SH
151
.B83
1995

BROODSTOCK MANAGEMENT
AND EGG AND LARVAL QUALITY

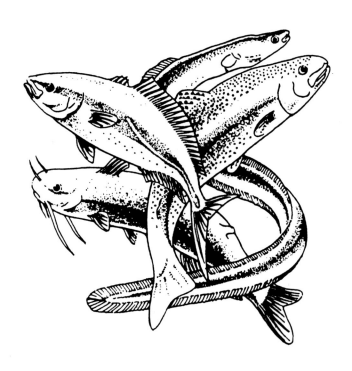

BROODSTOCK MANAGEMENT AND EGG AND LARVAL QUALITY

EDITED BY

NIALL R. BROMAGE
BSc, PhD
Professor of Reproductive Physiology and Endocrinology
Institute of Aquaculture, University of Stirling

RONALD J. ROBERTS
BVMS, PhD, FRCPath, FRCVS, FIBiol, FRSE
Director and Professor of Aquatic Pathobiology,
Institute of Aquaculture, University of Stirling

INSTITUTE OF AQUACULTURE

Blackwell Science

© 1995 by
Blackwell Science Ltd
Editorial Offices:
Osney Mead, Oxford OX2 0EL
25 John Street, London WC1N 2BL
23 Ainslie Place, Edinburgh EH3 6AJ
238 Main Street, Cambridge
  Massachusetts 02142, USA
54 University Street, Carlton
  Victoria 3053, Australia

Other Editorial Offices:
Arnette Blackwell SA
1, rue de Lille, 75007 Paris
France

Blackwell Wissenschafts-Verlag GmbH
Kurfürstendamm 57
10707 Berlin, Germany

Blackwell MZV
Feldgasse 13, A-1238 Wien
Austria

All rights reserved. No part of this publication may be reproduced, stored in a retrieval system, or transmitted, in any form or by any means, electronic, mechanical, photocopying, recording or otherwise, except as permitted by the UK Copyright, Designs and Patents Act 1988, without the prior permission of the copyright owner.

First published 1995

Set by Setrite Typesetters Ltd
Printed and bound in Great Britain by
  the University Press, Cambridge

DISTRIBUTORS

Marston Book Services Ltd
PO Box 87
Oxford OX2 0DT
(*Orders*: Tel: 01865 791155
        Fax: 01865 791927
        Telex: 837515)

USA
Blackwell Science, Inc.
238 Main Street
Cambridge, MA 02142
(*Orders*: Tel: 800 759-6102
        617 876-7000)

Canada
Oxford University Press
70 Wynford Drive
Don Mills
Ontario M3C 1J9
(*Orders*: Tel: 416 441 2941)

Australia
Blackwell Science Pty Ltd
54 University Street
Carlton, Victoria 3053
(*Orders*: Tel: 03 347-5552)

A catalogue record for this title is available from the British Library

ISBN 0-632-03591-9

Library of Congress
Cataloging-in-Publication Data

Broodstock management and egg & larval
  quality/[edited by] Niall R. Bromage and
  Ronald J. Roberts.
      p.   cm.
    'Institute of Aquaculture.'
    'This book arose out of the second of the
  science in aquaculture international
  conferences held at the University of Stirling,
  Scotland in June 1992' — Pref.
    Includes bibliographical references and
  index.
    ISBN 0-632-03591-9
    1. Fish hatcheries — Management.
  2. Fish-culture. I. Bromage, Niall R.
  II. Roberts, Ronald J. III. University of
  Stirling. Institute of Aquaculture.
  IV. Title: Broodstock management and egg
  and larval quality.
    SH151.B83 1995
    639.3 — dc20   94-19281
                                         CIP

Contents

	Contributors	vi
	Preface	viii
1	Broodstock Management and Seed Quality – General Considerations	1
	N. BROMAGE	
2	Sperm Physiology and Quality	25
	R. BILLARD, J. COSSON, L.W. CRIM, M. SUQUET	
3	Preservation of Gametes	53
	K. RANA	
4	Biotechnological Approaches to Broodstock Management	76
	G.H. THORGAARD	
5	Gilt-Head Sea Bream (*Sparus aurata*)	94
	Y. ZOHAR, M. HAREL, S. HASSIN, A. TANDLER	
6	Red Drum and Other Sciaenids	118
	P. THOMAS, C.R. ARNOLD, G.J. HOLT	
7	Sea Bass (*Dicentrarchus labrax*)	138
	M. CARRILLO, S. ZANUY, F. PRAT, J. CERDA, J. RAMOS, E. MANANOS, N. BROMAGE	
8	Atlantic Halibut (*Hippoglossus hippoglossus*) and Cod (*Gadus morhua*)	169
	E. KJØRSVIK, I. HOLMEFJORD	
9	Pacific Salmon (*Oncorhynchus* spp.)	197
	C.B. SCHRECK, M.S. FITZPATRICK, K.P. CURRENS	
10	Channel Catfish (*Ictalurus punctatus*)	220
	H. DUPREE	
11	African Catfish (*Clarias gariepinus*)	242
	C.J.J. RICHTER, E.H. EDING, J.A.J. VERRETH, W.L.G. FLEUREN	
12	Nile Tilapia (*Oreochromis niloticus*)	277
	D.J. MacINTOSH, D.C. LITTLE	
13	Carps (Cyprinidae)	321
	S. ROTHBARD, Z. YARON	
14	Origins and Functions of Egg Lipids: Nutritional Implications	353
	J.R. SARGENT	
15	Larval Foods	373
	P. LAVENS, P. SORGELOOS, P. DHERT, B. DEVRESSE	
16	Red Sea Bream (*Pagrus major*)	398
	T. WATANABE, V. KIRON	
	Index	414

Contributors

C.R. Arnold: Marine Science Institute, The University of Texas at Austin, PO Box 1267, Port Aransas, Texas 7837−1267, USA.

R. Billard: Museum d'Histoire Naturelle, Laboratoire d'Ichthylogy, IFREMER, 43 Rue Cuvier, 75231 Paris, France.

N. Bromage: Institute of Aquaculture, University of Stirling, Stirling, FK9 4LA, Scotland.

M. Carrillo: Instituto de Acuicultura de Torre de la Sal, 12595 Torre de la Sal, Castellon, Spain.

J. Cerda: Instituto de Acuicultura de Torre de la Sal, 12595 Torre de la Sal, Castellon, Spain.

J. Cosson: CNRS URA 671 06230, Villefranche Imer, France.

L.W. Crim: Memorial University, St Johns, Newfoundland, Canada.

K.P. Currens: Department of Fisheries and Wildlife, Oregon Cooperative Fishery Research Unit, Oregon State University, Corvallis, Oregon 97331−3803, USA.

B. Devresse: Artemia Systems NV, Oeverstraat 7, 3200 Baasrode, Belgium.

P. Dhert: Laboratory of Aquaculture, University of Gent, Rozier 44, 9000 Gent, Belgium.

H. Dupree: Fish Farming Experimental Laboratory, US Fish and Wildlife Service, Stuttgart, Arkansas, 72160 USA.

E.H. Eding: Department of Fish Culture and Fisheries, Agricultural University, PO Box 338, 6700 AH Wageningen, The Netherlands.

M.S. Fitzpatrick: Department of Fisheries and Wildlife, Oregon Cooperative Fishery Research Unit, Oregon State University, Corvallis, Oregon 97331−3803, USA.

W.L.G. Fleuren: Catfish Hatchery Fleuren, Zandstraat 86, 5712 XZ Someren, The Netherlands.

M. Harel: National Center for Mariculture, Israel Oceanographic and Limnology Research Institute, PO Box 1212, 8112 Eilat, Israel.

S. Hassin: National Center for Mariculture, Israel Oceanographic & Limnology Research Institute, PO Box 1212, 8112 Eilat, Israel; and Center of Marine Biotechnology and Agricultural Experiment Station, University of Maryland, 600 E, Lombard Street, Baltimore, MD 21202, USA.

I. Holmefjord: The Agricultural Research Council of Norway, Institute of Aquaculture Research, N−6600 Sunndalsora, Norway.

G.J. Holt: Marine Science Institute, The University of Texas at Austin, PO Box 1267, Port Aransas, Texas 7837, USA.

V. Kiron: Laboratory of Fish Nutrition, Department of Aquatic Biosciences, Tokyo University of Fisheries, Tokyo 108, Japan.

E. Kjørsvik: University of Trondheim, Brattora Research Station, Section for Aquaculture, N−7055 Dràgvoll, Norway.

P. Lavens: Laboratory of Aquaculture, University of Gent, Rozier 44, 9000 Gent, Belgium.
D.C. Little: School of Environment, Resources and Development, Asian Institute of Technology (AIT), GPO Box 2754, Bangkok 10501, Thailand.
D.J. Macintosh: Institute of Aquaculture, University of Stirling, Stirling, FK9 4LA, Scotland.
E. Mananos: Instituto de Acuicultura de Torre de la Sal, 12595 Torre de la Sal, Castellon, Spain.
F. Prat: Instituto de Acuicultura de Torre de la Sal, 12595 Torre de la Sal, Castellon, Spain.
J. Ramos: Instituto de Acuicultura de Torre de la Sal, 12595 Torre de la Sal, Castellon, Spain.
K. Rana: Institute of Aquaculture, University of Stirling, Stirling, FK9 4LA, Scotland.
C.J.J. Richter: Department of Fish Culture and Fisheries, Agricultural University, PO Box 338, 6700 AH Wageningen, The Netherlands.
S. Rothbard: YAFIT Laboratory, Fish Breeding Centre, Gan Shmuel 38810, Israel.
J.R. Sargent: Unit of Aquatic Biochemistry, School of Natural Sciences, University of Stirling, Stirling, FK9 4LA, Scotland.
C.B. Schreck: US Fish and Wildlife Service, Oregon Cooperative Fishery Research Unit, Oregon State University, Corvallis, Oregon 97331–3803; USA.
P. Sorgeloos: Laboratory of Aquaculture, University of Gent, Rozier 44, 9000 Gent, Belgium.
M. Suquet: IFREMER, Laboratoire de Physiologie et Zootechnie des poissons, 29280 Plaouzane, France.
A. Tandler: National Center for Mariculture, Israel Oceanographic and Limnology Research Institute, PO Box 1212, 8112 Eilat, Israel.
P. Thomas: Marine Science Institute, The University of Texas at Austin, PO Box 1267, Port Aranas, Texas 78373–1267, USA.
G.H. Thorgaard: Departments of Zoology and Genetics and Cell Biology, Washington State University, Pullman, Washington 99164–4236, USA.
J.A.J. Verreth: Department of Fish Culture and Fisheries, Agricultural University, PO Box 338, 6700 AH Wageningen, The Netherlands.
T. Watanabe: Laboratory of Fish Nutrition, Department of Aquatic Biosciences, Tokyo University of Fisheries, Tokyo 108, Japan.
Z. Yaron: Department of Zoology, Tel Aviv University, Tel Aviv 69987, Israel.
S. Zanuy: Instituto de Acuicultura de Torre de la Sal, 12595 Torre de la Sal, Castellon, Spain.
Y. Zohar: National Center for Mariculture, Israel Oceanographic and Limnology Research Institute, PO Box 1212, 8112 Eilat, Israel; and Center of Marine Biotechnology and Agricultural Experiment Station, University of Maryland, 600 E Lombard Street, Baltimore, MD 21202, USA.

Preface

This book arose out of the second of the 'Science in Aquaculture' international conferences held at the University of Stirling, Scotland, in June 1992. These biennial meetings are arranged by the Institute of Aquaculture in conjunction with BP Nutrition Aquaculture UK. The EC 'FAR' Programme of support for aquaculture research also helped sponsor the 1992 conference which was entitled 'Broodstock Management and Egg and Larval Quality'. This subject was chosen because the supply of seed is arguably the most important constraint on existing and future aquacultural production particularly that relating to marine fish.

The book has been written as a reference text for students of aquaculture, hatchery managers and fishery scientists whose remit involves the management of broodstock fish and the provision of eggs and fry for grow-out farms. It is also of considerable relevance to those concerned with the conservation of wild stocks of fish and those considering exploiting new species of fish for aquaculture.

The keynote contributors to the conference, which involves many of the most well-known workers in the field, were asked to provide individual chapters for the text based on their papers presented at the conference and in addition to include material of a general nature for the species or group of fish they were considering. The chapters have been edited with the aim of ensuring a reasonable homogeneity of the text, but without detracting from the individual nature of the contributions.

Emphasis has been placed on the more important groups of cultured finfish and includes details on the problems of broodstock management and seed production for that group, the induction of spawning, factors affecting egg and larval quality, rearing systems and conservation. In addition to this species or group approach there are more general chapters on egg and sperm quality, biotechnology and genetic manipulation, the importance of lipids and the provision of larval feeds. This is the first time that a single text has collected together all known information on broodstock management and seed production.

The editors hope that this volume, the second in the series, will prove as useful to its readers as the first and in particular will provide general information on methods of broodstock management and the production of seed which can be applied and successfully used for the exploitation of new species in aquaculture.

Chapter 1
Broodstock Management and Seed Quality – General Considerations

1.1 Control of reproduction
1.2 Induced spawning or hypophysation
1.3 Environmental manipulation
1.4 Cryopreservation
1.5 Gene banks and conservation
1.6 Fecundity and egg size
1.7 Egg quality
1.8 Nutritional status of broodfish
1.9 Broodstock husbandry and stress
1.10 Microbial influences
1.11 Overripening and egg quality
1.12 Conclusions
     Acknowledgements
     References

Currently, difficulties in the supply of eggs and fry, sometimes collectively known as seed, are amongst the most important constraints to further and more effective aquacultural development. For many farmed species, production is totally dependent on harvests of broodstock or seed from wild populations. Little is known about the control of reproduction of many farmed fish or about the detailed nutrient, metabolic and husbandry requirements of either broodstock or larvae or the ways in which these factors might be better managed to improve the numbers and continuity of eggs and larvae supply for farmers concerned with growing fish up to table size. Highlighted in this chapter are introductions to many of the more important areas of broodstock management and egg and larval quality, subjects which are fundamental to any considerations of seed quality. Some of these are considered further by other authors in this text.

1.1 Control of reproduction

A primary requirement as far as broodstock management and good farming practice are concerned is an ability to control fully the sexual maturation and spawning of the species under cultivation. Without this control farmers have to rely on collections of wild broodstock, larvae or fry in order to complete the cycle of production. Current methods of achieving this control are shown in Fig. 1.1. Many carps and catfish and other species like the milkfish, breams, eels and mullet do not complete maturation under farm conditions. In some like the carps, bream and catfish, spawning can be induced using a variety of

Fig. 1.1 Methods of management used to supply seed (gametes and fry) for aquacultural production.

hormonal preparations. In contrast, in others, e.g. the eel, yellowtail and milkfish, reproduction cannot be artificially induced and culture is totally dependent on harvests of mature broodstock, larvae or fry. Other species including the salmonids, bass and flatfish, whilst completing much of their maturation in culture, provide considerable difficulties in management.

However, for the vast majority of the 20 000 plus teleost species and of these the 1000 or so which are cultivated we have little or no information on management. Furthermore, only with those fish for which we have information on the detailed control of reproduction is it possible to programme the maturation of gametes and supply of seed and thus ensure an all-year-round production of table fish of consistent size and quality for retail markets.

Amongst the methods of controlled reproduction in use at present one would include induced spawning or hypophysation, environmental manipulation and cryopreservation.

1.2 Induced spawning or hypophysation

Induced spawning or hypophysation has been used for almost 60 years to bring about the final maturation or release of eggs and sperm. In many countries, pituitary extracts and HCG are still used extensively although there are periodic problems of purity, specificity, continuity of supply, methods of administration and potency, particularly when these materials are used in the field.

Increasingly, synthetic hypothalamic releasing hormones are being used to induce final maturation, ovulation and spawning because these hormones have a number of advantages over the more widely-used pituitary-gonadotropic materials. Releasing hormones are low molecular weight peptides which are secreted by the brain to control the synthesis and release of pituitary hormones. Gonadotropin-releasing hormone or GnRH, which is composed of a linear chain of 10 amino acids, is a potent inducer of pituitary gonadotropic hormone (GtH) release. GtH exerts overall control of many aspects of gonadal development, including oocyte growth and maturation, ovulation and spawning. Consequently, injections of GnRH or related peptides provide an effective alternative to pituitary extracts and HCG in the induction of spawning of a range of fish species. In practice, synthetic analogues of GnRH, known as LHRHa/GnRHa, are used because they have longer-lasting actions than the naturally-occurring hormone (Peter 1982, Zohar *et al.* this volume Chapter 5).

In addition to the induction of release of GtH by GnRH, there is evidence, at least as far as the cyprinids are concerned, that GtH secretion is also under the control of an inhibitory hormone (Peter *et al.* 1988). This release-inhibitory hormone, which is also produced in the hypothalamus, has been identified as dopamine. Treatment with antagonists of dopamine, like pimozide or domperidone, together with injections of LHRHa, leads to an enhanced release of GtH when compared with LHRHa alone, thus allowing for the dosage of LHRHa and the risks of excessive stimulation to be minimized. Recently, a combined preparation containing both LHRHa and a dopamine antagonist has become available. The simplicity and effectiveness of such combined treatments together with the absence of any special storage conditions and a good shelf life even under tropical conditions promises to extend considerably use of this method to new species and geographical areas.

Although methods of hypophysation enable seed to be produced from fish which currently do not spawn in captivity, at present the methods only bring about final maturation and the release or spawning of fully mature or ripened gametes and are of little use in radically altering the timing of maturation. In contrast techniques involving the artificial control of various environmental factors, like light and temperature, are able to control all stages of development and as a consequence are effective in achieving significant alterations in spawning time.

1.3 Environmental manipulation

Carp were probably the first fish on which the environmental control of spawning was practised. Farmers concerned with the production of carp eggs and fry are well aware of the importance of temperature in the control of maturation. In the common carp, maturation is stimulated by temperatures in excess of

17°C and preferably in the range of 21–25°C (see also Section 13.4.2). Grass carp and catfish have somewhat higher temperature requirements. All subsequently require final maturation and spawning to be hormonally induced using pituitary or hypothalamic hormones.

Of greater practical application for altering the rate of maturation and the time of spawning is the use of modified light or photoperiod regimes. All fish whose reproduction is cued by light, spawn at specific phases of the annually-changing cycle of day length. Most salmonids in the northern hemisphere spawn in the autumn and winter months under decreasing or short day length. In contrast other photoperiodic species spawn at different times of the year; for instance, most flatfish spawn in the Spring under long days, whereas the gilthead bream and sea bass reach maturity earlier in the year under the influence of short but increasing day lengths (Bromage et al. 1984, 1990a, Bromage & Duston 1986, Carrillo et al. 1993, Bromage et al. 1993).

In salmonids and sea and smallmouth bass, adjustments of the seasonal rate of change of photoperiod into shorter and longer periods than a year have produced respectively advances and delays of 3–4 months in the timing of spawning with no adverse effects on the quality of the eggs or fry (Carrillo et al. 1989, 1993, Cantin & Bromage 1991, Bromage et al. 1990a, 1993). Constant long days followed by constant short days produce similar 3–4 month advances in spawning time. Delays of equivalent duration have been produced by exposing fish to constant short days throughout the year or short day lengths followed by constant long days.

These constant photoperiod regimes are obviously less complex than modified seasonal light cycles and thus easier to manage and use on commercial farms. Once the spawning time of a stock has been modified in this way, the fish must be maintained permanently under light-proof covers and controlled light, otherwise spawning will revert to its timing under natural ambient conditions. Usually, after the spawning of a stock has been advanced or delayed by 3–4 months, then the fish are maintained on a light cycle which will give out-of-phase (advanced or delayed) yearly spawnings. Holding 3–4 separate broodstocks under different yearly light cycles ensures that eggs are available throughout the year.

The use of environmental manipulation to change spawning time is a particularly attractive proposition for flatfish, bass, bream and mullet because the very high fecundities of these fish mean that only small numbers of broodstock need to be maintained under artificial conditions.

Using photoperiod control it is also possible to spawn fish more than once a year. Thus, rainbow trout maintained under constant illumination (LL or LD 18;6) mature again after an interval of only 6 months (Bromage et al. 1984, Bromage & Duston 1986). There is, however, considerable variability in this response, with some fish spawning after 5 months, others after 6, 7 or 8 months, and still others failing to mature until the normal spawning time 1 year later. Exposure of fish to only 1 month of continuous light gives much more precise control, and recently methods using such pulses of light in an otherwise ambient photoperiod have been developed which are able to both advance and delay spawning (Bromage et al. 1992).

At present many farms and hatcheries in Europe and North America are successfully using photoperiod techniques to spread the spawning of rainbow trout. Similar techniques are also being used to alter the timing of spawning of the brown trout, sea and smallmouth bass, gilthead bream, some flatfish and cyprinids and more recently the Atlantic salmon. In Atlantic salmon in particular only a modest 4–6 week advancement in spawning constitutes a considerable commercial advance because fry are available earlier and hence there is a much greater likelihood of the fish becoming S1s (1 year smolts). For commercial farms it is necessary to combine this work on controlled spawning of the broodstock with associated photoperiodic alterations in the timing of smoltification (Thrush et al. 1994). Only by having eggs freely available throughout the year and a capability to stock smolts in the sea at the times prescribed by the grow-out farms will the 3–4 kg salmon that the market requires become available in every month of the year.

1.4 Cryopreservation

Short-term preservation or storage of gametes ranging from a few hours to a few weeks, usually at low temperatures, can also be used in hatcheries to overcome temporary shortages of gametes, asynchrony in artificial and natural spawning, transportation of gametes, disease control requirements and selective breeding. Thus, trout and salmon sperm can be stored successfully at −4°C for 3–4 weeks using a suitable extender with added antibiotic and eggs at 4°C for 2–3 days. Generally, storage times for sperm can be increased a little more by using oxygen-enriched environments.

Cryopreservation, the storage of biological material in a viable form at ultra-low temperature (−196°C), however, opens up the possibilities of storing sperm for infinitely long periods. However, despite significant progress having been made since the first reports of successful cryopreservation of fish sperm in the 1950s, there are still many contradictions in the literature. Much of this variability may be attributed to seasonality, varying sperm quality, collection techniques, diluents used, to precise freezing and thawing regimes, storage conditions, and to post-thaw fertilization techniques. Notwithstanding these difficulties, methods are now available to successfully cryopreserve milt from some 30 species of freshwater and marine fish including the brown and rainbow trout, Atlantic and Pacific salmon, the Atlantic halibut and several species of tilapia (McAndrew et al. 1993).

1.5 Gene banks and conservation

The development of successful protocols for long-term freezing will allow the storage of disease-free gametes of desirable strains and species for future use and also facilitate the comparison and evaluation of new strains of fish with original parental stock at minimum cost. In addition, frozen gene banks may be used in conjunction with live gene banks. A major advantage of this approach is that the effective population size can be dramatically increased at relatively low cost by storing spermatozoa from a large number of individual males.

At present cryogenic banking is possible for fish spermatozoa but not for ova or embryos. The fundamental problems of insufficient dehydration during the cooling of eggs and embryos, due to their relatively large volume and the presence of two or more membranes with different permabilities to water, have as yet not been overcome. In view of the current limitations to freezing eggs and embryos, other procedures for conserving the female genome are being evaluated, including the cryopreservation of masculined gynogenetic lines and their subsequent genetic reconstitution in an egg using the techniques of androgenesis, nuclear transplantation and heat or pressure shocks (Bromage *et al.* 1990b, McAndrew *et al.* 1993).

1.6 Fecundity and egg size

A major difference between fish and many of man's other domestic animals is their high fecundity, i.e. the large numbers of eggs produced. There are also significant differences between fish species with regards to fecundity of individual broodfish of flatfish and other marine species producing millions of eggs at a single spawning whilst other species e.g. the salmonids produce only thousands (Bromage, 1988). In addition many marine fish produce multiple batches of eggs at daily to weekly intervals over spawning seasons which may last 2–3 months.

In contrast salmonids produce a single batch of eggs each year which is usually stripped on a single occasion. These species differences are of profound importance to the planning and management of broodstock facilities with the less fecund salmonids requiring far more parental fish and in turn more extensive and costly installations to produce the same number of eggs as their marine counterparts. On the 'down side' the much smaller numbers of broodstock required for many marine hatcheries carries with it serious risks of inbreeding and genetic drift.

In addition to egg number or fecundity, the other parameter which is often considered in assessing the egg production capability of broodfish is the size of egg produced. This is because egg size is considered by some workers, at least, to be a determinant of egg quality (see Bromage *et al.* 1992, Richter *et al.* this volume Chapter 11, for discussion).

A number of biotic and environmental factors have been shown to influence fecundity and egg size, opening up the possibilities of artificial improvements in the production capability of broodstock fish.

One of the most important factors influencing both the number and size of eggs is the size of the broodfish. Generally, as fish size increases so does fecundity and the diameter of the eggs produced. This is particularly well seen in the salmonids (Fig. 1.2). There is, however, in the rainbow trout at least, a gradually diminishing increase in fecundity with increasing fish size (Bromage et al. 1990c, 1992); this means that larger broodfish have lower relative fecundities (i.e. the number of eggs produced per unit of post-spawned broodfish weight) and as a consequence 1 t of smaller broodfish trout would produce up to twice as many eggs as the same overall weight of individually-larger fish. Similar weight–fecundity relationships also seem to be in operation in sea bass (Carrillo et al. this volume Chapter 7). This has important implications for the optimization of egg production from limited water supplies, enclosures of specific size or where effluent or stocking density constraints may be of significance.

In contrast with fish size, fish age appears to be much less important in determining fecundity, with reported increases in egg number with age probably being due to concomitant increases in fish size (Bromage & Cumaranatunga 1988).

In addition to fish size a further important determinant of fecundity and egg size is the genotype of the broodfish. Bromage et al. (1990c) working with 12 commercial stocks of rainbow trout and Abée-Lund & Hindar (1990) with nine populations of wild brown trout showed that the most fecund of the stocks/strains produced twice as many eggs as the least fecund even after the potentially conflicting influences of fish size were partitioned statistically by co-variance analyses (ANCOVA). Regression analyses showed that all the stocks had similar slopes of the regression of fecundity on fish size but differed significantly in the elevation of their regressions (Fig. 1.3, adapted from Bromage et al. 1990c). Differences in egg size between the stocks were modest in comparison, with only a 10% difference in size between the smallest and largest eggs (Fig. 1.4, adapted from Bromage et al. 1990c). Collectively these data demonstrate that choosing strains for potential broodstock can have profound effects on the number and sizes of eggs produced by a farm.

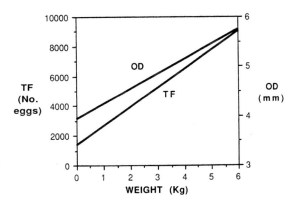

Fig. 1.2 Regression of total fecundity (TF) and egg diameter on post-stripped fish weight. (From Bromage et al. 1992.)

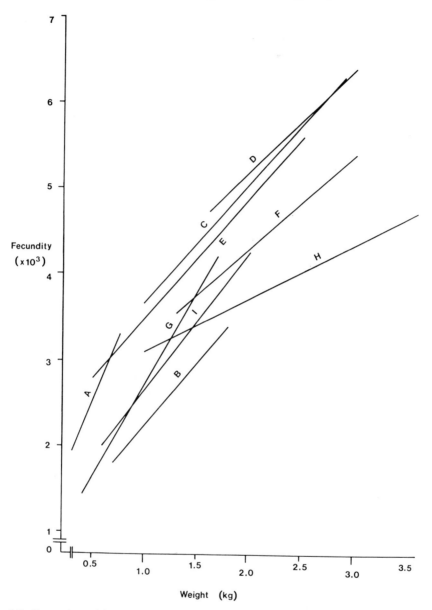

Fig. 1.3 Regressions of fecundity on post-stripped fish weight in nine strains (A-I) of rainbow trout. (Adapted from Bromage *et al.* 1990c.)

In addition to indirect effects on fecundity (through influences on fish size), the daily and seasonal rates of feeding (i.e. ration) of broodstock diets have direct effects on fecundity and egg size (Springate *et al.* 1985, Jones & Bromage 1987, Bromage & Cumaranatunga 1988, Bromage *et al.* 1992). Thus, feeding broodstock trout at half or three-quarters of their recommended daily ration of food throughout the year resulted in up to 25% fewer eggs even after differ-

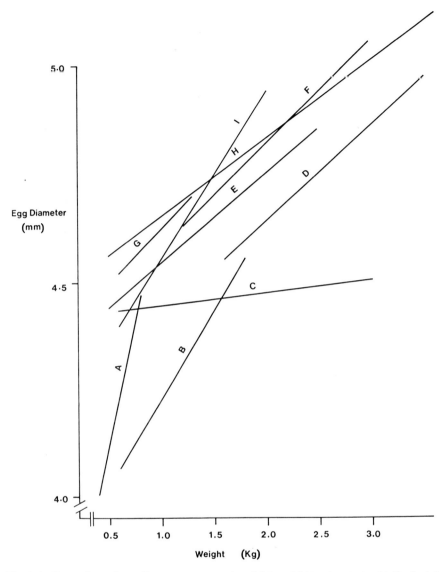

Fig. 1.4 Regressions of egg diameter on post-stripped fish weight in nine strains (A-I) of rainbow trout. (Adapted from Bromage *et al.* 1990c.)

ences in weight were taken account of by ANCOVA (Jones & Bromage, 1987). Significant numbers of the fish on low rations also failed to spawn, an important consideration when calculating potential egg outputs from specified numbers of broodfish.

Results from trials involving seasonal changes in ration also indicate that there are 'windows of opportunity' during which high and low feed rates produce significant effects on fecundity and rates of maturation. By splitting the annual reproductive cycle of the rainbow trout into three 4-month periods and feeding at either high (1.0% of body weight per day) or low (0.4%) rates for different periods (see Fig. 1.5 for protocol), Bromage et al. (1992) showed that those groups of fish fed on high rations for the first 4 months of the cycle had higher fecundities (Fig. 1.6) and included a higher percentage of spawned

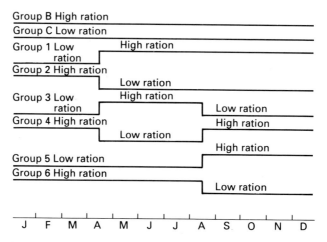

Fig. 1.5 Experimental protocol for feeding eight groups of rainbow trout with high and low rations for different periods of the year. Groups B and C received high and low rations throughout the year respectively whereas for example Group 3 received low ration for the period January to mid-April, followed by high ration until mid-August and low ration until spawning.

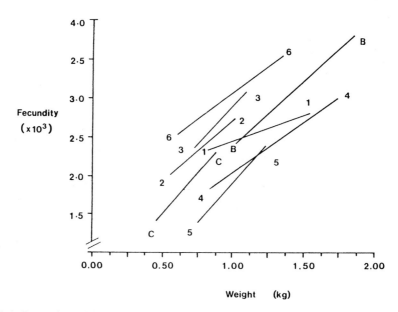

Fig. 1.6 Regressions of fecundity on post-stripped fish weight for eight groups of rainbow trout fed for different periods on high and low rations. The experimental protocol is shown in Fig. 1.5.

Table 1.1 Effects of feeding high and low rations for different periods of the year on percentage spawning rate, fecundity and weight of groups of rainbow trout (see Fig. 1.5 for experimental protocol for different groups of fish). The last two columns show the number and weight of fish required to produce 1 million eggs

Group	% Spawning	Mean fecundity	Mean weight (kg)	No. of fish*	Total fish biomass (kg)*
B	68.2	3036	1.571	483	759
C	34.5	1864	0.775	1555	1205
1	48.2	2562	1.321	810	1070
2	63.6	2355	0.903	668	603
3	47.3	2693	1.047	785	822
4	68.2	2268	1.305	646	843
5	40.9	1865	1.081	1311	1417
6	70.0	3060	1.163	467	543

* Number and total weight of fish required to produce 1 million eggs

fish (Table 1.1) than those which had been maintained on low rations over the same period.

In contrast feeding high rations during the latter stages of the cycle did not appear to affect the numbers of eggs produced by each individual fish but did increase the weight of the broodfish, i.e. they had lower relative fecundities than fish fed at low rations over the last 4 months of the reproductive cycle. These differences again have implications with regard to the number of broodstock fish required to produce a specific number of eggs. By optimizing the seasonal rates of feeding, broodstock numbers may be reduced by two-thirds and still produce the same number of eggs (see Table 1.1).

Although much of the work on fecundity determinants has been carried out with salmonid fish and the trout in particular, it is likely that similar complex relationships exist between fish size, genotype and feed rate and fecundity and egg size in other species of fish.

1.7 Egg quality

Given that it is possible for fish to spawn naturally or to be spawned artificially in captivity, it is essential that the eggs and fry produced are of the highest quality. Egg quality defined as those characteristics of the egg which determine its capacity to survive is a significant problem for many of the species currently being farmed and is almost certain to be a problem for the culture of any new species. In general for many fish e.g. bass, bream, turbot and halibut, mortality rates for eggs are very high, with survivals post-weaning often less than 5%. Only the salmonids exhibit better egg and larval quality but even here two-thirds of the eggs and fry may be lost during the first few months in the hatchery (Bromage et al. 1992).

Little is known about the determinants of egg quality although many factors have been implicated as possible causative agents (Fig. 1.7). There is also little

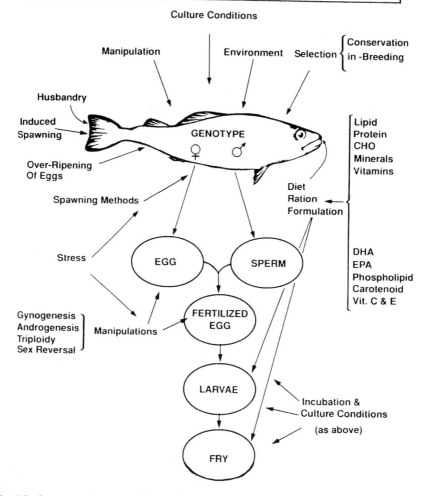

Fig. 1.7 Summary of suggested determinants of egg and larval quality.

agreement regarding reliable methods for the assessment of quality, an essential prerequisite if any firm conclusions regarding the factors which determine egg and larval quality are to be reached. Furthermore, in order to be of value to the farmer, methods of assessment must be simple to perform and should be capable of being carried out early in development to avoid occupying hatchery facilities and staff time on what may turn out to be unproductive batches of eggs.

In rainbow trout the percentage rate of fertilization, assessed either at 12 h, at the time of the first or second cleavage division or after 7 days with the formation of the primitive streak, provides a reliable indicator of subsequent performance, with survivors from batches of eggs with poor fertilization rates generally performing badly at all subsequent stages of development (Springate et al. 1984, Bromage & Cumaranatunga 1988, Bromage et al. 1992). As a result of the good correlation between eying and fertilization rates on the one hand and performance on the other, many trout farms use eying rates as indicators of quality (Bromage & Cumaranatunga 1988, Bromage et al. 1992).

There is, however, far less agreement regarding reliable methods of egg quality assessment for marine fish. Many hatcheries culturing marine species distinguish 'good' from 'poor quality' eggs by virtue of the eggs' ability to float or sink in sea water respectively (McEvoy 1984, Carrillo et al. 1989, Kjørsvik et al. 1990; Fig. 1.8). Other authors have suggested that the appearance of the chorion, the shape of the egg, its transparency and distribution of oil globules can be related to quality (see review of Kjørsvik et al. 1990). However, for a number of species including for example the halibut, no such correlations of buoyancy or of any other morphological characteristic, with quality has been consistently reported (Bromage et al. 1994).

Fertilization rate may prove to be as good an index of quality in marine species as it is for the salmonids, although it is reported not to correlate well with later patterns of survival in a number of marine fish (Kjørsvik et al. 1990). One problem may be the definition of fertilization rate. Correlations between

Fig. 1.8 Procedure used to measure volumes of good (floating) and poor (sinking) quality eggs in European sea bass. (Photograph kindly supplied by M. Carrillo and S. Zanuy.)

quality and overall rates of fertilization may be obscured in marine species by simply recording the total percentage of eggs fertilized. Supplemental information on, e.g., chromosome appearance and cell symmetry has proved useful for a number of species (Bromage et al. 1994, see also review by Kjørsvik et al. 1990 and Kjørsvik & Holmefjord this volume Chapter 8). Recent work on halibut has distinguished five categories of development immediately after fertilization but of these only the proportion of eggs which are spherical in shape and show symmetrical cleavages is thought to provide an assessment of quality which is indicative of subsequent performance (Bromage et al. 1994).

Another requirement which is important in all assessments of quality is to include data for all batches of eggs and not just record mean (pooled) development rates. Wherever possible individual egg batch survival figures should be included for each broodstock fish. The reasons for this approach become a little clearer when the full range of individual survival values are examined in populations of broodstock. Figure 1.9 shows the mean values together with the range of fertilization rates and survival levels for rainbow trout eggs until day 120. This clearly shows that included in the average values are batches of eggs, spawned by individual fish, with almost 0% and 100% survivals. Mean values, even with appropriate confidence intervals, can never truly represent the full range of egg quality in a stock of broodfish. Furthermore, some hatcheries and also some scientific studies exclude individual batches of eggs

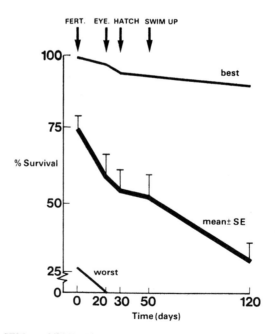

Fig. 1.9 Mean (± SEM; n = 15) fertilization rates and survival rates at eying, hatch and swim-up (% ages) of batches of rainbow trout eggs. Included are the highest and lowest individual values. (From N. Bromage & J.R.C. Springate unpublished data.)

with 100% mortalities (i.e. blanks) from their data sets. In many ways an improved understanding of why all the eggs from an individual fish have died may be more instructive than trying to attribute cause to less pronounced differences in quality.

Many factors have been suggested as possible determinants of egg quality including the nutrition, husbandry conditions and genetic make-up of the broodfish and fertilized zygote, and the size, chemical composition, microbial colonisation and also the overripening of the egg (Bromage & Cumaranatunga 1988, Kjørsvik et al. 1990, Bromage et al. 1992). However, of these only the nutritional status of the broodfish (Watanabe et al. 1985, Watanabe & Kiron this volume Chapter 16; see also Kjørsvik et al. 1990) and the extent to which they are exposed to stress (Campbell et al. 1992), the bacterial colonization of the surfaces of the eggs (Barker et al. 1989, 1991, Hansen & Olafsen 1989) and the overripening of eggs, which is the ageing process which occurs in eggs in the period following ovulation up to fertilization (see reviews in Kjørsvik et al. 1990, Bromage & Cumaranatunga 1988, Bromage et al. 1992), have been clearly shown to influence egg quality. These are considered in more detail in the following sections.

1.8 Nutritional status of broodfish

We have seen above (section 1.6) that modifications in the ration of food can significantly affect the fecundity of broodfish and their ability to mature as well as the sizes of egg produced. Although gross nutrient availability is clearly important in influencing various aspects of reproductive physiology, there is little evidence that such changes can affect egg and larval quality (see reviews in Hardy 1985, Watanabe 1985, Bromage et al. 1992) and increasingly attention is being paid to the role of individual components of the diets. These include the micronutrients, e.g. the essential polyunsaturated fatty acids (PUFA) particularly the (n-3) series including docosahexaenoic acid (DHA), 22−6 (n-3) and eicosapentaenoic acid (EPA), 20−5 (n-3), and their derivatives, the vitamins and particularly vitamins C and E, the carotenoids and specifically astaxanthin and various trace elements.

Pioneering work by Watanabe and colleagues (Watanabe 1985, Watanabe & Kiron this volume Chapter 16) showed that diets lacking in supplementary trace elements or with reduced levels of either lipid (PUFA) or vitamin E produced eggs of poorer quality than those produced by fish receiving more balanced formulations. Subsequent work by the same authors (Watanabe & Kiron this volume Chapter 16) have identified astaxanthin and in particular its mesoisomer 3S, 3'R, vitamin E and phospholipids containing DHA and EPA as being the most important determinants of egg quality.

Astaxanthin, along with other carotenoids and also vitamin E, are thought to act as quenchers or scavengers of singlet oxygens and also free radicals, i.e. they absorb the energy of these compounds within their extensive double bond structure, effectively preventing reactive damage to other molecules, particularly the polyunsaturated lipids. Sargent and colleagues (this volume Chapter 14) have suggested that the anti-oxidant effectiveness of carotenoids in protecting against light-induced free radical production may explain their high concentrations in the retina, a tissue known to contain high levels of DHA.

A number of studies, both on broodstock and larvae, have specifically indicated the importance of PUFA in egg and larval development. Studies by Watanabe & Kiron (this volume) and Zohar and colleagues in Eilat (Zohar et al. this volume Chapter 5) on the red sea and gilthead breams respectively have shown clearly that egg and larvae quality are profoundly affected by the levels of (n-3) PUFA in the diets of the broodfish. Sorgeloos and Lavens and colleagues (Lavens et al. this volume Chapter 15) also provide complementary evidence for similar requirements for PUFA in the diets of first feeding larvae. These requirements are hardly surprising, for the (n-3) PUFA and their derivatives, particularly the phospholipids, have a generalized role in the cell membranes of fish. Very high levels of phospholipid are found in all fish tissues and neural cell membranes in particular; in all cases DHA is the principal fatty acid.

Sargent and colleagues (Sargent et al. 1990, Sargent this volume Chapter 14) have suggested that the high levels of phosphatidylethanolamine and its constituent fatty acid DHA in the brain and eye assemblage of fish indicates a primary role for these compounds during embryonic and larval development which might involve both structural and behavioural functions. The other important function of the (n-3) PUFA is their role in the production of eicosanoids, substances which are important in stress reactions and adaptation. It is now accepted that EPA modulates the formation of eicosanoids from 20:4 (n-6) by competing with the enzyme systems responsible for controlling this pathway. Diets rich in (n-6) PUFA or deficient in (n-3) PUFA may result in an elevated production of eicosanoids during embryogenesis and larval development, which in turn will be reflected in an increased susceptibility of the resulting larvae to external stress (Sargent, this volume Chapter 14).

It is evident not only that the absolute levels of DHA and EPA are important but that due regard must be paid to the ratios of DHA:EPA and (n-3):(n-6) PUFA. It is also probable that these requirements and ratios will be different in different species and certainly between freshwater and marine fish. This area, however, is likely to be one of the most important constraints on future aquacultural developments.

1.9 Broodstock husbandry and stress

Ideally broodfish should be maintained under controlled conditions which as far as possible match or improve upon those to which the fish will have been exposed in the wild. In practice, however, it may not be possible to manage all of the rearing conditions. Water quality, feeding regime and diet, stocking density, exposure to pathogens and handling stress parameters may be optimized by appropriate management and husbandry practices, although such improvements may be difficult for species of fish which have only recently become farmed, mainly because the establishment of best husbandry practices requires a number of years of development and experimentation.

We are also seeing, even with species like the trout, where there is a long history of farming, the introduction of many new techniques and procedures in the hatchery particularly with regard to sex control, induced spawning and disease treatment. Fig. 1.10 shows the range of operations and treatment currently available in the modern trout hatchery. Inevitably, however sympathetic the husbandry and management, each of the procedures are likely to constitute a stress, and intuitively one would expect that this would be reflected in the health of the broodfish and the viability of the progeny.

The effects of stress on broodfish are somewhat of a paradox. In many ways, broodfish, possibly by virtue of their age, size and metabolic requirements and reserves, are far more tolerant of stress than fry or juvenile fish. Stresses which result in minimal apparent or measurable effects in broodfish often cause acute stress and mortality in young animals; e.g. broodfish are far more tolerant of poor quality waters than juvenile fish. However, one aspect of the biology of broodfish which clearly has a much lower threshold of effect of stress is reproduction. Because of this sensitivity, reproductive bioassays are sometimes used in preference to LD50s to assess the effects of environmental toxicants (Eaton 1973, Gerking 1982). No doubt chronic stress imposed by even modestly inappropriate rearing conditions is an important contributory factor in the failure of many fish to spawn or fully mature under culture conditions.

Regarding mechanisms, there is now evidence that acute handling and confinement stress reduce sex steroid levels in both male and female trout and that this effect is mediated by the action of cortisol on the hypothalamus, pituitary and gonad (Sumpter et al. 1987). In a subsequent paper from the same laboratories, Campbell et al. (1992) reported that ovulation was delayed in stressed fish and the fertilized eggs were smaller in size and of poorer quality. Spermatocrits were also reduced in stressed males.

Clearly, these data indicate that as far as reproduction is concerned husbandry conditions are likely to impose limitations on the quality of the eggs and sperm, which will almost certainly be reflected in a poorer productivity of the hatchery.

1.10 Microbial influences

A characteristic failure of most aquatic environments, both freshwater and marine, is the large numbers of micro-organisms which are present in the waters. There are further significant increases in these numbers following the culture of eggs and larvae of fish in the same environment. This enhancement of microbial growth probably occurs as a result of the increased amounts of nutrients from the metabolic by-products of the fish and the presence of

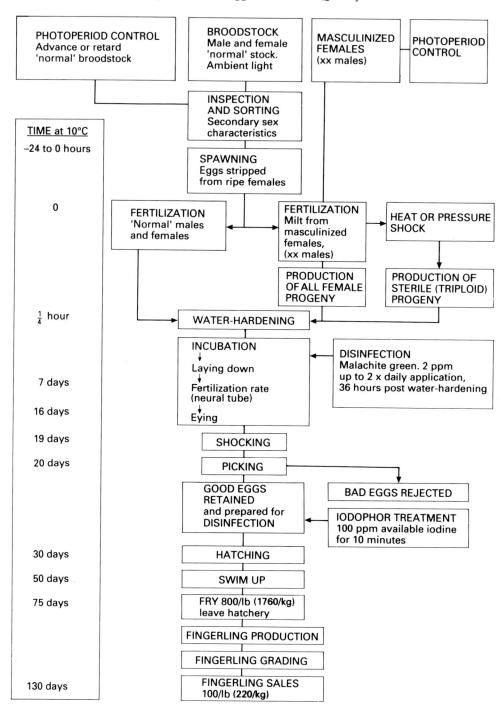

Fig. 1.10 Flow diagram showing procedures and methods used in a modern rainbow trout hatchery, together with timing of developmental stages at 10°C.

increased numbers of surfaces for colonization by micro-organisms and entrapment of organic debris. The various lipid and protein components of fish eggs also provide an excellent source of nutrients and their glycoprotein-rich chorion or shell encourages microbial attachment.

Despite the ubiquity of micro-organisms in culture systems, surprisingly, there have been few studies of the influences of these on the survival characteristics of the eggs and larvae of the various species under cultivation. The effects of the fungus *Saprolegnia* on the eggs of freshwater fish is a well-described problem in the salmonid hatchery and daily fungicidal treatments or the removal of dead eggs by 'picking' are techniques often employed to prevent fungal hyphae spreading from dead to healthy eggs.

The possible role of other micro-organisms, e.g. bacteria, is, however, less clear. A number of studies have reported the presence of large and species-specific bacterial communities in association with the surfaces of both freshwater and marine fish eggs (Hansen & Olafsen 1989, Barker *et al.* 1989, 1991); see also Kjørsvik & Holmefjord this volume Chapter 8).

Overall the most frequently isolated bacteria from the surfaces of live fish eggs are members of the genera *Cytophaga, Pseudomonas, Alteromonas* and *Flavobacterium* as well as *Aeromonas hydrophila*. In addition to being present in most water supplies, these bacteria are often present in the coelomic fluid of maturing female fish. Hence, the expulsion of eggs during natural spawning or 'stripping' will inevitably involve the contamination of the newly-spawned eggs. All of these bacteria have been identified as secondary opportunistic pathogens of fish. It has also been suggested that their presence may lead to increased mortalities of eggs due to the release of bacterial toxins or enzymes and/or increased competition for available oxygen supplies.

The addition of antibiotics to hatchery waters or immersion of eggs in disinfectants like glutaraldehyde have been shown to reduce mortalities in the developing eggs of several marine species (Jelmert & Mangor-Jensen 1987, Kjørsvik and Holmefjord this volume Chapter 8). Increases in or optimisation of the flow rates of water supplies to incubating eggs have also been shown to reduce surface colonization of eggs by bacteria (Barker *et al.* 1991). However, under commercial condition it may be advisable to prevent initial bacterial colonization by water treatment methods such as ozonization or UV sterilization rather than use chemicals or antibiotics to reduce the numbers of already established colonies. It remains to be seen whether overall egg and larval quality can be enhanced by long-term use of such methods for disinfection. However, it is clear that contamination of egg and larval culture systems may be an important factor in determining overall survival levels. It is essential that husbandry practices ensure that culture systems are kept as clean as possible.

1.11 Overripening and egg quality

Following ovulation unfertilized eggs undergo a process of ageing, commonly described as overripening (Nomura *et al.* 1974, Sakai *et al.* 1975, Springate *et al.* 1984, Kjørsvik *et al.* 1990, Bromage *et al.* 1994). During overripening eggs undergo a series of morphological and compositional changes and also a progressive loss in quality or viability (Sakai *et al.* 1975, Springate *et al.* 1984). In all species so far investigated this loss in quality would appear to occur irrespective of whether the eggs are retained in the body of the female following ovulation and before stripping, or whether they are stripped and stored *in vitro* in ovarian fluid. At present it is not clear whether the structural and other changes, which occur during overripening, are responsible, *per se*, for the deterioration in egg quality or merely gross symptoms of the overripening process. In many instances viability often appears to decline before any structural changes are visible.

The period of optimum ripeness (i.e. the time when the highest fertilization and egg and larval survival rates are achieved) varies with different species of fish (Table 1.2); the rate of overripening is also temperature dependent, with higher and lower temperatures curtailing and extending the period of optimum ripeness, respectively, for most species of fish (Gillet 1991). Most of the studies of overripening and egg quality have been carried out with fish in which oviposition does not naturally follow ovulation under laboratory or hatchery conditions. By leaving eggs in the same or series of different female broodfish

Table 1.2 Time after ovulation at which optimum egg quality is achieved in different species of finfish. (From Bromage *et al.* 1994.)

Species	Time	Reference
Oreochromis niloticus	1 h	Rana (unpublished data)
Prochilodus platensis	1 h	Fortuny *et al.* (1988)
Roccus saxatilis	1 h	Stevens (1966)
Carassius auratus	2–3 h	Formacion & Lam in Formacion (1991)
Macculochella peeli	2–3 h	Rowland (1988)
Misgurnus anguilicaudatus	3–8 h	Suzuki (1975)
Hippoglossus hippoglossus	4–6 h	Kjørsvik *et al.* (1990), Bromage *et al.* (1994), Holmefjord (1991), Norberg *et al.* (1991)
Rhamdia sapo	5–9 h	Espinach Ros *et al.* (1984)
Gadus morhua	9–12 h	Kjørsvik & Lonning (1983)
Clarias macrocephalus	10 h	Mollah & Tan (1983)
Scophthalmus maximus	10–20 h	McEvoy (1984), Howell & Scott (1989)
Plecoglossus altivelis	1–2 days	Hirose *et al.* (1977)
Limanda yokahamae	2–3 days	Hirose *et al.* (1979)
Salvelinus alpinus	5 days	Gillet (1991)
Oncorhynchus mykiss	4–6 days	Springate *et al.* (1984)
Clupea harengus	14 days	Hay (1986)
Oncorhynchus kisutch	20 days	Fitpatrick *et al.* (1987)

for varying periods of time, the effects of overripening on egg quality can be investigated.

In the rainbow trout the period of optimum ripeness occurs 4−6 days after ovulation at 10°C. Eggs fertilized immediately after and up to 3 days after ovulation show slightly reduced survivals compared to those stripped after 4−6 days. Although it is not clear whether this constitutes a definite period of under-ripeness, a similar phenomenon has been reported for the ayu, *Plecoglossus altivelis* (Hirose *et al*. 1977) and a Japanese flatfish, *Limanda yokahamae* (Hirose *et al*. 1979). In the trout, survivals to eying, hatch and swim-up broadly parallel the changes in fertilization rates seen with time (Springate *et al*. 1984) although eggs retained in the broodfish for longer than 12 days can often be fertilized but subsequently show no further development. With insufficient checking and stripping of broodfish, overripening has been shown to be a significant cause of losses in rainbow trout hatcheries (Bromage & Cumaranatunga 1988). In contrast much shorter periods of optimum ripeness are found in *Oreochromis* (K. Rana, unpublished data) and *Carassius auratus* (Formacion & Lam in Formacion 1991) whereas flatfish generally show periods of optimum ripeness which are intermediate in duration (Bromage *et al*. 1994 and see Table 1.2).

Overall, it would appear that overripening is a significant determinant of egg quality in a wide range of fish species. Further understanding of the process of overripening together with an ability to time ovulation accurately, particularly in batch spawners, should enable considerable improvements to be made in the supply of hatchery-reared juvenile fish.

1.12 Conclusions

Problems of seed supply together with the quality of the eggs and fry produced constitute one of the most important constraints on current and future aquacultural developments. Improvements in our understanding of the appropriate culture conditions and management procedures for the broodfish are essential if we are to programme reproductive development to produce reliably the numbers of eggs and fry required by grow-out farms. Research progress has been slow in these areas, principally because of the relatively short history of work on reproduction in many species, particularly marine fish, and because of the multiplicity of interrelated factors which are thought to act as egg and larval quality determinants in fish. Only by improving our understanding and control of all these processes will a continuity of finished product of consistent size and quality be available for retail sale. Inevitably, it is the stability of these supplies and in turn the maintenance of stable market prices for the product on which the success of all aquacultural enterprises depend.

Acknowledgements

The author is grateful for the information and comments of the other contributors to this volume and of the many other collaborators with whom he has worked over the years.

References

Abée-Lund, J.H. & Hindar, K. (1990) Interpopulation variation in reproductive traits of anadromous female brown trout, *Salmo trutta* L. *Journal of Fish Biology*, **37**, 755–63.

Barker, G.A., Smith, S.N. & Bromage, N. (1989) The bacterial flora of rainbow trout, *Salmo gairdneri* and brown trout, *Salmo trutta*, eggs and its relationship to developmental success. *Journal of Fish Diseases*, **12**, 281–93.

Barker, G.A., Smith, S.N. & Bromage, N. (1991) Commensal bacteria and their possible relationship to the mortality of incubating salmonid eggs. *Journal of Fish Diseases*, **14**, 199–210.

Bromage, N. (1988) Propagation and stock improvement. In *Intensive Fish Farming*, (eds J. Shepherd & N. Bromage), pp. 103–50. Blackwell Science, Oxford.

Bromage, N. & Cumaranatunga, R. (1988) Egg production in the rainbow trout. In *Recent Advances in Aquaculture*, Vol. IV (eds J.F. Muir & R.J. Roberts), pp. 63–138. Croom Helm/Timber Press, London and Sydney/Portland, Oregon.

Bromage, N. & Duston, J. (1986) The control of spawning in the rainbow trout using photoperiod techniques. *Report of the Institute of Freshwater Research Drottningholm*, **63**, 26–35.

Bromage, N., Elliot, J.A., Springate, J. & Whitehead, C. (1984) The effects of constant photoperiods on the timing of spawning in the rainbow trout. *Aquaculture*, **43**, 213–23.

Bromage, N., Duston, J., Randall, C., Brook, A., Thrush, M., Carrillo, M. & Zanuy, S. (1990a) Photoperiodic control of teleost reproduction. In *Progress in Comparative Endocrinology*, (eds A. Epple, C. Scancs & M. Stetson), pp. 620–6. Wiley-Liss, New York.

Bromage, N., Randall, C., Jones, J., McAndrew, B. & Rana, K. (1990b) New developments in the control of reproduction of farmed fish. In *Fisheries in the Year 2000. Proceedings of the 21st Anniversary. Conference IFM 10–14 September 1990*, (eds K. O'Grady, A. Butterworth, P. Spillet & J. Domaniewski), pp. 109–18. Institute of Fisheries Management, Nottingham.

Bromage, N., Hardiman, P., Jones, J., Springate, J. & Bye, V. (1990c) Fecundity, egg size and total egg volume differences in 12 stocks of rainbow trout. *Aquaculture and Fisheries Management*, **21**, 269–84.

Bromage, N., Jones, J., Randall, C., Thrush, M., Springate, J., Duston, J. & Barker, G. (1992) Broodstock management, fecundity, egg quality and the timing of egg production in the rainbow trout (*Oncorhynchus mykiss*). *Aquaculture*, **100**, 141–66.

Bromage, N., Randall, C., Duston, J., Thrush, M. & Jones, J. (1993) Environmental control of reproduction in salmonids. In *Recent Advances in Aquaculture*, Vol. IV (eds J. Muir & R. Roberts), pp. 55–66. Blackwell Science, Oxford.

Bromage, N., Shields, R., Basavaraja, N., Bruce, M., Young, C., Dye, J., Smith, P., Gillespie, M., Gamble, J. & Rana, K. (1994) Egg quality determinants in finfish: the role of overripening with special reference to the timing of stripping in the Atlantic halibut, *Hippoglossus hippoglossus*. *Journal of the World Aquaculture Society*, **25** (in press).

Cantin, M.C. & Bromage, N.R. (1991) Environmental control of the timing of smallmouth bass reproduction. *Proceedings of the First International Smallmouth Bass Symposium*, 73–5.

Campbell, P.M., Pottinger, T.G. & Sumpter, J.P. (1992) Stress reduces the quality of gametes produced by rainbow trout. *Biology of Reproduction*, **47**, 1140–50.

Carrillo, M., Bromage, N., Zanuy, S., Serrano, R. & Prat, F. (1989) The effects of modifications in photoperiod on spawning time, ovarian development and egg quality in the sea bass (*Dicentrarchus labrax*). *Aquaculture*, **81**, 351–65.

Carrillo, M., Zanuy, S., Prat, F., Serrano, R. & Bromage, N. (1993) Environmental and hormonal control of reproduction in sea bass. In *Recent Advances in Aquaculture*, Vol. IV (eds J.F. Muir & R.J. Roberts), pp. 43–54. Blackwell Science, Oxford.

Eaton, J.G. (1973) Chronic toxicity of copper, cadmium and zinc to the jarhead minnow (*Pimephales promelas* Ralinesque). *Water Research*, 7, 1723–36.

Espinach Ros, A., Amutio, V.G., Mestre Arceredillo, J.P., Orti, G. & Nani, A. (1984) Induced breeding of the South American catfish, *Rhamdia sapo* (C & V). *Aquaculture*, 37, 141–6.

Fitzpatrick, M.S., Schreck, C.B., Ratti, F. & Chitwood, R. (1987) Viabilities of eggs stripped from coho salmon at various times after ovulation. *Progressive Fish-Culturist*, 49, 177–80.

Formacion, M.J. (1991) *Overripening of ovulated eggs in goldfish*, Carassius auratus. Ph.D. thesis, National University of Singapore, Singapore.

Fortuny, A., Espinach Ros, A. & Amutio, V.G. (1988) Hormonal induction of final maturation and ovulation in the sabalo, *Prochilodus platensis* Homberg, latency and incubation times and viability of ovules retained in the ovary after ovulation. *Aquaculture*, 73, 373–81.

Gerking, S. (1982) The sensitivity of reproduction in fish to stressful environmental condition. In *Reproductive Physiology of Fish*, (compiled by C. Richer & H. Th. Goos), pp. 224–8. Pudoc Press, The Netherlands.

Gillet, C. (1991) Egg production in an Arctic charr (*Salvelinus alpinus*) broodstock: effects of temperature on the timing of spawning and the quality of the eggs. *Aquatic Living Resources*, 4, 109–16.

Hansen, G.M. & Olafsen, J.A. (1989) Bacterial colonisation of cod *Gadus morhua* and halibut (*Hippoglossus hippoglossus*) eggs in marine aquaculture. *Applied and Environmental Microbiology*, 55, 1435–46.

Hardy, R. (1985) Salmonid broodstock nutrition. In *Salmonid Reproduction*, (eds R. Iwamoto & S. Sower), pp. 98–108. Washington Sea Grant Programme, University of Washington, Seattle.

Hay, D.E. (1986) Effects of delayed spawning on viability of eggs and larvae of Pacific herring. *Transactions of the American Fisheries Society*, 115, 155–61.

Hirose, K., Ishida, R. & Sakai, K. (1977) Induced ovulation of ayu using HCG, with special reference to changes in several characteristics of eggs retained in the body cavity after ovulation. *Bulletin of the Japanese Society of Scientific Fisheries*, 43, 409–16.

Hirose, K., Machida, Y. & Donaldson, E.M. (1979) Induced ovulation of Japanese flounder (*Limanda yokohama*) with HCG and salmon gonadotropin, with special references to changes in the quality of eggs retained in the ovarian cavity after ovulation. *Bulletin of the Japanese Society of Scientific Fisheries*, 45, 31–6.

Holmefjord, I. (1991) Timing of stripping relative to spawning rhythms of individual females of Atlantic halibut (*Hippoglossus hippoglossus* L). In *Larvi 91. Fish and Crustacean Larviculture Symposium*, (eds P. Lavens, P. Sorgeloos, E. Jaspers & F. Ollevier), pp. 203–4. European Aquaculture Society Special Publication No. 15.

Howell, B. & Scott, A.P. (1989) Ovulation cycles and post-ovulating deterioration of eggs of the turbot (*Scophthalmus maximus*). *Rapports et Procès-Verbaux des Réunions, Conseil Permanent International pour l'Exploration de la Mer*, 191, 21–6.

Jelmert, A. & Mangor-Jensen, A. (1987) Antibiotic treatment and dose response of bacterial activity associated with flatfish eggs. *ICES CM 1980/F*: 19.

Jones, J. & Bromage, N. (1987) The influence of ration size on the reproductive performance of female rainbow trout. In *Reproductive Physiology of Fish 1987*, (eds D. Idler, L. Crim & J. Walsh), p. 202. Marine Science Laboratory, St Johns, Newfoundland.

Kjørsvik, E. & Lønning, S. (1983) Effects of egg quality on normal fertilization and early development of the cod, *Gadus morhua* L. *Journal of Fish Biology*, 23, 1–12.

Kjørsvik, E., Mangor-Jensen, A. & Holmefjord, I. (1990) Egg quality in fishes. *Advances in Marine Biology*, 26, 71–113.

McAndrew, B.J., Rana, K. & Penman, D.J. (1993) Conservation and preservation of genetic variation in aquatic organisms. In *Recent Advances in Aquaculture*, Vol. IV (eds J.F. Muir & R.J. Roberts), pp. 295–336. Blackwell Science, Oxford.

McEvoy, L.-A. (1984) Ovulatory rhythms and over-ripening of eggs in cultivated turbot, *Scophthalmus maximum* L. *Journal of Fish Biology*, 24, 437–48.

Mollah, M.F.A. & Tan, E.S.P. (1983) Viability of catfish (*Clarias macrocephallus*, Gunther) eggs fertilized at varying post-ovulation times. *Journal of Fish Biology*, 22, 563–6.

Nomura, M., Sakai, K. & Takashima, F. (1974) The overripening of rainbow trout – I. Temporal morphological changes of eggs retained in the body cavity after ovulation. *Bulletin of the Japanese Society of Scientific Fisheries*, 40, 977–84.

Norberg, B., Valkner, V., Huse, J., Karlsen, I. & Grung, G.L. (1991) Ovulatory rhythms and egg viability in the Atlantic halibut (*Hippoglossus hippoglossus* L). *Aquaculture*, **97**, 365–71.

Peter, R., Lin, H. & Van der Kraak, G. (1988) Induced ovulation and spawning of cultured freshwater fish in China: Advances in application of GnRH analogues and dopamine antagonists. *Aquaculture*, **74**, 1–10.

Peter, R.E. (1982) Neuroendocrine control of reproduction in teleosts. *Canadian Journal of Fisheries and Aquatic Science*, **39**, 48–56.

Rowland, S.J. (1988) Hormone-induced spawning of the Australian freshwater fish Murray cod, *Macculochella peeli* (Michell) Percichthyidae. *Aquaculture*, **70**, 371–89.

Sakai, K., Nomura, M., Takashima, F. & Oto, H. (1975) The over-ripening phenomenon of rainbow trout, II. Changes in the percentages of eyed eggs, hatching rate and incidence of abnormal alevins during the process of over-ripening. *Bulletin of the Japanese Society of Scientific Fisheries*, **41**, 855–60.

Sargent, J.R., Bell, M.V., Henderson, D.R. & Tocher, D.R. (1990) Polyunsaturated fatty acids in marine and terrestrial food webs. In *Comparative Physiology*, Vol. 5 (eds R. Kinne *et al.*), pp. 11–23. S. Karger, Basel, Switzerland.

Springate, J., Bromage, N., Elliott, J.A.K. & Hudson, D.L. (1984) The timing of ovulation and stripping and the effects on the rates of fertilization and survival to eying, hatch and swim-up in the rainbow trout (*Salmo gairdneri* L.). *Aquaculture*, **43**, 313–22.

Springate, J., Bromage, N. & Cumaranatunga, R. (1985) The effects of different rations on fecundity and egg quality in the rainbow trout (*Salmo gairdneri*). In *Nutrition and Feeding in Fish*, (eds C.B. Cowey, A.M. Mackie & J.A. Bell), pp. 371–91. Academic Press, London.

Stevens, R.E. (1966) Hormone-induced spawning of striped bass for reservoir stocking. *Progressive Fish Culturist*, **28**, 19–28.

Sumpter, J.P., Carragher, J., Pottinger, T.G. & Pickering, A.D. (1987) The interaction of stress and reproduction in trout. *Proceedings of the 3rd International Symposium on the Reproductive Physiology of Fish, Newfoundland*, 299–302.

Suzuki, R. (1975) Duration of development capacity of eggs after ovulation in the loach cyprinid fish. *Aquaculture*, **23**, 93–9.

Thrush, M., Duncan, N. & Bromage, N. (1994) The use of photoperiod in the production of out-of-season Atlantic salmon (*Salmo salar*) smolts. *Aquaculture*, **121**, 29–44.

Watanabe, T. (1985) Importance of the study of broodstock nutrition for further development of aquaculture. In *Nutrition and Feeding of Fish*, (eds C. Cowey, A. Mackie & J. Bell), pp. 395–414. Academic Press, London.

Watanabe, T., Koizumi, T., Suzuki, H., Satoh, S., Takeuchi, T., Yoshida, N., Kitada, T. & Tsukashima, Y. (1985) Improvement of quality of red sea bream eggs by feeding broodstock on a diet containing cuttlefish meal or on raw krill shortly before spawning. *Bulletin of the Japanese Society of Scientific Fisheries*, **51**, 1511–21.

Chapter 2
Sperm Physiology and Quality

2.1 Introduction
2.2 Morphology
2.3 Sperm production
2.4 Chemical composition of sperm and seminal fluid
    2.4.1 Sperm
    2.4.2 Seminal fluid
2.5 Evaluation of sperm quality
    2.5.1 Motility
    2.5.2 Fertilizing capacity
    2.5.3 Sperm concentration
2.6 Sperm physiology
    2.6.1 Medium for sperm activation
    2.6.2 Analyses of motility
    2.6.3 Mechanisms involved in the initiation of movement
    2.6.4 Mechanisms controlling motility
    2.6.5 Physiology of intra-testicular sperm
2.7 Short-term preservation
2.8 Cryopreservation
2.9 Recommendation on handling of brood males and artificial insemination
2.10 Conclusions
    Acknowledgements
    References

In order to consider all aspects relating to the physiology and the quality of sperm in fish, it is essential that detailed information is available about sperm when it is in the testis and the genital tract (i.e. the *in vivo* situation) and after collection during storage and dilution/fertilization (the *in vitro* situation). The most common parameters employed to study sperm physiology and evaluate its quality are the acquisition of and capacity for motility, the respiration and energetics of motility (intracellular energy store and endogenous and exogenous substrate), the biochemistry of sperm and seminal plasma, the structure of the sperm, various criteria for motility and survival during storage and freezing/thawing and the percentage of fertilization obtained at a known rate of dilution and ratio of sperm:egg.

The quality of sperm is highly variable and depends on various external factors such as the feeding regime, the quality of feed and the rearing temperature of the males. It also varies throughout its development right up to spawning. Sperm survival is better *in vivo* than *in vitro*. Large variability is found between males and also within the same individual. The reasons for such variability are not clear.

It is possible that *in vivo* sperm in contact with or close to the Sertoli cell layer or the sperm duct epithelium are more protected from ageing processes. In some species oxygen supply to sperm stored *in vitro* improves survival probably as a result of the maintenance of respiration and endogenous stores of ATP.

2.1 Introduction

One of the first scientists to study the biology of fish sperm was Spallanzani (quoted by de Quatrefages 1853) who observed fish sperm and showed that the duration of motility was much shorter in carp (*Cyprinus carpio*) (15 min) than in man (8 h). Further work on sperm biology was related to the development of techniques for the artificial reproduction of trout. In the eighteenth century Jacobi in Germany set up a method for artificial reproduction and the incubation of trout eggs (reported by Duhamel du Monceau 1773). The pioneering work of Jacobi remained unused for 70 years until it was rediscovered by two fishermen, Remy and Gehin, in France in the middle of the nineteenth century; this initiated the development of trout farming in Europe (see reviews by Wilkins 1989 and Thibault 1989). At this time several scientists were deeply involved in research on gamete biology and artificial fertilization. Coste (1853) published a treatise on pisciculture, and Vrasski 1856 (probably after Coste 1853 and de Quatrefages 1853, quoted by Wilkins (1989), defined the dry method of fertilization, mixing eggs and semen first and then adding water. Henneguy (1877) noticed that sperm were immotile in the semen and also observed that the duration of motility was very short. The seminal fluid was also studied (Kolliker 1855).

Early in the twentieth century more work was devoted to salmonid sperm, including the role of ions (Scheuring 1924, Gaschott 1925, Schlenk & Kahmann 1938) and osmotic pressure in motility (Huxley 1930, Ellis & Jones 1939). As the culture of fish, especially of salmonids, progressed in North America and Japan, further work was carried out in these countries and extended to other species (Reisenbichler 1882, Rutter 1902, Nakano & Nazawa 1925).

In the 1940s and 1950s some work on fish sperm concentrated on gamones (Hartmann *et al.* 1947, Medem *et al.* 1949), on the fertilizing capacity of the sperm (Hey 1939, Smith & Quistorff 1943, Shuman 1950), and on sperm motility in various media, (Smirnov 1963). This earlier literature was reviewed by Mann (1964), Scott (1981) and Stoss (1983).

The foregoing account shows clearly that more research was carried out as fish culture expanded. It was, however, mainly limited to salmonids which were artificially propagated. Cyprinids were also widely cultivated, but reproduction was allowed to occur naturally in ponds without artificial insemination and there was no study of the semen of cyprinids. It should also be pointed out that milt is more easily removed from salmonids than from cyprinids. Since the 1950s research on fish sperm has considerably expanded, dealing with a much wider range of species and with many aspects of sperm biology, morphology, motility, energetics of movement, and biochemistry of the seminal fluid. However, most of these works have been carried out on freshwater species and information on the sperm of marine species is still limited.

Research has also been conducted using model laboratory species such as the guppy *Poecilia reticulata*, which are easily and cheaply reared in the laboratory. However, most of the time the species studied are of economic interest (i.e. cultivated) with female broodstock available to test the fertilizing capacity of the sperm. This work was generally undertaken to set up techniques of artificial insemination and short term preservation and cryopreservation of sperm. These results were reviewed by Scott & Baynes (1980), Scott (1981) and Stoss (1983) and will be discussed below with more recent studies concentrating on the teleost fish species together with the sturgeons, which are also commonly studied owing to the interests of fish culturists.

2.2 Morphology

Knowledge of the morphology of sperm has progressed considerably since the development of the technique of electron microscopy. Pioneering work was carried out by Geiger (1955) and Mattei (1970). These were exhaustively reviewed by Jamieson (1991), who also imparted an evolutionary perspective. An acrosome is present in agnathans (hagfish and lampreys) and in all groups of fish except the teleosts. However, in this group temporary acrosome-like structures have been reported for example in *Lepadogaster lepadogoster* (Mattei & Mattei 1978), rainbow trout, *Oncorhynchus mykiss* (Billard 1983a), in *Gambusia affinis* and others (see Jamieson 1991).

As often noticed, the acrosome may not be necessary for fertilization in teleosts because of the presence of a micropyle, but one would expect specialized structures on the plasma membrane at the top of the sperm head to allow cell fusion during fertilization. The shape of the nucleus is highly variable and is related to the complexity of spermatogenesis and especially spermiogenesis (Grier 1981, Billard *et al.* 1982, Billard 1986, 1990a).

It is elongated, for example, in the guppy, in its final form after complex morphogenetic movement; slightly elongated in salmonids with a quasi-symmetry of revolution; and very primitive in cyprinids and many other species such as tilapia (Fig. 2.1), mullet and turbot, with the flagellum inserted laterally on the head. The nucleus is highly polymorphic, filiform, spherical or blade like (Jamieson 1991). The chromatin is highly condensed in the guppy (Fig. 2.1) and trout sperm where the histones have been replaced by protamines.

In tilapia (Fig. 2.1), as in cyprinids, where the replacement of histones by protamines is limited, the chromatin is less condensed. The mid-piece is well developed in the guppy and much reduced in size in salmonid, cyprinid, mullet and turbot sperm. It is usually located at the posterior part of the nucleus but it is sometimes found in the anterior part (elopomorph). In some primitive sperm it may be nearly as big as the nucleus, e.g. in tilapia (Fig. 2.1).

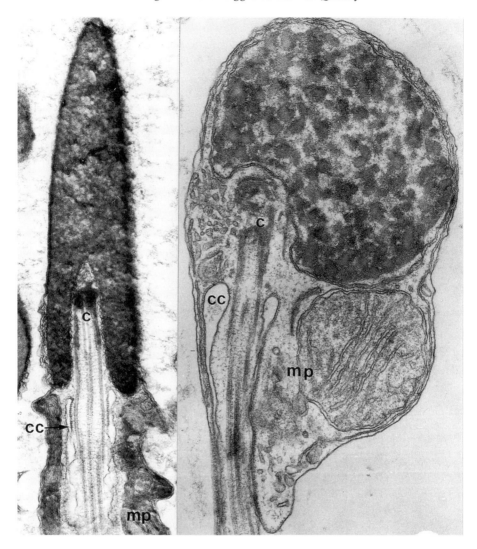

Fig. 2.1 Photomicrographs of fish sperm: left, *Poecilia reticulata* × 25 000 (evolved spermatozoon); right, *Oreochromis mossambicus* × 40 000 (primitive spermatozoon): c, centrioles; cc, cytoplasmic canal; mp, mid-piece.

The number of mitochondria is highly variable between species, up to 70 in the chondrichthyan *Hydrolagus collei* (Jamieson 1991). The mitochondria also often shows a collar-like appearance. The size of the mid-piece has been sometimes related to the duration of motility (see below). Remnants of cytoplasm are often present in the primitive sperm type (Fig. 2.1). In teleost fish the mid-piece is never in close contact with the axoneme, from which it is separated by the cytoplasmic canal (Fig. 2.1). The axoneme is inserted in the nucleus in a basal fossa which may be very deep in some species.

More or less complex structures for attachment (microtubules derived from the centrioles, filaments, osmiophilic rings, various satellites and rays) can be seen (Billard 1970a). These determine the solidity of anchorage of the flagellum. The attachment is, for instance, weaker in carp than in trout and the carp sperm is very sensitive to strong agitation. In fish there is usually one flagellum present, sometimes two. The axoneme is 9 + 2, but 9 + 0 is found in the groups of elopomorph which include the eel *Anguilla anguilla*. Accessory fibres and structures, sometimes attached to the axoneme and lateral expansions of the plasma membrane, are commonly found.

2.3 Sperm production

The production of sperm has been quantified in several species. In the guppy, with continuous spermatogenesis, the number of sperm released is 150×10^6 sperm per g of testis per day (or 55×10^9 per year) (Billard 1969). In fish species having a seasonal reproduction the total production is found in the testis just before the onset of spermiation. Testis size is a good indicator of the efficiency of spermatogenesis and is quite variable according to the species, with sizes varying from 0.2 to 10% of the body weight (Billard 1986, 1987). Total production in rainbow trout is 58×10^9 sperm per g testis per year, but the amount released during spermiation range from 20 to 50% (Billard 1990a).

From comparison between species, sperm production can be expressed in 10^9 sperm per g body weight; it was 7 for rainbow trout, 4 for carp, 2.7 for guppy, 0.6 for pike and 0.1 for *Leporinus* (Billard 1986). In practice only a part of this production can be collected and, especially in the case of oligospermic species, care should be taken to conserve and save sperm in order to manage properly the production available during the reproductive season. The number of sperm per ml of semen is highly variable from 2×10^6 to 6.5×10^{10} according to a review by Leung & Jamieson (1991).

2.4 Chemical composition of sperm and seminal fluid

2.4.1 Sperm

The biochemistry of sperm was reviewed recently by Billard & Cosson (1990) and Linhart *et al.* (1991).

The ionic composition has been reported for some species in mMol l^{-1}:

	Na^+	K^+	Mg^{++}	Ca^{++}	
Atlantic salmon	36.5	76.2	0.8	0.03	Hwang & Idler (1969)
Cod	77.4	60.6	0.2	0.4	Hwang & Idler (1969)
Grass carp	35.7	2.1	1.3	2.6	Gosh (1983)
Turbot	133.0	3.8			Suquet (1992)

The organic composition of sperm of several cyprinids was described by Belova (1982), who showed that the water content was in the range of 71–82% of fresh weight, whereas lipids varied from 2.5 to 3.6%, phospholipids represented 36–40% of the total lipid fraction, cholesterol 26–32% and triglycerides 0.6–2.2%. Labbé et al. (1991) showed that the fatty acid composition of the rainbow trout sperm is directly influenced by the composition of the diet.

Glycogen has been detected by histochemistry but only in species exhibiting internal fertilization, e.g. in the mid-piece of the sperm of the guppy (Billard & Jalabert 1973). Glycogen was also found in the toadfish (*Opsanus tau*) sperm but not in the rainbow trout (Billard & Cosson 1990). Jamieson (1991) reported the presence of glycogen in the flagellum of *Hydrolagus collei*. Enzymes involved in the energetic metabolism were identified, ATPases, phosphatases, lipases, esterase, oxidases, several uricolytic enzymes, LDH, NAD and NADP malicdehydrogenase (Billard & Cosson 1990). King et al. (1990) identified ATPase in the outer-arm dynein in trout sperm. ATP, ADP, cAMP, lactate, pyruvate, creatine phosphate and glucose 6 phosphate were also found by Dzuba & Cherapanov (1992).

2.4.2 Seminal fluid

The various data available show considerable intra- and inter-species variability in the composition of the seminal fluid of oviparous fish. The ionic composition is in the range of 103–140 mM Na^+, 20–66 mM K^+, 0.8–3.6 mM Mg^{++} and 0.3–2.6 mM Ca^{++} in salmonids. In cyprinids the range is 94–107 mM Na^+, 39–78 mM K^+, 0.02–1.2 mM Mg^{++}, 0.3–12.5 mM Ca^{++}. The ionic composition may change during the reproductive season (Linhart et al. 1992). Organic constituents are also variable; interspecies range (in mg l^{-1}) was 8–220 for glucose, 0–218 for fructose, 0–40 for cholesterol, 0–1316 for lipids, 35–391 for glycerol, 0.4–2800 for proteins [up to 12 000 in turbot (Suquet 1992)], 84–136 for amino acids, 12–136 for urea (Billard & Cosson 1990). In turbot osmotic pressure ranges from 232 to 401 mosmols kg^{-1} and pH from 7 to 8.3 (Suquet 1992). Linhart et al. (1991) reported in carp seminal fluid 1.7 nM pyruvate, 243 nM lactate, 15 nM malate, 36 nM citrate and 2.8 nM alpha-ketoglutarate. Similar values were found in grass carp. High levels of vitamin C were reported in the seminal fluid of several species of freshwater fish (K. Dabrowski, pers. comm.).

Enzymatic activities have been identified in seminal fluid, including LDH and MDH, acetate and butyrate esterases, alanyl and leucine aminopeptidase in carp and tilapia (Billard & Cosson 1990) and proteinase inhibitors in rainbow trout, whitefish (*Coregonus*) and yellow perch (*Perca flavescens*) (Dabrowski & Ciereszko 1992).

2.5 Evaluation of sperm quality

2.5.1 Motility

Motility is the most commonly used parameter to evaluate sperm quality. This parameter is acceptable, as in general sperm must be motile to achieve fertilization, although in some cases sperm which exhibit some motility may not be fertile (e.g. frozen/thawed carp sperm, Cognie et al. 1989). Sperm are usually immotile in the genital tract and are activated after dilution in the external medium. To test motility, semen is diluted in an appropriate diluent and examined under a microscope.

Evaluation of motility requires some care. A relatively high dilution is necessary (at least 1:1000) to initiate synchronously the motility of all the sperm. At lower dilutions the sperm are not all activated and initiation occurs progressively over a few minutes after dilution. It is therefore difficult to assess accurately the intensity and duration of sperm motility. Insufficient and irregular levels of dilution may explain many of the discrepancies in the literature. However, increasing dilution rate decreases sperm motility in sea bream (Chambeyron & Zohar 1990) and in turbot (Suquet 1992).

Fish semen is viscous and difficult to mix instantaneously with diluent unless a high dilution is made or if the sample is strongly agitated, which is harmful to the semen. After high dilution, a homogenous sperm suspension is obtained which is suitable for observation of synchronous motility and for the studies of the biochemical changes which occur during and after activation. A suitable procedure based on a two step dilution is used. First a 100 fold dilution is made in a test tube in a medium which does not activate sperm, usually with a diluent of the same osmotic pressure as the seminal fluid or with additional K^+ in required amounts as in salmonids (see below). Sperm activation is triggered at the second dilution step, which is carried out directly on the microscope stage. One μl of the prediluted sperm solution is mixed with 19 μl of the activating solution of the appropriate osmotic pressure (depending on the species) previously placed on the glass slide (Billard & Cosson 1992).

Several methods are used to measure motility. The most commonly used in the past was the estimation of the global movement of the sperm according to an arbitrary scale, usually 0 to 5 units (Goryczko & Tomasik 1975, Sanchez-Rodriguez & Billard 1977). The duration of motility refers to the total duration of the motility, including agitation of the flagellum without displacement (Morisawa et al. 1983, Billard & Cosson 1986) or to the duration of forward motion (Linhart et al. 1992) and the survival time of 50% of the sperm (Hines & Yashow 1971). The duration of motility is sometimes combined with estimates of swimming intensity (Baynes et al. 1981). More recent studies refer to the percentage of motile sperm (Billard et al. 1987, Cosson et al. (1991), Boitano & Omoto 1991, Miura et al. 1992).

More quantitive approaches to studies of sperm motility are the measurement of the beat frequency of the flagellum (Cosson *et al.* 1985) and photographic or video records of the head and/or the tail movement tracks allowing the measurement of the speed and the total displacement of the sperm during its life span, which is a parameter integrating most of the factors influencing motility (Billard & Cosson 1989, 1992). Motility is also studied experimentally with the technique of demembranation and reactivation *in vitro* (Cosson *et al.* 1991).

2.5.2 Fertilizing capacity

The fertilizing capacity is the most conclusive test of sperm quality. It is currently used in most studies on artificial insemination and sperm preservation. However, it integrates an independent factor which is the 'quality' of eggs and the interaction between gametes and between seminal and ovarian fluids. Care must be taken when trying to establish the fertilizing capacity and make comparisons between male or sperm treatments. Insemination must be carried out at different dilution rates in order to establish the minimum number of sperm to fertilize all fertilizable eggs in a well defined volume of diluent (Fig. 2.2).

The diluent used must be appropriate for both male and female gametes. For instance in carp (*Cyprinus carpio*) a good diluent for sperm is a KCl solution (50 mM) (Fribourgh 1965, Saad & Billard 1987) but the eggs are not fertilized when they are diluted in saline solution which contains more than 5–10 mM K^+ (Saad & Billard 1987). The percentage fertilization, which usually refers to the percentage of eyed eggs or hatched larvae, is also critical, but evaluating the percentage of developing larvae to first feeding may be a better index of gamete quality.

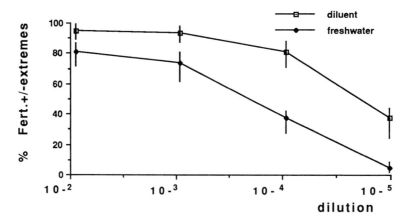

Fig. 2.2 Change in the percentage of fertilization in rainbow trout eggs (measured as the percentage of eyed eggs) with the sperm dilution ratio and with a saline solution (diluent) of fresh water; average with extreme values. (After Billard *et al.* 1974.)

2.5.3 Sperm concentration

The concentration of sperm in the milt is often used to characterize semen. An accurate way is to count the number of sperm with an haemocytometer or by spectrophotometry (Suquet *et al.* 1992). Spermatocrit (volume of sperm:volume of milt) is established after semen centrifugation in microtubes at 10 000 g for at least 5–10 min (Bouck & Jacobson 1976). However, this does not necessarily measure the number of sperm, as shown in turbot (Suquet *et al.* 1992), since individual sperm volumes may change during the year as in sea bass (R. Billard, unpublished data) or halibut (L.W. Crim, unpublished data).

2.6 Sperm physiology

2.6.1 Medium for sperm activation

Osmotic pressure, ionic composition and pH are the most important factors determining the activation of sperm. Motility occurs in a wider range of osmotic pressures in marine fish than in freshwater species (Fig. 2.3). Changes in osmotic pressure are the factor most commonly known to trigger the initiation of sperm motility, rising in comparison with seminal plasma in marine fish or decreasing in freshwater fish. However, factors other than osmotic pressure might be involved in triggering sperm motility in some marine fish (Chambeyron & Zohar 1990).

In salmonids, the immobility of sperm is due to high concentrations of K^+, and its decrease during dilution in freshwater initiates motility (Schlenk & Kahmann 1938). Ca^{++} is also required for the initiation of motility in salmonids (Morisawa & Morisawa 1990, Cosson *et al.* 1989). pH has been reported as a major sperm activation factor in other species such as the mullet (Hines and

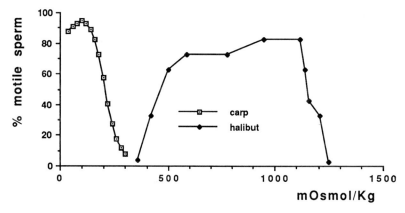

Fig. 2.3 Percentage of motile sperm observed after semen dilution in saline solutions of increasing osmotic pressure for a freshwater (carp) and a marine (halibut) species. (After Redondo *et al.* 1991 and Billard *et al.* 1993.)

Yashow 1971), turbot (C. Fauvel, unpublished data) and halibut (Billard et al. 1993) (Fig. 2.4).

The fertilizing capacity of sperm is also modified by the pH of the diluent as in carp, (optimum 7−8) (Saad & Billard 1987) and 7.5−8 in the grouper (*Epinephelus malabaricus*) (Chao *et al.* 1992). In pike (*Esox lucius*) the optimum value has been reported as pH 8 or more (Duplinsky 1982) or 8.5 (J. Marcel & R. Billard, unpublished data), but only at high osmotic pressures (Fig. 2.5). In the white sucker (*Catostomus commersoni*), Mohr & Chalanchuk (1985) observed sperm motility over a wide range of pH (3−7). Appropriate media were used instead of water for sperm and egg dilution and artificial insemination (Nomura 1964, Billard 1983b, 1985, 1990b).

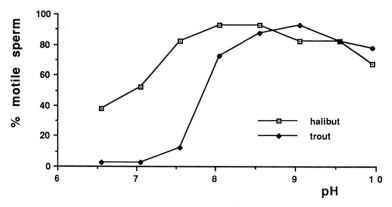

Fig. 2.4 Percentage of motile sperm after dilution in balanced NaCl solution (1 mM Ca^{++} supplemented) of various pH in halibut (Billard *et al.* 1993) and rainbow trout (M.P. Cosson *et al.*, unpublished data).

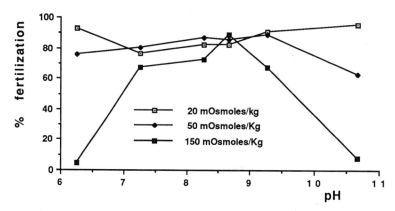

Fig. 2.5 Changes in fertilizing capacity of pike sperm diluted in medium for artificial insemination of increasing salinity (NaCl) and pH. (After J. Marcel & R. Billard, unpublished data.)

2.6.2 Analysis of motility

In most teleost species with external fertilization studied so far, the activity of sperm is brief and its intensity, which is maximal immediately after dilution, declines during the period of motion. Measurements of beat frequency on three species (Fig. 2.6) show that the duration of motility is very short in trout (20–25 s) and lasts slightly longer than 1 min in carp and halibut. The beat frequency of the majority of sperm declines progressively within 20–25 and 80–90 s respectively in trout and carp. In halibut it remains rather stable over the first 55 s but then drops suddenly to values around 10–15 Hz; this activity lasts for periods of up to 15–20 min in a limited number of sperm and involves only the anterior part of the flagellum.

The velocity of the spermatozoa analysed from photographs and video tape recordings declines in a way similar to the beat frequency (Fig. 2.7) and the total displacement can be estimated at 250–300 µm for trout and around 1 mm for halibut. The pattern of motility depends on the temperature. The total duration of the mass progressive movement decreases and the beat frequency increases as the temperature is raised from 5 to 25°C in trout (Fig. 2.8). Motility was shown to be highly variable between males and successive samplings for the same male (Tomasik 1973) (Fig. 2.9) and during the reproductive season in trout and sea bass (Fig. 2.10) and halibut (Methwen & Crim 1991). This was related to a decline of fertilizing capacity in salmonids (Aas *et al.* 1991).

2.6.3 Mechanisms involved in the initiation of movement

This question was studied in salmonids, where sperm motility is inhibited by high extracellular [K$^+$] concentration and can be activated by dilution of

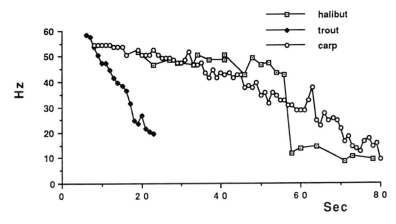

Fig. 2.6 Changes in the beat frequency (Hz) of trout, halibut and carp sperm after dilution in the specific activating (NaCl) solution: 250 mOsm kg^{-1} and pH 9 for rainbow trout, 120 mOsm kg^{-1} and pH 7 for carp, 750 mOsm kg^{-1} and pH 8 for halibut.

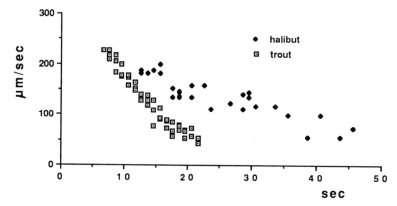

Fig. 2.7 Changes in the sperm velocity during the period of mass progressive movement in halibut (Billard *et al.* 1993) and in rainbow trout (Billard & Cosson 1992).

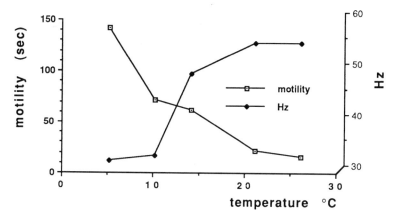

Fig. 2.8 Inverse relationships between the mass progressive motility (total duration at pH 9) and the beat frequency of the flagellum (Hz) with the increase of temperature in rainbow trout. (After Billard & Cosson 1992.)

extracellular [K^+]. Boitano & Omoto (1991) have shown that this dilution induces a membrane hyperpolarization which triggers activation, without increases in intracellular pH, although an internal pH increase was reported some time after dilution by Robitaille *et al.* (1987) and Gatti *et al.* (1990).

Considering that specific voltage-dependent inhibitors of Ca^{++} channels such as verapamil (Tanimoto & Morisawa 1988), especially when in its desmetoxy form, inhibit motility (Fig. 2.11) and that an influx of Ca^{++} is observed in the cell after dilution (Cosson *et al.* 1989, Boitano & Omoto 1992, Boitano *et al.* 1991, Okuno 1991), it is suggested that the entry of Ca^{++} is involved in the activation of motility. Ca^{++}, like other divalent ions, may have an extracellular effect by masking the charges and contributing to the hyperpolarization of the membrane (Boitano & Omoto 1991).

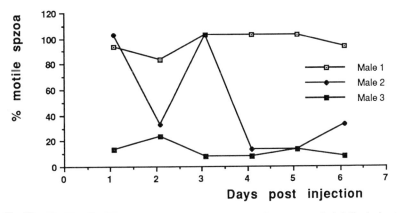

Fig. 2.9 Changes in the motility of sperm in three selected male carp sampled daily during 6 days after stimulation of spermiation by carp pituitary extract, $3\,\mathrm{mg\,kg^{-1}}$ body weight given once at day 0. (After Redondo et al. 1991.)

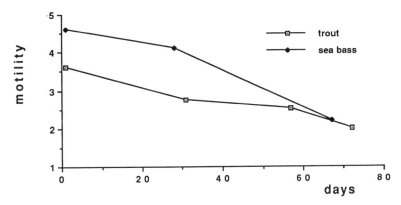

Fig. 2.10 Changes in intensity of motility (arbitrary units) during the reproductive season (days after the beginning of spontanaeous spermiation in rainbow trout (after Yazbeck Chemayel 1975) and sea bass (Billard et al. 1977).

Morisawa & Morisawa (1990) have proposed that this Ca^{++} influx increases the intracellular level of cAMP and initiates motility. However, measurement of cAMP, carried out on sperm activated at 2°C, shows a slow rise of cAMP which reaches a peak long after 100% of the spermatoza are activated (Fig. 2.12). There is also some interaction between cAMP and ATP (Cosson et al. 1991).

The capacity for movement of salmonid sperm is acquired only a few weeks after the completion of spermatogenesis. Muira et al. (1992) have suggested that a progestagen (17α, 20β dihydroxy-4-pregnen-3-one) increases the pH in the sperm duct, which in turn increases the intracellular cAMP in sperm, allowing the acquisition of motility (see also Morisawa et al. 1991). It was

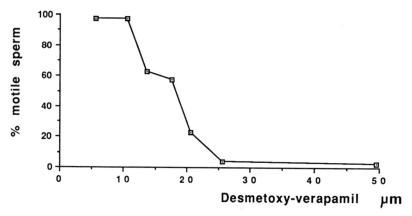

Fig. 2.11 Inhibition of motility of trout sperm by addition of increasing amounts of desmetoxyverapamil in the incubation medium. (After Cosson *et al.* 1989.)

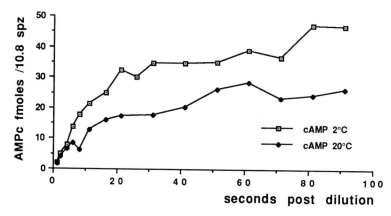

Fig. 2.12 Changes in cAMP in fmoles per 10^8 sperm) during period of motility in rainbow trout sperm activated and incubated at 2°C and 20°C. (After F. André *et al.*, unpublished data.)

reported by Billard & Cosson (1992) that the motility of intratesticular sperm is not inhibited by [K$^+$] but can be activated by addition of Ca^{++}.

Although some progress has been made recently, the mechanism of initiation of motility in salmonids is not completely elucidated, especially the events in the intracellular environment. In other species where sperm activation is due to changes in osmotic pressure, the mechanism remains entirely unknown.

2.6.4 Mechanisms controlling motility

In some species the oxidative capacity of the mitochondria is low and the energy (ATP) required for motility seems to be mainly pre-accumulated in the sperm before its release from the testis. Subsequently, this reserve becomes exhausted during motility. Fig. 2.13 shows this for trout (Christen *et al.* 1987)

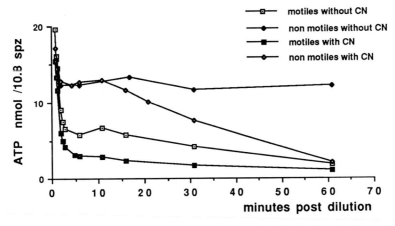

Fig. 2.13 Changes in the intracellular ATP in carp sperm after dilution in activating solution (45 mM NaCl, 5 mM KCl, 30 mM Tris, pH 7) or immobilizing solution (KCl 200 mM, Tris 30 mM, pH 7) alone or with 10 mM KCN added. ATP was measured by the technique of Christen *et al.* (1987) and expressed in nmoles per 10^8 cells. (After G. Perchec *et al.*, unpublished data.)

and for carp. In both species the blockage of respiration by KCN does not block motility but in trout it prevents the reaccumulation of ATP which occurs spontaneously within 15 min (Christen *et al.* 1987). This recovery was not observed in carp sperm left in the initial dilution medium (Fig. 2.13).

The situation also seems different in other species; for instance, in halibut, the blockage of respiration by KCN or NaN_3 inhibits motility (Fig. 2.14), suggesting that the ATP is not pre-accumulated but elaborated by the mitochondria, whose oxidative capabilities remain high whilst the sperm is in

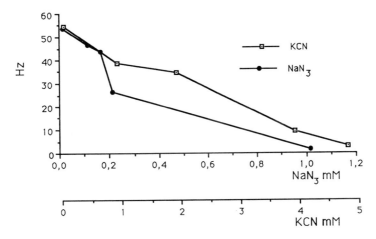

Fig. 2.14 Progressive inhibition of motility of halibut sperm measured by flagellum beat frequency (Hz) after dilution in a saline solution (0.6 M NaCl, 30 mM Tris, pH 8.4) with increasing amounts of KCN or NaN_3 (Billard *et al.* 1993).

motion. This may explain why the beat frequency remains at the same level during the time of motility (Fig. 2.6). However, reasons for the sudden drop in frequency are not known although they may be related to membrane alteration.

Respiration occurs in sperm as shown by the consumption of O_2:110–140 µl O_2 per 10^{10} cells per h in sucker, 20–40 µl in Atlantic salmon and cod and 25–29 µl in trout (quoted in Robitaille *et al.* 1987). Respiration is at least necessary for the maintenance of the ATP levels for these species, which during motility rely on endogenous energy stores. Experiments show clearly that the blockage of respiration leads to a depletion of ATP in salmonids (Christen *et al.* 1987, Robitaille *et al.* 1987).

The energetics of sperm motility was reviewed by Billard & Cosson (1990). A glycolytic mechanism exists in spermatids during spermatogenesis in teleost fish but becomes less significant in sperm. It remains efficient in poecilid sperm and probably in most species with internal fertilization but seems to be partly lost in sperm of species such as salmonids. This is due to the impermeability of the sperm membrane to glucose and to a reduced glycolytic activity.

The presence of LDH in seminal fluid suggests the contribution of lactic acid fermentation. The Krebs cycle is obviously involved in providing energy as specific enzymes such as MDH and pyruvate kinase have been identified (see [2.4]). Energetic substrates, monosaccharides, lipids and amino acids are found in the seminal fluid. In the sperm they probably contribute by providing energy for the maintenance of ATP and other nucleotide triphosphates. Substrates in the mid-piece are more likely to be involved in the energetics of motility of sperm, which seem to elaborate their ATP as they move, e.g. in the halibut. The exhaustion of such substrates may be the reason for the short duration of motility of the halibut sperm. The size of the mid-piece (see Fig. 2.1) may then have some significance and may be related to the duration of movement or its metabolism.

Another explanation for the sudden drop in beat frequency of halibut may be the lack of a shuttle molecule such as phosphocreatine (Christen *et al.* 1987, Robitaille *et al.* 1987) for the transportation of ATP to the distal part of the flagellum. The fact that the wave is seen only in the anterior part of the flagellum close to the mid-piece after the drop of the beat frequency (Fig. 2.6) would indicate that ATP is distributed only by diffusion, thus favouring a hypothesis involving deficiency of phosphocreatine or a related molecule. Research efforts in this field of energetics at the moment remain limited.

2.6.5 Physiology of intra-testicular sperm

In salmonids the sperm of sex-reversed males cannot be collected by stripping due to the abnormalities in the duct. Consequently, it has to be surgically removed. The sperm from the testis and its quality is highly variable. In this case there is no contribution of the sperm ducts to the formation of seminal fluid and the physiology of such sperm remains unknown.

2.7 Short-term preservation

Sperm is often collected from males and stored *in vitro* for short periods of time. Sperm survival is highly variable between species, between individual males and over time within individuals. In salmonids sperm *in vitro* in full semen survives for 1 to several days at temperatures of around 1–4°C (Carpentier & Billard 1978). This period can be prolonged from several weeks to 1 month by the addition of antibiotics and oxygen (Billard 1981, Stoss 1983). A similar situation is reported in the grouper (Chao *et al.* 1992) and carp (Saad *et al.* 1988) but there can be a large variability in the sperm quality (Fig. 2.15 and Redondo *et al.* 1991).

In all cases sperm survival is much shorter *in vitro* than *in vivo*, i.e. when the sperm is left in the sperm duct (Billard *et al.* 1981 and Fig. 2.15). One reason for the longer survival in the genital tract may be the better availability of oxygen, which contributes to keeping a higher level of ATP at least for species depending on endogenous stores of nucleotide triphosphates. Work by Christen *et al.* (1987) (see also Fig. 2.13) and Robitaille *et al.* (1987) indicates that when inhibitors of respiration are added to immotile sperm the level of ATP declines rapidly.

Sperm stored in the genital tract may benefit from an environment which provides ions, various substrates and steroids which are metabolized by the sperm. This has been as shown to be true for the trout (Ueda *et al.* 1984) and carp (Asahina *et al.* 1990, Barry *et al.* 1990). The seminal plasma is not, however, a good extender for *in vitro* storage because of the presence of a large variety of proteases which may induce some sperm alteration. Washed sperm may sometimes show a better capacity for short term storage (Cosson *et al.* 1985, Saad *et al.* 1988, Redondo *et al.* 1991). However, seminal fluid

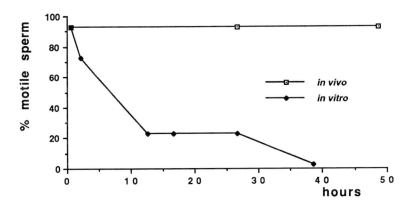

Fig. 2.15 Comparison of survival (percentage of motile sperm) of halibut sperm sampled from the sperm duct every day (*in vivo*) or sperm collected and stored *in vitro* 2 days on ice. (After L.W. Crim *et al.*, unpublished data.)

freshly taken from the sperm duct contains proteinase inhibitors (Dabrowski & Ciereszko 1992) and high concentrations of vitamin C which may protect membranes from oxidation (K. Dabrowski, pers. comm.).

In the genital tract, sperm may be more protected owing to their attachment or close proximity to the epithelium, as has been reported in mammals (Smith & Yanagimachi 1990, Raychoudhury & Suarez 1991). Close contact between sperm heads and somatic cells may exist in some fish species; it is common in the guppy, in the Sertoli cells during spermiogenesis (Billard 1970b,c) and in the epithelial cells of a pouch in the ovarian cavity of the female, where sperm can survive for several months (Jalabert & Billard 1969).

The function of such attachments is not known; the somatic cells may provide O_2, substrates or materials to restore sperm membranes which, because of the quiescent stage of the nucleus and the small size of the mid-piece, may have only a limited turnover. Such a favourable situation may apply only to a limited number of spermatozoa, which as a consequence may survive longer. This may explain in part the heterogeneity in sperm quality which is seen as the reproductive season progresses. In mammals, Smith & Yanagimachi (1990) considered that the only sperm which survive in the isthmus are the ones in close contact with the epithelium.

2.8 Cryopreservation

The present state of the art regarding the cryopreservation of fish sperm was reviewed by Stoss (1983), Leung & Jamieson (1991) and Billard (1992) for trout. It is also considered further in the present volume (see Chapter 3). Leung (1991) also summarized the principles of cryopreservation. Numerous works have been devoted to this subject and H. Stein (pers. comm.) has estimated that sperm from 200 fish species have been cryopreserved. Despite this attention few practical applications have been achieved so far. One was reported by Chao et al. (1992) for grouper in hatcheries in Taiwan. Some sperm banks have been established for cultivated salmonids (Rana this volume Chapter 3) and there are some attempts to establish sperm banks for saving some endangered species. Most of the work carried out so far is empirical, based on a trial and error approach, leading to heterogenous results.

A large number of parameters have to be taken into account, such as the extenders, the cryoprotectants, the dilution ratio, the freezing and thawing rates and the extenders for insemination. Extenders are usually based on saline or sucrose solutions at an osmolarity which inhibits motility with various compounds added (milk, proteins, amino-acids, hen yolk). A range of permeating cryoprotectants have been used but the most common are DMSO (7–10%) glycerol (up to 20%), methanol (15%) ethylene glycol (7%), and propane-diol (7–10%).

The dilution ratio is highly variable; one volume of semen is diluted with three to ten volumes of extender. Usually no equilibration time is allowed.

Freezing is accomplished on dry ice (pellet) or in straws in liquid nitrogen vapour at a rate of 10 to 45°C per min until the temperature reaches −70°C. Afterwards, the pellets and straws are plunged directly into liquid nitrogen. Thawing rate must be fast and pellets or straws are placed in a water bath at 30–40°C for 10–15 s, after which sperm are put directly onto the eggs previously mixed with the insemination diluent (activating solution) usually with one volume of diluent for each volume of eggs.

Post-thaw motility is very short in salmonids (5–10 s). It is nearly impossible to evaluate the percentage of motile cells or the beat frequency (J. Cosson et al. personal communication). The relatively high percentage of fertilization reported for salmonids after thawing (up to 80%) is due to the large amount of sperm inseminated (10–100 times more than the required number of fresh sperm). In fact in the grayling Thymallus thymallus, Lahnsteiner et al. (1992) have shown that only 10% of the sperm are intact after thawing. Thawed halibut sperm show a relatively good percentage of motile sperm (Fig. 2.16) and velocity compared with controls (Fig. 2.7). This raises again the problem of sperm quality and heterogeneity. The capacity of the sperm to be frozen and thawed depends on a large variety of factors. As mentioned before it depends on the species (Amann 1991), on individual males and on the advancement of the reproductive season.

The state of the sperm membrane is probably an important determinant of the success of cryopreservation. Membrane degradation, identified by the presence of a 42 KD protein in the seminal plasma, is an indicator of poor quality sperm in trout (Maisse et al. 1988). Membrane resistance to swelling, bursting

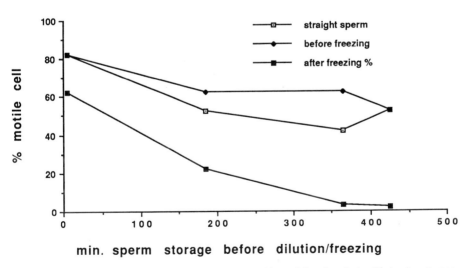

Fig. 2.16 Change in the capacity of halibut sperm to withstand freezing during 7 h *in vitro* storage. Semen was sampled at zero time and stored on ice; control test (straight sperm and diluted sperm without freezing) and freezing test were carried out after sampling, 180–360 and 420 minutes later. Survival was measured by the percentage of motile sperm (Billard et al. 1993).

and stainability of DNA with propidium iodine in trout sperm exposed to a hypotonic solution correlated with a good freezing capacity (Malejac et al. 1990). The ratio of cholesterol: phopholipids in sperm (mostly membranes) is 2–3 times bigger than in freshwater fish species and would explain why sperm of marine fish are easier to freeze (Drokin & Kopeika 1992). Sperm quality and survival after freezing in trout is also influenced by the rearing temperature of the males during the period of spermatogenesis (Labbé et al. 1991).

It has been reported that trout sperm frozen immediately after sampling have a better post-thaw fertilizing capacity than sperm left *in vitro* for a few hours (Stoss & Holtz 1983). Halibut sperm sampled and stored on ice gradually lose their capacity to withstand freezing over a 6–7 h period while control sperm stored *in vivo* maintain a good motility, even when diluted with the extender without freezing (Fig. 2.16).

2.9 Recommendation on handling of brood males and artificial insemination

In the fish farm the male may be reared in the same condition as for the female with good quality food and density usually less than for growing-on fish. Males and females may be put in the same raceway or pond at least during the early stages of the reproductive cycle but they must be separated at the time of spawning to avoid *in situ* spontaneous reproduction and the male agressive behaviour.

In males the secondary sexual characters such as a dark or red color and the hook in the lower and upper jaws are sometimes used to assess the proximity of spermiation. The establishment of spermiation is usually progressive. The very beginning can be detected by pressing the abdomen and checking the extrusion of minute amounts of milt. Such a low production may last for a few weeks. If necessary such sperm may be used for fertilization although the volume is low. The quality must also be checked (percentage of motile sperm and intensity of motility). During stripping the semen is often contaminated with urine which induces some alteration in the composition of the seminal fluid and partly initiates motility of spermatozoa. When the volume of semen collected is low the production of urine is higher and the consequent effects on sperm quality are more significant. This may be avoided by collecting the semen in a tube containing a sperm immobilizing solution.

Broodfish are generally of a large size and handling is facilitated by anaesthesia. Special care is required when handling males. It is essential to avoid wounds on which fungus will rapidly develop. The use of dip nets made of rough materials (e.g. nets with knots or metallic nets) should be avoided. Ideally fish should be kept in some water during handling, for instance in a deep net with a long pouch made of impermeable material and open at the distal end. When out of the water the fish must be kept on same form of moist support, e.g. a wet towel or wet polyurethane foam pad. Drying fish with a dry towel should be avoided because this results in mucus removal and skin abrasion.

In the traditional technique of artificial insemination semen is stripped directly on top of the eggs; eggs and sperm are mixed together and water is added. This does not allow a precise dosage of sperm and water is not necessarily the best diluent. It is advisable to strip males and females separately; sperm may be collected first and sperm stored until eggs are stripped. Sperm is collected directly into a dry dish, pan or test tube. Semen of oligospermic males may be collected in a syringe (Fig. 2.17). Eggs are stripped directly in a dish or pan (Fig. 2.18). The detailed procedure of artificial insemination varies according to the species (see Billard 1990b, 1992).

The question of diluent was discussed above (see [2.5.1]); saline solutions of various composition are used; 125 mM NaCl in salmonids and in pike, 45 mM NaCl + 5 mM KCl in cyprinids, 17 mM NaCl in *Silurus glanis*, full or slightly diluted sea water for marine species (turbot, halibut, sea bass, etc.). Usually diluents are buffered (20–30 mM Tris or a combination tris-glycine in salmonids): pH 8 in carp, *Silurus glanis* and halibut, 9 in salmonids. The optimum, ratio sperm, egg and diluent as defined for salmonids is 10–20 eggs/ml of sperm which has been diluted 1:1000. Practically, diluent is added to the container so that the upper level just covers the top of the eggs and 1 to 3 ml of sperm are used for 1 l of eggs (on average 1 l of diluent is required for 3 l of eggs in trout). Eggs and sperm are mixed by pouring them twice between two containers. The mixture is left for 15 min, rinsed with water for 10 s and the eggs transfered into incubators (Fig. 2.19). A similar procedure is used for some other species. In carp and *Silurus glanis* the sticky layer of eggs must be removed before transfer to incubators.

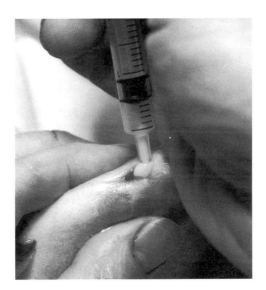

Fig. 2.17 Technique for collection of minute amounts of semen using a syringe.

46   *Broodstock Management and Egg and Larval Quality*

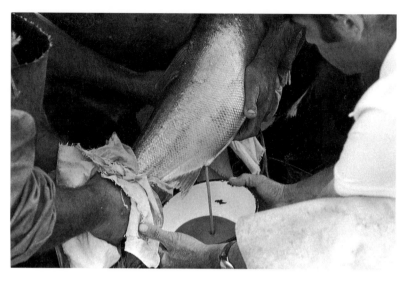

Fig. 2.18 Stripping eggs of a silver carp.

The number of spawners which should be used, in order to keep the required genetic variability in the population, was discussed by Chevassus (1989). As a rule of thumb a minimum of 30 males and 30 females is required. The number of females may exceed the number of males provided that each male contributes equally to the fertilization of each female.

2.10 Conclusions

Despite the equivalent genetic contributions of sperm and egg, little attention has been focused on the male and sperm in fish culture operations. The availability of good quality sperm in sufficient amounts at the right time ultimately determines the success of artificial reproduction in the fish farm. Care must be taken during the period of spermatogenesis, and several factors are of importance, including food, temperature, social interaction and the overall environment. Sperm can be now be efficiently used in artificial insemination in freshwater species with the help of an appropriate diluent. However, such methods are not as easy for marine fish. Care must be also taken during preservation to provide the right environment, in particular the energetic substrate and oxygen.

Many problems remain to be investigated and more work is needed on the environment to be offered to males during spermatogenesis and during the spawning season. We also need to know how to improve and manipulate endogenous stores of substrate and cAMP, ATP and other nucleotide triphosphates, and membrane cryoresistance, because these are very important in the determination of sperm quality.

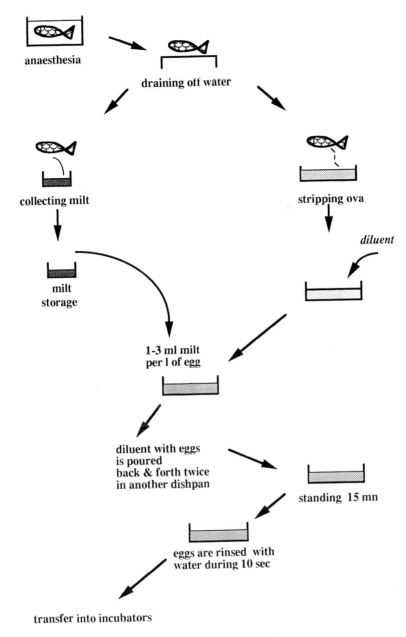

Fig. 2.19 Procedure for the artificial insemination of salmonids.

Acknowledgements

This work is part of a cooperative research project with IFREMER in France. Thanks are due to F. Andre, C. Jeulin and G. Perchec for help in the ATP measurement reported here. The manuscript was typed by J. Barthelemy.

References

Aas, G.H., Refstie, T. & Gjerde B. (1991) Evaluation of milt quality of Atlantic salmon. *Aquaculture*, **95**, 125–32.

Amann, R.P. (1991) Pourquoi les spermatozoïdes de toutes les espèces ne donnent pas un taux élevé de fertilité après congélation? Evaluation et amélioration de la fertilité du mâle; 30e *Réunion de la Société Française pour l'Étude de la Fertilité, 3–5/10/1991, Paris*, 134–42.

Asahina, K., Barry, P., Aida, K., Fususetani, N. & Hanyu, I. (1990) Biosynthesis of 17α, 20β dihydroxy-4-pregnen-3-one from 17α hydroxyprogesterone by spermatozoa of common carp, *Cyprinus carpio*. *Journal of Experimental Zoology*, **255**, 244–9.

Barry, P., Aida, K., Okumura, T. & Hanyu, I. (1990) The shift from C-19 to C-21 steroid synthesis in spawning male common carp, *Cyprinus carpio*, is regulated by the inhibition of androgen production by progestogens produced by spermatozoa. *Biology of Reproduction*, **43**, 105–12.

Baynes, S.M., Scott, A.P. & Dawson, A.P. (1981) Rainbow trout, *Salmo gairdneri* Richardson, spermatozoa: effects of cations and pH on motility. *Journal of Fish Biology*, **19**, 259–67.

Belova, N.V. (1982) Ecological and physiological properties of the semen of pond cyprinids III. Physiological-biochemical parameters of the semen of some cyprinid species. *Journal of Ichtyology*, **22**, 63–77.

Billard, R. (1969) Spermatogenèse de *Poecilia reticulata* II. La production spermatogénétique. *Annales Biologie Animale Biochimie Biophysique*, **9**, 307–13.

Billard, R. (1970a) Ultrastructure comparée de spermatozoïdes de poissons téléostéens. In *Comparative Endocrinology*, (ed B. Baccetti), pp. 71–9. Academic Press, London and Academia dei Lincei, Roma Quaderno, 137.

Billard, R. (1970b) La spermatogénèse de *Poecilia reticulata* III. Ultrastructure des cellules de Sertoli. *Annales Biologie Animale Biochimie Biophysique*, **10**, 37–50.

Billard, R. (1970c) La spermatogenèse de *Poecilia reticulata* IV. La spermiogenèse, étude ultrastructurale. *Annales Biologie Animale Biochimie Biophysique*, **10**, 493–510.

Billard, R. (1981) Short-term preservation of sperm under oxygen atmosphere in rainbow trout *Salmo gairdneri*. *Aquaculture*, **23**, 287–93.

Billard, R. (1983a) Spermiogenesis in the rainbow trout *Salmo gairdneri*; an ultrastructural study. *Cell Tissue. Research*, **233**, 265–84.

Billard, R. (1983b) Effects of coelomic and seminal fluids and various saline diluents on the fertilizing ability of spermatozoa in the rainbow trout *Salmo gairdneri*. *Journal of Reproduction and Fertility*, **68**, 77–84.

Billard, R. (1985) Artificial insemination in salmonids. In *Salmonid Reproduction*, (eds R.N. Iwamoto & S. Sower), pp. 116–28. Washington Sea Grant Program, Seattle.

Billard, R. (1986) Spermatogenesis and spermatology of some teleost fish species. *Reproduction, Nutrition and Development*, **26**, 877–920.

Billard, R. (1987) Testis growth and spermatogenesis in teleost fish; the problem of the large interspecies variability in testis size. *Proceedings of the 3rd International Symposium on Reproductive Physiology of Fish, St John's, Newfoundland*, 183–6.

Billard, R. (1990a) Spermatogenesis in teleost fish. In *Marshall's Physiology of Reproduction*, Vol. 2 (ed. G.E. Lamming), pp. 183–212. Churchill Livingstone, Edinburgh.

Billard, R. (1990b) Artificial insemination in fish. In *Marshall's Physiology of Reproduction*, Vol. 2 (ed. G.E. Lamming), pp. 870–88. Churchill Livingstone, Edinburgh.

Billard, R. (1992) Reproduction in rainbow trout: sex differentiation, dynamics of gametogenesis, biology and preservation of gametes. *Aquaculture*, **100**, 263–98.

Billard, R. & Cosson, M.P. (1986) Sperm motility in rainbow trout, *Parasalmo mykiss*; effect of pH and temperature; reproduction in fish. Basic and applied aspects in endocrinology and genetics. *INRA, Paris, les Colloques INRA*, **44**, 161–76.

Billard, R. & Cosson, M.P. (1989) Measurement of sperm motility in trout and carp. In *Aquaculture and Biotechnology in Progress*, (eds N. De Pauw, E. Jaspers, H. Ackefors & N. Wilkins), pp. 499–503. European Aquaculture Society, Bredene, Belgium.

Billard, R. & Cosson, M.P. (1990) The energetics aspects of fish sperm motility. In *Control of Sperm Motility, Biological and Clinical Aspects*, (ed. C. Gagnon), pp. 153–73. CRC Press, Boca Raton, FL.

Billard, R. & Cosson, M.P. (1992) Some problems related to the assessment of sperm motility in freshwater fish. *Journal of Experimental Zoology*, **261**, 122–31.

Billard, R. & Jalabert, B. (1973) Le glycogène au cours de la formation des spermatozoïdes et de leur transit dans le tractus génital femelle chez le guppy (poisson poecilidae). *Annales Biologie Animale Biochimie Biophysique*, **13**, 313–20.

Billard, R., Petit, J., Jalabert, B. & Szollosi, D. (1974) Artificial insemination in trout using a sperm diluent. In *Symposium on the Early Life History of Fish, Oban*, (ed. J.H.S. Blaxter), pp. 715–23, Springer-Verlag, Berlin.

Billard, R., Dupont, J. & Barnabé, G. (1977) Diminution de la motilité et de la durée de conservation du sperme de *Dicentrarchus labrax* (poisson téléostéen) pendant le période de spermiation. *Aquaculture*, **11**, 363–70.

Billard, R., Marcel, J. & Matei, D. (1981) Survie post mortem des gamètes de truite fario, *Salmo trutta fario*. *Canadian Journal of Zoology*, **59**, 29–33.

Billard, R., Fostier, A., Weil, C. & Breton, B. (1982) Endocrine control of spermatogenesis in teleost fish. *Canadian Journal of Fisheries and Aquatic Sciences*, **39**, 65–79.

Billard, R., Cosson, M.P. & Christen R. (1987) Some recent data on the biology on trout spermatozoa. *Proceedings of the 3rd International Symposium on Reproductive Physiology of Fish, St John's, Newfoundland*, 187–90.

Billard, R., Cosson, J. & Crim L.W. (1993) Motility of fresh and aged halibut sperm. *Aquatic Living Resources*, **6**, 67–75.

Boitano, S. & Omoto, K.C. (1991) Membrane hyperpolarization activates trout sperm without an increase in intracellular pH. *Journal of Cell Science*, **98**, 346–9.

Boitano, S. & Omoto, C.K. (1992) Trout sperm swimming patterns and role of intracellular Ca^{++}. *Cell Motility and the Cytoskeleton*, **21**, 74–82.

Boitano, S., Ashley, R. & Omoto, C.K. (1991) Mechanism of application to trout sperm motility. In *Comparative Spermatology 20 Years After* (ed. B. Baccetti), pp. 451–6. Raven Press, New York.

Bouck, R. & Jacobson, J. (1976) Estimation of salmonid sperm concentration by microhematocrit technique. *Transactions of the American Fisheries Society*, **108**, 534–5.

Carpentier, P. & Billard, R. (1978) Conservation à court terme des gamètes de salmonidés à des températures voisines de 0°C. *Annales Biologie Animale Biochimie Biophysique*, **18**, 1083–8.

Chambeyron, F. & Zohar Y. (1990) A diluent for sperm cryopreservation of gilthead sea bream *Sparus aurata*. *Aquaculture*, **90**, 345–52.

Chao, N.H., Tsai, H.P. & Liao, I.C. (1992) Short-term and long-term cryopreservation of sperm and sperm suspension of the grouper, *Epinephelus malabaricus* (Bloch & Schneider). *Asian Fisheries Science*, **5**(1), 3–116.

Chevassus, B. (1989) Contribution of aquacultural stocks: genetic aspects. *Advances in Tropical Aquaculture, AQUACOP-IFREMER, Actes de Colloque*, **9**, 569–92.

Christen, R., Gatti, J.L. & Billard, R. (1987) Trout sperm motility. The transient movement of trout sperm is related to changes in the concentration of ATP following the activation of the flagellar movement. *European Journal of Biochemiistry*, **166**, 667–71.

Cognie, F., Billard, R. & Chao, N.H. (1989) La cryopreservation de la laitance de la carpe *Cyprinus carpio*. *Journal of Applied Ichtyology*, **5**, 165–76.

Cosson, M.P., Billard, R., Gatti, J.L. & Christen, R. (1985) Rapid and quantitative assessment of trout spermatozoa motility using stroboscopy. *Aquaculture*, **46**, 71–5.

Cosson, M.P., Billard, R. & Letellier, L. (1989) Rise of internal Ca^{2+} accompanies the initiation of trout sperm motility. *Cell Motility and the Cytoskeleton*, **14**, 424–34.

Cosson, M.P., Cosson, J. & Billard, R. (1991) cAMP dependence of movement initiation in intact and demembranated trout spermatozoa. *Proceeding of the Fourth International Symposium on the Reproductive Physiology of Fish. University of East Anglia, Norwich, UK, 7–12/07/1991* (eds A.P. Scott, J.P. Sumpter, D.E. Kime & M.S. Rolfe), pp. 262–4.

Coste, J.M. (1853) *Instructions Pratiques sur la Pisciculture*. Librairie Victor Masson, Paris.

Dabrowski, K. & Ciereszko, A. (1992) Seminal plasma proteinase inhibitor(s) in fish. *Workshop on Gamete and Embryo Storage and Cryopreservation in Aquatic Organisms*, Marly le Roi, France, 30/03–2/04/1992, Abstract, p. 11.

Drokin, S.I. & Kopeika, E.F. (1992) Cryoresistance and lipid characteristics of sperm of different fish species. *Workshop on Gamete and Embryo Storage and Cryopreservation in Aquatic*

Organisms, Marly le Roi, France, 30/03–2/04/1992, Abstract, p. 50.

Duhamel du Monceau, H. (1773) Traité général des pêches, seconde partie, seconde section. *Academie Royale des Sciences*, **2**, 209–13.

Duplinsky, D. (1982) Sperm motility of northern pike and chain pikerel at various pH values. *Transactions of the American Fisheries Society*, **11**, 768–71.

Dzuba, B.B. & Cherepanov, V.V. (1992) A content of some intermediates of energetic exchange in activated carp sperm prior to and after cryopreservation. *Workshop on Gamete and Embryo Storage and Cryopreservation in Aquatic Organisms*, Marly le Roi, France, 30/03–2/04/1992, Abstract, p. 51.

Ellis, W.G. & Jones, J.W. (1939) The activity of the spermatozoa of *Salmo salar* in relation to osmotic pressure. *Journal of Experimental Biology*, **15**, 530–4.

Fribourgh, J.H. (1965) The effects of potassium tellurite on goldfish spermatozoan activity. *Transactions of the American Fisheries Society*, **94**, 399–402.

Gaschott, O. (1925) Contributions to the physiology of movement of trout sperm. *Archives Hydrobiologie*, supplement 4, 441–78 (in German).

Gatti, J.L., Billard, R. & Christen, R. (1990) Ionic regulation of the plasma membrane potential of rainbow trout *Salmo gairdneri* spermatozoa: role in the initiation of sperm motility. *Journal of Cell Physiology*, **143**, 546–54.

Geiger, N. (1955) Electron microscope studies on salmonoid spermatozoa. *Revue Suisse Zoologique*, **62**, 325–34 (in German).

Goryczko, K. & Tomasik, L. (1975) An influence of males on the variability and fertilization degree of trout eggs. *Acta Ichthyology & Pisciculture*, **5**, 3–11.

Gosh, R.I. (1983) Chemical composition of the cavity fluid and ova of the grass carp and the carp. *Hydrobiology Journal*, **19**, 84–7.

Grier, H.J. (1981) Cellular organization of the testis and spermatogenesis in fishes. *American Zoologist*, **21**, 345–57.

Hartmann, M., Medem, F.G., Kuhn, R. & Bielig, H.J. (1947) Investigations on the gamones of the rainbow trout. *Zeitscrift Naturforschung*, **2**, 330–49, (in German).

Henneguy, L.F. (1877) Recherches sur la vitalité des spermatozoïdes de la truite. *Compte Rendus Academie de Science, Paris*, **84**, 1333–5.

Hey, D. (1939) A preliminary report concerning the causes of the low fertility of trout eggs at the trout hatchery. *Annale University Stellenboschen*, **17**, 1–20.

Hines, R. & Yashouw, A. (1971) Some environmental factors influencing the activity of spermatozoa of *Mugil capito* Cuvier, a grey mullet. *Journal of Fish Biology*, **3**, 123–7.

Huxley, J.S. (1930) The maladaptation of trout spermatozoa to fresh water. *Nature*, **125**, 194.

Hwang, P.C. & Idler, D.R. (1969) A study of major cations, osmotic pressure and pH in seminal components of Atlantic salmon. *Journal of the Fisheries Research Board of Canada*, **26**, 413–19.

Jalabert, B. & Billard, R. (1969) Etude ultrastructurale du site de conservation des spermatozoïdes dans l'ovaire de *Poecilia reticulata* (poisson téléostéen). *Annales Biologie Animale Biochimie Biophysique*, **9**, 273–80.

Jamieson, B.G.M. (1991) *Fish Evolution and Systematics: Evidence from Spermatozoa*. Cambridge University Press.

King, M., Gatti, J.L., Moss, G. & Witman, B. (1990) Outer-arm dynein from trout spermatozoa: substructural organization. *Cell Motility and Cytoskeleton*, **16**, 266–78.

Kolliker, A. (1855) Physiological studies on seminal fluid. *Zeitschrift Wissenschaftliche Zoologie*, **7**, 201–72 (in German).

Labbé, C., Loir, M., Kaushik, S. & Maisse, G. (1991) The influence of both rearing temperature and dietary lipid origin of fatty and composition of spermatozoan phospholipids in rainbow trout *Oncorhynchus mykiss*. Effect on sperm conservation tolerance. In *Proceedings of Fish Nutrition in Practice, June 1991, Biarritz, France*, pp. 49–59.

Lahnsteiner, F., Weismann, T. & Patzner, R.A. (1992) Fine structural changes in spermatozoa of the grayling, *Thymallus thymallus* (Pisces: Teleostei), during routine cryopreservation. *Aquaculture*, **103**, 73–84.

Leung, L.K.P. (1991) Principles of biological cryopreservation. In *Fish Evolution and Systematics: Evidence from Spermatozoa*, (ed. B.G.M. Jamieson), pp. 231–44. Cambridge University Press.

Leung, L.K.P. & Jamieson, B.G.M. (1991) Live preservation of fish gametes. In *Fish Evolution and Systematics: Evidence from Spermatozoa*, (ed. B.G.M. Jamieson), pp. 245–95. Cambridge University Press.

Linhart, O., Slechta, V. & Slavik, T. (1991) Fish sperm composition and biochemistry. *Bulletin International Zoology, Academia Sinica Monograph*, **16**, 285–311.

Linhart, O., Billard, R. & Proteau, J.P. (1992) Cryopreservation of European catfish *Silurus glanis* L. spermatozoa. *Aquaculture*, **115**, 347–59.

Maisse, G., Pinson, A. & Loir, M. (1988) Characterisation de l'aptitude à la congélation du sperme de truite arc-en-ciel *Salmo gairdneri* par des critères physico-chimiques. *Aquatic Living Resources*, **1**, 45–51.

Malejac, M.L., Loir, M. & Maisse, G. (1990) Qualité de la membrane des spermatozoïdes de truite arc en ciel *Oncorhynchus mykiss*; relation avec l'aptitude du sperme à la congélation. *Aquatic Living Resources*, **3**, 43–54.

Mann, T. (1964) *The Biochemistry of Semen and the Male Reproductive Tract*. Methuen, London.

Mattei, X. (1970) Spermiogenèse comparée des poissons. In *Comparative Spermatology*, (ed. B. Baccetti), pp. 57–70. Academic Press, New York.

Mattei, C. & Mattei, X. (1978) La spermatogénèse d'un poisson téléostéen *Lepadogaster lepadogaster*. La spermatide. *Biologie Cellulaire*, **32**, 257–66.

Medem, F., Rotheli, A. & Roth, H. (1949) Biological evidence for gamones in whitefish and brook trout. *Schweizerische Zeitschrift Hydrobiologie*, **11**, 361–77 (in German).

Methwen, D.A. & Crim, L.W. (1991) Seasonal changes in spermatocrit, plasma sex steroids, and motility of sperm from Atlantic habitut *Hippoglossus hippoglossus*. *Proceeding of the Fourth International Symposium on the Reproductive Physiology of Fish*, (eds A.P. Scott, J.P. Sumpter, D.E. Kime & M.S. Rolfe), p. 170. University of East Anglia, Norwich, UK, 7–12 July 1991.

Miura, T., Yamauchi, K., Takahashi, H. & Nagahama, Y. (1992) The role of hormones in the acquisition of sperm motility in salmonid fish. *Journal of Experimental Zoology*, **261**, 359–63.

Mohr, C. & Chalanchuk, M. (1985) The effect of pH on sperm motility of white suckers, *Catostomus commersoni*, in the experimental Lakes Area. *Environmental Biology of Fishes*, **14**, 309–14.

Morisawa, M. & Morisawa, S. (1990) Acquisition and initiation of sperm motility. In *Controls of Sperm Motility: Biological and Clinical Aspects*, (ed. C. Gagnon), pp. 137–51. CRC Press, Boca Raton, FL.

Morisawa, M., Suzuki, K., Shimizu, H., Morisawa, S. & Yasuda, K. (1983) Effects of osmolality and potassium on motility of spermatozoa from freshwater cyprinid fishes. *Journal of Experimental Biology*, **107**, 95–103.

Morisawa, S., Ishida, K. & Morisawa, M. (1991) The role of extracellular pH on the acquisition of sperm motility in chum salmon. In *Comparative Spermatology 20 Years After*, (ed. B. Baccetti), pp. 503–505. Raven Press, New York.

Nakano, S. & Nozawa, A. (1925) On the viability of the egg and sperm of *O. masou* (Landlocked). *Journal of the Imperial Fisheries Institute, Tokyo*, **21**, 17–18.

Nomura, M. (1964) Studies on reproduction of rainbow trout *Salmo gairdneri* with special reference to egg taking – VI. The activities of spermatozoa in different media and preservation of semen. *Bulletin of the Japanese Society of Scientific Fisheries*, **30**, 723–33.

Okuno, M. (1991) Initiation of salmonid fish sperm motility with calcium. In *Comparative Spermatology 20 Years After*, (ed. B. Baccetti), pp. 517–20. Raven Press, New York.

de Quatrefages, A. (1853) Vitalité des spermatozoïdes de quelques poissons d'eau douce. *Annales des Sciences Naturelles*, **19**, 341–69.

Raychoudhury, S.S. & Suarez, S.S. (1991) Porcine sperm binding to oviductal explant in culture. *Theriogenology*, **36**, 1059–70.

Redondo, M., Cosson, M.P., Cosson, J. & Billard, R. (1991). *In vitro* maturation of the potential for movement of carp spermatozoa. *Molecular Reproduction and Development*, **29**, 259–70.

Reisenbichler, G. (1882) The thick or thin fertilisation of eggs. *Report of the US Commission on Fisheries, 1879*, **7**, 633–5.

Robitaille, P.M.L., Munford, K.G. & Brown, G. (1987) 31P. nuclear magnetic resonance study of trout spermatozoa at rest, after motility and during short-term storage. *Biochemistry and Cell Biology*, **65**, 474–85.

Rutter, C. (1902) The natural history of the quinnat salmon. *Bulletin of the Bureau of Fisheries, Washington*, **22**, 65–141.

Saad, A. & Billard, R. (1987) Composition et emploi d'un dilueur d'insémination chez la carpe *Cyprinus carpio*. *Aquaculture*, **66**, 329–45.

Saad, A., Billard, R., Théron, M.C. & Hollebecq, M.G. (1988) Short-term preservation of carp *Cyprinus carpio* semen. *Aquaculture*, **71**, 133–50.

Sanchez-Rodriguez, M. & Billard, R. (1977) Conservation de la motilité et du pouvoir fécondant du sperme de la truite arc en ciel maintenu à des températures voisines de 0°C. *Bulletin Française Pisciculture*, **265**, 143–52.

Scheuring, L. (1924) Biological and physiological studies on salmonid sperm. *Archives Hydrobiologie*, supplement **4**, 181–318 (in German).

Schlenk, W. (1938) Spermatozoo motility and hydrogen ion concentration. Investigation with rainbow trout sperm. *Biochemical Zoology*, **265**, 29–35 (in German).

Schlenk, W. & Kahmann (1938) The chemical composition of seminal fluids and their physiological importance study with trout sperm. *Biochemical Zoology*, **295**, 283–301 (in German).

Scott, A.P. (1981) Bibliography on biology, handling and storage of fish spermatozoa. *Bibliography of Physiology and Reproduction*, **37**, 6 pp.

Scott, A.P. & Baynes S.M. (1980) A review of the biology, handling and storage of salmonid spermatozoa. *Journal of Fish Biology*, **17**, 707–39.

Shuman, R.F. (1950) On the effectiveness of spermatozoa of the pink salmon, *Oncorhynchus gorbuscha*, at varying distances from the point of dispersal. *Fishery Bulletin Fish and Wildlife Service US*, **51**, 359–63.

Smirnov, A.I. (1963) The fertilizability of the eggs and spermatozoa of pink salmon, *Oncorhynchus gorbuscha*, when held in water. *Nauchnye Doklady Vysshei Shkoly Biologischeskie Nauki*, **3**, 37–41 (in Russian). Translation available: Fisheries Research Board of Canada Translation Series, 544.

Smith, R.T. & Quistorff, E. (1943) Experiments with the spermatozoa of the steelhead trout, *Salmo gairdnerii*, and the chinook salmon. *Onchorhynchus tshawystscha*. *Copeia*, **3**, 164–7.

Smith, T. & Yanagimachi, R. (1990) The viability of hamster spermatozoa stored in the isthmus of the oviduct: the importance of sperm-epithelium contact for sperm survival. *Biology of Reproduction*, **42**, 450–7.

Stoss, J. (1983) Fish gamete preservation and spermatozoan physiology. In *Fish Physiology*, Vol. IX B (eds W.S. Hoar, J.D. Randall & E.M. Donalson), pp. 305–50. Academic Press, New York.

Stoss, J. & Holtz, W. (1983) Successful storage of chilled rainbow trout (*Salmo gairdneri*) spermatozoa for up to 34 days. *Aquaculture*, **31**, 269–74.

Suquet, M. (1992) La production de sperme chez le turbot; aspects descriptifs et experimentaux. Mémoire EPHE, Paris.

Suquet, M., Omnes, M.H., Normant, Y. & Fauvel, C. (1992) Assessment of sperm concentration and motility in turbot *Scophthalmus maximus*. *Aquaculture*, **101**, 177–85.

Tanimoto, S. & Morisawa, M. (1988) Roles for potassium and calcium channels in the initiation of sperm motility in rainbow trout. *Develop. Growth & Differ*, **30**, 117–24.

Thibault, M. (1989) La redécouverte de la fécondation artificielle de la truite en France en milieu du XIXème siècle; les raisons de l'engouement et ses conséquences. In *Colloque Homme, Animal et Société, 13–16 mai 1987. 3. Histoire et Animal*. Toulouse Institute, Etudes Politiques, 205–31.

Tomasik, L. (1973) Specific and individual differences in motility between salmonid spermatozoa. *Acta Ichthyol. Piscat*, **3**, 11–17.

Ueda, H., Kambegawa, A. & Nagahama, Y. (1984) In vitro 11–ketotestosterone and 17α, 20β-dihydroxy-4-pregnen-3-one, production by testicular fragments and isolated sperm of rainbow trout, *Salmo gairdneri*. *Journal of Experimental Zoology*, **231**, 435–9.

Wilkins, N.P. (1989) *Ponds, Passes and Parcs. Aquaculture in Victorian Ireland*. Glendale, Dublin.

Chapter 3
Preservation of Gametes

3.1 Introduction
3.2 Role of gamete storage in aquaculture and conservation
    3.2.1 Short-term storage
    3.2.2 Cryopreservation
3.3 Short term storage of gametes
    3.3.1 Biological variation in gamete quality
    3.3.2 Preservation conditions
3.4 Cryopreservation of gametes
    3.4.1 Major physico-chemical objectives during cryopreservation
    3.4.2 Cryo-injuries
    3.4.3 Components of milt cryopreservation
3.5 Conclusions
    Acknowledgements
    References

In vitro preservation of gametes is assuming an increasing role in seed production and genetic management of broodstock. Preservation techniques, however, have largely been established for spermatozoa. Gametes, notably spermatozoa, may be stored for several weeks to facilitate the mass fertilization of eggs and genetic management of broodstock or may be cryopreserved for longer periods of storage to conserve genetic resources. To date, the gametes of over 50 freshwater and marine species, particularly those of commercial interest, have been stored with varying degrees of success. The reported post-storage viability, however, even for the same species, is very variable.

The reasons for the variability under *in vitro* storage conditions are unclear and our understanding of the interacting factors influencing storage success has been hampered by the non-standardized approaches in evaluating post-storage success. Several factors are likely to contribute and these largely relate to the quality of the gametes and preservation technology used. For spermatozoa, although the quality is widely reported to vary during the season, the collection techniques employed to date make it difficult to compare such variation and establish its importance for *in vitro* storage. In addition, for shorter term preservation, storage vessels, media and environment will influence the duration of successful storage.

The problems associated with the cryopreservation of eggs and embryos are more complex. To date, although some success is reported for invertebrates, viable cryopreservation of fish eggs or embryos has not been demonstrated. Therefore much of the reported literature relates to spermatozoa.

The processes of cooling and thawing during cryopreservation are traumatic to gametes. These effects can, however, be ameliorated by optimizing the interacting factors such as gamete quality, diluents, cooling and warming protocols, freezing and storage method.

This chapter addresses the possible roles of *in vitro* storage of gametes in aquaculture and conservation and discusses some of the major factors, protocols and possible reasons for the variability of published results.

3.1 Introduction

The preservation of fish gametes, particularly spermatozoa, has been widely reported for a number of fish species. To date over 50 species of freshwater and marine species have been preserved (see reviews by Scott & Baynes, 1980, Stoss 1983, Leung & Jamieson 1991, McAndrew *et al.* 1993, Rana 1994). The majority of publications, however, relate to three groups of fish of aquacultural importance; the salmonids, tilapias and carps.

The aim of all preservation techniques is to increase the longevity of gametes by lowering the storage temperature and so reducing their metabolic burden. The duration of successful storage of gametes will depend on the final storage temperature and the choice of the storage method will depend on the purpose of storage. Gametes may be stored to facilitate the practical management of seed production, for research or lately for the conservation or preservation of genetic biodiversity of aquatic resources. The storage of fish eggs has been more problematic and to date storage of ova or embryos is limited to chilled storage. The storage of spermatozoa is more commonly practised.

Under chilled or near zero temperature conditions, gametes can be preserved for up to several weeks in an unfrozen form. To maintain their viability for longer periods gametes need to be cryopreserved. Under correct cryopreservation and storage conditions, the viability of the gametes can be preserved for up to 32 000 years (Ashwood-Smith 1980).

Although numerous studies have been published since the late 1960s on the cryopreservation of milt, the incompatibility of experimental design between several of these studies has led to heterogenous results for the same species and similar protocols. Consequently the uptake of this technology in aquaculture and conservation has been negligible.

3.2 Role of gamete storage in aquaculture and conservation

3.2.1 Short-term storage

The need to store gametes for short periods depends on several factors: the country in which the hatchery is located, the size and type of aquacultural or conservational operation, the management of broodstock, methods of seed production and the principal purpose and markets for the seed. All these factors can influence gamete management and the need to preserve gametes.

Asynchrony in spawning state between sexes can often result in a shortfall of gametes of the required quality and quantity. Depending on species and biogeography, male and female fish can mature at different times in a spawning season (Bromage & Cumaranatunga 1988). Under these asynchronous conditions, especially during the beginning and end of the spawning season, gametes can be collected and held under appropriate storage conditions for later use. Similarly, if milt or ova of special interest are acquired for hybridization, they may need to be stored until fertilization can be carried out.

Depending on the species cultured and the number and size of broodstock used for seed production, gametes, particularly milt, may also need to be stored to facilitate mass fertilization. Such an application may be of particular interest to salmonid and carp seed producers, who may need to spend considerable time screening broodstock for ovulated females. In such circumstances milt can be independently stripped and stored for later use; in this way the yield of eyed eggs is maximized.

In cases where the disease-free status of seed is critical and a farm is known to be infected with a disease, e.g. IPN, storage of milt in an unfrozen state can be used to minimize financial loss by reducing the number of egg batches that need to be discarded. In such circumstances milt can be collected from labelled males and stored. The males can then be tested for the presence of relevant disease(s) and only milt originating from disease-free stock used for mass fertilization of ova.

3.2.2 Cryopreservation

The genetic information represented by a species can be maintained by conserving fish in wet or live gene banks. Such banks, however, suffer from two major drawbacks. Firstly, they offer no long-term guarantee of the genetic stability of the population or species over time and therefore captive stocks cannot be maintained in a pristine condition (Tave 1986, McAndrew *et al.* 1993). Secondly, given the high natural genetic variation, it will be prohibitively expensive to maintain the desired representative genetic information of a population or species in a live gene bank.

To overcome these shortfalls genetic information can be preserved in the form of frozen gametes and embryos. Cryobanks can play a crucial role in the genetic management and conservation of aquatic resources. At present the basic technologies are available for the cryopreservation of spermatozoa of several fish species and it is possible to establish frozen sperm banks. Frozen banks can be used to hold spermatozoa from a substantial number of males, thus increasing the effective breeding number (N_e) and thereby increasing the efficacy of conservation by enabling greater intraspecific variation to be conserved.

Frozen sperm banks may also have additional applications in aquaculture. Genes from valuable individuals or stocks which have desirable characteristics, e.g. for use in inbred lines and selection programmes (Tave 1986), can be preserved for future use and development. In such cases representative genes in the form of sperm can be selected for cryopreservation. Sperm cryobanks can also be used to manage the genetic integrity of farmed stocks. For fish species, especially those which have a short life cycle, such banks are of particular importance to protect against inbreeding, founder effects and random drift of captive stocks. To minimize these detrimental effects milt from the representative stocks can be cryopreserved and genetic material periodically reintroduced to increase the heterozygosity in stocks or to increase the contribution of desired genes in the offspring.

Cryopreservation can also play a crucial role in genetic selection programmes for fish, especially in those instances where the life span of the species is relatively long. Since selection programmes typically take several generations to evaluate, the interpretation of any selection advantage can be confounded by environmental change and management practice. In such cases milt from original stocks can be cryopreserved and used to make half-sibling comparisons at a later date. Such evaluations will provide a more accurate prognosis of selection benefits.

The disease-free status of parent fish is an integral component of broodstock management. During disease outbreaks entire stocks can be affected. Therefore, periodic freezing of milt from disease-free males can help to buffer farms against such catastrophes. The use of post-thawed milt to fertilize ova from similar species or strains can aid the rate of recovery of farms.

In instances of scarcity of males, e.g. protandrous hermaphrodite species of commercial interest, such as groupers, milt can be stored for later use. The value of cryopreserved milt is even greater in species such as bass, *Epinephelus malabraicus* which can take up to 8–10 years to mature (N.C. Chao, pers. comm.).

Cryobanks can facilitate more efficient use of hatchery space for live genebanks and seed production. In instances where a farm or institute houses multiple species and strains, a large part of the genetic representation of the stocks can be held in the form of a frozen sperm bank.

3.3 Short term storage of gametes

Preservation of fish gametes below their culture or ambient temperature can be used to prolong their viability. The majority of the publications relate to storage of milt under chilled conditions, usually in domestic refrigerators (Table 3.1). Although ova can be successfully stored, most of the studies and applications of liquid storage of gametes relate to spermatozoa and to salmonids (Scott & Baynes 1980, McAndrew et al. 1993).

The viability of the gametes at the end of the storage period depends on several, interrelated factors. The success of gamete storage, particularly of sperm, will be influenced by the intrinsic biological variability of collected gametes, collection techniques and contamination, storage procedures such as storage temperature and container configuration, composition of gamete diluting medium, gamete:diluent ratio and post-collection contamination.

The predominance of sperm preservation in the reported literature is reflected in the following sections.

3.3.1 Biological variation in gamete quality

The evaluation of milt quality is a function of the criteria used to define its quality. The percentage of spermatozoa activated, the duration of motility, sperm concentration and spermatocrit, and seminal plasma composition have been used to quantify milt quality. Studies on the composition of milt suggest large intraspecific and interspecific variations in spermatozoa concentration and fitness and seminal plasma composition. These variations have been attributed to genetic variability, intratesticular aging of spermatozoa, seasonality (Billard *et al.* 1977, Benau & Terner 1980, Scott & Baynes 1980, Piironen & Hyvarinen 1983, Kruger *et al.* 1984, Munkittrick & Moccia 1987), breeding state and strategy (McAndrew *et al.* 1993).

Change in milt quality during the spawning season has been reported for salmonids and bass. The significance of these changes for *in vitro* gamete storage, however, is still nuclear. For example, in rainbow trout (*Oncorhynchus mykiss*) Billard *et al.* (1977), Buyukhatipoglu & Holtz (1984) and Munkittrick & Moccia (1987) reported that sperm density declined as the season advanced, whereas Sanchez-Rodriguez *et al.* (1978) and Piironen & Hyvarinen (1983) noted that the spermatocrit increased over the stripping season.

Studies on rainbow trout, brown trout (*Salmo trutta*), brook trout (*Salvelinus fontanalis*) and Atlantic salmon (*Salmo salar*) have shown that duration of motility also exhibits seasonal variation. During the peak spawning season activated rainbow trout spermatozoa remained motile for 30–55 s. By the end of the spawning season the duration of the motility declined to 15 s (Benau & Terner 1980). Similarly, for sea bass (*Dicentrarchus labrax* L.) the duration of motility decreased from 5 min at the start of the season to 30 s at the end of the season (Billard *et al.* 1977). In addition, in rainbow trout, the proportion of spermatozoa that are activated may also gradually decrease as the spawning season progresses (Buyukhatipoglu & Holtz 1984, Munkittrick & Moccia 1987).

Variations in the inorganic and organic composition of seminal plasma (Benau & Terner 1980, Piironen & Hyvarinen 1983, Kruger *et al.* 1984, Munkittrick & Moccia 1987) may also affect the preservation properties of milt. The concentration and the ratio of ions such as potassium and sodium (Morisawa & Suzuki 1980) which are implicated in the initiation of sperm motility decreases as the season progresses (Munkittrick & Moccia, 1987).

The interpretation of published data on intraspecific and interspecific variation of milt quality, however, may be confounded by collection techniques. Almost without exception, all previous studies on milt quality evaluation were conducted on samples collected by abdominal massage. Consequently, even with the

Table 3.1 Viability of unfrozen fish spermatozoa during short term storage at near zero temperatures

Species	Storage conditions			Milt:eggs (ml:no) used	Fertilization rate (% of control)	References
	°C	Form	Duration			
Brown trout (*Salvelinus fontinalis*)	2	neat	18.5–20.5 h	1:2000	69–92	Plosila & Keller 1974
	2	neat	18.5–20.5 h	1:2000	83–95	Plosila & Keller 1974
	10	1:63[1] dil	24 h	0.1:4 ml	90–95	Erdahl & Graham 1987
Rainbow trout (*Oncorhynchus mykiss*)	0	neat[2]	34 days	excess:25	93	Stoss & Holtz 1983
	0	neat	7 days	motility test	12–13 activation	Stoss et al. 1987
	4	neat[2]	5 days	0.1:200	80	Billard 1981
	4	1:1 dil[3]	37 days	1.4×10^6 sperm per egg (300 eggs)	81	McNiven et al. 1993
	10	1:31 dil[1]	24 h	0.1:4 eggs	90–95	Erdahl & Graham 1987
Pink salmon (*Oncorhynchus gorbuscha*)	10	1:255 dil[1]	30 min	0.1:4 ml eggs	85–90	Erdahl & Graham 1987
	8–9	neat	46 h	NG	90	Withler & Humphreys 1967
	3.2	neat	160 h	NG	50	Withler & Humphreys 1967
Sockeye salmon (*Oncorhynchus nerka*)	3	neat	180 h	NG	50	Withler & Humphreys 1967
	10	neat	180 h	NG	50	Withler & Humphreys 1967
Chum salmon (*Oncorhynchus keta*)	3	neat	8 days	0.75:150	50	Jensen & Alderdice 1984
	9	neat	88 h	0.75:150	50	Jensen & Alderdice 1984
Atlantic salmon (*Salmo salar*)	−4	1:1 dil	21 days	0.4:1000	93	Rana unpublished data

Species	Temp	Dilution	Duration	Result	Reference
Halibut (*Hippoglossus hippoglossus*)	−4	neat	29 days	75	Edwardes 1991
Carp (*Cyprinus carpio*)	0–5 4	neat neat	45 h 13 days	90 83	Hulata & Rothbard 1979 Saad *et al.* 1988
Walleye (*Stizostedion vitreum vitreum*)	1	1:2 dil[4]	14 days	91	Moore 1987
Tilapia					
(*Oreochromis aureus*)	4	1:19 dil[5]	6 days	50% motile	Muiruri 1988
(*Oreochromis niloticus*)	4	1:19 dil[5]	6 days	50% motile	Muiruri 1988
(*Oreochromis andersoni*)	4	1:19 dil[5]	5 days	50% motile	Muiruri 1988
(*Oreochromis mossambicus*)	4 5	1:19 dil[5] neat	4 days 27 h	50% motile 41% motile	Muiruri 1988 Harvey 1983
Sarotherodon galilaeus	4	1:19 dil[5]	5 days	50% motile	Muiruri 1988
Tilapia zilli	4	1:19 dil[5]	4 days	50% motile	Muiruri 1988

NG = Not given

[1] Extender = $CaCl_2$ 0.1 g l^{-1}, $MgCl_2$ 0.2 g l^{-1}, Na_2HPO_4 0.25 g l^{-1}, KCl 2.55 g l^{-1}, NaCl 5.85 g l^{-1}, Glucose 10 g, KOH 10 ml at 1.27 g/100 ml, Bicine 10 ml at 5.3 g/100 ml, deionised water 980 ml
[2] Stored under pure oxygen
[3] Extender = Fluorocarbon, FC-77
[4] Extender = $CaCl_2.2H_2O$ 0.243 g l^{-1}, $MgCl_2.6H_2O$ 0.267 g l^{-1}, Na_2HPO_4 0.472 g l^{-1}, NaCl 13.2 g l^{-1}, Glucose 20 g l^{-1}, Citric acid 0.2 g l^{-1}, NaOH 40 ml at 1.27 g 100 ml^{-1}, Bicine at 5.3 g 100 ml^{-1}, distilled water 1920 ml
[5] Extender = NaCl 6.5 g l^{-1}, KCl 3.0 g l^{-1}, $NaHCO_3$ 0.2 g l^{-1}, $CaCl_2.6H_2O$ 0.30 g l^{-1}, distilled water 1000 ml

utmost care urine contamination is inevitable owing to the proximity of sperm duct and ureter. Therefore variation in sperm density and seminal plasma composition can be misinterpreted (Rana 1995).

Recent studies with Atlantic salmon at the Institute of Aquaculture, University of Stirling, aimed at elucidating the effect of contamination, have shown that urine is a major contaminant of collected milt (Rana *et al.* 1992). The potential significance of urine contamination on milt quality can be seen by comparing milt samples collected by catheterization of the sperm duct with that from abdominal massage after clearing the bladder of urine from the same individuals (Fig. 3.1). These studies suggest that urine contamination of milt after clearing of the bladder can dilute milt by up to 80%, and this can significantly increase the variability (Fig. 3.1) and reduce the mean osmolality of milt (Fig. 3.2). Such contamination will in turn alter the ionic composition of the seminal plasma. In Atlantic salmon urine, contamination also increased the variability and reduced the mean concentration of potassium in seminal plasma from $1382\,mg\,l^{-1}$ to $797\,mg\,l^{-1}$ (Fig. 3.3).

The effect of urine contamination on seminal plasma composition and hence on milt preservation qualities can be further complicated by possible intraspecific differences in urine composition. In Atlantic salmon, for example, urine protein concentration and osmolality can vary by as much as 145 and 72%, respectively (Table 3.2).

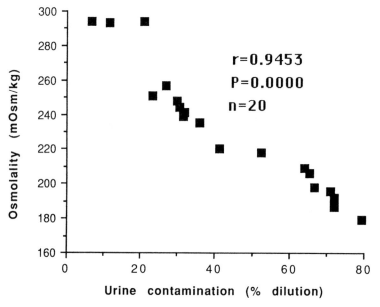

Fig. 3.1 Effect of urine contamination on the osmolality of Atlantic salmon (*Salmo salar*) milt. Urine contamination expressed as dilution estimated from the relative decrease in sperm density between non-catheterized and catheterized milt samples from same individuals.

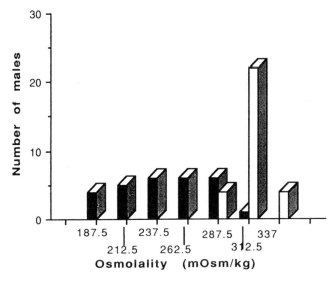

Fig. 3.2 Variation in osmolality of Atlantic salmon milt taken from non-catheterized (■,NC) and catheterized (□,C) fish. Means and medians for NC and C were 233 and 237 and 313 and 311, respectively.

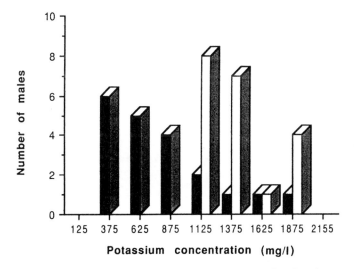

Fig. 3.3 Variation in potassium concentration of Atlantic salmon milt taken from non-catheterized (■,NC) and catheterized (□,C) fish. Means and medians for NC and C were 797 and 631 and 1382 and 1322, respectively.

3.3.2 Preservation conditions

Storage temperature

Storage temperature is a major factor affecting the viability of gametes during *in vitro* storage. Viability can be prolonged by maintaining gametes at near

Table 3.2 Mean[a] variations in urine[b] characteristics of Atlantic salmon

Fish No.	Osmolality (mOsm kg^{-1})	Protein (µg ml^{-1})
1	62.3 (1.21)	601 (9.70)
2	72.3 (0.87)	306 (6.10)
3	69.7 (0.87)	375 (6.99)
4	107.0 (0.64)	245 (9.35)

[a] Mean (± SEM) based on triplicate measurements
[b] Urine catheterized from bladder

zero temperatures to reduce their metabolic burden. The ability of gametes and embryos to tolerate low temperature may vary between temperate and tropical species (Leung & Jamieson 1991).

The viability of salmon ova maintained at −1°C in ovarian fluid was prolonged for 20 days. (Harvey et al. 1983). Earlier studies by Withler & Morley (1968) on sockeye salmon (*Oncorhynchus nerka*) and pink salmon (*O. gorbuscha*), however, report no fertility with ova stored at 8.5°C. A further reduction of storage temperature to 2.9°C resulted in 12% of the ova being viable after 14 days. Later studies on chum salmon (*O. keta*) suggest that 50% of the ova remained fertilizable after 9−10 days of storage at 3°C (Jensen & Alderdice 1984). When stored milt and ova were used together, however, fertilization success was zero after a similar storage period.

The temperature and duration for the storage of ova of sub-tropical and tropical species such as carp (Zlabek & Linart 1987) and tilapia (Harvey & Kelley 1984, Carrilho 1990, Aboyewa 1991) are higher and shorter, respectively, than those reported for salmonids. In grass carp, the fertilization rates for ova stored at 9−10°C and 22°C for 1 h was reduced from between 83 and 87% to between 6.5 and 17% after 5 h of storage. A similar trend was also reported for silverhead carp after 1 h of storage at 10 and 25°C (Zlabek & Linart 1987). Similarly, in *Oreochromis mossambicus* (Harvey & Kelley 1984) and *O. niloticus* (Adam 1988), storage of unfertilized eggs with 200 µg ml^{-1} kanamycin at temperatures of above and below 20−22°C reduced fertility and increased the incidence of embryonic abnormalities (Harvey & Kelley, 1984).

The sensitivity of eggs to *in vitro* storage may also vary between species, individuals and spawning pattern. In halibut (*Hippoglossus hippoglussus*), for example, where batches of ova are shed every four or so days, *in vitro* storage of ova in ovarian fluid at 4°C beyond 12 h resulted in a rapid decline in fertility rates and an increase in abnormalities of the embryos. In addition, the storage properties of ova varied between females and between each successive stripping (Basavaraja 1991).

In comparison, the viability of milt stored at chilled temperatures can be maintained for longer periods (Table 3.1). For the salmonids the storage of milt at 4−8°C can prolong their viability up to 37 days. Lowering the storaǥ

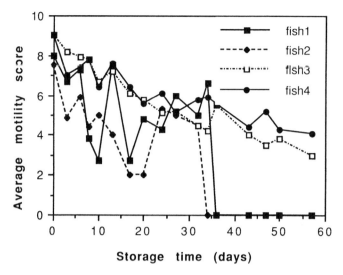

Fig. 3.4 Viability of spermatozoa from individual Atlantic halibut *Hippoglossus hippoglossus* (L) held at −4°C. Motility index based on a score of 0–10, 0 being no and 10 total post-activation motility. (Data adapted from Edwardes 1991.)

temperature of milt to −4°C and adding cryoprotectants such as DMSO and methanol has been shown to increase the viability to about 2 months. Recent studies on halibut suggest that undiluted milt containing 5% methanol can be successfully stored for over 60 days at −4°C, but this varied with individual males (Fig. 3.4).

The shelf life may be prolonged by using other storage procedures in conjunction with low temperature. These are considered in the following sections.

Oxygen enriched environments

Maintaining sperm cells in an aerobic environment is a prerequisite for *in vitro* preservation (Scott & Baynes 1980, Stoss 1983, Billard 1988, Billard *et al.* this volume Chapter 2). Studies on rainbow trout (Buyukhatipoglu & Holtz 1978, Billard 1981, Stoss & Holtz 1983) suggest that the fertility of spermatozoa can be prolonged when preserved under oxygen compared with air. To ensure high oxygen availability and distribution to the cells several different approaches have been reported. Milt has been stored in polythene bags (Stoss & Holtz 1983, Billard 1988) or continuously flushed in a moisture-saturated desiccator (Stoss & Holtz 1983, McNiven *et al.* 1993). By combining this technique with the use of antibiotics and lowering the storage temperature to 0°C rainbow trout milt has been successfully stored for 34 days (Stoss & Holtz 1983).

The use of perfluorocarbon emulsions (PFC) such as fluosol and FC-77, which were originally used for respiratory gas transport in human medicine and cell culture (King *et al.* 1989, Lowe 1991) has increased the longevity of poultry semen under chilled conditions (Rogoff 1985). The use of such inert organic gas carriers, which have a very high affinity for oxygen, to prolong the viability of fish milt was recently reported for rainbow trout (McNiven *et al.* 1993). In these studies rainbow trout milt held over a non-aqueous layer of PFC (FC-77) in a moisture laden atmosphere at 0°C remained viable for 37 days. Similar studies on Atlantic salmon using fluosol, however, at 4°C and −4°C showed that although milt could be stored for up to 29 and 69 days, respectively, at each temperature there was no significant advantage over storage in air at either temperature (K. Rana, unpublished data).

The depth of milt in the storage container, and hence gaseous diffusion, is also reported to influence the fertility of milt after storage (Stoss *et al.* 1987). By sampling rainbow trout milt at various depths in a test tube the authors demonstrated a pronounced decrease in post-activation motility at depths below 5 mm.

Dilution media

Eggs are generally stored in the ovarian fluid and often not diluted prior to storage; the only additives are generally antibiotics (Stoss 1983).

Spermatozoa concentrations in fish can range from 2×10^6 to 5.3×10^{10} cells per ml (Leung & Jamieson 1991). To minimize the demand on dissolved oxygen the cell density can be reduced with diluents. Ideally these diluents should be isotonic with the seminal plasma, maintain the spermatozoa in an immotile state and be able to sustain all the metabolic activities and needs of the cells (see Billard *et al.* this volume Chapter 2). Unlike spermatozoa from viviparous fish species and mammals, the duration of motility of fish spermatozoa ranges from 30 to 300 s (Stoss 1983). The efficacy of solutions used for gamete dilution during preservation will depend on the ability of such solutions, notably potassium ion concentration and osmolality, to maintain spermatozoa in a quiescent state until required.

Diluents which are commonly used have been reviewed recently by several authors (Scott & Baynes 1980, Stoss 1983, Leung & Jamieson 1991, McAndrew *et al.* 1993) and, therefore, their composition, which is designed to simulate fish blood or seminal plasma (Randall & Hoar 1971), is not covered in this chapter.

Current collection methods for milt inevitably result in microbial contamination. To suppress the subsequent proliferation of bacteria and fungi during storage, antibiotics and anti-mycotics are commonly added to diluents (Scott & Baynes 1980, Stoss 1983, Leung & Jamieson 1991). Penicillin and streptomycin at 125–6000 IU or mg ml^{-1} milt are most widely used for salmonids (Scott & Baynes 1980) but success is also reported with kanamycin (Harvey & Kelley 1984, Adam 1988) and ampicillin (Moore, 1987). The advantages of antibiotics and antimycotics are still unclear as high post-storage viability can also be obtained without the use of such additives (Fig. 3.4).

By optimizing all the above factors the longevity of gametes during *in vitro* storage may be prolonged for several weeks. For longer periods of preservation, gametes need to be cryopreserved.

3.4 Cryopreservation of gametes

The science of cryopreservation relates to the long-term preservation of biological material in a frozen but viable form at ultra-low temperatures, usually below −130°C. At these temperatures cellular viability can be maintained in a genetically stable form and is affected only by background radiation (Ashwood-Smith 1980).

At present, attempts to freeze fish ova and embryos have been unsuccessful. Inadequate dehydration and cryoprotectant penetration of ova and embryos, which contain relatively large volume of yolk and impermeable perivitelline membranes, are challenging problems that as yet have not been overcome. Recently, however, some success has been cited with Australian bass (*Macquaria novemaculeata*) ova, which were first equilibrated under vacuum for 1 and 5 min before being frozen in methanol/dry ice bath for 5 min (Leung & Jamieson 1991). Using this technique up to 21% fertile ova were obtained. Several attempts to repeat these studies with rosy barb (*Puntius conhonius*), zebra fish (*Brachydanio reio*) and *Oreochromis niloticus* ova have been unsuccessful (K. Rana *et al.* unpublished data). Nevertheless, cryopreserved spermatozoa can make an important contribution to the conservation and genetic management of fish resources.

3.4.1 Major physico-chemical objectives during cryopreservation

Cryopreservation of gametes encompasses a set of complex physical and chemical events, with 'cryosuccess' depending on a number of interrelated factors (Franks 1985). The relevant physico-chemical principles and events are reviewed elsewhere (Franks 1985, Grout & Morris 1987, Leung & Jamieson 1991, Rana 1993). The overall objective during cooling and warming is to prevent or minimize the formation of damaging intracellular ice crystals (Franks 1985).

When heat is removed from a cell suspension, its temperature decreases past its freezing point and undercools before ice formation or freezing is initiated, usually in the extracellular medium (Grout & Morris 1987). The freezing process, which is accompanied by ice formation and accumulation, results in an increase in solute concentration and a rise in temperature to the freezing point. Further cooling past the eutectic point, the freezing point of the solutes, results in the solidification of the solution and solutes.

The rise in extracellular solute concentration of the unfrozen fraction during freezing causes cellular water to be removed and cells to dehydrate. The rate and extent of dehydration is dependent on the size of the organism being frozen and the cooling rate (Grout & Morris 1987). With successful cryopreservation procedures the dehydrated cells suffer minimal damage from intracellular freezing of water.

As the temperature decreases further molecular motion is reduced. At temperatures below −130°C, the glass transition temperature, all cellular molecular motion ceases. It is this cessation of cellular activity in frozen cells that forms the basis for indefinite storage of cryopreserved material. Biological material can be successfully stored at −150°C (Mazur 1964) but for practical reasons samples are usually stored at −196°C, the temperature of liquid nitrogen.

3.4.2 Cryo-injuries

During cooling and warming, cells are subject to considerable physical and chemical trauma. The cell membrane is the primary barrier between the cytoplasm and the environment and therefore the prime site for cryo-injury (Franks 1985). These injuries may result from chill damage or hypertonic chemical effects during undercooling (Lovelock 1957, Morris & Watson 1984), change in membrane composition and structure (Morris & Watson 1984), speed shape and size of ice crystal formation (Fujikawa 1978), volume changes due to cell dehydration, compression and rehydration (Renard 1986, Schneider 1986) and reduction in buffering capacity of the freezing medium (van den Berg & Soliman 1969).

In fish, the degeneration of the sperm membrane during cooling and thawing is regarded as the principal cause of reduced post-thaw viability (Stoss & Donaldson 1982, Billard 1983, Maisse et al. 1988, Lahnsteiner et al. 1992, J. Lawrence & K. Rana unpublished data). Several approaches have been used to evaluate fish sperm membrane integrity. High levels of proteins (Yoo et al. 1987, Maisse et al. 1988) and intracellular enzymes (Gallant & McNiven 1991) in the seminal plasma, changes in flagellar beat frequency (Billard et al. 1993) sperm speed (Lawrence 1992) and staining of nuclear DNA with vital dyes (Fribourgh 1966) fluorescent probes (J. Lawrence & K. Rana (unpublished data)) have all been used to detect cryo-injury. All these approaches confirm that cooling and thawing reduce the integrity of fish sperm membrane but evidence to support a clear relationship between sperm fitness and motility is still lacking.

3.4.3 Components of milt cryopreservation

Some of the above deleterious effects can be ameliorated by the selective use of cryoprotectants and by optimization of all of the interrelated components that encompass cryobanking.

The process of cryobanking includes milt physiology, collection procedures, pre-freezing milt handling and storage, storage containers, cooling method and rate, thawing rate and the method of post-thaw evaluation. The variability of these factors, particularly the cooling methods and post-thaw evaluation for the same species between studies, have resulted in heterogenous results (Scott & Baynes 1980, McAndrew *et al.* 1993).

Pre-cooling stage

This phase includes the pre-freeze quality of milt, collection procedures and the dilution of milt with suitable extenders and cryoprotectants.

Milt quality

The quality of milt used for cryopreservation is crucial for optimizing post-thaw viability. The factors that influence biological variability considered earlier for chilled storage apply equally to cryopreservation. The ionic composition of the suspending medium will affect the freezing characteristics (Franks 1985) and, therefore, variability in seminal plasma composition and contamination can both have detrimental effects on post-thaw viability.

The cryo-resistance of spermatozoa may also be influenced by the rearing temperature and diet of males, both of which can affect the fatty acid composition and, thereby, the fluidity of membranes (Kruuv *et al.* 1978). Studies by Baynes & Scott (1987) on the effects of sprat diets high in $w3$ on the post-thaw viability of rainbow spermatozoa trout, however, suggest that sperm fitness can at best be marginally improved by high dietary level of $w3$ and furthermore that this can be compensated for by modifying the dilution medium.

The duration of storage of milt prior to cryopreservation may also influence post-thaw cryosuccess, although this may vary with species and cryopreservation protocol. It is widely recommended that gametes should be frozen immediately after collection to avoid a decline in post-thaw fertility (Scott & Baynes 1980, Stoss 1983, Harvey 1983, Schmidt-Baulain & Holtz 1989, Leung & Jamieson 1991). The detrimental effects of pre-cooling storage, however, are unclear. Studies on rainbow trout by Schmidt-Baulain & Holtz (1989) report that a delay of 60 min between milt collection and cryopreservation significantly decreases post-thaw viability. In contrast, Baynes & Scott (1987) noted significantly higher post-thaw fertility when milt of the same species was held at 0°C for between 8 and 26 h. In Atlantic salmon, milt stored for 21 days at −4°C before freezing showed similar viability to milt frozen 4 h after storage (K. Rana, unpublished data). Similarly, in *Oreochromis niloticus*, milt stored for up to 6 days at 4°C showed no significant loss in post-thaw viability (Rana *et al.* 1990).

Diluents

To enhance post-thaw viability milt should be extended in suitable diluents (Scott & Baynes 1980). A number of diluents of varying chemical composition and complexity have been successfully used. Details of their composition are reviewed by Scott & Baynes (1980) Leung & Jamieson (1991) and McAndrew et al. (1993). Most diluents used for cryopreservation are based on fish salines or sugars but published information shows no clear advantage of complex diluents over simple ones such as those containing only 0.3 M sucrose and 10% DMSO (Stoss & Refstie 1983) or 0.6 M glucose and 10% glycerol (Holtz et al. 1991). Typically most diluents include $0.25-7.5\,g\,l^{-1}$ KCl, $1.9-14.0\,g\,l^{-1}$ NaCl, $0.1-0.4\,g\,l^{-1}$ $CaCl_2$ $0.2-7.5\,g\,l^{-1}$ $NaHCO_3$, $0.09-0.23\,g\,l^{-1}$ $MgSO_4$, $0.6-10\,g\,l^{-1}$ of either fructose, lactose, sucrose, glucose. In addition, hens egg yolk, BSA, promine D, and lecithin have also been included in selected diluents.

To improve the yield of post-thaw viable cells, milt is usually diluted prior to cryopreservation (Scott & Baynes 1980). Several pre-freezing milt dilution ratios have been investigated but the variable sperm:egg ratio used between these studies to evaluate treatments often confounds any dilution effect. For salmonids a ratio of milt to diluents of between 1:1 and 1:19 had no significant effect on post-thaw viability (Truscott & Idler 1969, Ott & Horton 1971, Buyukhatipoglu & Holtz 1978). Studies on rainbow trout, however, suggest that a dilution ratio of 1:3 was superior to either 1:1 or 1:9 (Legendre & Billard 1980). Milt of O. niloticus can be diluted 20 times without any loss of viability (Rana & McAndrew 1989) whereas in O. mossambicus no advantage was reported for dilutions greater than 1:5 (Harvey 1983). For most species dilution ratios of 1:3 to 1:6 are commonly used (McAndrew et al. 1993).

Cryoprotection

Cryo-injuries associated with cooling and thawing can be ameliorated by the addition of non-permeating and permeating organic compounds to the diluents. Non-permeating compounds such as sugars (glucose, trehalose sucrose), proteins (milk powder, hens egg yolk, calf serum, promine D, glycoproteins) interact with permeating cryoprotectants in diluents to suppress the freezing point and raise the glass transition temperature (Shlafer 1981).

Permeating cryoprotectants such as glycerol, DMSO and methanol are widely used to depress the freezing point of the extracellular medium, ameliorate the damaging effects of ice crystals and regulate the rate of cellular dehydration (Shlafer 1981). Although a large selection of cryo-additives are available (see Shlafer 1981), glycerol, DMSO and methanol are most widely used for fish gametes preservation studies (McAndrew et al. 1993). New cryo-additives are being continually tested for fish milt cryopreservation. Recent studies have shown that dimethlyacetamide (10%) can confer good cryoprotection to rainbow trout spermatozoa (Gallant & McNiven 1991). The final choice of cryoprotectant may, however, depend on availability, quality safety considerations and cost.

Cryoprotectants are toxic to gametes. At high concentrations and during prolonged equilibration cellular proteins can be denatured (Shlafer 1981) and this can reduce pre-freezing viability. Recent studies with O. niloticus spermatozoa suggest that DMSO is more toxic than methanol (Rana & McAndrew 1989).

For the commonly used cryoprotectants, final concentrations of between 7 and 15% have been successfully used. The optimum concentration may, however, vary between cryoprotectants, species, equilibration time used and criteria used for evaluation of post-thaw viability. For *O. niloticus* 10% methanol was demonstrated to be optimal (Fig. 3.5) whereas using post-thaw motility 5% methanol was optimal for *O. mossambicus* (Harvey 1883).

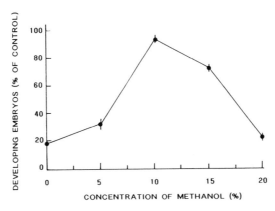

Fig. 3.5 Effect of cryoprotectant concentration on the viability of post-thawed spermatozoa frozen in modified fish Ringers. Means given with SEM, n = 4. (Adapted from Rana & McAndrew 1989.)

Equilibration time

For effective protection during cooling sufficient time must be allowed to facilitate the penetration of cryoprotectants into cells (Grout & Morris 1987). The rate of uptake may depend on the size of biological material and molecular size of the cryoprotectant. Smaller molecules such as methanol will enter cells at a faster rate than larger compounds such as glycerol (Harvey & Ashwood-Smith 1982). Recent studies on a range of fish spermatozoa suggests that an equilibration time of up to 2h (Truscott & Idler 1969, Ott & Horton 1971) may not be needed to facilitate cryoprotectant penetration. Harvey (1983) advocates immediate cooling after the addition of diluent containing the cryoprotectant. Good post-thaw survival has been reported for *O. niloticus* spermatozoa following an equilibration of 15 min, though no deleterious effects were noted for equilibration times up to 90 min (Rana *et al.* 1990). Equilibration time may be influenced by diluents used. Studies on *O. niloticus* using two fish Ringer-based diluents suggest that equilibration time and diluents are significantly related (Table 3.3).

Table 3.3 The effect of diluent and equilibration time on post-thaw motility of *Oreochromis niloticus* spermatozoa. (Data based on video analysis)

Equilibration time (min)	Mean post-thaw motility (%)*	
	D_1^{**}	D_2^{**}
15	33(4.0)^{a1}	14(1.8)^{b3}
30	56(4.0)^{a2}	11(2.2)^{b3}
45	54(3.7)^{a2}	4(1.3)^{b4}
60	48(4.9)^{a2}	0^{b5}
90	50(4.4)^{a2}	0^{b5}

* Means (n = 4) bearing different letters and numbers between columns and within rows, respectively, are significantly ($P < 0.05$) different
** D_1 and D_2 are diluents used by Rana *et al.* (1990) and Harvey (1983), respectively

Cooling and warming rates

Cooling and thawing rates are regarded as the most critical phases of cryopreservation (Franks 1985, Grout & Morris 1987). However, in most studies relating to the preservation of gametes these factors are the least standardized of variables, thus frustrating attempts to evaluate published protocols. The cooling rates generated in practice vary with the method of cooling (Leung & Jamieson 1991, McAndrew *et al.* 1993, Rana 1994).

Fish milt can be cooled as pellets on dry ice blocks (Buyukhatipoglu & Holtz 1978, Stein & Bayrle 1978, Legendre & Billard 1980, Stoss & Holtz 1981, Stoss & Refstie 1983). This method, though convenient for field application, can only generate one cooling rate, approximately, 30–35°C min^{-1} and is commonly used for cryopreserving trout milt. Cooling milt in liquid nitrogen vapour in a polystyrene box or in the neck of a liquid nitrogen dewar is more widely used, but the cooling rates are unpredictable and variable. Moreover, studies using this technique do not determine the cooling rates within the milt vessels and therefore are of little value in developing a reliable protocol. The use of programmable control rate coolers provides reliable methods to simulate consistent cooling and warming rates for protocol evaluation.

The limited data on cooling rates suggest an optimum cooling rate for spermatozoa which may vary between species (Scott & Baynes 1980, Stoss 1983, Leung & Jamieson 1991, McAndrew *et al.* 1993). For rainbow trout, cooling milt containing 7–12.5% DMSO at 30–35°C min^{-1} resulted in 0–98% eyed eggs (Scott & Baynes 1980, McAndrew *et al.* 1993) though Baynes & Scott (1987) reports 67% eyed eggs for the same species at cooling rates of 80°C min^{-1}. Lower cooling rates have been more successfully achieved in subtropical and tropical species. Sea bass and sea bream milt cooled at 10°C min^{-1} gave highest post-thaw activation (Billard *et al.* 1977). In *O. niloticus* cooling rates of 5–50°C min^{-1} had no significant effect on fertilization rates (Rana & McAndrew 1989).

To avoid recrystallization rapid warming rates are used. In salmonids (Stoss 1983) and tilapia (Harvey 1983) warming rates of 100–1500°C min^{-1} and 90–320°C min^{-1} have been used (Stoss 1983). In practice samples are generally thawed in a 30–80°C water bath prior to insemination.

Insemination of thawed cryopreserved milt

The injuries sustained during the process of cooling and thawing are difficult to isolate. Cryo-injuries can result in spontaneous activation of spermatozoa and reduce the proportion of activated spermatozoa upon thawing. In addition, the speed and duration of motility can be reduced when compared with fresh unfrozen spermatozoa (Stoss 1983, Billard 1988). In view of such deterioration, post-thaw milt should be used immediately after thawing. For rainbow trout a delay of 30 s decreased the eyed egg yield by 16% (Stoss & Holtz 1981).

To prolong the duration of motility, minimize osmotic shock during fertilization and disperse the sperm cells around the ova, inseminating solutions are advocated. The compositions of such solutions are similar to ovarian fluid (Ginsberg 1963, Scott & Baynes 1980) or have a salinity of 5% and 20% for fresh and sea water species, respectively. For salmonids, buffered 0.1–0.15 M sodium bicarbonate or chloride is commonly used (Stoss 1983). The addition of compounds such as isobutyl-1-methylxanthine (Benau & Terner 1980) and theophylline (Scheerer & Thorgaard 1989, Wheeler & Thorgaard 1991) to buffered fertilizing medium may also help to reduce osmotic swelling and prolong the duration of sperm motility.

3.5 Conclusions

The majority of *in vitro* preservation studies relate to *in vitro* storage of spermatozoa and are biased towards three commercial groups of fish of aquacultural importance, the salmonids, tilapines and carps.

To date, preservation of fish ova is limited to short-term storage. *In vitro* storage of unfrozen spermatozoa has been widely reported since the 1960s, but often the interrelated factors influencing post-storage viability are poorly controlled or reported, thus resulting in variable and non reproducible results.

The period of *in vitro* preservation is influenced by storage temperature. For short-term storage of unfrozen gametes, pre-storage gamete quality, stability of desired storage temperature, aerobic conditions and desiccation are principal variables that need rigorous control.

For cryopreservation the quality of milt is of crucial importance. Diluents must maintain cells in the immotile state and contain optimal levels of cryoprotectant. Cryoprotectants such as DMSO, glycerol and methanol are commonly used at final concentrations of 7–15%. Cooling rates of 1–80°C have been tested but methods of cooling and reproducibility of cooling rates are variable. Post-thaw spermatozoa have a reduced duration of motility and should be used immediately to inseminate ova. Although post-thaw motility is often used as a criterion for cell viability, it should be accompanied by a standardized fertilization method. The number of eggs used should be similar, and more importantly the sperm:egg ratio during insemination must be kept constant. In addition, since cryoprotectants can affect pre-freezing activity, an additional control containing fresh milt and diluent for a similar equilibration time should be used to establish pre-freezing milt viability.

Cryopreservation technology is likely to play an increasing role in the conservation of aquatic species and in aquaculture. Current preservation technologies can be used to minimize the cost of wet gene banks and to preserve valuable aquacultural stocks. It may not be necessary to store female gametes to conserve the nuclear genome. A combination of sperm cryobank and androgenesis or nuclear transfer techniques could be used to conserve a large gene pool for different species (McAndrew *et al.* 1993). The application of such techniques is, however, dependent on the reliability of protocols developed.

Acknowledgements

Part of the studies referred to in this chapter have been conducted under a grant and facilities provided by the British Overseas Development Administration at the Institute of Aquaculture, University of Stirling. The author thanks Ann Gilmour for her assistance in compiling the reference list and tables.

References

Aboyewa, M.F. (1991) *Effects of storage medium and temperature on the short-term storage of* Oreochromis niloticus *(L.) eggs*. M.Sc. thesis, Institute of Aquaculture, University of Stirling.
Adam, M.M. (1988) *Short-term storage of* Oreochromis niloticus *(L.) ova*. M.Sc. thesis, Institute of Aquaculture, University of Stirling.
Ashwood-Smith, M.J. (1980) *Low Temperature Preservation Cells, Tissues and Organs: Low Temperature Preservation in Biology*. Pitman Medical, Tunbridge Wells.
Basavaraja, N. (1991) *Effects of stripping and fertilisation methodology on egg quality in the Atlantic halibut*, Hippoglossus hippoglossus *(L.)* M.Sc. thesis, Institute of Aquaculture, University of Stirling.
Baynes, S.M. & Scott, A.P. (1987) Cryopreservation of rainbow trout spermatozoa: The influence of sperm quality, egg quality and extender composition on post-thaw fertility. *Aquaculture*, **66**, 53–67.
Benau, D. & Terner, C. (1980) Initiation, prolongation and reactivation of the motility of salmonid spermatozoa. *Gamete Research*, **3**, 247–57.
van den Berg, L. & Soliman, F.S. (1969) Effect of glycerol and dimethyl sulfoxide on changes in composition and pH of buffer salt solutions during freezing. *Cryobiology*, **6**, 93–7.
Billard, R. (1981) Short-term preservation of sperm under oxygen atmosphere in rainbow trout. *Aquaculture*, **23**, 287–93.
Billard, R. (1983) A quantitative analysis of spermatogenesis in the trout *Salmo truta fario. Cell Tissue Research*, **230**, 495–502.

Billard, R. (1988) Artificial insemination and gamete management in fish. *Marine Behavior and Physiology*, **14**, 3–21.

Billard, R., Dupont, J. & Barnabé, G. (1977) Diminution de la motilité et de la durée de conservation du sperme de *Dicentrarchus labrax* (poisson téléostéen) pendant la période de spermiation. *Aquaculture*, **11**, 363–7.

Billard, R., Casson, J. & Crim, L.W. (1993) Motility of fresh and aged halibut sperm. *Aquatic Living Resources*, **6**, 67–75.

Bromage, N. & Cumaranatunga, R. (1988) Egg production in rainbow trout. In *Recent Advances in Aquaculture*, Vol. IV (eds J.F. Muir & R.J. Roberts), pp. 63–138. Blackwell Science, Oxford.

Buyukhatipoglu, B. & Holtz, W. (1978) Preservation of trout sperm in liquid or frozen state. *Aquaculture*, **14**, 49–56.

Buyukhatipoglu, B. & Holtz, W. (1984) Sperm output in rainbow trout – effect of age, timing and frequency of stripping and presence of females. *Aquaculture*, **37**, 63–71.

Carrilho, M.C. (1990) *Influence of storage medium, temperature and storage time on the viability of unfertilised* Oreochromis niloticus *(L) eggs*. M.Sc. thesis, Institute of Aquaculture, University of Stirling.

Edwardes, S. (1991) *A preliminary investigation into the seasonal variation and short term storage of milt from the Atlantic halibut* Hippoglossus hippoglossus. M.Sc. thesis, Institute of Aquaculture, University of Stirling.

Erdahl, A.W. & Graham, E.F. (1987) Fertility of teleost semen as affected by dilution and storage in a seminal plasma-mimicking medium. *Aquaculture*, **60**, 311–21.

Franks, F. (1985) *Biophysics and Biochemistry at Low Temperatures*. Cambridge University Press, Cambridge.

Fribourgh, J.H. (1966) Application of a differential staining method to low-temperature studies on goldfish spermatozoa. *Progressive Fish Culturist*, **28**, 227–31.

Fujikawa, S. (1978) Morphology evidence of membrane damage caused by intracellular ice crystals. *Cryobiology*, **15**, 707.

Gallant, R.K. & McNiven, M.A. (1991) Cryopreservation of rainbow trout spermatozoa. *Bulletin of the Aquaculture Association of Canada*, **91**, 25–7.

Ginsberg, A.S. (1963) Sperm – egg association and its relationship to the activation of the egg in salmonid fishes. *Journal of Embryology and Experimental Morphology*, **11**, 13–33.

Grout, B.W. & Morris, G.J. (1987) *The Effect of Low Temperature on Biological Systems*. Edward Arnold, London.

Harvey, B. (1983) Cryopreservation of (*Sarotherodon mossambicus*) spermatozoa. *Aquaculture*, **32**, 313–20.

Harvey, B. & Ashwood-Smith, M.J. (1982) Cryoprotectant penetration and super cooling in the eggs of salmonid fish. *Cryobiology*, **19**, 29–40.

Harvey, B., Stoss, J. & Butchart, W. (1983) Supercooled storage of salmonid ova. *Canadian Technical Report of Fisheries and Aquatic Sciences*, **1222**, Series i, 1–9.

Harvey, B. & Kelley, R.N. (1984) Short-term storage of *Sarotherodon mossambicus* ova. *Aquaculture*, **37**, 391–5.

Holtz, W., Schmidt-Baulain, R. & Meiners-Gefken, M. (1991) Cryopreservation of rainbow trout (*Oncorhynchus mykiss*) semen in a sucrose/glycerol extender. *Fourth International Symposium on Reproductive Physiology of Fish*, Abstract 63.

Hulata, G. & Rothbard, S. (1979) Cold storage of carp semen for short periods. *Aquaculture*, **16**, 267–9.

Jensen, J.O.T. & Alderdice, D.F. (1984) Effect of temperature on short term storage of eggs and sperm of chum salmon (*Oncoryhynchus keta*). *Aquaculture*, **37**, 251–65.

King, A.T., Mulligan, B.J. & Lowe, K.C. (1989) Perfluorochemicals and cell culture. *Biotechnology*, **7**, 1037–41.

Kruger, J.C. de W., Smit, G.L., Van Vuren, J.H.J. & Ferreira, J.T. (1984) Some chemical and physical characteristics of the semen of *Cyprinus carpio* (L.) and *Oreochromis mossambicus* (Peters). *Journal of Fish Biology*, **24**, 263–72.

Kruuv, J., Lepock, J.R. & Keith, A.D. (1978) The effect of fluidity of membrane lipids on freeze thaw survival of yeast. *Cryobiology*, **15**, 73–9.

Lahnsteiner, F., Weisman, T. & Patzner, R.A. (1992) Fine structural changes in spermatozoa of the grayling, (*Pisces Teleostei*), during routine cryopreservation *Aquaculture*, **103**, 73–84.

Lawrence, C. (1992) *Development of computer image analysis and fluorometry to assess the fitness of cryopreserved tilapia spermatozoa*. M.Sc. thesis, Institute of Aquaculture, University of Stirling.

Legendre, M. & Billard, R. (1980) Cryopreservation of rainbow trout sperm by deep freezing. *Reproduction, Nutrition, Development*, **20**, 1859–68.

Leung, L.K.-P. & Jamieson, B.G.M. (1991) Live preservation of fish gametes. In *Fish Evolution and Systematics: Evidence from Spermatozoa*, (ed. B.G.M. Jamieson), pp. 245–69. Cambridge University Press, Cambridge.

Lovelock, J.E. (1957) The denaturation of lipid–protein complexes as a cause of damage by freezing. *Proceedings of the Royal Society B*, **14**, 427–33.

Lowe, K.C. (1991) Synthetic oxygen transport fluids based on perfluorochemicals: applications in medicine and biology. *Vox Sanguinis*, **60**, 129–40.

McAndrew, B.J., Rana, K.J. & Penman, D.J. (1993) Conservation and preservation of genetic variation in aquatic organisms. In *Recent Advances in Aquaculture*, Vol. IV (eds J.F. Muir & R.J. Roberts), pp. 295–336. Blackwell Science, Oxford.

McNiven, M., Gallant, R.K. & Richardson, G.F. (1993) Fresh storage of rainbow trout (*Oncorhyncus mykiss*) semen using a non-aqueous medium. *Aquaculture*, **109**, 71–82.

Maisse, G., Pinson, A. & Loir, M. (1988) Characterisation of the fitness for cryopreservation of milt from rainbow trout together with physico chemical criteria. *Aquatic Living Resources*, **1**, 45–51.

Mazur, P. (1964) Basic problems in cryobiology. *Advances in Cryogenic Engineering*, **9**, 28–37.

Moore, A. (1987) Short term storage and cryopreservation of walleye sperm. *The Progressive Fish Culturist*, **49**, 40–3.

Morisawa, M. & Suzuki, K. (1980) Osmolality and potassium ion: their roles in initiation of sperm motility in teleosts. *Science*, **210**, 1145–6.

Morris, G.J. & Watson, P.F. (1984) Cold shock injury – a comprehensive bibliography. *Cryoletters*, **5**, 352–72.

Muiruri, R.M. (1988) *Chilled and cryogenic preservation of tilapia spermatozoa*. M.Sc. thesis, Institute of Aquaculture, University of Stirling.

Munkittrick, K.R. & Moccia, R.D. (1987) Seasonal changes in the quality of rainbow trout (*Salmo gairdneri*) semen: effect of delay in stripping on spermatocrit, motility, volume and seminal plasma constituents. *Aquaculture*, **64**, 147–56.

Ott, A.G. & Horton, H.F. (1971) Fertilisation of steelhead trout (*Salmo gairdneri*) eggs with cryopreserved sperm. *Journal of the Fisheries Research Board of Canada*, **28**, 1915–18.

Piironen, J. & Hyvarinen, H.H. (1983) Cryopreservation of spermatozoa of the whitefish *Coregonus muksun* Pallas. *Journal Fish Biology*, **22**, 159–64.

Plosila, D.S. & Keller, W.T. (1974) Effects of quantity of sperm and water on fertilization on brook trout eggs. *The Progressive Fish Culturist*, **36**, 42–5.

Rana, K.J. (1993) Cryopreservation of fish spermatozoa. In *Workshop Proceedings Genetics in Aquaculture and Fisheries Management, University of Stirling, 31 August–4 September 1992*, (eds D. Penman, N. Roongratri & B. McAndrew), pp. 49–53.

Rana, K.J. (1995) Cryopreservation of fish spermatozoa. In *Cryopreservation and Freeze Drying Protocols*, (eds D.G. Day & M.R. McLellan) (in press). The Humana Press Inc., New Jersey.

Rana, K.J. & McAndrew, B.J. (1989) The viability of cryopreserved tilapia spermatozoa. *Aquaculture*, **76**, 335–45.

Rana, K.J., Muiruri, R.M., McAndrew, B.J. & Gilmour, A. (1990) The influence of diluents, equilibration, time and pre-freezing storage time on the viability of cryopreserved *Oreochromis niloticus* (L.) spermatozoa. *Aquaculture and Fisheries Management*, **21**, 25–30.

Rana, K.J., Gupta, S.D. & McAndrew, B.J. (1992) The relevance of collection techniques on the quality of manually stripped Atlantic salmon (*Samnio salar*) milt. *Workshop on Gamete and Embryo Storage and Cryopreservation in Aquatic Organisms*, 30 March–2 April 1992, 4.

Randall, D.J. & Hoar, W.S. (1971) *Special Techniques in Fish Physiology*, Vol. 6. Academic Press, New York.

Renard, J.P. (1986) Cryopreservation of embryos. *Cryobiology*, **23**, 547.

Rogoff, M.S. (1985) *The effects of aeration and diluent composition on the fecundity of cold stored turkey and chicken semen*. M.Sc. thesis, Clemson University, Clemson SC.

Saad, A., Billard, R., Theron, M.C. & Hollebecq, M.G. (1988) Short term preservation of carp (*Cyprinus carpio*) semen. *Aquaculture*, **71**, 133–50.

Sanchez-Rodriguez, H., Escaffre, A.M., Marlot, S. & Reinaud, P. (1978) The spermiation period in rainbow trout (*Salmo gairdnerii*). Plasma gonadotropin and androgen levels, sperm production and biochemical changes in the seminal fluid. *Annales Biologie Animale Biochimie et Biophysique*, **18**, 943–8.

Scheerer, P. & Thorgaard, G. (1989) Improved fertilisation by cryopreserved rainbow trout semen treated with Theophyline. *The Progressive Fish Culturist*, **51**, 179–82.

Schmidt-Baulain, R. & Holtz, W. (1989) Deep freezing of rainbow trout sperm at varying intervals after collection. *Theriogenology*, **32**, 439–43.

Schneider, U. (1986) Cryobiological principles of embryo freezing. *Journal of In Vitro Fertilisation and Embryo Transfer*, **3**, 3–9.

Scott, A.P. & Baynes, S.M. (1980) A review of the biology, handling and storage of salmonid spermatozoa. *Journal of Fish Biology*, **17**, 707–39.

Shlafer, M. (1981) Pharmacological considerations in cryopreservation. Chapter 10. In *Organ Preservation for Transplantation*, (eds A.M. Karow & D.E. Pegg) 2nd edn, pp. 177–212. Marcell Dexuer Inc., NY.

Stein, H. & Bayrle, H. (1978) Cryopreservation of the sperm of some freshwater teleosts. *Annales Biologie Animale Biochimie et Biophysique*, **18**, 1073–6.

Stoss, J. (1983) Fish gamete preservation and spermatozoan physiology. In *Fish Physiology*, Vol. IXB (eds W.S. Hoar, D.J. Randall & E.M. Donaldson), pp. 305–50. Academic Press, New York.

Stoss, J. & Holtz, W. (1981) Cryopreservation of rainbow trout (*Salmo gairdneri*) sperm, 1. Effect of thawing solutions, sperm density and interval between thawing and insemination. *Aquaculture*, **22**, 97–104.

Stoss, J. & Holtz, W. (1983) Successful storage of chilled rainbow trout (*Salmo gairdneri*) spermatozoa for up to 34 days. *Aquaculture*, **31**, 269–74.

Stoss, J. & Reftsie, T. (1983) Short term storage and cryopreservation of milt from Atlantic salmon and sea trout. *Aquaculture*, **30**, 229–36.

Stoss, J., Geries, L. & Holtz, W. (1987) The role of sample depth in storing chilled rainbow trout (*Salmo gairdneri*) semen under oxygen. *Aquaculture*, **61**, 275–9.

Stoss, J. & Donaldson, E.M. (1982) Preservation of fish gametes. In *Proceedings of the International Symposium on Reproductive Physiology of Fish, Pudoc Netherlands*, (eds C.J.J. Richter & H.J.Th. Goos).

Tave, D. (1986) *Genetics for Hatchery Managers*. Avi Publishing Co Inc., Westport, Connecticut.

Truscott, B. & Idler, D.R. (1969) An improved extender for freezing Atlantic salmon spermatozoa. *Journal of the Fisheries Research Board of Canada*, **26**, 3254–8.

Wheeler, P.A. & Thorgaard, G.H. (1991) Cryopreservation of rainbow trout semen in large straws. *Aquaculture*, **93**, 95–100.

Withler, F.C. & Humphreys, R.M. (1967) Duration of fertility of ova and sperm of sockeye (*Oncorhyncus nerka*) and pink (*O. gorbuscha*) salmon. *Journal of the Fisheries Research Board of Canada*, **24**, 1573–8.

Withler, F.C. & Morley, R.B. (1968) Effects of chilled storage on viability of stored ova and sperm of sockeye and pink salmon. *Journal of the Fisheries Research Board of Canada*, **25**, 2695–9.

Yoo, B.Y., Ryan, M.A. & Wiggs, A.J. (1987) Loss of protein from spermatozoa of Atlantic salmon (*Salar salar L.*) because of cryopreservation. *Canadian Journal of Zoology*, **65**, 9–13.

Zlabek, A. & Linhart, O. (1987) Short term storage of non-inseminated and unfertilised eggs of the common carp, grass carp and silver carp. *Bulletin VURH Vodney*, **23**, 3–11.

Chapter 4
Biotechnological Approaches to Broodstock Management

4.1 Introduction
4.2 Qualities of fish facilitating biotechnological approaches
4.3 Biotechnologies available for broodstock improvement
    4.3.1 Chromosome set manipulation
    4.3.2 Gene transfer
    4.3.3 Marker-assisted selection
4.4 Major breeding goals for aquaculture
    4.4.1 Sex control
    4.4.2 Extended environmental tolerance
    4.4.3 Disease resistance
    4.4.4 Growth rate or meat quality
4.5 Biotechnology and possible broodstock improvement
    Acknowledgements
    References

Biotechnological approaches with implications for broodstock improvement in aquaculture include chromosome set manipulation, gene transfer and marker-assisted selection. Polyploidy can be induced either directly with heat or pressure treatment of fertilized eggs or indirectly by hybridizing diploids and tetraploids. Triploids are advantageous because of their sterility and the increased viability of triploid interspecific hybrids. Gynogenesis (all-maternal inheritance) and androgenesis (all-paternal inheritance) can also be induced and have significant research and practical applications. Gene transfer is being widely applied to fish species; the technology is rapidly improving and transgenic fish with traits such as faster growth and the likelihood of improved disease resistance will be available in the near future. However, environmental concerns may limit their use. Numerous new genetic markers are becoming available through a variety of DNA technologies. Such markers should be useful in improving the efficiency of selection for aquacultural traits such as environmental tolerance, disease resistance and growth rate.

4.1 Introduction

Biotechnology is a relatively new term, which has been used to describe a variety of manipulations of biological systems for human benefit. In the context of broodstock improvement, biotechnology can be used to describe the application of new genetic technologies to improving broodstock more rapidly or in ways which would not be possible using conventional selective breeding techniques.

By highlighting such approaches in this chapter, the intent is not to detract from more classical approaches. Such approaches have been the basis for the dramatic improvement which has already been made in some aquaculture stocks (Gjedrem 1983, Refstie 1990) and will continue to be central to genetic improvement in the future. Animal breeders do indeed have a valid claim to be considered the original genetic engineers (Dickerson 1986). One of the exciting challenges will be the integration of the newer biotechnological approaches with the established selective breeding approaches for genetic improvement. Successful integration of biotechnology and selective breeding will allow the effects of changes to be better evaluated, accelerate the rate of improvement and allow changes outside the range of what could be readily accomplished by selective breeding alone.

Although finfish, shellfish and aquatic plants share many potential biotechnological approaches in common and useful ideas can often be gained by noting approaches used in different systems, this review will be confined to biotechnological applications to finfish. The major topics to be covered include:

(1) the qualities of fish facilitating biotechnology applications;
(2) biotechnologies available and how they can be applied;
(3) major breeding goals in aquaculture and selected examples of biotechnological applications to these goals; and
(4) opportunities for combining these technologies to maximize the efficiency of improvement.

4.2 Qualities of fish facilitating biotechnological approaches

Fish have a number of qualities which facilitate applying biotechnological approaches to their genetic improvement. Although many initial breakthroughs (e.g. development of improved methods for handling and analysing DNA) will often be done by investigators studying other groups of organisms, investigators studying fish have significant advantages for applying the breakthroughs in a living animal. Fish are certainly the most experimentally tractable group of vertebrates for applying a full range of biotechnological approaches to genetic improvement.

The reproductive qualities of fish are advantageous for biotechnological applications. Fish typically produce large numbers of eggs and in some species can be induced to spawn more than once per year. Their gametes are easy to handle and sperm can generally be stored for a considerable length of time before use in fertilization and can be cryopreserved (Stoss 1983). Eggs can sometimes also be stored for extended periods after removal from the female and before fertilization, a significant advantage for many biotechnological applications. Many of the approaches which will be discussed involve planned matings which can be most readily done in a controlled fashion over a period of time. External fertilization facilitates such studies by allowing controlled crosses to be readily made to a timetable.

The low cost per individual of rearing fish (relative to even small mammals and poultry) is a substantial advantage for biotechnological applications. As will be discussed, many of these applications are not yet at the commercial stage and considerable research is needed before they can be commercially applied. The range of approaches which can be investigated is obviously limited by the cost of such applications, and in many domestic livestock species the cost per individual of rearing the animals can be a significant obstacle to progress. However, because of the large number of individual fish which can be reared on a reasonable budget, the potential for using fish in fundamental biotechnology studies is being recognized. Small fish species have similarly been recognized as excellent model vertebrates for development and cancer research because of the low cost of rearing per individual (Hoover 1984).

Fish can tolerate a range of ploidy changes which are lethal or subvital in many other vertebrates; such alterations have significant implications for biotechnology. Ploidy-manipulated fish which can be produced include viable triploids and tetraploids and inbred fish with both chromosome sets from the female or male parent.

The range of genetic variability available for research is another significant advantage in applying biotechnology to fish. All fish species currently being cultured still have progenitor stocks available in the wild which can provide a source of genetic material for future improvement. Wild progenitors have been a major source of raw material for the genetic improvement of plants (Wilkes 1983) but have seen less application in domestic animals. However, the advantageous properties of fish (which make them similar to plants) and the opportunities afforded by the new biotechnologies suggest that variation within species may be a significant resource for future genetic improvement in fish species.

Using biotechnological approaches, it may be possible to select particular attributes from wild populations without excessively diluting the improvement which has been made in the cultured stocks. Similarly, there are typically a number of closely related species for any given cultured species which can serve as a 'genetic reservoir' for the future improvement of the cultured species. We are likely to see increased use of related species for their desired attributes as the ability to identify and transfer genes associated with particular traits is enhanced.

4.3 Biotechnologies available for broodstock improvement

Biotechnologies which have significant potential application for broodstock improvement include chromosome set manipulation, gene transfer and marker-assisted selection. These approaches have in some cases been applied for many years with fish, but all have seen their broadest application only in recent years. With the expanded use of recombinant DNA technology and the range of DNA technologies available, gene transfer in particular has seen a dramatic

4.3.1 Chromosome set manipulation

Chromosomal manipulations which can be tolerated in fish include induced polyploidy, gynogenesis and androgenesis (reviewed by Purdom 1983, Thorgaard 1986, Ihssen *et al.* 1990). These manipulations have a number of potential applications (Fig. 4.1).

Polyploidy can be tolerated in fish, unlike mammals, and has significant applications for genetic improvement. Triploids are of primary interest because female triploids have been shown to be sterile and to not develop eggs or secondary sexual characters. This is advantageous in many culture situations.

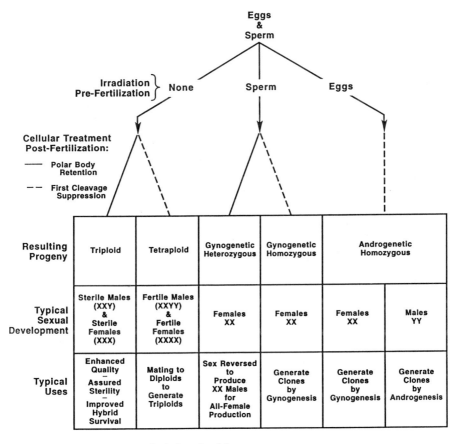

Fig. 4.1 Chromosome set manipulations for fish.

Male triploid fish do develop the secondary sexual characters associated with sexual maturation and are not generally desirable (Lincoln & Scott 1984). However, the use of sex-reversed XX males can allow all-female triploid populations to be produced.

Another benefit of induced triploidy has been the improved survival that has often been observed in interspecific triploid hybrids relative to the corresponding diploid hybrids (Chevassus *et al.* 1983, Scheerer & Thorgaard 1983). Some triploid hybrids such as the rainbow trout (*Oncorhynchus mykiss*) × brook trout (*Salvelinus fontinalis*) hybrid (Fig. 4.2) may survive when the corresponding diploid hybrid is inviable. Triploid hybrids may be of interest in their own right or as research animals to characterize genetic differences between species and the potential for transferring useful traits from one species to another.

Fig. 4.2 Appearance of triploid rainbow trout (*Oncorhynchus mykiss*) × brook trout (*Salvelinus fontinalis*) interspecific hybrids. This hybrid is normally inviable as a diploid. The triploid hybrid shows resistance to several viruses which rainbow trout are sensitive to (Dorson *et al.* 1991). The top fish is a male showing secondary sexual characteristics; the bottom fish is a female lacking secondary sexual characteristics, as expected. The fish are approximately 3 years old and 450 cm long.

Triploidy can be induced either directly or indirectly through a tetraploid parent in fish. Direct induction of triploids typically involves treating the eggs with a heat or pressure shock treatment shortly after fertilization to induce retention of the second polar body. Such fish have two maternal chromosome sets and generally show good viability. The heat and pressure treatments have been successfully scaled up for large-scale application and are being commercially applied with several species (see e.g. Bye & Lincoln 1986).

The alternative approach of inducing triploidy using tetraploid parents is attractive because it avoids the need to apply potentially damaging treatments to fertilized eggs. To date, however, this alternative has been demonstrated only in the rainbow trout (Chourrout et al. 1986, Myers & Hershberger 1991).

The principal obstacles to the tetraploid × diploid alternative are that tetraploids consistently show poor viability, at least in the initial generation after induction, and that it appears to be significantly more difficult to induce tetraploidy than it is to induce triploidy. An additional obstacle to the use of tetraploids to generate triploids is the poor fertility of tetraploid males when crossed with diploid females, apparently because their large sperm makes it difficult to pass through the small micropyle of the egg from normal diploid fish (Chourrout et al. 1986).

Gynogenesis and androgenesis are two techniques which also involve suppression of cell divisions in generating viable offspring. However, in these techniques, the gametes are typically irradiated before fertilization so that only one parent contributes to the next generation. In gynogenesis, only the female parent contributes; in androgenesis, only the male contributes.

In gynogenesis, sperm is irradiated before fertilization most often with ultraviolet light. By then applying the same sorts of treatments which have been applied to induce triploidy and tetraploidy, partially inbred in the case of second polar body retention or fully homozygous fish in the case of suppression of first cleavage can be produced.

One application of such fish is that in female homogametic species, which most fish species appear to be, the progeny after gynogenesis will all be females. A second application of gynogenesis lies in the ability to map genes relative to their centromeres in fish after retention of the second polar body (e.g. Allendorf et al. 1986). A third application is the generation of homozygous lines by applying a second cycle of gynogenesis to the homozygous fish produced initially by gynogenesis involving suppression of the first cleavage division (Streisinger et al. 1981, Naruse et al. 1985, Komen et al. 1991). Such fish are likely to be very valuable tools in basic research involving fish and could be useful in evaluating some biotechnologies.

Homozygous lines have proved to be useful in mice and in plants to evaluate the genetic basis of quantitative characters when crossed and used to generate new recombinant inbred lines (Festing 1979). Such lines might similarly be used to correlate DNA markers to quantitative traits in fish.

A fourth application of gynogenesis is the generation of fish with two maternal chromosome sets plus chromosome fragments from the male parent if the sperm is irradiated with gamma radiation at a dose which does not completely inactivate the sperm chromosomes (Thorgaard et al. 1985). This could provide a means to transfer genes from one species into the background of another if lines which stably transmit the introduced gene/fragment to the progeny can be developed (Disney et al. 1988, Thorgaard et al. 1992).

Androgenesis has not been as widely investigated in fish as gynogenesis. Like gynogenesis, androgenesis can be used to generate clonal lines (Scheerer et al. 1991). An advantage of androgenesis is the option of storing and regenerating lines from cryopreserved sperm by androgenesis; eggs have not yet been successfully cryopreserved in fish (Stoss 1983) and this precludes the storage of gynogenetic lines in that way.

4.3.2 Gene transfer

Gene transfer has become a very active topic of research in fish in recent years (reviewed by Chen & Powers 1990, Maclean & Penman 1990, Houdebine & Chourrout 1991). This is probably because of the considerable interest in transgenic technology and the recognition that, for the reasons discussed above, fish are ideal vertebrates in which to pursue such studies. Much of the research to date has focused on developing basic techniques for gene transfer in fish, but the time when transgenic fish will be ready for commercial application is rapidly approaching.

The principal approach used to date to transfer genes into fish eggs has been microinjection. Cloned DNA sequences are typically injected into the eggs shortly after fertilization. Transfer of the genes can then be monitored by the presence of the foreign DNA in the progeny or expression of the transferred genes. Another critical question is transfer of the introduced genes from the initial (microinjected) generation to their progeny; this has also been confirmed in several studies.

Alternative approaches to microinjection are being investigated. These include electroporation of DNA into the eggs shortly after fertilization (see e.g. Buono & Linser 1991) and the transfer of DNA on the sperm itself (see e.g. Mueller et al. 1991). These techniques could facilitate larger-scale production of transgenic fish.

Gene transfer currently has a number of limitations. Although it is easier to accomplish and afford in fish than in mammals, the preparatory steps of gene isolation and constructing the vector carrying the gene, in a way that insures consistent and advantageous expression, still need to be carried out. Such studies tend to be costly and require a laboratory with specialized skills. One emerging impression is that DNA sequences which include fish control sequences give the best gene expression (Houdebine & Chourrout 1991).

Another significant constraint to the application of gene transfer is concerns which have been expressed about hazards associated with escape of transgenic animals into the wild (Kapuscinski & Hallerman 1991). Could such fish readily become established and create unforeseen problems? We are in fact already impacting wild fish populations with hatchery fish which in many cases are substantially different genetically from the wild populations (Hindar *et al.* 1991) and perhaps transgenic fish should be considered as a subset of this broader issue. However, public and consumer perceptions are important and could represent a real constraint to the use of transgenics in production aquaculture situations in the near future.

4.3.3 Marker-assisted selection

Marker-assisted selection represents a third broad biotechnology approach. The basic concept is that, by identifying underlying genes or markers associated with a particular phenotype, it may be possible to practise more efficient selection (Soller & Beckmann 1983, Haley 1991). This has always been a possibility for investigators using protein markers (Utter & Seeb 1990), but with the development of DNA technologies the number of potential markers grows dramatically and marker-assisted selection becomes more feasible.

A range of DNA-based methods are available for developing genetic markers (reviewed by Hallerman & Beckmann 1988, Phillips & Ihssen 1990). The most widely used to date have been restriction fragment length polymorphisms (RFLPs), which detect polymorphisms for a single gene or DNA region by cutting the total genomic DNA with a restriction enzyme, separating the DNA by size on a gel, blotting the DNA onto a membrane, probing with the gene or sequence of interest, and visualizing the polymorphisms as bands on a gel. Methods for detecting polymorphisms based on the new techniques of polymerase chain reaction (PCR) are being developed and have already shown considerable application to plant breeding (Michelmore *et al.* 1991, Williams *et al.* 1991).

The RAPD (random amplified polymorphic DNA) technique involves the use of short primers of arbitrary sequence combined with PCR to generate bands which show variations both within and between species (Fig. 4.3). PCR-based methods have significant advantages in that only a small quantity of DNA is needed, it may not need to be of high molecular weight, and fewer steps are needed in the analysis. Finally, direct sequencing of the DNA, often in combination with the PCR technique, is another direct approach for detecting genetic differences (see e.g. Carr & Marshall 1991, Bartlett & Davidson 1992).

Two basic approaches in marker-assisted selection are the candidate gene approach and the correlation of random markers to the trait of interest. In the candidate gene approach, a gene which is suspected to show genetic variation related to the trait is investigated to identify polymorphisms (e.g. using protein electrophoresis or one of the above DNA analysis approaches). When a polymorphism is detected, it can be tested for correlation to the trait and, if a

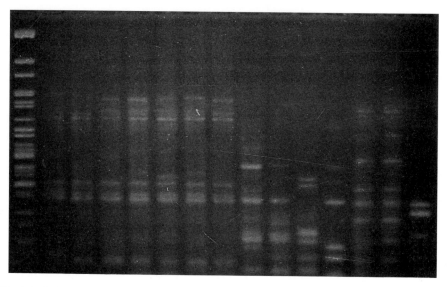

Fig. 4.3 RAPD pattern for four species of fish produced after PCR amplification with the primer GCGACGCCTA. The primer was added to genomic DNA and characteristic patterns of bands were produced after numerous cycles of PCR (polymerase chain reaction). Lanes, from left: (1) molecular weight marker, bacteriophage lambda digested with restriction enzyme BstN1 producing bands from 5 kb to 200 bp in size; (2–8) rainbow trout (*Oncorhynchus mykiss*); (9–11) coho salmo (*Oncorhynchus kisutch*); (12) walleye (*Stizostedion vitreum*); (13–14) yellow perch (*Perca flavescens*); (15) no genomic DNA control. Polymorphisms within rainbow trout and coho salmon and marked differences between species are apparent. However, the observation of bands where no genomic DNA was added illustrates the sensitivity and potential for confusion with this technique. (Courtesy of Andy Peek, Washington State University).

positive correlation is detected, become the direct object of selection. Examples of candidate genes that might be used in such an approach are discussed in later sections.

In the random markers approach, populations segregating for the trait of interest are analysed to identify random markers which might correlate with the trait. If such markers are identified, they can then be used as direct subjects for selection. A particular type of RFLP marker known as DNA fingerprint or VNTR sequences may show particular attributes for the random marker approach, as large numbers of markers are available and there is a high level of genetic variation for such markers (Taggart & Ferguson 1990, Castelli *et al.* 1990, Turner *et al.* 1991). Salmonid fishes in particular appear to show high numbers of bands and high levels of polymorphism for such markers (Fig. 4.4).

In addition, methods based on DNA mixing are being applied which will allow markers to be more readily screened for association with phenotypes in large numbers of individuals (Dunnington *et al.* 1991, Michelmore *et al.* 1991).

The marker-assisted selection approach makes the assumption that there are major gene effects underlying many of the quantitative traits which have conventionally been selected on the basis of phenotype (Haley 1991). The

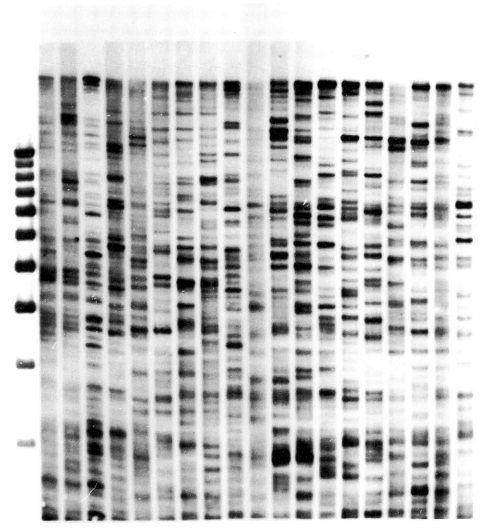

Fig. 4.4 DNA fingerprint patterns of rainbow trout (*Oncorhynchus mykiss*). Genomic DNA was digested with the restriction enzyme Hae III and subjected to electrophoresis in an agarose gel. The 33.6 probe (sequence TGGAGGAGGGC × 2) was hybridized to the DNA after it was transferred to a nylon membrane and the resulting complex pattern of bands was detected. Lanes, from left; (1) Molecular weight marker (top band 12 kb, bottom band 3 kb); (2) pooled DNA from eight anadromous rainbow trout from the Skamania, Washington, strain; (3–10) DNA from eight individuals from the Skamania strain; (11) pooled DNA from nine anadromous rainbow trout from the Chambers Creek, Washington, strain; (12–20) DNA from nine individuals from the Chambers Creek strain. A high level of polymorphism among individuals is evident. (Courtesy of Paul Spruell, Washington State University).

extent to which this is true will obviously be a major outcome of research in this field, but results in other systems to date are encouraging. Marker-assisted selection may have particular benefits for traits such as disease resistance, where it would be advantageous never to expose broodstock, whose progeny are intended for use in production facilities, to pathogens. By using marker-based methods, it may be possible to determine which individuals would be resistant without ever having to expose them to a pathogen.

4.4 Major breeding goals for aquaculture

Four major breeding goals for aquaculture include sex control (generation of monosex or sterile fish), extended environmental tolerances for aquaculture species, improved disease resistance and enhanced growth rate or meat quality. Examples of current or potential biotechnological approaches to these goals will be discussed. The number of cases in which biotechnology has successfully contributed toward these goals is fairly limited to date. It is hoped that the considerable promise of these approaches will be realized and that there will be additional examples of commercial application of these biotechnological approaches in the near future.

4.4.1 Sex control

The production of monosex or sterile populations can improve the efficiency of aquaculture for some species (Donaldson & Hunter 1982). Females often mature sexually at a later age than males or produce eggs with a high commercial value; making them more desirable than males. In some species, e.g. channel catfish, males grow more rapidly than females, making them preferable (Simco et al. 1989). Sterile fish can be a desirable goal when overpopulation is a problem or when growth and quality slow when sexual maturity occurs before harvest size.

The production of all-female salmonid populations of several species (rainbow trout, Atlantic salmon, chinook salmon) is a very successful example of applying biotechnology in a commercial setting (Bye & Lincoln 1986, Devlin et al. 1991). Chromosome set manipulation has contributed toward this goal because gynogenesis has sometimes been used to generate the XX population, which is hormonally sex-reversed, into XX males, which in turn generate all-female progeny. Induced triploidy, particularly in combination with all-female production, has been used in salmonids to generate sterile populations for aquaculture (Bye & Lincoln 1986).

All-male populations have been produced by fertilizing eggs with sperm from YY males in tilapia (Scott et al. 1989) and rainbow trout (Chevassus et al. 1988, Scheerer et al. 1991). YY males have been produced by crossing XY individuals after hormonal (Chevassus et al. 1988) or spontaneous (Scott et al. 1989) sex reversal or by androgenesis (Scheerer et al. 1991). Although there is no commercial demand for all-male salmonid populations, such approaches could be useful for other species (e.g. channel catfish, tilapia).

Gene transfer might potentially be applied to sterilize fish, but the need for such an application is not as immediate as some other applications of transgenic technology, given the available alternatives, triploidy and hormonal sterilization.

The isolation of a Y-specific DNA probe from chinook salmon and development of a PCR-based test for sex testing in that species (Devlin *et al.* 1991) is the best example to date of the use of a DNA marker in commercial aquaculture. This test has been used to identify rare XY fish among an XX male population being used to produce all-female chinook salmon populations in British Columbia, Canada.

Major genes are known which control the timing of sexual maturation in fish species (e.g. Kallman 1983), indicating that this trait could be a productive one in which to use marker-assisted selection.

4.4.2 Extended environmental tolerance

Extending the range of environmental tolerance of aquaculture species can be an important breeding goal. By producing fish which can tolerate a wider range of environmental conditions, e.g. temperature, salinity and pH, at critical life stages, the environments in which a given species can be successfully cultured can be extended.

The production of triploid hybrids can help extend environmental tolerance. By hybridizing a cultured species to another with a wider environmental tolerance, new environments might be exploited. The early sea water tolerance of triploid rainbow trout × pink salmon hybrids (Thorgaard *et al.* 1992) and of diploid masu salmon × pink salmon hybrids (Ma & Yamazaki 1986) contributed by the pink salmon male parents, although not useful commercially, demonstrates this concept. Undoubtedly, other similar opportunities are available through hybridization.

A current example of applying gene transfer to extending environmental tolerance is the effort to transfer a gene for an anti-freeze protein (found in some sea fish living in subzero waters) into Atlantic salmon (Shears *et al.* 1991). Atlantic salmon, which normally lack such a protein, are subject to mortality in some cases in the winter when sea water on the Atlantic coast of Canada drops below 0°C. Indications to date are that the gene has been successfully transferred into Atlantic salmon but that the level of expression is not yet adequate for protection.

Marker-assisted selection has not yet been applied to the problem of extending environmental tolerance but is likely to contribute in the future. For example, cold tolerance in tilapia appears to have a genetic basis (Behrends *et al.* 1990), and through appropriate studies it may be possible to identify DNA markers linked to major genes for cold tolerance and to use these markers in selection programmes to combine cold tolerance with rapid growth. The use of homozygous lines and the recombinant inbred approach to analysis of quantitative traits might facilitate isolating such a gene.

4.4.3 Disease resistance

Disease resistance is obviously a major issue in aquaculture breeding (Chevassus & Dorson 1990) and several biotechnology approaches have been used to address this issue.

Triploid hybrids could contribute toward disease resistance breeding in much the same way that they could contribute toward extended environmental tolerance. Triploid hybrids of rainbow trout with three other species (coho salmon, brook trout and Arctic char) have been demonstrated to show increased viral resistance relative to rainbow trout (Dorson & Chevassus 1985, Parsons *et al.* 1986, Dorson *et al.* 1991); all three hybrids are resistant to viral haemorrhagic septicaemia virus (VHSV) and the coho salmon has shown resistance to infectious haematopoietic necrosis virus (IHNV). The culture potential of these hybrids needs further study. Even if these hybrids do not prove useful in aquaculture, they could aid in the identification of disease resistance genes which might be useful in gene transfer studies.

Two approaches which show promise at the cell culture level for gene transfer of viral disease resistance are the transfer of 'anti-sense' virus DNA sequences (reviewed by Chen & Powers 1990) and the introduction of viral nucleocapsid protein genes (Anderson & Leong 1991). Such efforts are initially being applied to cell cultures but will probably be extended to whole fish in the near future.

Breeding for disease resistance provides an ideal situation for applying marker-assisted selection because in this way disease-resistant broodstocks could be developed which had never been exposed to the pathogen. Alleles of the transferrin gene have been associated with differential disease resistance in coho salmon (Suzumoto *et al.* 1977), indicating that this might be an excellent candidate locus for disease resistance breeding. The genes of the major histocompatibility complex (MHC, termed HLA in humans) have been associated with differential disease resistance in humans and other animals (Thomson 1988) and indications that a homologous gene complex exists in fish (Hashimoto *et al.* 1990) suggest that breeding based on alleles at that locus will probably be a fruitful approach for disease resistance breeding in the future. Efforts to screen for random markers associated with disease resistance could also be productive.

4.4.4 Growth rate or meat quality

Selection for improved growth rate or meat quality is an important aquacultural goal and another area where biotechnology is already being used to develop improved strains.

Triploid sterility can provide a means to enhance the growth rate and quality of mature fish in salmonids (Bye & Lincoln 1986, Thorgaard 1986) although it does not appear to be helpful in commercial trials in channel catfish (Wolters *et al.* 1991). Triploidy might also provide a means to enhance quality in new types of hybrids; there are indications that the triploid hybrids of rainbow trout with coho and chinook salmon may show desirable taste and texture qualities (J. Parsons, pers. comm.).

Efforts to transfer growth hormone genes into commercial species represent the greatest area of focus for fish gene transfer research (Chen & Powers 1990). Studies with growth hormone supplementation (Agellon *et al.* 1988, Schulte *et al.* 1989) and promising early results with gene transfer (Chen & Powers 1990) suggest that this approach is likely to be successful commercially. Other genes in the pathway might also provide opportunities for enhanced growth if transferred. Whether the advocates of transgenic fish with extra growth hormone genes can overcome regulatory (and potentially consumer) resistance remains to be seen. Growers themselves who are concerned about the natural image of their products may initially be hesitant to utilize engineered fish and this concern can extend to other biotechnology approaches as well.

Marker-assisted selection on growth rate is likely to be productive. A major gene which affects the early rate of development and markedly affects the size of rainbow trout has been identified (Allendorf *et al.* 1983) and similar loci are likely to be identified in other species in the future. Variation in trypsin-like isozymes has been correlated with growth rate in Atlantic salmon, indicating that this should be a suitable marker in breeding programs (Torrissen 1991). The candidate gene approach also might be used for alleles of the growth hormone gene and related genes, as well as screening of random markers for association with growth rate.

4.5 Biotechnology and possible broodstock improvement

Progress towards broodstock improvement is likely to be made most effectively by integrating the available biotechnologies and using them in a complementary fashion. For example, the clonal lines which can be generated through chromosome set manipulation could provide excellent materials with which to identify major genes for important traits through the recombinant inbred approach, as previously discussed. This could improve the efficiency of marker-assisted selection approaches.

Similarly, the use of standard lines or hybrids between lines might facilitate the evaluation of transgenic fish. Transgenic approaches and marker-assisted selection might interact in that major genes identified through the selection studies might subsequently be candidates for gene transfer. Induced triploidy might interface with gene transfer in the identification of species differences in disease resistance, which might subsequently be transferred through single gene transfer or the transfer of chromosome fragments by incomplete gynogenesis.

It is important to emphasize that the success of all these biotechnologies will depend on the ability to integrate them with the traditional tools used by animal breeders; careful, objective statistical evaluation of performance and maintenance of selected and control lines. The full range of skills to bring a technology from the laboratory to the farm are rarely found in a single research group, and team approaches to production and evaluation of improved fish are likely to become increasingly common and important.

Acknowledgements

I thank our current laboratory group (Patrick Gibbs, Paul Spruell, Shawn Cummings, Peter Galbreath, Andy Peek, Bill Young and Paul Wheeler) for thought-provoking discussions on many of the topics in this review. Our laboratory was supported by grants from the National Institutes of Health (NIEHS PO1 ESO4766 and BMMRP #1 RO1-RR06654-01), the US Department of Agriculture, the USDA Western Regional Aquaculture Consortium, and the Washington Sea Grant Program (projects R/A-62 and R/B-4) while this paper was written.

References

Agellon, L.B., Emery, C.J., Jones, J.M., Davies, S.L., Dingle, A.D. & Chen, T.T. (1988) Promotion of rapid growth of rainbow trout (*Salmo gairdneri*) by a recombinant fish growth hormone. *Canadian Journal of Fisheries and Aquatic Science*, **45**, 146–51.

Allendorf, F.W., Knudsen, K.L. & Leary, R.F. (1983) Adaptive significance of differences in the tissue-specific expression of a phosphoglucomutase gene in rainbow trout. *Proceedings of the National Academy of Science of the USA*, **80**, 1397–400.

Allendorf, F.W., Seeb, J.E., Knudsen, K.L., Thorgaard, G.H. & Leary, R.F. (1986) Gene-centromere mapping of 25 loci in rainbow trout. *Journal of Heredity*, **77**, 307–12.

Anderson, E.D. & Leong, J.C. (1991) Reduction of IHNV induced CPE by transgene expression of the nucleocapsid gene. *Second International Marine Biotechnology Conference (IMBC '91) (abstract)*.

Bartlett, S.E. & Davidson, W.S. (1992) FINS (Forensically Informative Nucleotide Sequencing): a procedure for identifying the animal origin of biological specimens. *BioTechniques*, **12**, 408–11.

Behrends, L.L., Kingsley, J.B. & Bulls, M.J. (1990) Cold tolerance in maternal mouthbrooding tilapias: phenotypic variation among species and hybrids. *Aquaculture*, **85**, 271–80.

Buono, R.J. & Linser, P.J. (1991) Transgenic zebrafish made by electroporation. *Second International Marine Biotechnology Conference (IMBC '91) (abstract)*.

Bye, V. & Lincoln, R.F. (1986) Commercial methods for the control of sexual maturation in rainbow trout (*Salmo gairdneri* R.). *Aquaculture*, **57**, 299–309.

Carr, S.M. & Marshall, H.D. (1991) A direct approach to the measurement of genetic variation in fish populations: applications of the polymerase chain reaction to studies of Atlantic cod, *Gadus morhua* L. *Journal of Fish Biology*, **39** (supplement A), 101–7.

Castelli, M., Philippart, J.-C., Vassart, G. & Georges, M. (1990) DNA fingerprinting in fish: a new generation of genetic markers. In *Fish-Marking Techniques. American Fisheries Society Symposium*, 7 (eds N.C. Parker, A.E. Giorgi, R.C. Heidinger, D.B. Jester, Jr, E.D. Prince & G.A. Winans), pp. 514–20. American Fisheries Society, Bethesda, Maryland, USA.

Chen, T.T. & Powers, D.A. (1990) Transgenic fish. *Trends in Biotechnology*, **8**(8) 209–15.

Chevassus, B. & Dorson, M. (1990) Genetics of resistance to disease in fishes. *Aquaculture*, **85**, 83–107.

Chevassus, B., Guyomard, R., Chourrout, D. & Quillet, E. (1983) Production of viable hybrids in salmonids by triploidization. *Genetics, Selection and Evolution*, **15**, 519–32.

Chevassus, B., Devaux, A., Chourrout, D. & Jalabert, B. (1988) Production of YY rainbow trout by self-fertilization of induced hermaphrodites. *Journal of Heredity*, **79**, 89–92.

Chourrout, D., Chevassus, B., Krieg, F., Happe, A., Burger, G. & Renard, P. (1986) Production of second generation triploid and tetraploid rainbow trout by mating tetraploid males and diploid female — Potential of tetraploid fish. *Theoretical and Applied Genetics*, **72**, 193–206.

Devlin, R.H., McNeil, B.K., Groves, T.D.D. & Donaldson, E.M. (1991) Isolation of a Y-chromosomal DNA probe capable of determining genetic sex in chinook salmon (*Oncorhynchus tshawytscha*). *Canadian Journal of Fisheries and Aquatic Science*, **48**, 1606–12.

Dickerson, G.E. (1986) Genetic engineering of livestock improvement. In *Third World Congress on Genetics Applied to Livestock Production*, (eds G.E. Dickerson & R.K. Johnson), pp. 3–8. University of Nebraska, Lincoln, Nebraska, USA.

Disney, J.E., Johnson, K.R., Banks, D.K. & Thorgaard, G.H. (1988) Maintenance of foreign gene expression in adult transgenic rainbow trout and their offspring. *Journal of Experimental Zoology*, **248**, 335–44.

Donaldson, E.M. & Hunter, G.A. (1982) Sex control in fish with particular reference to salmonids. *Canadian Journal of Fisheries and Aquatic Sciences*, **39**, 99–110.

Dorson, M. & Chevassus, B. (1985) Etude de la réceptivité d'hybrides triploides truite arc-en-ciel × saumon coho à la necrose pancréatique et à la septicemie hemorragique virale. *Bulletin Francaise Pisciculture*, **296**, 29–34.

Dorson, M., Chevassus, B. & Torhy, C. (1991) Comparative susceptibility of three species of char and of rainbow trout × char triploid hybrids to several pathogenic salmonid viruses. *Dis. Aquat. Org.*, **11**, 217–24.

Dunnington, E.A., Gal, O., Siegel, P.B., Haberfeld, A., Cahaner, A., Lavi, U., Plotsky, Y. & Hillel, J. (1991) Deoxyribonucleic acid fingerprint comparisons between selected populations of chickens. *Poultry Science*, **70**, 463–7.

Festing, M.F.W. (1979) *Inbred Strains in Biomedical Research*. Macmillan, London.

Gjedrem, T. (1983) Genetic variation in quantitative traits and selective breeding in fish and shellfish. *Aquaculture*, **33**, 51–72.

Haley, C.S. (1991) Use of DNA fingerprints for the detection of major genes for quantitative traits in domestic species. *Animal Genetics*, **22**, 259–77.

Hallerman, E.M. & Beckmann, J.S. (1988) DNA-level polymorphism as a tool in fisheries science. *Canadian Journal of Fisheries and Aquatic Sciences*, **45**, 1075–87.

Hashimoto, K., Nakanishi, T. & Kurosawa, Y. (1990) Isolation of carp genes encoding major histocompatibility complex antigens. *Proceedings of the National Academy of Science, USA*, **87**, 6863–7.

Hindar, K., Ryman, N. & Utter, F. (1991) Genetic effects of cultured fish on natural fish populations. *Canadian Journal of Fisheries and Aquatic Sciences*, **48**, 945–57.

Hoover, K.L. (ed.) (1984) *Use of Small Fish Species in Carcinogenicity Testing*. National Cancer Institute (USA), Monograph 65.

Houdebine, L.M. & Chourrout, D. (1991) Transgenesis in fish. *Experientia*, **47**, 891–7.

Ihssen, P.E., McKay, L.R., McMillan, I. & Phillips, R.B. (1990) Ploidy manipulation and gynogenesis in fishes: cytogenetic and fisheries applications. *Transactions of the American Fisheries Society*, **119**, 698–717.

Kallman, K.D. (1983) The sex determining mechanism of the poeciliid fish, *Xiphophorus montezumae*, and the genetic control of the sexual maturation process and adult size. *Copeia*, **1983**, 755–69.

Kapuscinski, A.R. & Hallerman, E.M. (1991) Implications of introduction of transgenic fish into natural ecosystems. *Canadian Journal of Fisheries and Aquatic Sciences*, **48**, 97–107.

Komen, J., Bongers, A.B.J., Richter, C.J.J., van Muiswinkel, W.B. & Huisman, E.A. (1991) Gynogenesis in common carp (*Cyprinus carpio* L.). 2. The production of homozygous gynogenetic clones and F sub(1) hybrids. *Aquaculture*, **92**, 127–42.

Lincoln, R.F. & Scott, A.P. (1984) Sexual maturation in triploid rainbow trout, *Salmo gairdneri* Richardson. *Journal of Fish Biology*, **25**, 385–92.

Ma, H. & Yamazaki, F. (1986) Characteristics of the hybrid F_1 juveniles between female masu salmon, *Oncorhynchus masou*, and male pink salmon, *Oncorhynchus gorbuscha*. *Bulletin of the Faculty of Science, Hokkaido University*, **37**(1), 6–16.

Maclean, N. & Penman, D. (1990) The application of gene manipulation to aquaculture. *Aquaculture*, **85**, 1–20.

Michelmore, R.W., Paran I. & Kesseli, R.V. (1991) Identification of markers linked to disease-resistance genes by bulked segregant analysis: a rapid method to detect markers in specific genomic regions by using segregating populations. *Proceedings of the National Academy of Science, USA*, **88**, 9828–32.

Mueller, F., Ivics, Z., Erdelyi, F., Varadi, L., Horvath, L., MacLean, N. & Orban, L. (1991) Introducing foreign genes into fish eggs using electroporated sperm as a carrier. *Second International Marine Biotechnology Conference (IMBC '91) (abstract)*.

Myers, J.M. & Hershberger, W.K. (1991) Early growth and survival of heat-shocked and tetraploid-derived triploid rainbow trout (*Oncorhynchus mykiss*). *Aquaculture*, **96**, 97–107.

Naruse, K., Ijiri, K., Shima, A. & Egami, N. (1985) The production of cloned fish in the medaka (*Oryzias latipes*). *Journal of Experimental Zoology*, **236**, 335–41.

Parsons, J.E., Busch, R.A., Thorgaard, G.H. & Scheerer, P.D. (1986) Increased resistance of triploid rainbow trout × coho salmon hybrids to infectious haematopoietic necrosis virus. *Aquaculture*, **57**, 337–43.

Phillips, R.B. & Ihssen, P.E. (1990) Genetic marking of fish by use of variability in chromosomes and nuclear DNA. In *Fish-Marking Techniques. American Fisheries Society Symposium*, 7 (eds N.C. Parker, A.E. Giorgi, R.C. Heidinger, D.B. Jester, Jr, E.D. Prince & G.A. Winans), pp. 499–513. American Fisheries Society, Bethesda, Maryland, USA.

Purdom, C.E. (1983) Genetic engineering by the manipulation of chromosomes. *Aquaculture*, **33**, 287–300.

Refstie, T. (1990) Application of breeding schemes. *Aquaculture*, **85**, 163–9.

Scheerer, P.D. & Thorgaard, G.H. (1983) Increased survival in salmonid hybrids by induced triploidy. *Canadian Journal of Fisheries Aquatic Sciences*, **40**, 2040–4.

Scheerer, P.D., Thorgaard, G.H. & Allendorf, F.W. (1991) Genetic analysis of androgenetic rainbow trout. *Journal of Experimental Zoology*, **260**, 382–90.

Schulte, P.M., Down, N.E., Donaldson, E.M. & Souza, L.M. (1989) Experimental administration of recombinant bovine growth hormone to juvenile rainbow trout (*Salmo gairdneri*) by injection or immersion. *Aquaculture*, **76**, 145–56.

Scott, A.G., Penman, D.J., Beardmore, J.A. & Skibinski, D.O.F. (1989) The 'YY' supermale in *Oreochromis niloticus* (L.) and its potential in aquaculture. *Aquaculture*, **78**, 237–51.

Shears, M.A., Fletcher, G.L., Hew, C.L., Gauthier, S. & Davies, P.L. (1991) Transfer, expression, and stable inheritance of antifreeze protein genes in Atlantic salmon (*Salmo salar*). *Molecular Marine Biology and Biotechnology*, **1**, 58–63.

Simco, B.A., Goudie, C.A., Klar, G.T., Parker, N.C. & Davis, K.B. (1989) Influence of sex on growth of channel catfish. *Transactions of the American Fisheries Society*, **118**, 427–34.

Soller, M. & Beckmann, J.S. (1983) Genetic polymorphism in varietal identification and genetic improvement. *Theoretical Applied Genetics*, **67**, 25–33.

Stoss, J. (1983) Fish gamete preservation and spermatozoan physiology. In *Fish Physiology*, Vol. 9B (eds W.S. Hoar, D.J. Randall & E.M. Donaldson) pp. 305–50. Academic Press, New York.

Streisinger, G., Walker, C., Dower, N., Knauber, D. & Singer, F. (1981) Production of clones of homozygous diploid zebra fish (*Brachydanio rerio*). *Nature*, **291**, 293–6.

Suzumoto, B.K., Schreck, C.B. & McIntyre, J.D. (1977) Relative resistances of three transferrin genotypes of coho salmon (*Oncorhynchus kisutch*) and their hematological responses to bacterial kidney disease. *Journal of the Fisheries Research Board of Canada*, **34**, 1–8.

Taggart, J.B. & Ferguson, A. (1990) Minisatellite DNA fingerprints of salmonid fishes. *Animal Genetics*, **21**, 377–89.

Thomson, G. (1988) HLA disease associations: models for insulin dependent diabetes mellitus and the study of complex genetic disorders. *Annual Review of Genetics*, **22**, 31–50.

Thorgaard, G.H. (1986) Ploidy manipulation and performance. *Aquaculture*, **57**, 57–64.

Thorgaard, G.H., Scheerer, P.D. & Parsons, J.E. (1985) Residual paternal inheritance in gynogenetic rainbow trout: implications for gene transfer. *Theoretical and Applied Genetics*, **71**, 119–21.

Thorgaard, G.H., Scheerer, P.D. & Zhang, J.J. (1992) Integration of chromosome set manipulation and transgenic technologies for fishes. *Molecular Marine Biology and Biotechnology*, **1**, 251–6.

Torrissen, K. (1991) Genetic variation in growth rate of Atlantic salmon with different trypsin-like isozyme patterns. *Aquaculture*, **93**, 299–312.

Turner, B.J., Elder, J.F. & Laughlin, T.F. (1991) Repetitive DNA and the divergence of fish populations: some hopeful beginnings. *Journal of Fish Biology*, **39** (supplement A), 131–42.

Utter, F.M. & Seeb, J.E. (1990) Genetic marking of fishes: overview focusing on protein variation. In *Fish-Marking Techniques. American Fisheries Society Symposium*, 7 (eds N.C. Parker, A.E. Giorgi, R.C. Heidinger, D.B. Jester, Jr, E.D. Prince & G.A. Winans), pp. 426–38. American Fisheries Society, Bethesda, Maryland, USA.

Wilkes, G. (1983) Current status of crop plant germplasm. *CRC Critical Reviews in Plant Science*, **1**(2), 133–81.

Williams, J.G.K., Kubelik, A.R., Livak, K.J., Rafalski, J.A. & Tingey, S.V. (1991) DNA polymorphisms amplified by arbitrary primers are useful as genetic markers. *Nucleic Acids Research*, **18**, 6531–5.

Wolters, W.R., Lilyestrom, C.G. & Craig, R.J. (1991) Growth, yield, and dress-out percentage of diploid and triploid channel catfish in earthen ponds. *The Progressive Fish Culturist*, **53**, 33–6.

Chapter 5
Gilt-Head Sea Bream (*Sparus aurata*)

5.1 Introduction
5.2 Taxonomy
5.3 Sex reversal and sex ratio
5.4 Induction of spawning
    5.4.1 Structure – activity relationships of GnRH and its analogues
    5.4.2 The use of dopamine antagonists
    5.4.3 Delivery systems for GnRH
5.5 Year-round egg production
5.6 Broodstock diet and egg and larval quality
5.7 Recommendations on broodstock management
    5.7.1 Broodstock and maintenance
    5.7.2 Spawning
    5.7.3 Egg collection
    5.7.4 Larval rearing
Acknowledgements
References

A basic requirement of intensive farming of any fish species is a constant supply of good quality eggs and fry from captive broodstocks. Commercial growers should be able to programme the timing and the magnitude of spawning to fit with their hatchery and growout requirements. However, many sea bream broodstock populations in European hatcheries either do not spawn spontaneously in captivity, or their spawning is unpredictable. Also, like many other fish of temperate zones, gilt-head sea bream is a seasonal spawner. Depending on the geographical location, the spawning season of this species extends from December to May. Broodstock should, however, be managed so as to secure a year-round supply of eggs.

In addition, the gilt-head sea bream displays some unique reproductive characteristics. It is a hermaphroditic species which undergoes sex reversal, has non-synchronous ovarian development and spawns repeatedly over a long period of time. These characteristics should be taken into consideration when designing broodstock management programmes for gilt-head sea bream, so as to ensure the desired sex ratio and a continuous production of good quality eggs and larvae.

Survival of larvae is another major problem in gilt-head sea bream, with common levels of only 15% from hatching to 1 g fish. This percentage represents a dramatic improvement from levels of less than 1% which were reported by Barnabe (1976) and Alessio *et al.* (1976). Such progress was the result of intensive research on environmental and biotic factors that regulate growth and survival, e.g. photoperiod, light intensity (Tandler & Mason 1983, 1984), salinity (Anav 1991) and larval age-related-rotifer-size preference (Helps 1982). The larval rearing procedure of sea bream was further improved by developing live food enrichment regimes aimed at modifying the composition of live food to suit the larval requirements (Koven *et al.* 1989, 1990, 1992).

As a result of these research and development efforts, survival rates of larvae at 32 days (15 mg wet weight) presently surpass 30% and survival rates at 1 g exceed 20%. However, in species like gilt-head sea bream or red sea bream (Watanabe & Kiron this volume Chapter 16) which spawn daily large masses of eggs (see below), the quality of the eggs, embryo and larvae is also very dependent on the diet fed to the broodstock. Understanding the nutritional requirements of the spawners will help in formulating broodstock diets that will optimize survival of offspring.

5.1 Introduction

The gilt-head sea bream, *Sparus aurata*, has traditionally been considered as a high value fishery product of the Mediterranean and Atlantic waters, with present annual landings of 3000–5000 t. The demand and high value of this species led to efforts to farm this fish under extensive conditions in Valli Cultura in Italy during the early part of the century (Kirk 1987) with documented reports in the 1950s (D'Ancona 1954). The reproductive cycles of gilt-head sea bream and of other Mediterranean sparids were studied as early as 1941 by the Italian biologists Pasquali (1941) and D'Ancona (1941). Attempts to induce gilt-head sea bream to spawn in captivity and to raise their larvae were first reported by Lumare & Villani (1970), who were able to grow a small number of larvae but only to a very early stage of development. Barnabe (1976) and Alessio *et al.* (1976) proposed the first larval rearing protocols for this species. Mass production of sea bream in captivity has gradually evolved since then (Person-Le Ruyet & Verillaud 1980, Tandler & Helps 1985).

At present, gilt-head sea bream is one of the major farmed fish in Europe. The production rates of this species increase annually with reported production of 1855 metric tons in 1989 (FAO 1991). Unofficial production figures for 1994 are in excess of 10 000 metric tons, at a wholesale price of 12–15 US$ per kg depending on the season, size of fish and overall supply. It is thus clear that the production level of gilt-head sea bream is rapidly growing. In order for this production to stay competitive, the farming of this species has to intensify and become more efficient. Planned broodstock management programmes will significantly contribute to achieving these goals.

5.2 Taxonomy

Saprus aurata L. is in the family Sparidae, a member of the Perciformes. Its body is oval, rather deep and compressed. The head profile is regularly curved with small eyes, scaly cheeks, scaleless pre-operculum, a low, slightly oblique mouth and thick lips. It has four to six canine-like teeth anterior in each jaw, followed posteriorly by blunter teeth which become progressively molar-like and are arranged in two to four rows (teeth in the two outer rows are stronger). Its long dorsal fin has 11 spines and 13–14 soft rays and it has 73–85 scales along the lateral line. Its colour is silvery grey with a large black blotch at the origin of the lateral line extending to the upper margin of the operculum where it is edged below by a yellowish area. It has a typical golden band between eyes edged by two dark areas (not well defined in young individuals). Dark longitudinal lines are often present on the sides of its body and a dark band on the dorsal fin. The forks and tips of the caudal fin are edged with black.

5.3 Sex reversal and sex ratio

Gilt-head sea bream is a protandrous hermaphroditic species (D'Ancona 1949, Zohar et al. 1978). In captivity, all fish function as males during the first year of their life. Under certain environmental conditions all fish remain males during their second year. From the first or second year onwards, a certain percentage of the males undergo sex reversal, so that over succeeding years the ratio of females in the population increases (Zohar et al. 1978, 1984). All males start to develop ovaries at the end of the spawning season (May), when their gonads become ambisexual, with the final commitment to changing sex into mature females made by September (Zohar et al. 1978). Males that enter into the ambisexual stage but ultimately do not become females, develop testes once more and become functional males again. Our studies showed that the proportion of males reversing sex is socially controlled (Happe & Zohar 1988, Table 5.1).

However, it is only during the period of May through September that the final sex of the ambisexual fish can be influenced by social and hormonal factors. We thus refer to this period as the sensitive period for sex determination. The presence or addition of young fish (potential males) during the sensitive period increases the number of older fish that change sex into females (Table 5.1, compare lines 1, 2 and 3). On the other hand, the presence or addition of

Table 5.1 The effect of sex ratio and age of males on the sex reversal process in sea bream broodstock. The four groups of fish were established in May (end of spawning season) and the sex ratio was determined again in the next spawning season (6 months later, in December). At the beginning of the experiment young males were around 15 months old and old males were around 27 months old. Each group consisted of 16–20 fish

Sex ratio in May (%)			Sex ratio in December (%)		
Young males	Old males	Females	Young males	Old males	Females
100	0	0	50	0	50
0	100	0	0	60	40
50	50	0	50	10	40
0	50	50	0	50	50

older females will inhibit sex reversal in younger fish and cause them to develop back into males (Table 5.1, compare lines 2 and 4).

The above considerations suggest that careful attention should be paid to establishing groups of broodstock, or modifying their composition, otherwise it might lead to a sex ratio that is detrimental to spawning. For instance, if younger fish are added to a group of spawners during the sensitive period for sex reversal, the older males will change sex into females. The younger males produce less sperm (Zohar et al. 1984) and the population will be driven into a situation where there are excess big females and shortages of sperm. As a consequence, the quantity and quality of the spawned eggs will be affected. This is especially critical in situations in which the growers are unable to add fish to their broodstock, for instance, when the broodstock is manipulated to spawn out of season or when it includes genetically-selected fish.

5.4 Induction of spawning

When the sea bream broodstock management programme was begun in Eilat in the 1970s, spontaneous spawnings were obtained only rarely from our captive broodstock (Gordin & Zohar 1978). While spermatogenesis is completed in males held in captivity, oocytes in the females develop only to the final stages of vitellogenesis and then undergo rapid atresia (degeneration). Thus, final oocyte maturation, ovulation and spawning do not occur. At present, a certain proportion of the females (that may be as high as 50%) spawn spontaneously. This is probably a result of domestication, a process that naturally occurs in captive broodstocks over generations. However, the proportion of females that spawn spontaneously and, more importantly, the timing of their spawning are highly unpredictable. This is unacceptable in intensive hatcheries, that require precise planning of year-round egg and fry production. Therefore, even in situations in which spontaneous spawnings do occur, it is important to be able to control totally the number of eggs produced and the timing of their production.

As mentioned above, female sea bream are sequential spawners. They can spawn almost daily for a period of up to 3–4 months an average of 20 000–30 000 eggs per kg body weight (BW) per day (Gordin & Zohar 1978, Zohar & Gordin 1979, Devauchelle 1984, Zohar et al. 1989a). The fecundity of female sea bream thus reaches 2–3 million eggs per kg BW per season. Obviously, this is the desired spawning pattern for farmed sea bream. However, early efforts to induce spawning in sea bream (Alessio et al. 1976, Arias 1976, Barnabe 1976, Villani 1976) failed to obtain such long-term daily spawnings. In these studies, female sea bream were injected with high doses of human chorionic gonadotropin (hCG, at 800–15 000 IU per kg BW) and then stripped of their eggs two to six times. Lowering the dose of hCG to 100–200 IU per kg BW resulted in natural spawning (no stripping necessary) at daily intervals for periods ranging from 4 to 100 days (Gordin & Zohar 1978, Zohar & Gordin 1979, Zohar et al. 1984).

Recently, we have developed a homologous radioimmunoassay (RIA) for sea bream maturational gonadotropin (GtH-II, referred to as GtH in this review, Zohar et al. 1990a). Using this RIA, it has been shown that female sea bream that fail to spawn in captivity do so because the GtH accumulates in their pituitary but is not released in to the blood stream (Zohar 1988). This finding led us to investigate the use of the gonadotropin-releasing hormones (GnRH), instead of hCG, for spawning induction.

In addition to stimulating the release of the fish's own GtH, GnRHs are advantageous over hCG because they are small and non-immunogenic peptides that are simple to synthesize and because super-active analogues of GnRH can be selected specifically for the fish of interest. A single injection of the mammalian GnRH analogue (GnRHa) [D-Ala^6, Pro^{9}NEt]-LHRH was shown to be as effective as an injection of hCG in inducing spawning in sea bream (Zohar et al. 1989a). However, a single injection of either hCG or GnRHa induced long-term daily spawning in only 25–35% of the treated females (Zohar et al. 1989a and unpublished data). The other injected females spawned for 1–6 days only and then stopped, while the rest of the vitellogenic oocytes underwent rapid atresia. This type of result is a failure as far as the fish farmer is concerned.

It was clear that the efficiency of the spawning induction treatment in sea bream had to be drastically improved. Thus, our subsequent research towards understanding and improving the GnRHa-based spawning induction technology in sea bream focused on three areas:

(1) studying structure – activity relationships of GnRH and analogues with the aim of selecting super-active GnRH analogues;
(2) examining the need for the use of dopamine antagonists and
(3) developing delivery systems for the efficient administration of GnRH analogues.

Our studies in each of these areas are detailed below.

5.4.1 Structure – activity relationships of GnRH and its analogues

A single injection of a native form of GnRH to female sea bream undergoing the final stages of vitellogenesis induces a very short-term surge of GtH release. This surge, lasting for only 6 h, is not sufficient to induce ovulation and spawning. We have demonstrated that the low bioactivity of the native forms of GnRH is the result of their rapid degradation by specific peptidases located in the pituitary, kidney and liver of the female (Fig. 5.1, Zohar et al. 1989b, Goren et al. 1990). Immediately after the injection of the native peptides (which are decapeptides) into the fish, these enzymes cleave the peptides in positions 5–6 and 9–10, resulting in a number of inactive fragments. Substituting the amino-acids at positions 6 and 10 of the native peptide, respectively, with D-amino acids and a NEt residue, renders the cleavage sites resistant to

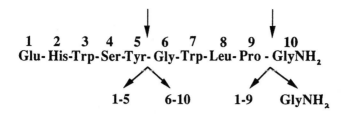

Fig. 5.1 The pattern of salmon GnRH degradation by pituitary, kidney and liver-bound enzymes in the gilt-head sea bream. The upper arrows indicate sites of cleavage, the lower part indicates the resulting fragments.

degradation, resulting in long-lasting, super-potent analogs of GnRH (Zohar et al. 1989b, 1990b).

Table 5.2 shows that the GnRH analogues that are resistant to degradation are also much more active *in vivo* compared with the native peptides, in terms of stimulating GtH release in female sea bream. Obviously, increased resistance to enzymatic degradation contributes to the *in vivo* super-activity of the GnRH analogues. However, as is shown in Table 5.2, there is no direct relationship between resistance of GnRH analogues to degradation and their relative bioactivities. This indicates that other factors might be involved in regulating the bioactivity of GnRH and its analogues in sea bream. In fact, studies in other species showed that some of the superactive GnRH analogues have a higher affinity to the pituitary GnRH receptors, as compared with the native peptides (Crim *et al.* 1988a, de Leeuw *et al.* 1988, Andersson *et al.* 1989, Habibi *et al.* 1989). Recently, we have characterized the GnRH receptors in the pituitary of the gilt-head sea bream (Pagelson & Zohar 1992) and demonstrated that the GnRH analogues that are superactive in inducing GtH release display higher binding affinities to these receptors (Table 5.2).

Therefore, increased binding affinity of some GnRH analogues to the pituitary receptors contributes also to their increased bioactivity. In the goldfish, a specific GnRH binding protein has been demonstrated in the blood (Huang & Peter 1988). This binding protein was shown to have a higher binding affinity to the superactive GnRHa than it has to the native peptide. Thus, better

Table 5.2 Biological properties of native GnRH and some GnRH analogues. *In vivo* bioactivity, resistance to degradation and receptor affinity are relative to salmon-GnRH. Data obtained for all analogues are significantly different (*, $P < 0.05$) from those obtained for the native peptide

Peptide		*In vivo* bioactivity		Resistance to degradation		Receptor affinity		Half life time *in vivo* [min]
salmon-GnRH		1		1		1		5.5
[D-Arg^6,Pro^{9}NEt]-sGnRH		14		10.3		29		12.5
[D-Trp^6]-LHRH	*	23	*	5.0	*	8	*	17.7
[D-Ala^6,Pro^{9}NEt]-LHRH		23		9.5		7.5		22.1

protection of GnRH analogues from clearance by a binding protein might also contribute to their superactivity in fish.

On the basis of the above-described studies a number of GnRH analogues were selected, including analogues of mammalian and salmon GnRH, and their *in vivo* biological activity compared in terms of inducing GtH release and ovulation in the female sea bream (Zohar *et al.* 1989b). The analogues were tested at a dose range of $0.2-20\,\mu g\,kg^{-1}\,BW$. This study showed that when higher doses of GnRHa are injected (above $2.5\,\mu g\,kg^{-1}\,BW$), all tested GnRH analogues are equipotent. However, at lower doses (below $2.5\,\mu g\,kg^{-1}\,BW$), surprisingly the analogues of mammalian GnRH were found to be more potent than analogues of salmon GnRH (see also Table 5.2).

This finding has an important applied implication, in that it suggests that in sea bream there is no need to use the costly analogues of salmon GnRH for spawning induction treatments. Low-cost analogues of mammalian GnRH, such as [D-Ala^6, Pro^{9}NEt]-LHRH, are as potent as, or even more potent than, analogues of salmon GnRH in inducing GtH secretion and spawning. In some fish species, mainly cyprinids, the salmon GnRH analogue [D-Arg^6, Pro^{9}NEt]-sGnRH has been shown to be more potent than analogues of mammalian GnRH (Peter *et al.* 1987, Lin *et al.* 1988). However, in other species, such as trout, salmon and winter flounder, findings tend to be similar to our results in sea bream, namely [D-Ala^6, Pro^{9}NEt]-LHRH and other analogues of mammalian GnRH are as potent as analogues of salmon GnRH in inducing GtH secretion and gonadal development (Crim *et al.* 1988a).

The above-described studies led us to identify analogues of GnRH that are highly resistant to degradation, have a high affinity to the pituitary GnRH receptors and are thus superactive in inducing GtH secretion and ovulation in the female sea bream. However, a single injection of such analogues to females undergoing the final stages of vitellogenesis still induces only a short-term surge of GtH secretion into the circulation (Fig. 5.2). The GtH surge induced by such a GnRHa, [D-Ala^6, Pro^{9}NEt]-LHRH, is much more intensive than that induced by the native peptide, in terms of both the amplitude and the duration of the GtH release. However, the duration of the GtH surge induced by the analogue is only about 24 h (Fig. 5.2). This short-term effect of even the most potent analogues of GnRH probably underlies the low efficiency of a single injection of such analogues in inducing long-term daily spawning in sea bream.

The short-term effect of the GnRH analogues in inducing GtH release was surprising in view of their resistance to degradation by cleaving enzymes. In order to get a better understanding of the bioavailability of the injected GnRH analogues, the *in vivo* clearance rates of native and modified forms of GnRH from the circulation of female sea bream were studied (Gothilf & Zohar 1991). As can be seen in Table 5.2, all GnRH analogues that are resistant to enzymatic degradation disappear from the fish circulation significantly more slowly than the native peptide. However, their *in vivo* half-life times are still short, in the range of 12–22 min. This means that even the highly-resistant GnRH analogues

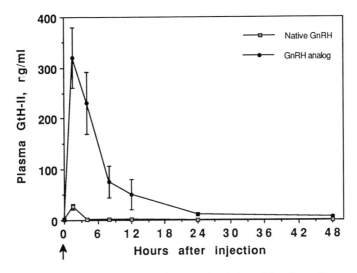

Fig. 5.2 GtH-II levels (mean ± SEM, n = 8) in the circulation of female sea bream undergoing the final stages of vitellogenesis before and at various times after a single injection of a native GnRH (LHRH) or of a GnRH analog ([D-Ala^6, Pro^{9}NEt]-LHRH). Both peptides were injected at a dose of $10\,\mu g\,kg^{-1}$.

disappear from the sea bream circulation relatively quickly after their injection, which explains their low efficiency. Similarly, GnRH and their analogues have been shown to disappear relatively quickly from the circulation of goldfish (Sherwood & Harvey 1986) and rainbow trout (Crim et al. 1988b).

Therefore, we concluded that selecting the most resistant and potent GnRH analogues is only the first step toward the development of a reliable spawning induction treatment in sea bream. In order for such a treatment to succeed, there is a need for a much more prolonged effect of the GnRH analogues. Two approaches were thus taken with the aim of increasing the efficiency of the GnRHa spawning induction therapy. The first involved assessment of possible dopaminergic inhibition of GtH release and the second the development of efficient delivery modes for the GnRH analogues. They are considered in more detail below.

5.4.2 The use of dopamine antagonists

The inhibitory effect of dopamine on GtH secretion and on GnRH-induced GtH release is well established in cyprinids (Peter et al. 1986, 1991) and catfish (Goos et al. 1987). Therefore, GnRHa-based spawning induction methods for cyprinids and some other freshwater species include the injection of a dopamine antagonist, such as domperidone or pimozide, in addition to the injection of the GnRHa (Lin et al. 1988, Peter et al. 1988). In order to examine whether in gilt-head sea bream a dopaminergic inhibition of GtH release has also to be overcome, we examined whether treating female sea bream with dopamine antagonists will potentiate the GnRHa effect on GtH secretion. Figure 5.3 shows the results of one of these experiments, using pimozide as a dopamine antagonist. The experiment included four groups of eight female sea bream undergoing final stages of vitellogenesis.

At the beginning of the experiment, two groups of females were injected with pimozide (10 mg kg^{-1} BW) and two other groups were injected with the vehicle solution with no drug. Three hours later, one of the pimozide injected groups and one of the vehicle injected groups were treated with [D-Ala^6, Pro^{9}NEt]-LHRH, at a dose of 7.5 µg kg^{-1} BW. The other pimozide and vehicle injected groups were treated with saline. Blood samples were taken from all treated females at 6, 24, 48 and 72 h following the second injection. GtH levels in the blood were determined using RIA. As can be seen in Fig. 5.3, the pimozide treatment alone did not induce any GtH secretion. Moreover, injecting pimozide did not enhance the effect of GnRHa on GtH release. GtH levels in females treated with GnRHa alone or with GnRHa and pimozide were not significantly different. The amplitude and the duration of the GtH surge induced in the two groups were also similar. Comparable results were obtained with the dopamine antagonist domperidone.

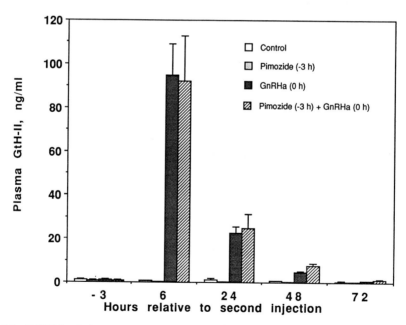

Fig. 5.3 GtH-II levels (mean ± SEM, n = 8) in the circulation of female sea bream undergoing the final stages of vitellogenesis before and at various times after the injection of the dopamine antagonist pimozide alone (10 mg kg^{-1}), GnRHa ([D-Ala^6, Pro^{9}NEt]-LHRH) alone (7.5 µg kg^{-1}) or a combination of both pimozide and the GnRHa. The control group was treated with both the vehicle of the pimozide and saline. Pimozide or its vehicle were injected 3 h before the injection of GnRHa or saline.

Therefore, one must conclude that the pituitary of female sea bream is highly sensitive to GnRHa alone, and that injecting even low doses of GnRHa alone is sufficient to induce intensive GtH release. Dopamine antagonists fail to prolong the GtH surge induced by GnRHa. Consequently, there is no need to include dopamine antagonists in the GnRHa spawning induction therapies for sea bream.

Copeland & Thomas (1989) demonstrated a similar lack of dopaminergic inhibition of GtH release in another marine fish, the Atlantic croaker. The data in sea bream and in Atlantic croaker, together with the fact that many marine fish are highly sensitive to GnRHa (see Zohar 1989 for review) may suggest that marine fish do not require dopamine antagonists in addition to GnRHa for their spawning induction treatments.

5.4.3 Delivery systems for GnRH

In an effort to increase the duration of GnRHa bioavailability in the circulation of the treated females, we started to develop systems that will deliver the GnRHa in a sustained manner. Such systems have been previously used to induce and synchronize spawning in a variety of fish. GnRHa-containing cholesterol and cholesterol/cellulose implants have been successfully applied in salmonids (Crim et al. 1983, Crim & Glebe 1984), milkfish (Lee et al. 1986), striped mullet (Tamaru et al. 1989), sea bass (Almendras et al. 1988) and winter flounder (Harmin & Crim 1992). Our studies focused on the use of novel polymer-based delivery systems that have been recently developed for sustained administration of polypeptides and proteins in biomedical applications. It has been demonstrated that such systems are highly efficient in inducing prolonged GtH secretion and successful spawning in sea bream and other species (Zohar 1988, 1989, Zohar et al. 1990c, Breton et al. 1990).

Figure 5.4 shows the result of an experiment in which we studied the efficiency of sustained administration of GnRHa compared with a single injection of the peptide in inducing GtH release, ovulation and spawning. Female sea bream undergoing the final stages of vitellogenesis were given either a single injection of GnRHa ($7.5\,\mu g\,kg^{-1}$ BW) or implants containing GnRHa (at $100\,\mu g\,kg^{-1}$ BW). The implants were made of the biodegradable copolymer of polylactic-polyglycolic acid (PLGA). The females were bled before and at various intervals after the administration of the hormone, and levels of GtH in the blood measured by specific RIA. Additional groups of fish were given the same treatments and monitored for spawning activities.

As can be seen in Fig. 5.4, the sustained administration of the GnRHa using the biodegradable implants induced constantly elevated circulating levels of GtH for at least 10 days. This and numerous additional experiments that we carried out showed that the GnRHa delivery systems (implants or microspheres) induced a highly predictable spawning pattern in over 80% of the treated females (as compared with 25% induced by a single injection) for periods

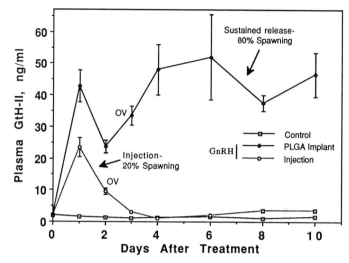

Fig. 5.4 Plasma levels of GtH-II (mean ± SEM) in female sea bream undergoing the final stages of vitellogenesis before and at various times after their treatment with either a single GnRHa injection ([D-Ala^6, Pro^{9}NEt]-LHRH, at 7.5 µg kg^{-1} BW) or with GnRHa implants made of polylactic-polyglycolic acid copolymer (PLGA, at 100 µg kg^{-1} BW). OV indicates the date of ovulation. Also shown is the percentage of females that responded to each of the treatments by long-term daily spawning.

ranging up to 4 months. No adverse effect of the GnRHa implants on fecundity or on quality of the spawned eggs was found, as compared with those parameters observed in females that spawn spontaneously (Table 5.3).

On the basis of the above-described work it was concluded that sustained administration of GnRHa is the most efficient spawning induction treatment in female sea bream. This treatment is routinely used in the National Center for Mariculture to produce sea bream eggs commercially. The GnRHa sustained release delivery systems can be prepared either in the form of implants (3 mm in diameter) that are inserted under the skin of the fish using a special implanter, or in the form of microspheres (50–150 µm in diameter) that are suspended in a vehicle solution and injected into the muscle. Females undergoing the final

Table 5.3 Spawning characteristics of a group of spontaneously spawning female sea bream and of a group of females induced to spawn using GnRHa implants. Each group was stocked in a 4 m^3 round tank. Eggs were collected and counted daily. Hatching rates were determined once a week, by incubating 3000–5000 eggs in 25 l tanks

	Control	GnRH implanted
No. of females	15	15
No. of spawning days	133	137
No. of viable eggs per day (Mean ± SEM)	203 075 ± 15 670	286 448 ± 20 100
% hatching (Mean ± SEM)	51 ± 3	57 ± 3

stages of vitellogenesis are selected for treatment according to Gordin & Zohar (1978) and Zohar & Gordin (1979).

Briefly, fish are anaesthetized and samples of oocytes biopsied by means of a haematocrit tube inserted into the ovary via the ovipore. The oocytes are dispersed in a 0.9% saline solution and examined under the microscope. Ten to twenty of the larger oocytes are measured. Only females in which the larger oocytes measure above 530 µm in diameter, and in which no more than 10% of the follicles are undergoing atresia are selected for hormonal treatment. Effective doses of GnRHa are $25-100 \mu \mathrm{kg}^{-1} \mathrm{BW}$, depending on the delivery system used. The treated females are returned to their tanks together with spermiating males. Females start to spawn 48–72 h following the hormonal treatment, and continue to spawn at daily intervals. No stripping is necessary. More details of managing broodstock and spawning are given in [5.7].

5.5 Year-round egg production

The GnRHa-based technology is efficient in inducing in-season spawning. However, intensive sea bream operations require a year-round availability of eggs. Out-of-season spawning is achieved by environmental phase-shifting of the gametogenic cycle. Photoperiod and temperature manipulation have been used to extend the spawning season in a variety of fish species (see Bromage & Cumaranatunga 1988, Zohar 1989). Attempts to prolong the spawning season in gilt-head sea bream by manipulating environmental factors were reported by Devauchelle (1984). In most of the studies of out-of-season spawning, compressed photoperiod and temperature cycles have been used. In these manipulations day length and water temperature cycles are changed faster than under the natural cycle; alternatively, fish which are under constant day length are exposed to brief periods of long or short days. Since we found that such regimes tend to harm egg quality in sea bream (unpublished data), our approach to phase-shifting spawning time has been different.

Broodstocks are exposed to phase-shifted environmental conditions; the photoperiod and temperature regimes are 12 months long. Three groups of broodstock are exposed to year-long environmental regimes that are shifted by 3, 6 and 9 months in relation to the natural regime. This results in groups of fish ready to be spawned at 3-month intervals, in winter (fish exposed to natural conditions) and in spring, summer and autumn (the shifted fish). Approximately 3 weeks after the shortest day of the year, females in each group are treated with GnRHa sustained-release delivery systems (implants or microspheres) to initiate spawning. As female sea bream spawn daily for periods ranging up to 4 months, using this approach it is possible to obtain eggs every day of the year, all year round (Table 5.4).

A major concern when using environmental and hormonal manipulation of broodstock is the quality of the eggs and larvae produced. Figure 5.5 shows the hatching rates and 32-day survival rates of larvae derived from eggs produced

Table 5.4 Year-round egg production by four groups of sea bream broodstock. One group (winter) is exposed to natural environmental conditions, while the other three groups are exposed to phase-shifted photoperiod and temperature regimes. Each group consisted of 30–50 fish

Group	Shortest day	Spawning dates	No. of eggs
Winter	21 December	14.01.87–24.05.87	92 146 000
Spring	21 March	20.04.87–05.08.87	45 153 000
Summer	21 June	14.07.87–22.11.87	57 221 000
Autumn	21 September	15.10.87–13.02.88	76 969 000
Winter	21 December	13.01.88–19.05.88	89 598 000
Spring	21 March	15.04.88–11.08.88	63 164 000

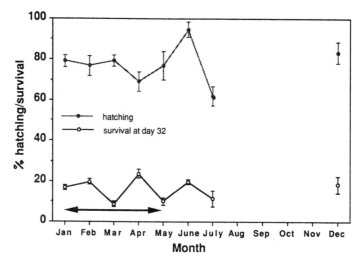

Fig. 5.5 Quality of sea bream eggs and larvae produced during the natural spawning season (indicated by the black arrow) and at a few dates out-of-season (by broodstock exposed to phase-shifted environmental conditions). Hatching rates and survival rates of larvae at day 32 (mean ± SEM) are shown. The data represent a mean of 3 years (1987–9).

during the natural spawning season (arrow) and from a series of out-of-season spawnings. It is obvious that both hatching rates and survival rates of larvae remained relatively stable, with moderate fluctuations, throughout the year. This indicates that eggs and larvae produced out-of-season are of the same quality as those produced within the natural spawning season. Therefore, by combining the manipulation of the external factors to which the fish are exposed, with the use of GnRHa devices, one can obtain commercial quantities of high quality gilt-head sea bream eggs all year round.

5.6 Broodstock diet and egg and larval quality

Final ovarian development in fish involves major physiological and biochemical changes which result in a massive incorporation of lipids and proteins into the growing oocytes. In a relatively short period of time, oocytes grow rapidly to constitute 20–40% of the female's body weight. Many fish species also decrease their food intake throughout final ovarian development. Therefore, nutrients and energy necessary for ovarian growth and other functions must be drawn from body stores. In Atlantic salmon, for instance, final oogenesis is accompanied by a depletion in muscle proteins and lipids and by an increase in muscle water content (Aksnes et al. 1986). In rainbow trout, ovarian development is associated with a drastic mobilization of carcass and visceral lipids (Nassour & Leger 1989).

As mentioned earlier, sea bream display a non-synchronous ovarian development and spawn daily for a period of 3–4 months per year. During the spawning season, at any one time some of the oocytes are undergoing the final stages of vitellogenesis while others are only starting the process. Therefore, yolk material is continuously deposited in the ovaries over many months of the year. Moreover, during the 3–4 month spawning period female sea bream spawn a total of 0.5–2 kg eggs per kg BW, which is the equivalent of 0.5–2 times their body weight. This massive egg production can be supported only by dietary nutrients and energy. Indeed, female sea bream continue to feed throughout their long spawning season.

It is clear that the composition of the diet fed to sea bream broodstock will have a major impact on the quality of their eggs and larvae. In fact, a series of studies conducted on the red sea bream, *Pagrus major*, which is a close relative of the gilt-head and displays a similar spawning pattern, showed that fecundity and egg quality in terms of the number of buoyant (viable) eggs, number of oil globules, hatching rates and the percentage of normal larvae, are improved by feeding broodstock with frozen raw krill or cuttlefish meal (Watanabe et al. 1984a,b,c,d, 1985a,b).

Based on the above considerations, we started to study the effect of the diet fed to gilt-head sea bream broodstock on the quality of their eggs. Figure 5.6 shows the results of an experiment in which five groups of broodstock female sea bream were fed with commercial pellets that were supplemented with:

(1) 0.1% vitamin C, 0.2% vitamin E and 5% capelin oil (a source of n-3 highly unsaturated fatty acids- HUFA); these components were found to be effective in improving spawning performance and egg quality in red sea bream (Watanabe et al. 1984a,b);
(2) 0.1% vitamin C and 5% capelin oil;
(3) 0.2% vitamin E and 5% capelin oil;
(4) 0.1% vitamin C and 0.2% vitamin E (no HUFA); or
(5) squid meal.

In the diet fed to the last group all meals in the commercial diet (meat, fish and soybean) were removed and replaced by squid meal. Each group consisted of four females and six males, divided equally between two tanks. Broodfish were fed with these diets at a ration of 1.5% of their BW per day for 60 days

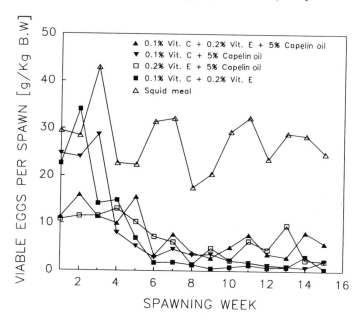

Fig. 5.6 Quantities of viable eggs spawned by five groups of broodstock sea bream fed with different diets. Each group consisted of four females and six males, devided equally between two tanks. Capelin oil was the source of n-3 HUFA. Feeding with the respective diets started 60 days prior to spawning, at a daily ratio of 1.5% of the BW.

prior to the initiation of spawning. Spawning was induced using GnRHa implants. As can be seen in Fig. 5.6, broodfish fed with the first four diets spawned poor quality eggs, as indicated by the small quantities of viable buoyant eggs collected from these groups. Only those females that were fed with the diet supplemented with squid meal spawned good quality eggs throughout their spawning season. These data indicate that squid meal contains components that are essential for good egg quality, in addition to vitamin C and E and n-3 HUFA.

Therefore, we carried out an experiment in which we tested the importance of squid protein and squid lipids in the diet for egg and larval quality (Fig. 5.7). Broodstock sea bream were fed with pellets supplemented with either squid meal extract or with squid meal in which the protein or lipid fractions were removed and replaced with either soybean protein or soy oil (commercial meal extract), or with 1:1 mixture of soybean and squid proteins or oils, respectively. Figure 5.7A demonstrates that replacing either the protein or the lipid fractions of the squid meal-based broodstock diet resulted in a drastic decrease in the hatching rates of the eggs. Larval survival rate at day 3 post hatching was also drastically decreased (Fig. 5.7B). These data indicate that both the protein and lipid fractions of the squid meal diet fed to sea bream broodstock contain components which are essential for the performances of sea bream eggs and larvae.

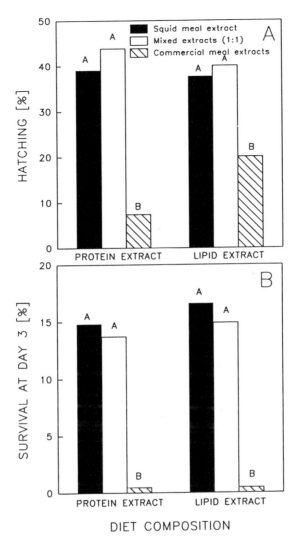

Fig. 5.7 The effect of the protein and lipid source in the diet fed to sea bream broodstock on the hatching rates of the eggs they spawned (A), and on the survival rates at day 3 of the hatched larvae (B). Broodstock were fed with pellets supplemented with either squid meal extract or with squid meal in which the protein or lipid fractions were removed and replaced with either soybean protein or soy oil (commercial meal extract), or with a 1:1 mixture of soybean and squid proteins or oils, respectively. Each group included six females stocked individually with two males. Spawning and hatching were monitored for individual females. Each bar represents the mean of the six spawning tanks. Bars with different letters are significantly different from each other ($P < 0.05$).

Despite the fact that sea bream females feed throughout spawning, we showed that during the long spawning season they lose 52% of their muscular lipids and 45% of the lipids associated with the adipose tissues. However, most of the mobilized lipids are saturated and mono-unsaturated fatty acids. Thus, it is clear that the essential highly-unsaturated fatty acids have to be supplied by the diet. We have also shown a strong correlation between the level of n-3 and n-6 essential fatty acids in the broodstock diet and the levels of these fatty acids in the spawned eggs.

Based on these findings, it was concluded that one essential component in the squid meal is the group of n-3 HUFA and that egg quality can be affected rapidly by a dietary shortage of these essential lipids. We thus carried out an experiment (Fig. 5.8) in which the effect of a sudden change in the broodstock diet on the quality of the spawned eggs was investigated. Eighteen female sea bream were induced to spawn using GnRHa implants and were divided into three treatment groups. Females were placed individually in 18 tanks and continuously monitored for spawning and egg quality. One group was fed with a full squid meal-based diet (diet 1) which contained squid oil as a source of n-3 HUFA. A control group was fed with a squid meal-based diet in which the lipid fraction was replaced by a 1:1 mixture of squid and soy oils (diet 2). Another group was fed with a squid meal-based diet in which the lipid fraction was replaced by soy oil, which does not contain n-3 HUFA (diet 3).

The left part of Fig. 5.8 shows that fish fed a full squid meal-based diet or a 1:1 mixture of squid and soy oils continuously produced a high percentage of

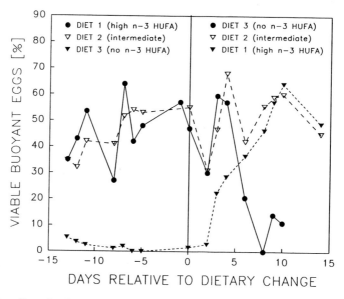

Fig. 5.8 The effect of a change in the broodstock diet on the percentage of viable eggs spawned. The experiment consisted of three groups of six females, each female stocked in a separate tank with two males. One group was fed with a diet rich in n-3 HUFA (containing squid oils, diet 1), a second group (control) was fed with a diet containing a 1:1 mixture of squid oils and soybean oils (diet 2) and a third group was fed with a diet with no n-3 HUFA (containing only soybean lipids, diet 3). Fifteen days after the beginning of the experiment, diets 1 and 3 were switched, the first group started to be fed with diet 3, whereas the third group started to be fed with diet 1. The second group (control) was fed with the same diet. Each data point represents the mean of six spawning tanks.

viable, buoyant eggs. However, when the full squid meal diet fed to the broodstock was changed to a diet which was deficient in squid oil (day 0), the percentage of viable eggs dropped continuously, reaching very low levels only 8–10 days after the diet change (the right portion of Fig. 5.8). Fish fed with the diet deficient in squid oil (diet 3) produced mostly dead eggs. When the squid oil-deficient diet was switched to a full squid meal-based diet, the percentage of viable, buoyant eggs increased dramatically and within 10 days of feeding reached a level equal to that of fish fed a full squid meal-based diet or a 1:1 mixture of squid and soy oils.

The above data emphasize the extreme importance of feeding gilt-head sea bream broodstock with a diet supplemented with frozen squid or squid meal. A change in the broodstock diet, even for short periods of time, will drastically affect the quality of the spawned eggs, and result in spawning of non-viable eggs. Efforts are currently being made to identify all the essential components in the squid meal which influence egg quality in sea bream. This will lead to the development of an optimal commercial diet for gilt-head sea bream broodstock.

5.7 Recommendations on broodstock management

5.7.1 Broodstock and maintenance

Sea bream used as broodstock are 2–6 years old. They are stocked in tanks of 4–20 m^3, at densities of 10–15 kg m^{-3}. In order to obtain out-of-season spawning, the tanks should be either indoors or covered to allow photoperiod manipulation. The sex ratio at stocking is 1:1, albeit the ratio changes over the years. Adding younger fish to older populations should be avoided, as it may induce all older males to change sex. Fish are fed 1–3 times daily with 1–1.5% per kg body weight, per day of squid meal-based dry pellets, composed of 50–55% protein and 10–15% marine-type lipid. The lipids have to contain at least 5% n-3 HUFA, mainly of the 22:6n-3 (DHA) type. This diet has to be fed to the broodstock starting at least 15 days before initiation of spawning. Alternatively, fish may be fed 1–1.5% of dry commercial pellets, supplemented with 2–3% of chopped frozen squid. Water temperatures for broodstock are kept in the range of 16–21°C.

5.7.2 Spawning

The natural spawning season in Eilat extends from 15 January through to 15 May. By manipulating the environment, eggs can be obtained all year round. Spawning is initiated by treating the females with GnRHa delivery systems, as described in [5.4.3]. Only females are treated. Males fertilize the eggs as soon as they are spawned. Spawning typically starts 48–72 h after the hormonal treatment. In the first few days of induced spawning, spawning may occur at different times of the day. Around 1 week after the onset of spawning, the timing of spawning of females within the population becomes synchronized. Spawning then usually occurs around sunset, at 24 h intervals.

The viable eggs, which are around 1 mm in diameter and transparent, and normally contain a single oil globule, tend to float at the surface of the tank. Non-viable eggs or eggs that are not fertilized for several hours turn opaque and sink to the bottom. Spawning is preceded and accompanied by characteristic behaviour during which fish change colour and chasing activity becomes widespread. Broodstock should be kept undisturbed during spawning or spawning may cease, as spawning populations are highly sensitive to stress. Spawning periods for each of the spawning tanks range from 3 to 5 months. If spawning activity decreases, as indicated by the numbers of eggs collected, females can be selected again, as described in [5.4.3] and treated with the GnRHa delivery systems to increase egg production again.

5.7.3 Egg collection

The viable (floating) eggs are collected by running the surface water through a collecting device (Fig. 5.9). An exit at the surface level of the spawning tank (A) directs the overflow water containing the floating eggs and other suspended materials into a 60 l settlement tank (B) where most of the suspended materials (mainly feed leftovers) sink. The overflow from tank B, containing the floating eggs, flows into a similar 60 l tank (C), into which an egg collector is fitted (D). The egg collector itself consists of 700 µm mesh stretched over a plastic (CPVC) frame. A small aerator set underneath the egg collector prevents blocking of the mesh by sinking eggs. During daytime the overflow water from the spawning tank is directed to the drain, without passing through the settlement tank or egg collector. This usually eliminates most of the dirt that might block the collector. One to two hours before spawning is expected to start, the overflow is directed into the settlement tank and egg collector.

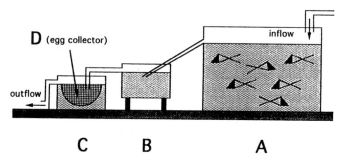

Fig. 5.9 Flow diagram showing procedure for egg collection. An exit at the surface level of the spawning tank (A) directs the overflow water containing the floating eggs and other suspended materials into a 60 l settlement tank (B) where most of the suspended materials (mainly feed leftovers) sink. The overflow from tank B, containing the floating eggs, flows into a similar 60 l tank (C), into which an egg collector (D) is fitted.

Since spawning occurs around sunset and is highly synchronized, all floating (i.e. viable) eggs are collected during the night. The next morning, eggs are gently scooped from the surface of the egg collector using a 1 l beaker. At 20°C, first cleavage of the embryo occurs about 45 min after fertilization. From this time on, fertilization rates can be determined by counting, under a dissecting microscope, the number of eggs with developing embryos. Eggs can be counted by draining them of water and weighing. There are 1800–2000 eggs per g. This number should be verified occasionally, and might differ, according to the size of the females, their strain, their feeding regime, etc.

5.7.4 Larval rearing

Larval rearing of *Sparus aurata* is performed in conical fibreglass tanks (600 l in volume) supplied with running filtered (sand filter of 10 µm) sea water. The incoming water is pumped directly from the gulf of Eilat (Aqaba). It has a salinity of 40–41‰, 0.1–0.2 µm NH_3-NH_4^+ and a pH of about 8.2. Before being used, the water is directed into four independent heat exchange systems designed to supply fresh sea water continuously at the following temperatures: 19.0 ± 0.5°C, 20.5 ± 0.5°C, 22.5 ± 0.5°C and 24.5 ± 0.5°C. Each tank is stocked with 60 000 eggs (100 eggs per litre) added by volume from a calibrated container.

After 48 h, the number of hatched larvae in each tank is estimated by counting at least five 250 ml samples taken from different areas of the tank at a depth of *c.* 80 cm. The mean of these counts is subsequently used as a basis to calculate the final survival rates. The following protocol is used for larval rearing. Larvae are reared in a flow-through system with a flow rate of 1 l per min (2.4 water exchanges per day) from incubation to the end of the 19.0°C rearing period, 1.5 l per min (3.6 exchanges per day) at the two intermediate temperatures, and 2 l per min (4.8 exchanges per day) at the 24.5°C rearing period.

At the initiation of larval pigmentation, 3 days after hatching, larvae are offered rotifers (*Brachionus plicatilis*) at a concentration of 20 animals per ml and algae (*Nannochloropsis* sp.) at a concentration of $5-7 \times 10^5$ cells per ml. The concentration of both rotifers and algae is maintained by continuous addition of both via a special dispensing system. However, these concentrations have to be adjusted every morning since the addition of rotifers and algae is limited to 18–20 h per day.

Fifteen days after hatching, the larvae are offered newly hatched *Artemia* nauplii for 3 days. Subsequently they are offered nauplii enriched for 24 h with Super Selco (*Artemia* systems). The number of *Artemia* offered daily increases gradually with demand, often reaching $4-5 \times 10^3$ per l in the 5th week of rearing. At the age of 30–35 days the post-larvae are counted and transferred to the nursery, where they are weaned from live food.

Acknowledgements

Parts of the research reported here were supported by Grants 3430 and I-1468-88 from BARD, the United States-Israel Binational Research and Development fund, and by Grant NA9OAA-D-FG-424 from the Massachusetts Institute of Technology SeaGrant Program.

References

Aksnes, A., Gjerde, B. & Roald, S.O. (1986) Biological, chemical and organoleptic changes during maturation of farmed Atlantic salmon, *Salmo salar. Aquaculture*, **53**, 7–20.

Alessio, G., Gandolfi, G. & Screiber, B. (1976) Induction de la ponte, élevage et alimentation des larves et des alevins des poissons euryhalins. *Etudes Revues, Conseil Générale des Pêches Méditerranéan, FAO*, **55**, 143–57.

Almendras, J.M., Duenas, C., Nacario, J., Sherwood, N. & Crim, L.W. (1988) Sustained hormone release. III. Use of gonadotropin releasing hormone analogues to induce multiple spawnings in sea bass, *Lates calcarifer. Aquaculture*, **74**, 97–111.

Anav, F. (1991) *The effect of salinity on growth, respiration and Na_+/K_+ ATPase activity in gilthead sea bream* (Sparus aurata) *larvae*. M.Sc. thesis, Tel-Aviv University.

D'Ancona, U. (1941) Ulteriori osservazioni e considerazioni sull'ermafroditismo ed il differenziamento sessuale dell'orata *Sparus auratus* L. *Pubblicazione della Stazione Zoologica di Napoli*, **18**, 313–36.

D'Ancona, U. (1949) Il differenziamento della gonade e l'inversione sessuale degli sparidi. *Arch. Oceanogr. Limnol.* **6**, 97–163.

D'Ancona U. (1954) Fishing and fish culture in brackish water lagoons. *FAO Fish Bulletin*, 4.

Andersson, E., Borg, B. & De Leeuw, R. (1989) Characterization of gonadotropin-releasing hormone binding sites in the pituitary of the three-spined stickleback, *Gasterosteus aculeatus. General and Comparative Endocrinology*, **76**, 41–5.

Arias, A.M. (1976) Reproduction artificielle de la daurade *Sparus aurata* (L.). *Etudes Revues, Conseil Générale des Pêches Méditerranéans, FAO*, **55**, 159–73.

Barnabe, G. (1976) Rapport technique sur la ponte induite et l'elevage des larves du loup *Dicentrarchus labrax* (L.) et de la daurade *Sparus aurata* (L.). *Etudes Revues, Conseil Générale des Pêches Méditerranéan, FAO*, **55**, 63–116.

Breton, B., Weil, C., Sambroni, E. & Zohar, Y. (1990) Effects of acute versus sustained administration of GnRHa on GtH release and ovulation in the rainbow trout, *Oncorhynchus mykiss. Aquaculture*, **91**, 373–83.

Bromage, N. & Cumaranatunga, R. (1988) Egg production in the rainbow trout. In *Recent Advances in Aquaculture* (eds J.F. Muir & R.J. Roberts, pp. 63–138. Croom Helm/Timber Press, London and Sydney/Portland, Oregon.

Copeland, P.A. & Thomas, T. (1989) Control of gonadotropin release in the Atlantic croaker (*Micropogonias undulatus*): Evidence for lack of dopaminergic inhibition. *General and Comparative Endocrinology*, **74**, 474–83.

Crim, L.W. & Glebe, B.D. (1984) Advancement and synchrony of ovulation in Atlantic salmon with pelleted LHRH analog. *Aquaculture*, **43**, 47–56.

Crim, L.W., Sutterlin, A.M., Evans, D.M. & Weil, C. (1983) Accelerated ovulation by pelleted LHRH analogue treatment of spring-spawning rainbow trout *Salmo gairdneri* held at low temperature. *Aquaculture*, **35**, 299–307.

Crim, L.W., Nestor, J.J. & Wilson, C.E. (1988a) Studies of the biological activity of LHRH analogs in the rainbow trout, landlocked salmon, and the winter flounder. *General and Comparative Endocrinology*, **72**, 372–82.

Crim, L.W., Sherwood, N.M. & Wilson, C.E. (1988b) Sustained hormone release. II. Effectiveness of LHRH analog (LHRHa) administration by either single time injection or cholesterol pellet implantation on plasma gonadotropin levels in a bioassay fish, the juvenile rainbow trout. *Aquaculture*, **74**, 87–95.

Devauchelle, N. (1984) Reproduction decalee du bar (*Dicentrarchus labrax*) et de la duarade (*Sparus aurata*). In *Aquaculture de Bar et des Sparides*, (eds R. Billard & G. Barnabe), pp. 53–63. INRA Press, Paris.

FAO (1991) Aquaculture Production (1986–1989). FAO *Fisheries Circular*. No. 815(3), Rome.

Goos, H.J.Th., Joy, K.P., DeLeeuw, R., Van Oordt, P.G.W.J., Van Delft, A.M.L. & Gielen, J.Th. (1987) The effect of luteinizing hormone-releasing hormone analogue (LHRHa) in combination with different drugs with anti-dopamine and anti-serotonin properties on gonadotropin release and ovulation in the African catfish, *Clarias gariepinus*. *Aquaculture*, **63**, 143–56.

Gordin, H. & Zohar, Y. (1978) Induced spawning of *Sparus aurata* (L.) by mean of hormonal treatments. *Annales Biologie Animale Biochimie Biophysique*, **18**, 985–90.

Goren, A., Zohar, Y., Fridkin, M., Elhanati, E. & Koch, Y. (1990) Degradation of gonadotropin releasing hormone in the gilthead seabream, *Sparus aurata*: I. Cleavage of native salmon GnRH, mammalian LHRH and their analogs in the pituitary. *General and Comparative Endocrinology*, **79**, 291–305.

Gothilf, Y. & Zohar, Y. (1991) Clearance of different forms of GnRH from the circulation of the gilthead seabream, *Sparus aurata*, in relation to their degradation and bioactivities. In *Reproductive Physiology of Fish*, (eds A.P. Scott, J.P. Sumpter, D.E. Kime & M.S. Rolfe), pp. 35–7. University of East Anglia Press, Norwich.

Habibi, H.R., Marchant, T.A., Nagorniak, C.S., Van Der Loo, H., Peter, R.E., Rivier, J.E. & Vale, W.W. (1989) Functional relationship between receptor binding and biological activity for analogs of mammalian and salmon gonadotropin-releasing hormones in the pituitary of the goldfish (*Carassius auratus*). *Biology and Reproduction*, **40**, 1152–61.

Happe, A. & Zohar, Y. (1988) Self-fertilization in the protandrous hermaphrodite *Sparus aurata*: Development of the technology. In *Reproduction in Fish – Basic and Applied Aspects in Endocrinology and Genetics*, (eds Y. Zohar & B. Breton), pp. 177–80. INRA Press, Paris.

Harmin, S. & Crim, L. (1992) Gonadotropic hormone-releasing hormone analog (GnRH-A) induced ovulation and spawning in female winter flounder, *Pseudopleuronectes americanus* (Walbaum). *Aquaculture*, **104**, 375–90.

Helps, S. (1982) An examination of prey size selection and its subsequent effect on survival and growth of larval gilthead seabream (*Sparus aurata*). M.Sc. thesis, Plymouth Polytechnic, Plymouth, UK.

Huang, Y.P. & Peter, R.E. (1988) Evidence for a gonadotropin releasing hormone binding protein in goldfish (*Carassius auratus*) serum. *General and Comparative Endocrinology*, **69**, 308–16.

Kirk, R. (1987) *A History of Marine Fish Culture in Europe and North America*. Fishing News Books, Oxford.

Koven, W.M., Kissil, G.Wm. & Tandler, A. (1989) Lipid and n-3 HUFA requirement of *Sparus aurata* larvae during starvation and feeding. *Aquaculture*, **79**, 185–91.

Koven, W.M., Tandler, A., Kissil, G.Wm., Sklan, D., Friezlander, O. & Harel, M. (1990) The effect of dietary (n-3) polyunsaturated fatty acids on growth, survival and swim bladder development in *Sparus aurata* larvae. *Aquaculture*, **91**, 131–41.

Koven, W.M., Tandler, A., Kissil, G.Wm. & Sklan D. (1992) The importance of n-3 highly unsaturated fatty acids for growth in larval *Sparus aurata* and their effect on survival, lipid composition and size distribution. *Aquaculture*, **104**, 91–104.

Lee, C.S., Tamaru, C.S., Kelley, C.D. & Banno, J.E. (1986) Induced spawning of milkfish, *Chanos chanos*, by a single application of LHRH-analogue. *Aquaculture*, **58**, 87–98.

De Leeuw, R., Van Der Veer, C., Smit-Van Dijk, W., Goos, H.J.Th. & Van Oordt, P.G.W.J. (1988) Binding affinity and biological activity of gonadotropin-releasing hormone analogs in the African catfish, *Clarias gariepinus*. *Aquaculture*, **71**, 119–31.

Lin, H.R., Van Der Kraak, G., Zhou, X.J., Liang, J.Y., Peter, R.E., Rivier, J.E. & Vale, W.W. (1988) Effects of [D-Arg^6, Trp^7, Leu^8, Pro^{9}NEt]-luteinizing hormone-releasing hormone (LHRH-A), in combination with pimozide or domperidone, on gonadotropin release and ovulation in the Chinese loach and common carp. *General and Comparative Endocrinology*, **69**, 31–40.

Lumare, F. & Villani, P. (1970) Contributo alla conoscenza delle uove e dei primi stadi larvali di *Sparus aurata* L. *Pubblicazione della Stazione Zoologica di Napoli*, **38**, 364–9.

Nassour, I. & Leger, C.L. (1989) Deposition and mobilization of body fat during sexual maturation in female trout *Salmo gairdneri* R. *Aquatic Living Resources*, **2**, 153–9.

Pagelson, G. & Zohar, Y. (1992) Characterization of gonadotropin-releasing hormone (GnRH) binding to pituitary receptors in the gilthead seabream (*Sparus aurata*). *Biology and Reproduction*, **47**, 1004–8.

Pasquali, A. (1941) Contributo allo studio dell'ermafroditismo e del diffeenziamento della gonada nell'orata *Sparus auratus* L. *Pubblicazione della Stazione Zoologica di Napoli*, **18**, 282–312.

Person-Le Ruyet, J. & Verillaud, P. (1980) Techniques d'élevage intensif de la daurade dorée (*Sparus aurata* L.) de la naissance à l'age de deux mois. *Aquaculture*, **20**, 351–70.

Peter, R.E., Chang, J.P., Nahorniak, C.S., Omeljaniuk, R.J., Sokolowska, M., Shih, S.H. & Billard, R. (1986) Interactions of catecholamines and GnRH in regulation of gonadotropin secretion in teleost fish. *Recent progress in Hormone Research*, **42**, 513–48.

Peter, R.E., Lin, H.R. & Van Der Kraak, G. (1987) Drug/hormone induced breeding of Chinese teleosts. In *Reproductive Physiology of Fish*, (eds D.R. Idler, L.W. Crim & J.M. Walsh), pp. 120–3. Memorial University Press. St John's, Newfoundland.

Peter, R.E., Lin, H.R. & Van Der Kraak, G. (1988) Induced ovulation and spawning of cultured freshwater fish in China: Advances in application of GnRH analogues and dopamine antagonists. *Aquaculture*, **74**, 1–10.

Peter, R.E., Trudeau, V.L., Sloley, B.D., Peng, C. & Nahorniak, C.S. (1991) Action of catecholamines, peptides and sex steroids in regulation of gonadotropin-II in the goldfish. In *Reproductive Physiology of Fish*, (eds A.P. Scott, J.P. Sumpter, D.E. Kime & M.S. Rolfe), pp. 30–4. University of East Anglia Press, Norwich.

Sherwood, N.M. & Harvey, B. (1986) Topical absorption of gonadotropin-releasing hormone (GnRH) in goldfish. *General and Comparative Endocrinology*, **61**, 13–19.

Tamaru, C.S., Kelley, C.D., Lee, C.S., Aida, K. & Hanyu, I. (1989) Effects of chronic LHRH-A +17α-methyltestosterone or LHRH-A+ testosterone therapy on egg growth in the striped mullet *Mugil cephalus*. *General and Comparative Endocrinology*, **76**, 114–27.

Tandler, A. & Helps, S. (1985) The effect of photoperiod and water exchange rate on growth and survival of gilthead seabream (*Sparus aurata* Linnaeus, Sparidae) from hatching to metamorphosis in a mass rearing system. *Aquaculture*, **48**, 71–82.

Tandler, A. & Mason C. (1983) Light and food density effects on growth and survival of larval gilthead seabream (*Sparus aurata* L., Sparidae). *Proceedings of the Warmwater Fish Culture Workshop, Special Publication*, **3**, 103–15.

Tandler, A. & Mason, C. (1984) The use of ^{14}C-labelled rotifers (*Brachionus plicatilis*) in the larvae of gilthead seabream (*Sparus aurata*): Measurements of the effect of rotifer concentration, the lighting regime and seabream larval age on their rate of rotifer ingestion. *European Mariculture Society*, **8**, 241–59.

Villani, P. (1976) Ponte induite et élevage des larves de poissons marins dans les conditions de laboratoire. *Etudes Revues Conseil Générale des Pêches Méditerraneán, FAO*, **55**, 117–32.

Watanabe, T., Arakawa, T., Kitajima, C. & Fujita, S. (1984a) Effect of nutritional quality of broodstock diet on reproduction of red seabream. *Bulletin of the Japanese Society of Scientific Fisheries*, **50**, 495–501.

Watanabe, T., Ohashi, S., Itoh, A., Kitajima, C. & Fujita, S. (1984b) Effect of nutritional composition of diets on chemical components of red seabream broodstock and egg produced. *Bulletin of the Japanese Society of Scientific Fisheries*, **50**, 503–15.

Watanabe, T., Itoh, A., Murakami, A., Tsukashima, Y., Kitajima, C. & Fujita, S. (1984c) Effect of dietary protein level on reproduction of red seabream. *Bulletin of the Japanese Society of Scientific Fisheries*, **50**, 1015–22.

Watanabe, T., Itoh, A., Murakami, A., Tsukashima, Y., Kitajima, C. & Fujita, S. (1984d) Effect of nutritional quality of diets given to broodstock on the verge of spawning on reproduction of red seabream. *Bulletin of the Japanese Society of Scientific Fisheries*, **50**, 1023–8.

Watanabe, T., Itoh, A., Satoh, S., Kitajima, C. & Fujita, S. (1985a) Effect of dietary protein levels and feeding period before spawning on chemical components of eggs produced by red seabream. *Bulletin of the Japanese Society of Scientific Fisheries*, **51**, 1501–9.

Watanabe, T., Koizumi, T., Suzuki, H., Satoh, S., Takeuchi, T., Yoshida, N., Kitada, T. & Tsukashima, Y. (1985b) Improvement of quality of red seabream eggs by feeding broodstock on a diet containing cuttlefish meal on a raw krill shortly before spawning. *Bulletin of the Japanese Society of Scientific Fisheries*, **51**, 1511–21.

Zohar, Y. (1988) Gonadotropin releasing hormone in spawning induction in teleosts: Basic and applied considerations. In *Reproduction in Fish – Basic and Applied Aspects in Endocrinology and Genetics*, (eds Y. Zohar & B. Breton), pp. 47–62. INRA Press, Paris.

Zohar, Y. (1989) Endocrinology and fish farming: Aspects in reproduction, growth and smoltification. *Fish Physiology and Biochemistry*, **7**, 395–405.

Zohar, Y. & Gordin H. (1979) Spawning kinetics in the gilthead seabream, *Sparus aurata* L. after low doses of human chorionic gonadotropin. *Journal of Fish Biology*, **15**, 665–70.

Zohar, Y., Abraham, M. & Gordin, H. (1978) The gonadal cycle of the captivity reared hermaphroditic teleost *Sparus aurata* (L.) during the first two years of life. *Annales Biologie Animale Biochimie Biophysique*, **18**, 877–82.

Zohar, Y., Billard, R. & Weil, C. (1984) La reproduction de la daurade et du bar: Le cycle sexuel et l'induction de la ponte. In *Aquaculture de Bar et des Sparides*, (eds R. Billard & G. Barnabe), pp. 3–24. INRA Press, Paris.

Zohar, Y., Tosky, M., Pagelson, G. & Finkelman, Y. (1989a) Induction of spawning in the gilthead seabream, *Sparus aurata*, using [D-Ala^6-Pro^{9}NET]-LHRH: Comparison with the use of hCG. *Israel Journal of Aquaculture*, **4**, 105–13.

Zohar, Y., Goren, A., Tosky, M., Pagelson, G., Liebovitz, D. & Koch, Y. (1989b) The bioactivity of gonadotropin-releasing hormones and its regulation in the gilthead seabream, *Sparus aurata*, in vivo and in vitro studies. *Fish Physiology and Biochemistry*, **7**, 59–67.

Zohar, Y., Breton, B., Sambroni, E., Fostier, E., Tosky, M., Pagelson, G. & Liebovitz, D. (1990a) Development of homologous radioimmunoassay for a gonadotropin of the gilthead seabream, *Sparus aurata*. *Aquaculture*, **88**, 189–204.

Zohar, Y., Goren, A., Fridkin, M., Elhanati, E. & Koch, Y. (1990b) Degradation of gonadotropin releasing hormone in the gilthead seabream, *Sparus aurata*: II. Cleavage of native salmon GnRH, mammalian LHRH and their analogs in the pituitary, kidney and liver. *General and Comparative Endocrinology*, **79**, 306–19.

Zohar, Y., Pagelson, G., Gothilf, Y., Dickhoff, W.W., Swanson, P., Duguary, S., Gombotz, W., Kost, J. & Langer, R. (1990c) Controlled release of gonadotropin releasing hormones for the manipulation of spawning in farmed fish. *Proc. Inter. Symp. Control. Rel. Bioact. Mater.*, **17**, 51–2.

Chapter 6
Red Drum and Other Sciaenids

6.1 Introduction
6.2 Reproduction
    6.2.1 Oocyte growth
    6.2.2 Final oocyte maturation and ovulation
6.3 Culture methods: broodstock
    6.3.1 Capture and care of broodstock
    6.3.2 Induction of precocious sexual maturation
    6.3.3 Induction of gonadal recrudescence
    6.3.4 Induction of spawning
    6.3.5 Maintenance of extended spawning
6.4 Culture methods: eggs and larvae
    6.4.1 Production of eggs
    6.4.2 Production of larvae
6.5 Conclusions
    References

Red drum (*Sciaenops ocellatus*) is the only sciaenid fish currently in commercial production. Interest in establishing red drum farms is greatest in the South Atlantic and Gulf regions of the United States. High market value and a ban on commercial fishing for red drum drive the industry, while pond overwintering problems constrain it. The aquaculture potential of red drum was significantly advanced when captive spawning with temperature and photoperiod was accomplished and larval rearing techniques developed (Arnold *et al.* 1977). This technology was promptly adapted by the Texas Parks and Wildlife Department to produce red drum juveniles for stocking.

Red drum is the only marine fish that is hatchery reared and released in large numbers for stock enhancement in the United States (McCarty *et al.* 1986). Several commercial hatcheries modelled on the research and stock enhancement facilities began successful operations in the mid 1980s. Currently, there are six red drum farms in Texas and others in Louisiana, Mississippi and Florida. Most of the red drum are grown in ponds. Severe fish losses caused by record low winter temperatures in the southern US in 1986 slowed down expansion of the industry. The potential for fast growth in the tropics (Sandifer 1991) and the availability of breeding technology has led to preliminary grow-out trials in Panama and Martinique (Soletchnik *et al.* 1988).

Red drum breeding technology has been successfully applied to other sciaenids, and several are being evaluated for stock enhancement programmes. The state of California is considering a marine fish hatchery and stock enhancement programme for the white sea bass, *Atractoscion nobilis* (Orhun 1989), while Texas is evaluating spotted seatrout (*Cynoscion nebulosus*) for stocking. Black drum (*Pogonias chromis*), black drum x red drum crosses (Henderson-Arzapalo *et al.* 1988) and orangemouth corvina, *Cynoscion xanthulus* (Prentice & Thomas 1987), have been examined for their aquaculture potential.

Effective care and management of red drum broodstock is rarely a major concern for fish hatcheries. Red drum (Fig. 6.1) adapt well to captivity and are relatively resistant to stressors and disease. Infestations with the gill parasite *Amyloodinium ocellatum* are the only problems frequently encountered and these can be treated effectively with copper. Some other species such as spotted sea trout and orangemouth corvina appear to be more susceptible to culture stressors and can sustain injuries when they are handled. However, once adapted to culture conditions all three species can be induced to undergo gonadal recrudescence by normal and shortened seasonal cycles of temperature and photoperiod. Moreover, spawning can be readily induced by several methods in these species and continued for many months without a decline in egg quality.

Suitable methods for rearing larvae of red drum and other sciaenid fishes in ponds and recirculating systems have been developed. However the production of live food organisms of the correct size and in sufficient quantities for large scale culture of red drum larvae remains a major technological problem, as it does for culturing larvae of many other marine fish. Red drum larvae, like the adults, are also susceptible to *Amyloodinium ocellatum* infections. High mortality of red drum

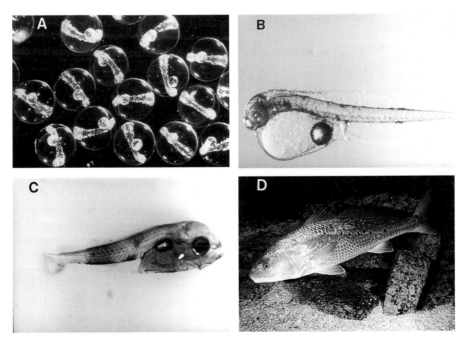

Fig. 6.1 Red drum at different stages of their life cycle: A eggs 12-hours old (25°C); B one-day old yolk-sac larva; C eight-day old larvae (25°C); D two-year old fish which has been reared in captivity. Red drum commonly possess a single large pigmented spot at the base of the caudal fin; individuals with multiple spots like the one shown here are rare.

larvae can however frequently be tolerated at the hatchery because of the remarkable fecundity of this species.

In conclusion, red drum have many of the ideal characteristics of an aquaculture species and there are few biological constraints for its widespread culture. Many of the current problems in the USA are related to the youth of the red drum aquaculture industry and to the scarcity of experienced fish culturists.

6.1 Introduction

The family Sciaenidae (drums) is composed of over 200 marine, estuarine and freshwater perciform fish species which inhabit most of the temperate zones of the world. There are 34 sciaenid species in North America, six of which are of commercial and/or recreational importance. Currently, red drum (*Sciaenops ocellatus*) is the most important aquaculture and recreational sciaenid fish. Red drum is also known by the common names of redfish, spottail bass, and channel bass, and ranges from Cape Cod in the north Atlantic into the Gulf of Mexico and west to Tampico, Mexico. Adult red drums spend most of their time in offshore waters, occasionally entering deeper bay areas. Spawning season begins in the autumn and occurs in offshore waters near the mouths of passes and inlets in the Gulf of Mexico. There is also evidence that red drum spawn in certain lagoons in southern Florida (Johnson & Funicelli 1991). Eggs and larvae move with currents through tidal passes into bays and estuaries where the larvae grow to juveniles. At the end of 4–5 years and at a weight of 4–5 kg, red drums migrate back into offshore waters, where they become part of the spawning stock.

Spotted sea trout (*Cynoscion nebulosus*), another sciaenid of recreational importance, is a euryhaline species which spawns and lives most of its life in bays. Spotted sea trout begin spawning in late spring and continue spawning until late summer, each individual spawning several times (Brown-Peterson *et al.* 1988). A close relative of the spotted sea trout, the weakfish (*Cynoscion regalis*), inhabits the nearshore area of the north-eastern coast of North America and spawns during the summer months (Epifanio *et al.* 1988). Current evidence suggests that sciaenid fish commonly spawn around dusk in the wild (Holt *et al.* 1985, Brown-Peterson *et al.* 1988) and also under laboratory conditions. Up to a million small pelagic eggs (approximately 1 mm in diameter) are released during each spawning and are passively transported to the nursery grounds inshore.

6.2 Reproduction

6.2.1 Oocyte growth

The endocrine control of the female reproductive cycle in sciaenid fish has been investigated most extensively in Atlantic croaker and spotted sea trout.

The identities of the pituitary hormones that regulate ovarian recrudescence have not been established. Although the pituitary content of maturational gonadotropin (GTH II) increases during ovarian recrudescence in croaker, the hormone remains barely detectable in the plasma (Copeland & Thomas 1989a).

A second gonadotropin which has recently been isolated from croaker, GTH I, is a more likely candidate for the vitellogenic GTH (Copeland & Thomas 1993). In addition, two other pituitary hormones purified from sciaenids, growth hormone and somatolactin, stimulate ovarian steroidogenesis and may have physiological roles during oocyte growth (Safford 1992, Singh & Thomas 1991). The hepatic production of vitellogenin, the precursor of the yolk proteins, appears to be regulated primarily by oestradiol in croaker, spotted sea trout and orangemouth corvina (Trant 1987, Copeland & Thomas 1988, Thomas *et al.* in press), although the possible involvement of other hormones has not been examined. Treatment of sea trout with oestradiol induces production of vitellogenin and also the hepatic oestrogen receptor, an intermediary in the oestrogen regulation of vitellogenesis (Copeland & Thomas 1988, Smith & Thomas 1990). The 2-month period of oocyte growth in sea trout coincides with seasonal increases in plasma concentrations of oestradiol and vitellogenin and hepatic concentrations of the oestrogen receptor (Thomas *et al.* 1987, Smith & Thomas 1991).

Although plasma testosterone levels are also elevated during this period, testosterone is unlikely to be directly involved in vitellogenesis because the oestrogen receptor is highly specific for oestrogens and shows negligible binding to androgens (Smith & Thomas 1990). Plasma oestradiol levels are maximal at the end of ovarian recrudescence ($1-2\,\text{ng}\,\text{ml}^{-1}$) in all three species and decline prior to spawning (Thomas *et al.* 1987, Trant 1987). However, plasma concentrations of the steriod are maintained at half maximal values and vitellogenesis persists in the multiple spawning sea trout throughout the remainder of its 6-month spawning season (Thomas *et al.* 1987, Smith & Thomas 1991). Oestradiol concentrations also show marked daily fluctuations in this species with peak values in the early morning (Thomas *et al.* 1987).

6.2.2 Final oocyte maturation and ovulation

Final oocyte maturation, ovulation and spawning are initiated by a surge in GTH II secretion from the pituitary. Plasma GTH II levels increase several fold in croaker within 1 h of injection with an ovulatory dose of des-Gly^{10} [D-Ala^6]-luteinizing hormone releasing hormone ethylamide (LHRHa, $20\,\mu\text{g}$ $\text{kg}^{-1}\,\text{BW}$) and remain elevated for at least another 24 h at which time the oocytes have completed maturation and are within a few hours of ovulation (Copeland & Thomas 1989b). Prolonged elevations of GTH II have also been observed in sea trout, red drum and orangemouth corvina following LHRHa or salmon gonadotropin releasing hormone analogue (sGnRHa) injection (Copeland & Thomas 1992). GTH II levels were elevated on the fifth day of spawning in corvina but had declined to undetectable levels 9 days later after spawning had ceased (Thomas *et al.* unpublished data). The persistent elevation of GTH II in sea trout and corvina after injection with 100 µg kg^{-1} LHRHa may account for the multiple spawning frequently observed. Preliminary results suggest that the elevation is more transient in sea trout during natural spawning.

Unlike all the other species investigated to date, the dopaminergic inhibitory system for the control of GTH II secretion is completely lacking in croaker, sea trout and red drum (Copeland & Thomas 1989b, and unpublished data), and therefore appears to be absent in sciaenids. A serotonergic system is present in croaker which augments LHRHa-induced GTH II secretion, but the involvement of this neuroendocrine pathway in the control of final oocyte maturation is uncertain (Khan & Thomas 1992).

GTH II performs two different functions during two temporally distinct stages of final oocyte maturation in croaker and sea trout. GTH II initially induces priming of the follicle-enclosed oocytes so that they become competent to undergo final maturation in response to the maturation-inducing steroid (MIS) (Patiño & Thomas 1990a,b, Thomas & Patiño 1991). During the GVBD phase, GTH II stimulates MIS synthesis, the traditional role ascribed to GTH II during final oocyte maturation (Trant & Thomas 1989a,b, Thomas & Trant 1989, Patiño & Thomas 1990c). Interestingly, one of the critical events during GTH II priming is induction of the ovarian membrane receptor for the MIS (Patiño & Thomas 1990d, Thomas & Patiño 1991).

Extensive evidence indicates that 17α,20β,21-trihydroxy-4-pregnen-3-one (20β-S), a MIS first identified in sciaenid fishes (Trant *et al.* 1986), is the major MIS in croaker and sea trout and that 17α,20β-dihydroxy-4-pregnen-3-one (17,20-P) is of minor importance. 20β-S is a potent inducer of final oocyte maturation *in vitro*, its production coincides with the onset of the GVBD phase of final oocyte maturation, and it is the only MIS produced in large quantities by the ovary *in vitro* (Trant & Thomas 1988, 1989a,b, Patiño & Thomas 1990c). Moreover the ovarian membrane receptor for 20β-S is highly specific and shows negligible binding to 17,20-P (Patiño & Thomas 1990d).

Plasma levels of immunoreactive 20β-S rise dramatically during the later stages of final oocyte maturation in sea trout and reach peak values (6–8 ng ml^{-1}) at ovulation whereas 17,20-P levels remain low (Thomas *et al.* 1987). Plasma levels of 17,20-P rise in parallel with those of 20β-S in red drum and corvina during final maturation so 17,20-P could conceivably have an important function during the periovulatory period in some sciaenid species (Thomas 1988). Possible pheromonal actions of 20β-S, 17,20-P or their metabolites have not been investigated, although the presence of ovulating female croaker in the tank was found to increase plasma GTH II levels in males (P. Thomas, unpublished data).

6.3 Culture methods: broodstock

6.3.1 Capture and care of broodstock

It is preferable to capture red drum and spotted sea trout by hook and line because it causes fewer abrasions and lower mortalities than netting. Occasionally the abdomen of large red drum will become filled with air during capture, presumably owing to distension of the air bladder, causing the fish to float upside down. The air can be removed by puncturing the abdomen with a hypodermic needle. Anaesthetics such as quinaldine sulphate have proved useful for handling and transporting red drum (Robertson *et al.* 1988). Typically four adult red drum (two males and two females) are placed in each 160 000 l recirculating tank equipped with biological filtration (Fig. 6.2). Ammonia and nitrite levels in the tanks should be kept below 0.5 ppm, pH should adjusted to 7.8 and salinity maintained at 30‰. The tanks are routinely treated with copper ($CuSO_4$, 0.3 ppm) at most red drum facilities whenever sea water is added to prevent infections of *Amyloodinium ocellatum*.

Most red drum acclimatize quickly to captivity and begin feeding within 2 weeks a diet of fresh shrimp, squid and fish. Spotted sea trout usually take longer to acclimatize to laboratory conditions and initially only accept live food. The broodfish are fed until they are satiated for the first 3 months and then fed approximately 3% of their body weight in food per day.

6.3.2 Induction of precocious sexual maturation

Currently most red drum eggs and fry are obtained from wild-caught broodstock and no selective breeding of this species for desirable traits such as rapid

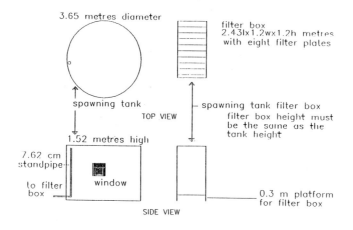

Fig. 6.2 Spawning tank system.

growth, resistance to stress and disease, and cold tolerance has been undertaken. The 4–5 years for red drum to reach sexual maturity and breed successive generations has been a major deterrent to establishing such a breeding programme.

Two methods have been investigated for inducing precocious maturation and spawning in red drum, environmental manipulation and hormonal treatments. Fish were raised at high temperatures and on long or natural photoperiods and fed high food rations to maximize growth. One group of 20 fish was reared under a natural photoperiod and at 24–26°C until they were 10 months old (February–December). The fish were then maintained on a 12L:12D photoperiod regime and at 26°C from December until spawning began the following October, when the fish were 19.5 months old and mean weight was 2.9 kg (Arnold 1991). The size of the spawns ranged from a few thousand to 200 000 eggs, and fertilization rates ranged from 0 to 80%. Altogether ten of these fish spawned 49 times during the first 3 months of spawning.

A collaborative study with Texas Parks and Wildlife Department investigated the effects of dietary administration of thyroxine and 11-ketotestosterone and long photoperiods (16 h light) on growth and sexual maturation. All fish were maintained at 26°C and fed a high food ration. Fish not fed hormones grew rapidly and this growth rate was not enhanced by addition of hormones to the diet. Sexual maturation occurred first in the androgen-treated group, but all the fish were phenotypic males (P. Thomas *et al.* unpublished data). The fish were subjected to an abbreviated annual photoperiod and temperature cycle (McCarty *et al.* 1986) when they were 18 months old and began spawning 6 months later when they were under autumn environmental conditions and weighed 4–5 kg.

It is concluded from these studies showed that sexual maturation in red drum is not an age-dependent phenomenon under strict genetic control but is also influenced by phenotypic characteristics such as size. Thus the potential exists for advancing sexual maturation in sciaenid fishes by increasing the growth rate.

Another valuable tool for selective breeding, cryopreservation of spermatozoa, has been developed for Atlantic croaker and is probably applicable to other sciaenids. Extensive studies showed that the extenders NaCl, glucose and sucrose resulted in high fertilization rates and that a mixture of semen to dimethyl sulphoxide to extender in the proportion of 10:15:75 gave optimal results (Gwo 1989). Freezing rates of -10 to $-15°C$ per min were effective for preserving viable spermatozoa.

6.3.3 Induction of gonadal recrudescence

Temperature and photoperiod are the most important environmental factors controlling gonadal recrudescence and spawning in all temperate and subtropical sciaenid species studied to date; red drum, spotted sea trout, Atlantic croaker and orangemouth corvina. Environmental manipulation techniques to induce gonadal recrudescence and spawning in captivity of red drum and spotted sea trout were developed by Arnold and co-workers (Arnold et al. 1976, 1977). Broodstock were acclimated to 30 000 l indoor recirculating spawning tanks containing seawater (salinity 25–35‰S) for 14–40 days prior to the beginning of temperature and photoperiod manipulation. Knowledge of the environmental conditions at each stage of the natural reproductive cycles of red drum and spotted sea trout was used in the design of temperature and photoperiod regimes to induce gonadal recrudescence and spawning of these species in captivity.

Red drum were exposed to a seasonal cycle of temperature and photoperiod changes, beginning with late autumn conditions (9L:15D, 20°C) for 1 month, then winter conditions (9L:15D, 18°C) for 2 months, spring conditions (12L:12D, 19°C) for 1 month, summer conditions (15L:9D, 25°C) for 2 months, later summer conditions (12L:12D, 26°C) for 2 months, and finally autumn conditions (9L:15D, 23°C). Spawning began after exposure for 1.5 months to autumn conditions (Arnold et al. 1977).

Spotted sea trout were first exposed to winter conditions (temperature 14–17°C, photoperiod 9L:15D) for 2 months, after which the photoperiod was changed to summer conditions (15L:9D) and the temperature was gradually raised to 26°C over a 2-month period (Arnold et al. 1976). This photoperiod/temperature regime caused gonadal recrudescence. Spawning began in mid April, 1 month after the water temperature reached 26°C and around the time that spotted sea trout begin spawning in the wild.

It was later found that condensed annual cycles of photoperiod and temperature were effective in inducing gonadal recrudescence in sciaenids. A 4-month annual cycle is now routinely used to induce gonadal recrudescence in red drum (Roberts 1987, Roberts et al. 1978, McCarty et al. 1986). A condensed (6-month) annual cycle has also been used to induce ovarian recrudescence in orangemouth corvina (Prentice & Thomas 1987).

Thus, exposure to a condensed annual photoperiod and temperature cycle is the preferred method of inducing gonadal recrudescence in all the sciaenid species investigated so far. No attempts have been made to induce gonadal growth with hormonal treatments.

6.3.4 Induction of spawning

Environmental

Special procedures are not usually required to induce spawning of red drum and spotted sea trout; once the gametes are fully developed the broodstock are maintained at the temperature and photoperiod conditions at which they spawn naturally. Salinity should be maintained at 30‰, and noise and vibration minimized to ensure spawning. Pre-spawning behaviour of red drum includes chasing the females and may continue for up to 30 days before spawning begins. If spawning does not commence by then it can sometimes be induced by rapidly lowering the water temperature 3–4°C for 48 h and then rapidly reverting to the original temperature (Roberts 1990, Arnold 1988).

Spawning in red drum is preceded by loud drumming and the development of a dark wine coloration above the lateral line in the males (Arnold et al. 1977). Pre-spawning behaviour of spotted sea trout is similar to that of red drum. However, the coloration of both male and female spotted sea trout changes, becoming almost black above the lateral line. Spawning occurs at dusk, with all the males engaging in the spawning act. The white sea bass, *Atractoscion nobilis*, has also recently been spawned in captivity by temperature-photoperiod manipulation (Orhun 1989).

Hormonal (by injection)

The induction of spawning of some sciaenid species such as orangemouth corvina by temperature and photoperiod manipulation has frequently been unsuccessful owing to a lack of knowledge of the environmental cues required to induce spawning (Prentice & Colura 1984). Hormone treatments have been shown to be reliable methods of inducing spawning of orangemouth corvina and other sciaenid species that do not respond predictably to environmental manipulation (Prentice & Thomas 1987, Thomas & Boyd 1988).

A single injection of a LHRHa at a dose of 100 µg kg^{-1} BW was effective in inducing spawning of orangemouth corvina, spotted sea trout, red drum and Atlantic croaker with fully grown oocytes (Thomas & Boyd 1988, Copeland & Thomas 1989b). Spawning of all four species occurred 30–36 h later with high rates of fertilization and hatching success. A second spawn was usually observed on the second day which may have been related to the sustained elevation of GTH II induced by this dose of LHRHa (Thomas & Boyd 1988). Subsequently, lower doses of LHRHA (50 µg kg^{-1} and 20 µg kg^{-1}) were shown to be equally effective in inducing spawning of croaker, sea trout and red drum (Copeland & Thomas 1989b, P. Thomas, unpublished data).

The salmon GnRH analog, des-Gly^{10} [D-Arg^6, Trp^7, Leu^8, Pro^9]-LHRH-ethylamide (sGnRHa), had a similar potency to LHRHa (P. Thomas, unpublished data). Repeated treatment with LHRHa or sGnRHa did not adversely affect subsequent spawning success. These studies demonstrate that a single injection of LHRHa or sGnRHa at a dose of 20–100 µg kg^{-1} BW is a practical method of inducing spawning of sciaenids, especially when predictable spawning is required. Red drum, spotted sea trout, orangemouth corvina and Atlantic croaker with mean oocyte diameters of 600 µm, 400 µm, 440 µm, and 550 µm, respectively (determined by ovarian biopsy), can be reliably induced to spawn by this single injection protocol. Simultaneous injection of a dopamine antagonist such as pimozide or domperidone is not required because the dopaminergic inhibitory system for GTH II secretion is lacking in these species.

Spawning has also been induced in sciaenids with injections of heterologous gonadotropins. Human chorionic gonadotropin (hCG) induced spawning of Atlantic croaker and red drum at doses of 125–250 IU per kg and 500–600 IU per kg, respectively (Middaugh & Yoakum 1974, Colura 1990), whereas injection of orangemouth corvina with 500 IU hCG caused massive swelling of the ovary due to simultaneous maturation and hydration of a large percentage of the oocytes which resulted in death (Prentice & Colura 1984). Thus orangemouth corvina and other fractional spawners such as sea trout, in which only a small batch of fully mature oocytes normally undergo maturation and hydration during each spawning cycle, may be particularly sensitive to over stimulation by high doses of mammalian gonadotropins. Similar problems have not been encountered when LHRHa or sGnRHa have been administered, so these are preferable hormonal treatments for spawning these species.

Oral administration of hormone

Certain sciaenid species, for example spotted sea trout and orangemouth corvina, are especially susceptible to handling stressors when they are in spawning condition, and may fail to ovulate or may die if they are not anesthetized prior to netting, handling and injection, particularly if the environmental conditions (e.g. temperature, salinity) are sub-optimal. A hormonal method of spawning sea trout in which LHRHa was administered in the diet was developed which eliminated the stress associated with capture and handling (Thomas & Boyd 1989). Oral administration of 1.0–2.5 mg LHRHa per kg BW, injected in acidified saline into the diet (dead shrimp), in four separate trials resulted in spawning 32–38 h later with high rates of fertilization and hatching success (Table 6.1). To our knowledge this was the first demonstration of successful spawning in any teleost species by dietary administration of hormones.

Subsequent trials showed that sGnRHa is equally effective in inducing spawn-

Table 6.1 Summary of spawning trials for spotted sea trout administered LHRHa orally[a]

Trial	Dose LHRHa Mean (mg kg^{-1} BW)	Dose LHRHa Range (mg kg^{-1} BW)	Results (h)	No. of eggs collected (×1000)	% fertilization	% hatch	% survival
1	2.0	1.0–4.0	spawn (32)[b]	1628	99.0	28.6	69.8
2	1.0	0.5–2.0	spawn (38)	100	93.6	67.0	25.0
3	2.0	1.0–6.0	spawn (38)	751	97.3	93.0	86.0
4	2.5	–	spawn (38)	901	92.2	90.2	81.5
5	0.2	0.1–0.4	no spawn	–	–	–	–
6	0.2	0.1–0.6	no spawn	–	–	–	–

[a] Data from Thomas & Boyd 1989 (with permission)
[b] Time from LHRHa feeding to spawning

ing by this route of administration. Doses of LHRHa and sGnRHa as low as 0.5 mg kg^{-1} are effective, although a dose of 1 mg kg^{-1} gives more consistent results. These results suggest that at least ten times more LHRHa or sGnRHa is required to induce ovulation in spotted sea trout by dietary administration than by injection. Slow release from the diet, enzymatic degradation, and only partial absorption by the gut could account for the lower potency of the orally administered hormone.

A disadvantage of this spawning method is that it is usually impossible to determine whether all the broodstock have received the same dose of the hormone. In practice, however, this is rarely a problem when spawning sea trout since all the females in the tank will ovulate even if only some of them have received the hormone. It is concluded, therefore, that dietary administration of LHRH analogues has great potential for spawning sea trout and other species susceptible to stress. Although more hormone is required for oral administration, the increased cost may be offset by substantial savings in effort and personnel over that required to capture and administer the hormone by injection and in reduction of handling stress for the broodstock.

6.3.5 Maintenance of extended spawning

Environmental

A remarkable characteristic of red drum and spotted sea trout reproduction in captivity is the ability to spawn continually over extended periods. Once spawning begins the photoperiod is kept constant and spawning frequency is regulated by minor changes in the temperature (Arnold *et al.* 1976, Arnold 1988). Spawning of four red drum broodfish (two males and two females) was monitored over a 7-year period. The fish were subjected to an abbreviated seasonal cycle of temperature and photoperiod until they reached autumn spawning conditions (12L:12D, 26°C) after which the photoperiod was held constant until spawning began (August 1980).

Initially spawning frequency was high (10–20 per month) but declined after the fourth month and averaged 5.8 spawns per month for the first 41 months of the trial (Fig. 6.3, see Arnold 1988). The broodfish failed to spawn in only one month (February 1982) of this 41-month period (Fig. 6.3). Spawning ceased following a dramatic fall in water temperature to 17°C in December 1983, and did not resume until March 1984, 1 month after the temperature had risen to 24°C. Spawning continued until the end of the trial (June 1987), although it was interrupted several times when the water temperature decreased and when the fish were transferred to new facilities.

Temperature had a profound effect on spawning frequency. Spawning decreased when water temperatures were below 23°C and ceased altogether below 20°C. For example, spawning was interrupted by a 3°C decrease in water temperature to 21°C and resumed 5 days later when the temperature rose

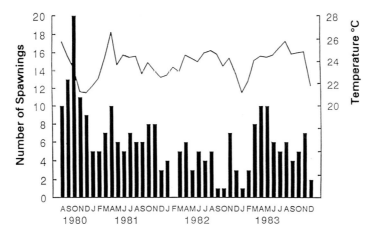

Fig. 6.3 Spawning frequency (black bars) of red drum broodstock (two females and two males) maintained on a constant photoperiod (12L:12D) for 41 months, and mean water temperature. (From Arnold 1988 with permission.)

again. This procedure is therefore, useful for limiting the production of eggs to periods when they are required.

Altogether the two female broodfish spawned 360 times during the 7-year period and produced approximately 250 million fertilized eggs. This remarkable reproductive output was achieved under a constant photoperiod, which suggests that red drum can be spawned indefinitely without going through a refractory period. The ability to produce large numbers of fertilized eggs throughout the year from a few broodstock without the need for complicated manipulations is undoubtedly one of the most attractive features of red drum as an aquaculture species.

Hormonal

Spawning techniques developed for red drum are not suitable for large scale production of eggs from multiple spawning species such as spotted sea trout, which release relatively few eggs during each spawning cycle (Brown-Peterson *et al.* 1988). Spawning is unpredictable so that many broodstock would be required to obtain enough eggs on a particular day for stocking hatchery ponds. Hormonal treatments can be used to synchronize spawning and obtain large numbers of eggs at a time from relatively few broodfish.

It has been shown that spotted sea trout broodfish can be induced to spawn several times by repeated dietary administration of sGnRHa (P. Thomas *et al.* unpublished data). Three female spotted sea trout were each induced to spawn four times during a 4.5 month period by repeated oral administration of 1–5 mg sGnRHa per kg BW. High doses of sGnRHa ($2.5-5.0\,\text{mg}\,\text{kg}^{-1}$) were used to initiate spawning and a dose of $1.0\,\text{mg}\,\text{kg}^{-1}$ was effective in subsequent trials. Females were induced to spawn every 20–30 days, the interval required for a new cohort of oocytes to complete vitellogenic growth. The batch fecundity, percentage fertilization, hatching success and larval survival did not decline during the serial spawning trials. Altogether 2–4 million viable eggs were produced by each female spotted sea trout during this period. Total egg production during the spawning period is considerably reduced if sGnRHa is administered by injection.

Recovery after the stress of capture and handling is slow in sea trout and both feeding and vitellogenic growth of the oocytes are interrupted after each injection. Moreover, the fish frequently develop bacterial infections after repeated handling. Repeated oral administration of a GnRH analogue is, therefore, the method of choice for large scale production of spotted sea trout eggs.

6.4 Culture methods: eggs and larvae

6.4.1 Production of eggs

Eggs of sciaenids are positively buoyant and spherical, contain a single oil globule and range in diameter from 0.6 mm in spotted sea trout and silver perch to 1.2 mm in black drum (Holt *et al.* 1988) and 1.3 mm in white sea bass (Orhun 1989). Red drum eggs range from 0.9 to 1.0 mm in diameter, are released and fertilized at night, and develop to a fully formed embryo ready to hatch in 18–29 h depending on water temperature (Holt *et al.* 1981a, Holt *et al.* 1985). Diel spawning patterns (Holt *et al.* 1985) and egg development are similar in many sciaenids, with early cell division essentially completed in 2 h and early embryo formation by 12 h (Fig. 6.1A). Lipids are the energy source for these small, rapidly-developing eggs. Wax esters and triglycerides together provide over 98% of the energy consumed, whereas carbohydrate, in the form of glycogen, provides less than 2% of the energy, and protein content actually increases during development (Vetter *et al.* 1983).

Development time is a function of incubation temperature while egg diameter is correlated with spawning salinity. Many euryhaline sciaenids spawn in estuarine and coastal waters where their eggs are exposed to a wide range of salinities. Spotted sea trout, and estuarine spawner, produces small diameter eggs (0.60 mm) in 45‰ and large diameter eggs (0.86 mm) in 21‰ salinity. A similar relationship is seen with red drum spawned in the laboratory at 24‰ producing 1.01 (SE.017) mm eggs, at 28‰ 0.95 (SE.012) mm eggs and at 37‰ 0.92 (SE.004) mm diameter eggs.

Sciaenid eggs are collected in 500 micron mesh bags submerged in the water of an external bio-filter box, or in special egg collection baskets placed below the outflow from the spawning tank. Volumetric methods are used to estimate the number of eggs spawned before they are transferred to incubators for 72–96 h (Henderson-Arzapalo 1990) or directly to rearing tanks, or hatched and then moved to rearing tanks after 24 h (Holt et al. 1990). In our laboratory, eggs are treated with a 10 ppm formalin bath for 1 h and counted, a sub-sample is taken for measurement of the hatch rate, and the remainder are placed in rearing tanks maintained at the same temperature and salinity as the spawning tank.

6.4.2 Production of larvae

Sciaenid larvae are small at hatching (1.5–2.5 mm long and 20–75 µg dry weight), develop quickly and within a few days deplete the yolk sac and require an external source of nutrients (Fig. 6.1B,C). During the yolk sac period red drum larvae utilize maternally-derived lipids and proteins, leading to a reduction in body weight, protein and RNA content by the time of first feeding (Holt & Sun 1991, Lee et al. 1984, Westerman & Holt 1988). Small live prey such as rotifers (*Brachionus plicatilis*) or wild zooplankton are required in high concentrations for first-feeding larvae.

Growth rates are temperature-dependent but are typically 0.2 mm per day during the first 2 weeks (Lee et al. 1984), increasing to nearly 1 mm per day thereafter under ideal conditions (Holt 1990). Temperature affects all aspects of larval development, including the rate of yolk utilization, size at first feeding, motor and sensory development, and growth (Fig. 6.4). The optimum temperature for red drum culture is between 25 and 30°C. Growth is arrested below 20°C (Holt et al. 1981b). In contrast larval white sea bass, a sciaenid that occurs off the southern California coast, was found to have an optimum culture temperature of 19°C (Orhun 1989).

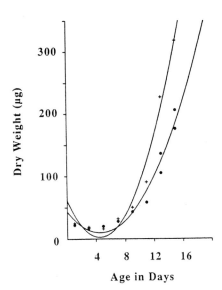

Fig. 6.4 Daily change in dry weight of larval red drum grown at two temperatures 24°C(●) and 28°C(+). (Data fit with quadratic equation from Lee et al. 1984.)

Early growth rates are not influenced by salinity but survival may be reduced if osmoregulatory competence is not developed. Zero mortality salinity limits vary from 15 to 35‰ for red drum larvae (Holt *et al.* 1981b) to 10–40‰ for spotted sea trout, but there is evidence that spawning salinity and larval acclimatization can change these limits (Banks *et al.* 1991, G.J. Holt, unpublished data). Red drum are initially stocked at a salinity near that of the spawning tank and then gradually adjusted to ambient salinities. A major inducement for aquaculture development is that red drum can be grown in fresh water after they are fully scaled (25 mm SL, Crocker *et al.* 1981).

Oxygen requirements are not unusual and maximum ammonia (1 ppm) and nitrite (10 ppm) levels for culture (Holt & Arnold 1983) are easily controlled. Most larvae produced commercially are grown in outdoor ponds where natural blooms of phytoplankton and zooplankton are induced by a fertilization programme. But variation associated with pond culture is great (Sturmer 1990) and production is limited to coastal areas where sea water is available. Moreover, pond production of red drum larvae is limited to about 6 months a year in southern regions of the United States because temperature is critical to growth, and survival is significantly reduced at 20°C or less (Holt *et al.* 1981b).

A technique for producing red drum in indoor recirculating water systems has been developed (Arnold *et al.* 1990) and at least two commercial red drum hatcheries are using this method. The major impediment to indoor production is supplying enough live prey to feed a million or so larvae during the pre-weaning period. Red drum are typically fed at rates of 3–5 rotifers per ml twice a day for 10 days, then *Artemia salina* nauplii are fed at 1–2 per ml until the larvae are weaned to non-living food. Though most of the biological issues related to culturing rotifers have been resolved, large scale production can be unreliable and expensive and the product may have low nutritional value. Moreover, it is difficult to wean larvae from live to inert feed. In addition, cannibalism can result in high mortality and severely reduce production.

In order to resolve these problems and advance the use of recirculating systems for hatchery production, a substitute diet was developed. Inert particles are readily accepted by first-feeding larvae if they are small (250 μg or less) and buoyant enough to be retained in the water column (Holt 1992). A 150 l conical, fibreglass larviculture system with an internal biological filter, aerated from the bottom to generate water currents to disperse food, was designed to augment diet development (Fig. 6.5). This design provided an efficient closed system for rearing several thousand larvae for the month-long feeding trials.

Although the alimentary canal is simple at first feeding in red drum larvae, it is divided into a short foregut, an expanded midgut and a hindgut, and the digestive glands, including liver, gall bladder and pancreas are developed (Holt 1992). Larvae fed commercial fry diets, compared in trials with controls fed live prey, grew significantly less than controls and none survived to metamorphosis (Holt 1992). The inability of larvae to derive sufficient nutrition from non-living food may result from low enzyme activity, implying a crucial

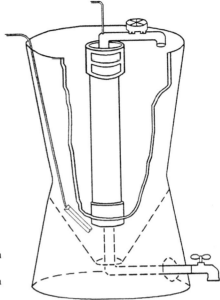

Fig. 6.5 A 150 l fibreglass cone tank with bottom aeration and an automatic feeder. Water recirculates through an internal biological filter via an air lift. (From Holt 1993.)

role for exogenous enzymes, supplied by living prey, in early feeding success. It has been shown that protease and lipase are present, but at low activity levels, and that these activities increase when larvae feed on live rotifers (Holt & Sun 1991).

This was further investigated by feeding larvae a combination of live rotifers and a microdiet for different lengths of time before weaning. Daily feeding rates of live food were reduced by half and a commercial microparticulate diet was substituted for 1–5 days, and then feeding live food was discontinued completely. The best growth was achieved on the 5 day combination of live food and microdiet, exceeding even that of controls fed live prey (Holt 1993). Besides acquiring excellent growth and survival, early weaning eliminated the need to feed artemia and the mortalities associated with later weaning.

A feeding protocol for red drum that requires live food for only 5 days, well before a functional stomach is formed, and a semi-purified diet for testing nutrient requirements in week-old larvae have been developed (G.J. Holt, unpublished data). The effectiveness of marine bacteria in digesting microdiets and for enhancing culture conditions is also being examined. This research is aimed toward developing an inert stable and nutritious diet for first feeding larvae, eliminating the need for live zooplankton cultures.

6.5 Conclusions

Reproduction of several sciaenid species can be induced in captivity by exposing them to natural or condensed seasonal cycles of photoperiod and temperature and by hormonal treatments. Environmental treatments are invariably used to stimulate gonadal recrudescence. Spawning is also induced routinely in red drum by environmental manipulation. This species can spawn continually for several years when held on a constant autumn photoperiod. Injection of LHRHa is the only reliable method of spawning certain species, such as orangemouth corvina. Repeated oral administration of sGnRHa is an effective method of synchronizing spawning and of obtaining large numbers of eggs from few spotted sea trout broodstock. Large numbers of sciaenid larvae can be cultured in ponds or in indoor high density recirculating systems. Suitable enviromental conditions for optimal survival and growth of red drum larvae can be readily achieved in recirculating systems. Red drum larvae can be weaned onto artificial dry diets at 1 week old, thereby eliminatinig the need to feed artemia.

References

Arnold, C.R. (1988) Controlled year-round spawning of red drum *Sciaenops ocellatus* in captivity. *Contributions in Marine Science*, **30** (supplement), 65–70.

Arnold, C.R. (1991) Precocious spawning of red drum. *The Progressive Fish Culturist*, **53**, 50–1.

Arnold, C.R., Lasswell, J.L., Bailey, W.H., Williams, T.D. & Fable, W.A. Jr (1976) Methods and techniques for spawning and rearing spotted seatrout in the laboratory. *Proceedings of the Annual Conference of the Southeastern Association of Game and Fish Commissioners*, **30**, 167–78.

Arnold, C.R., Bailey, W.H., Williams, T.D., Johnson, A. & Lasswell, J.L. (1977) Laboratory spawning and larval rearing of red drum and southern flounder. *Proceedings of the Annual Conference of the Southeastern Association of Game and Fish Commissioners*, **31**, 437–40.

Arnold, C.R., Reid, B. & Brawner, B. (1990) High density recirculating grow-out systems. In *Red Drum Aquaculture*, (eds G.W. Chamberlain, R.J. Miget & M.G. Haby), pp. 182–4. Texas A & M Sea Grant College Program No. TAMU-SG-90-603.

Banks, M.A., Holt, G.J. & Wakeman, J.M. (1991) Age-linked changes in salinity tolerance of larval spotted seatrout (*Cynoscion nebulosus*, Cuvier). *Journal of Fish Biology*, **39**(4), 505–14.

Brown-Peterson, N., Thomas, P. & Arnold, C.R. (1988) Reproductive biology of the spotted seatrout, *Cynoscion nebulosus*, in South Texas. *Fishery Bulletin*, **86**(2), 373–88.

Colura, R.L. (1990) Hormone-induced strip-spawning of red drum. In *Red Drum Aquaculture*, (eds G.W. Chamberlain, R.J. Miget & M.G. Haby), pp. 33–4. Texas A & M Sea Grant College Program No. TAMU-SG-90-603.

Copeland, P.A. & Thomas, P. (1988) The measurement of plasma vitellogenin levels in a marine teleost, the spotted seatrout (*Cynoscion nebulosus*) by homologous radioimmunoassay. *Comparative Biochemistry and Physiology B*, **91**(1), 17–23.

Copeland, P.A. & Thomas, P. (1989a) Purification of maturational gonadotropin from Atlantic croaker (*Micropogonias undulatus*) and development of a homologous radioimmunoassay. *General and Comparative Endocrinology*, **73**, 425–41.

Copeland, P.A. & Thomas, P. (1989b) Control of gonadotropin release in the Atlantic croaker: evidence for lack of dopaminergic inhibition. *General and Comparative Endocrinology*, **74**, 474–83.

Copeland, P.A. & Thomas, P. (1992) Isolation of maturational gonadotropin subunits form spotted seatrout (*Cynoscion nebulosus*) and development of a beta subunit-directed radioimmunoassay for gonadotropin measurement in sciaenid fish. *General and Comparative Endocrinology*, **88**, 100–10.

Copeland, P.A. & Thomas, P. (1993) Isolation of gonadotropin subunits and evidence for two distinct gonadotropins in Atlantic croaker (*Micropogonias undulatus*). *General and Comparative Endocrinology*, **91**, 115–25.

Crocker, P.A., Arnold, C.R., DeBoer, J.A. & Holt, J. (1981) Preliminary evaluation of survival and growth of juvenile red drum (*Sciaenops ocellata*), in fresh and salt water. *Journal of the World Mariculture Society*, **12**, 122–34.

Epifanio, C.E., Doshorn, D. & Targett, T.E. (1988) Induction of spawning in the weakfish, *Cynoscion regalis*. *Fishery Bulletin*, **86**(1), 168–71.

Gwo, J.-C. (1989) *Cryopreservation of Atlantic croaker spermatozoa: Optimization of procedures, evaluation of morphological changes, and assessment of motility*. PhD thesis, Texas A & M University.

Henderson-Arzapalo, A. (1990) Red drum egg and larval incubation. In *Red Drum Aquaculture*, (eds G.W. Chamberlain, R.J. Miget, M.G. Haby), pp. 51–2. Texas A & M Sea Grant College Program No. TAMU-SG-90-603.

Henderson-Arzapalo, A., Colura, R.L. & Maciorowski, A.F. (1988) A comparison of black drum red drum, and their hybrid in saltwater pond culture. *Contributions in Marine Science*, **30** (supplement), 195–6.

Holt, G.J. (1990) Growth and development of red drum eggs and larvae. In *Red Drum Aquaculture*, (eds G.W. Chamberlain, R.J. Miget & M.G. Haby), pp. 46–50. Texas A & M Sea Grant College Program No. TAMU-SG-90-603.

Holt, G.J. (1992) Experimental studies of feeding in larval red drum. *Journal of the World Aquaculture Society*, **23**(4), 265–70.

Holt, G.J. (1993) Feeding larval red drum on microparticulate diets in a closed recirculating water system. *Journal of the World Aquaculture Society*, **23**, 225–30.

Holt, G.J. & Arnold, C.R. (1983) Effects of ammonia and nitrite on growth and survival of red drum eggs and larvae. *Transactions of the American Fisheries Society*, **112**, 314–18.

Holt, G.J. & Sun, F. (1991) Lipase activity and total lipid content during early development of red drum *Sciaenops ocellatus*. *European Aquaculture Society Special Publication*, **15**, 30–3.

Holt, G.J., Johnson, A.G., Arnold, C.R., Fable, W.A. Jr & Williams, T.D. (1981a) Description of eggs and larvae of laboratory reared red drum, *Sciaenops ocellata*. *Copeia*, **1981**(4), 751–6.

Holt, G.J., Godbout, R. & Arnold, C.R. (1981b) Effects of temperature and salinity on egg hatching and larval survival of red drum, *Sciaenops ocellata*. *Fishery Bulletin*, **79**(3), 569–73.

Holt, G.J., Holt, S.A. & Arnold, C.R. (1985) Diel periodicity of spawning in sciaenids. *Marine Ecology – Progress Series*, **27**, 1–7.

Holt, S.A., Holt, G.J. & Young-Abel, L. (1988) A procedure for identifying sciaenid eggs. *Contributions in Marine Science*, **30** (supplement), 99–108.

Holt, G.J., Arnold, C.R. & Riley, C.M. (1990) Intensive culture of larval and post-larval red drum. In *Red Drum Aquaculture*, (eds G.W. Chamberlain, R.J. Miget & M.G. Haby), pp. 53–6. Texas A & M Sea Grant College Program No. TAMU-SG-90-603.

Johnson, D.R. & Funicelli, N.A. (1991) Spawning of the red drum in Mosquito Lagoon, east-central Florida. *Estuaries*, **14**(1), 74–9.

Khan, I.A. & Thomas, P. (1992) Stimulatory effects of serotonin on maturational gonadotropin release in the Atlantic croaker, *Micropogonias undulatus*. *General and Comparative Endocrinology*, **88**, 388–96.

Lee, W.Y., Holt, G.J. & Arnold, C.R. (1984) Growth of red drum larvae in the laboratory. *Transactions of the American Fisheries Society*, **113**, 243–6.

McCarty, C.E., Geiger, J.G., Sturmer, L.N., Gregg, B.A. & Rutledge, W.P. (1986) Marine finfish culture in Texas: A model for the future. In *Fish Culture in Fish Management*, (ed. R.H. Stroud), pp. 249–62. American Fisheries Society, Washington, DC.

Middaugh, D.P. & Yoakum, R.L. (1974) The use of chorionic gonadotropin to induce laboratory spawning of the Atlantic croaker, *Micropogonias undulatus*, with notes on subsequent embryonic development. *Chesapeake Science*, **15**, 110–23.

Orhum, M.R. (1989) *Early life history of white seabass* Atractoscion nobilis. M.Sc. thesis, San Diego State University.

Patiño, R. & Thomas, P. (1990a) Effects of gonadotropin on ovarian intrafollicular processes during the development of oocyte maturational competence in a teleost, the Atlantic croaker: Evidence for two distinct stages of gonadotropic control of final oocyte maturation. *Biology of Reproduction*, **43**, 818–27.

Patiño, R. & Thomas, P. (1990b) Induction of maturation of Atlantic croaker oocytes by 17α,20β, 21-trihydroxy-4-pregnen-3-one *in vitro*: consideration of some biological and experimental variables. *Journal of Experimental Zoology*, **255**, 97−109.

Patiño, R. & Thomas, P. (1990c) Gonadotropin stimulates 17α,20β,21-trihydroxy-4-pregnen-3-one production from endogenous substrates in Atlantic croaker ovarian follicles undergoing final maturation *in vitro*. *General and Comparative Endocrinology*, **78**, 474−8.

Patiño, R. & Thomas, P. (1990d) Characterization of membrane receptor activity for 17α,20β, 21-trihydroxy-4-pregnen-3-one in ovaries of spotted seatrout (*Cynoscion nebulosus*). *General and Comparative Endocrinology*, **78**, 204−17.

Prentice, J.A. & Colura, R.L. (1984) Preliminary observations of orangemouth corvina spawn inducement using photoperiod, temperature and salinity cycles. *Journal of the World Mariculture Society*, **15**, 162−72.

Prentice, J.A. & Thomas, P. (1987) Successful spawning of orangemouth corvina following des-Gly^{10}, [D-Ala^6]-luteinizing hormone-releasing hormone (1−9) ethylamide and pimozide injection. *Progressive Fish Culturist*, **49**, 66−9.

Roberts, D.E., Jr (1990) Photoperiod/temperature control in the commercial production of red drum (*Sciaenops ocellatus*) eggs. In *Red Drum Aquaculture*, (eds G.W. Chamberlain, R.J. Miget & M.G. Haby), pp. 35−43. Texas A & M University Sea Grant College Program No. TAMU-SG-90-603.

Roberts, D.E. Jr, Harpster, B.V. & Henderson, G.E. (1978) Conditioning and induced spawning of the red drum (*Sciaenops ocellatus*) under varied conditions of photoperiod and temperature. *Proceedings of the World Mariculture Society*, **9**, 311−32.

Robertson, L., Thomas, P. & Arnold, C.R. (1988) Plasma cortisol and secondary stress responses of cultured red drum (*Sciaenops ocellatus*) to several transportation procedures. *Aquaculture*, **68**, 115−30.

Safford, S. (1992) *Purification and chemical and biological characterization of prolactin and somatolactin, and partial chemical characterization of growth hormone from two marine teleosts, red drum* (Sciaenops ocellatus) *and the Atlantic croaker* (Micropogonias undulatus) PhD thesis, The University of Texas at Austin.

Sandifer, P.A. (1991) Species with aquaculture potential for the Caribbean. In *Status and Potential of Aquaculture in the Caribbean*, (eds A.A. Hargreaves & D.E. Alston), p. 274. Baton Rouge Press, Louisiana.

Singh, H. & Thomas, P. (1991) Mechanism of stimulatory action of growth hormone on ovarian steroidogenesis in spotted seatrout. In *Proceedings of the Fourth International Symposium on the Reproductive Physiology of Fish*, (eds A.P. Scott, J.P. Sumpter, D.E. Kime & M.S. Rolfe), p. 104. University of East Anglia, Norwich, UK.

Smith, J.S. & Thomas, P. (1990) Binding characteristics of the hepatic estrogen receptor of the spotted seatrout, *Cynoscion nebulosus*. *General and Comparative Endocrinology*, **77**, 29−42.

Smith, J.S. & Thomas, P. (1991) Changes in hepatic estrogen-receptor concentrations during the annual reproductive and ovarian cycles of a marine teleost, the spotted seatrout *Cynoscion nebulosus*. *General and Comparative Endocrinology*, **81**, 234−45.

Soletchnik, P., Thouard, E., Goyard, E., Yvon, C. & Baker, P. (1988) First larval rearing trials of red drum (*Sciaenops ocellatus*) in Martinique (French West Indies). *Contributions in Marine Science*, **30** (supplement), 125−8.

Sturmer, L.N. (1990) Zooplankton composition and dynamics in fingerling red drum rearing ponds. In *Red Drum Aquaculture*, (eds G.W. Chamberlain, R.J. Miget & M.G. Haby), pp. II 80−90. Texas A & M Sea Grant College Program No. TAMU-SG-90-603.

Thomas, P. (1988) Changes in the plasma levels of maturation-inducing steroids in several perciform fishes during induced ovulation. *American Zoologist*, **28**, 294.

Thomas, P. & Boyd, N.W. (1988) Induced spawning of spotted seatrout, red drum and orangemouth corvina (family:Sciaenidae) with luteinizing hormone-releasing hormone analog injection. *Contributions in Marine Science*, **30** (supplement), 43−7.

Thomas, P. & Boyd, N.W. (1989) Dietary administration of a LHRH analog induces successful spawning of spotted seatrout (*Cynoscion nebulosus*). *Aquaculture*, **80**, 363−70.

Thomas, P., Copeland, P.A. & Prentice, J.A. (in press) Preliminary observations on the reproductive physiology of female orangemouth corvina in captivity. *Journal of the World Aquaculture Society*.

Thomas, P. & Patiño, R. (1991) Changes in 17α,20β,21-trihydroxy-4-pregnen-3-one membrane receptor concentrations in ovaries of spotted seatrout during final oocyte maturation. In

Proceedings of the 4th International Symposium on Reproductive Physiology of Fish, (eds A.P. Scott, J. Sumpter, D. Kune & M.S. Rolfe), pp. 122−4. University of East Anglia, Norwich, UK.

Thomas, P. & Trant, J.M. (1989) Evidence that 17α,20β,21-trihydroxy-4-pregnen-3-one is a maturation-inducing steroid in spotted seatrout. *Fish Physiology and Biochemistry*, **7**, 185−91.

Thomas, P., Brown, N.J. & Trant, J.M. (1987) Plasma levels of gonadal steroids during the reproductive cycle of female spotted seatrout (*Cynoscion nebulosus*). In *Proceedings of the 3rd International Symposium on Reproductive Physiology of Fish* (eds D.R. Idler, C.W. Crim & J.M. Walsh), p. 219. Memorial University Press, St John's, Newfoundland.

Trant, J.M. (1987) *Synthesis of ovarian steroids during the reproductive cycle of a marine teleost, Micropogonias undulatus: evidence for a novel oocyte maturation-inducing steroid*. PhD thesis, The University of Texas at Austin.

Trant, J.M. & Thomas, P. (1988) Structure-activity relationships of steroids in inducing germinal vesicle breakdown of Atlantic croaker oocytes *in vitro*. *General and Comparative Endocrinology*, **71**, 307−17.

Trant, J.M. & Thomas, P. (1989a) Changes in ovarian steroidogenesis *in vitro* associated with final maturation of Atlantic croaker oocytes. *General and Comparative Endocrinology*, **75**, 405−12.

Trant, J.M. & Thomas, P. (1989b) Isolation of a novel maturation-inducing steroid produced *in vitro* by ovaries of Atlantic croaker. *General and Comparative Endocrinology*, **75**, 397−404.

Trant, J.M., Thomas, P. & Shackleton, C.H.L. (1986) Identification of 17α,20β,21-trihydroxy-4-pregnen-3-one as the major ovarian steroid produced by the teleost (*Micropogonias undulatus*) during final oocyte maturation. *Steroids*, **47**, 89−99.

Vetter, R.D., Hodson, R.E. & Arnold, C.R. (1983) Energy metabolism in rapidly developing marine fish egg, the red drum (*Sciaenops ocellata*). *Canadian Journal of Fisheries and Aquatic Sciences*, **40**(5), 627−34.

Westerman, M.E. & Holt, G.J. (1988) The RNA-DNA ratio: Measurement of nucleic acids in larval *Sciaenops ocellatus*. *Contributions in Marine Science*, **30** (supplement), 117−24.

Chapter 7
Sea Bass (*Dicentrarchus labrax*)

7.1 Introduction and taxonomy
7.2 Reproduction
    7.2.1 Reproductive strategies
    7.2.2 Seasonal endocrine cycles
7.3 Hormonal and environmental induction of spawning
    7.3.1 Hormonal induction of spawning
    7.3.2 Environmental control of reproduction
    7.3.3 Endogenous rhythms
7.4 Broodstock nutrition and effects on the quality of progeny
    7.4.1 Effects of ration
    7.4.2 Effects of quality of diet
7.5 Egg and larval quality
    7.5.1 Assessments of quality
    7.5.2 Factors affecting larval development
7.6 Holding tanks/facilities and management
    7.6.1 Broodstock
    7.6.2 Egg collection & larval rearing
7.7 Future prospects
Acknowledgements
References

The management of sea bass broodstock presents a number of problems that must be overcome in order to achieve spontaneous natural spawnings and the production of eggs and larvae of good quality. This work reviews these aspects and presents knowledge already attained on the reproductive physiology of broodstock sea bass and the quality of the eggs and larvae produced. More specifically, exogenous hormonal administration, environmental manipulation (i.e. photoperiod, temperature and salinity), diet composition and ration are considered with respect to their effects on the metabolism and physiology of broodfish and in turn on egg and larval quality. Reproductive cycles and endogenous rhythms of this species are reviewed as, together with nutritional aspects, these are considered fundamental to solving problems of inducing spawning of sea bass in captivity and of the quality of the seed produced. Finally, the problems in defining egg and larval quality are presented mainly with respect to the effects of nutritional deficiency.

7.1 Introduction and taxonomy

Sea bass is a much sought-after food fish in European markets because harvests from commercial fisheries are restricted both in terms of the quantity and seasonality of the supplies. It is also much prized by anglers. Present catches in no way meet the ever-increasing market demands for sea bass. As a result of its high value, its well-defined and buoyant market position and the restrictions in supplies, there has been much interest in the farming of sea bass (Bromage et al. 1988).

Sea bass is a teleost fish that belongs to the order Perciformes, Family Serranidae and genus Dicentrarchus. There are two species, *Dicentrarchus labrax* (Linne 1758) and *D. punctatus* (Bloch 1972), widely distributed throughout the Mediterranean and Atlantic coasts from latitude 30°N (Morocco) to 55°N (Ireland, Baltic and North Sea) (Kennedy & Fitzmaurice 1972, Tortonese 1986). Sea bass is a synchronous and gonochoristic species which spawns annually for a number of years during its life span. Spawning lasts from December to March in the Mediterranean area but becomes more restricted in length and delayed in northern latitudes (Bibliography in Bruslé & Roblin 1984). Sea bass is eurythermic (temperatures 5–28°C) and euryhaline (salinities 3‰ to full strength sea water) and spawns small (1.02–1.39 mm) pelagic eggs near the river mouths and estuaries or in littoral areas where salinity is high (30‰ or over) (Bibliography in Barnabé 1980, 1991).

At northern latitudes size and age at first maturity are increased. Puberty in females is attained at 3–5 years in the Mediterranean area (Barnabé 1973, Bou Ain 1977) and at 5–8 years in Irish waters (Kennedy & Fitzmaurice 1972) being delayed by 1–2 years in males. Although there is no clear sexual dimorphism, males are smaller than females at the time of first maturity.

Sea bass is a very fecund species, with absolute fecundities exceeding 2 million in the oldest fish (>18 years) (Mayer et al. 1990). Reports on relative fecundity vary from 293×10^3 to 358×10^3 eggs per kg body weight as determined by Kennedy & Fitzmaurice (1972) for Irish bass and from 492×10^3 to 955×10^3 eggs per kg as determined by Bou Ain (1977) for Mediterranean bass.

Despite optimistic production forecasts for sea bass of 27 000 t, at present total farmed production amounts only to a few hundred tonnes per annum (Bromage et al. 1988). Reasons for this shortfall include the lack of domestication of the species, the restricted seasonal and unpredictable nature of the spawnings of the broodstock, the higher proportion of males in many fish farm populations (sometimes attaining 70–90%), the poor quality of the eggs produced, the losses of larvae and fry during yolk sac resorption and also the slow growth of the younger fish up to table size.

As a result of these difficulties, growth in the commercial production of sea bass has been severely constrained. In addition to these problems, the poor quality and survival of eggs and larvae also appears to be affected by genetic drift in hatcheries resulting from reduced numbers of broodstock, uneven proportions of males and females and poor selection of broodstock fish. Even at this early stage of development of sea bass farming, Rodriguez (1991) has observed an important genetic erosion in a Spanish hatchery stock compared with natural populations.

Control of reproduction is essential for the success of fish farming, principally because it enables supplies of eggs and fry to be made available at precisely those times required for on-growing farms and not just in the 3–4 months of the year when natural spawning occurs. Moreover, the quality of the eggs and larvae produced depends to a great extent on how this control is performed, particularly with regard to broodstock management and nutrition. The control of reproduction also enables farmers to delay or inhibit sexual maturation, thus ensuring that metabolic activities and food inputs are channelled into somatic growth, thus avoiding the sometimes regressive changes in growth and flesh composition which accompany maturation. As a consequence there is a need for detailed knowledge of:

(1) the sexual cycle under culture conditions;
(2) the techniques required to induce and synchronize spawnings in mature animals, to alter the timing of spawning by environmental manipulation and to modify sex either by hormonal or genetic approaches; and
(3) the special husbandry and nutritional requirements of the broodstock fish and fry produced.

These subjects are considered further in the present chapter.

7.2 Reproduction

7.2.1 Reproductive strategies

The sea bass, like many teleosts inhabiting temperate latitudes, uses seasonal patterns of changing day length, modulated by temperature, to synchronize spawning, so that it occurs at the most favourable time of the year for survival of the offspring. During the first year of life the gonads of sea bass remain undifferentiated, with sexual differentiation occurring some time after this event (Roblin & Bruslé 1983). However, recently Blázquez et al. (1992) suggested that sex differentiation in captive sea bass may occur at an earlier age (9 months). In addition a 'sensitive period' for alteration of ovarian or testicular configuration by steroid administration has been identified between 126 and 226 days after hatching.

A high proportion of males (70–90%) has been observed in several Mediterranean hatcheries (S. Zanuy and M. Carrillo, unpublished results), although the possible environmental and/or social factors that produce such unequal sex ratios are not known.

Under culture conditions, puberty in males is reached near the second year of life, depending more on the body size attained than on the age of the animal. In contrast, puberty in females usually occurs 1 year later (Bruslé & Roblin 1984). These differences in reproductive strategy result in a higher

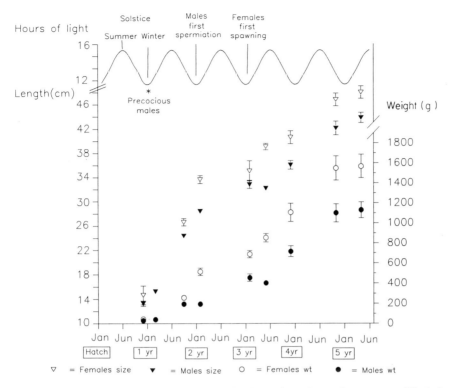

Fig. 7.1 Timing of spawning of cultured sea bass (both sexes) on the south-east coast of Spain in relation to day length and somatic growth. Vertical bars indicate standard error of the mean.

growth performance in females than males (Fig. 7.1). In captivity, under natural photoperiod and temperature, 2 year old immature female sea bass, attain 0.5 ± 0.04 kg in weight and 33.7 ± 0.6 cm in length compared with 0.3 ± 0.02 kg and 28.5 ± 0.6 cm for spermiating males of the same age. During the third year of life, the differences in weight and size between males and females are not as pronounced as in the previous year, probably as a result of the maturity of females. These facts favour the development of techniques for the monosex (all female) culture of sea bass.

As in other teleosts, gonadal maturation involves many months of preparation, whereas the latter stages of growth and gamete release are but brief interludes in the developmental process, i.e. vitellogenesis starts in October and lasts 4–5 months, whereas spawning time in the Mediterranean area is restricted to a month in the winter. In sea bass the different cellular types in the gonad and their change during the sexual cycle have already been described (Caporiccio & Connes 1977, Zohar et al. 1984, Zanuy et al. 1986, Mayer et al. 1988, Carrillo et al. 1989).

Alvariño et al. (1992a), studying the pattern of sea bass oocyte development after experimental ovarian stimulation, reported that sea bass show group-synchronous oocyte development and have ovaries which contain more than one group of developing oocytes. Thus, the sea bass in culture is a multiple spawner, confirming previous observations of Mayer et al. (1990) in wild stocks.

7.2.2 Seasonal endocrine cycles

Although the distribution of sGnRH in the brain and in the pituitary (Kah et al. 1991) and the inmunocytochemical identification of the cell types in the pituitary (Cambré et al. 1986) of the sea bass have already been completed, there has not been any attempt to purify the GtH(s) of this species. This explains why most of the work-carried out has concerned changes in steroid hormones.

Thus, Prat et al. (1990) described the seasonal changes in plasma levels of gonadal steroids of sea bass maintained under culture conditions in south-east Spain (40°N) and Hassin et al. (1991) for sea bass acclimatised to the warm water conditions of the Red Sea (29° 31′ N). A slight shift of the hormonal profiles and times of spawning were observed according to the latitude, but in general plasma levels of 17β-oestradiol (E2) and testosterone (T) attained high levels during vitellogenesis and spawning, concomitant with a dramatic increase in oocyte diameter (Fig. 7.2A). The significantly raised plasma levels of E2 during spawning can be accounted for by production of this steroid from the remaining groups of vitellogenic follicles, as is to be expected in a fractional spawner such as the sea bass.

As in other teleosts, E2 appears to be involved in vitellogenin synthesis. These statements are further supported by results derived from the development of a specific enzyme-linked immunosorbent assay (ELISA) for quantification of plasma vitellogenin (VTG) in sea bass (Mañanós et al. 1991a,b). Changes in plasma VTG levels correlate well with the presence of vitellogenic oocytes. In fact, a gradual increase in plasma VTG levels is observed during autumn, preceding the first appearance of vitellogenic oocytes; subsequently VTG and numbers of vitellogenic oocytes both peak during December at the time when E2 levels also attain high values (Figs 7.2A,B).

Over the following months, during the spawning period (SP) (January– March), the percentage of vitellogenic oocytes and E_2 plasma levels still remains elevated although plasma VTG levels were reduced to one half their previous values. Towards the end of SP and beginning of post-SP (March–April), the numbers of atretic oocytes increase, coinciding with low plasma levels of E_2 and T. A second minor peak of plasma VGT is observed during this period, possibly related to the reabsorption of yolk (Fig. 7.2B).

Measurements of 17α-hydroxy-20β-dihidroxyprogesterone (17α,20β-P), the maturation-inducing steroid (MIS) in salmonids, showed that there were only traces of the free form of this hormone in the blood of ovulating sea bass females (Prat et al. 1990, Hassin et al. 1991). Subsequently, high levels of this

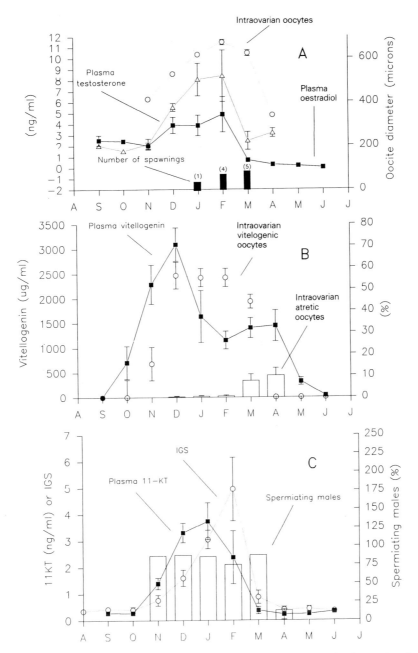

Fig. 7.2 Steroid and vitellogenin profiles in plasma in relation to oocyte development and spawning of sea bass. (A) Redrawn and modified from Prat *et al.* (1990). (B) After E. Mañanós, S. Zanuy, M. Carrillo, J. Nuñez & F. LeMenn, (unpublished results). (C) Redrawn and modified from Prat *et al.* (1990).

steroid were found in mature females, but in the conjugated rather than the free form, suggesting that the former may be the mediator of oocyte maturation in sea bass (Scott et al. 1990). However, it is also possible that other steroids may be acting as the MIS in this species.

Male fish commence spermiation 2 months in advance of the spawning time of female fish, coinciding with the first elevation of the gonadosomatic index (GSI). Spermiation ends up to 1 month after the spawning period of the corresponding group of females when the GSI is low. The first increase of plasma 11-ketotestosterone coincides with the presence of spermiating males, peaking at the beginning of the spawning period but decreasing immediately afterwards, suggesting that this hormone could be related to the initiation of spermiation in sea bass (Prat et al. 1990) (Fig. 7.2C).

Seasonal changes of plasma lipids (Fernández et al. 1989), insulin (Gutiérrez et al. 1987) and thyroid hormones (Carrillo et al. 1991a) have been described and their relationship to the reproductive cycle of sea bass has been reviewed (Zanuy et al. 1993). In general, plasma lipids and insulin levels are increased during vitellogenesis (autumn) whereas thyroid hormones are raised during spawning. Different seasonal patterns were observed for plasma cortisol (Planas et al. 1990), coinciding with natural sea water temperatures and photoperiod cycles, being highest during summer and lowest during winter. As a result, levels of plasma cortisol decrease during vitellogenesis, attaining their lowest values during spawning. High levels of plasma cortisol during spawning have been associated with thermic stress and low quality spawnings (M. Carrillo, S. Zanuy, J. Planas, F. Prat and N. Bromage, unpublished results).

Collectively these data suggest that the reproductive process in sea bass is a complex mechanism involving the participation of many hormones and that the interrelationships of nutrition and reproduction and stress and reproduction are very important for the reproductive success of these species.

7.3 Hormonal and environmental induction spawning

7.3.1 Hormonal induction of spawning

Sea bass breed normally in captivity, but methods for inducing spawning have been developed in order to synchronize the production of eggs and sperm. Most of the methods for inducing spawning in sea bass involve injections of hCG (human chorionic gonadotropin) or LHRHa (luteinizing hormone releasing hormone analogue). The first studies of hormonal synchronization of spawning in sea bass used hCG (Alessio et al. 1976, Villani 1976, Barnabé 1980, Barnabé & Paris 1984, Zohar et al. 1984). In general at 11–13°C, two injections of hCG, 6 h apart, at doses of 800–1000 IU per kg (or even lower) produce a strong oocyte hydration 1 or 2 days after the injection. Spawning occurs 48–78 h after the first injection.

However, there are a number of problems with this treatment. Only 50% of the females respond to the treatment, especially if only one injection is administered; secondly, no response occurs at water temperatures above 17°C. There are also immunological reactions of the broodfish to hCG and often the eggs produced are of low variable quality (<80% of viable eggs).

These difficulties and those relating to the induction of out-of-season spawning in photoperiodically-treated animals (see section on the use of constant day lengths) have been largely overcome by the use of LHRHa (Barnabé & Barnabé-Quet 1985, Zanuy et al. 1986, Bouget & De La Gándara 1987, Devauchelle & Coves 1988, Carrillo et al. 1989, Carrillo et al. 1991b). However, LHRHa, although effective in accelerating final maturation and inducing spontaneous egg release in photo-induced animals at high temperatures (Carrillo et al. 1991b), is unable to stimulate the earlier stages of gametogenesis without prior use of photoperiod (Devauchelle & Coves 1988). More detailed studies are needed to investigate and confirm this last aspect, including the possible use of LHRHa implants (Zohar et al. this volume Chapter 5).

An optimized procedure for spawning induction with LHRHa during the natural reproductive period involves two intraperitoneal injections of LHRHa ($5\,\mu g\,kg^{-1}$ and $10\,\mu g\,kg^{-1}$), 4–6 h apart, administered to females with ovaries in which the egg diameters are at least 650 μm (Figs 7.3, 7.4). Spawning generally

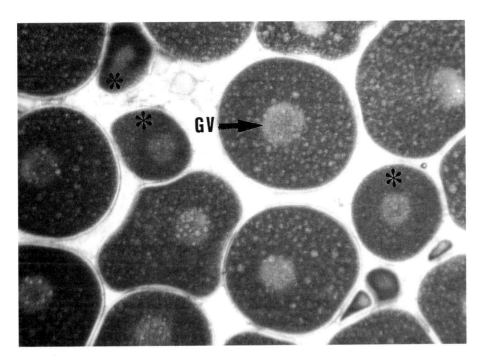

Fig. 7.3 Photomicrograph of late-vitellogenic oocytes (700 μm diameter) ready to be artificially induced for maturation, ovulation and spawning using LHRHa administration. Oocytes were obtained by intraovarian cannulation and then observed under the microscope. A round and clear germinal vesicle (GV) appears in the centre of the cell. Oocytes in earlier stages of vitellogenesis can also be observed (*).

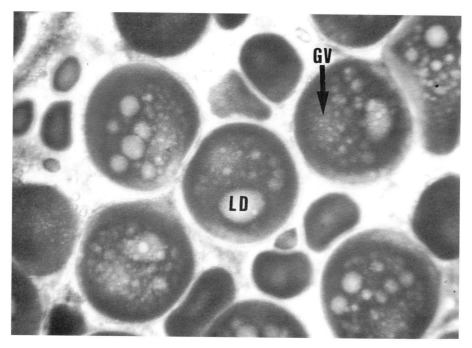

Fig. 7.4 Photomicrograph of maturing oocytes (900 μm diameter) obtained by intraovarian cannulation and then observed under the microscope after LHRHa administration. The germinal vesicle (GV) is near the animal pole and appears less conspicuous than the lipid droplets (LD) which have started to fuse.

occurs 72 h after the first injection (S. Zanuy and M. Carrillo, unpublished results). The influence of time of day on the ovarian activity of LHRHa-treated sea bass has also been assessed in order to optimize the induction of ovulation by LHRHa (Alvariño et al. 1992b). Generally, almost 100% of the injected fish respond to LHRHa administration and the quality of the eggs and larvae obtained is similar to that in the controls; this could explain why anti-dopaminergic drugs, often used in association with LHRHa in the spawning induction of other fish, have only rarely been investigated in sea bass.

Notwithstanding this, Prat et al. (1987) studied the physiological effects of LHRHa administered alone or combined with pimozide (PIM). When used with LHRHa, peaks of E2 and T appeared 12 h and 24 h respectively after the first injection. More delayed peaks were observed in fish that received pimozide in combination with LHRHa, either in the first or in the second injection (Fig. 7.5A,B). The percentage of maturing oocytes exibiting germinal vesicle migration was also reduced. In addition, the increase in percentage of oocytes with germinal vesicle breakdown (GVBD) occurred earlier and at a faster rate in fish that received LHRHa alone than in those receiving a combination of LHRHa and pimozide (Fig. 7.5C). Spawnings were obtained between 84 h and 108 h after the first injection.

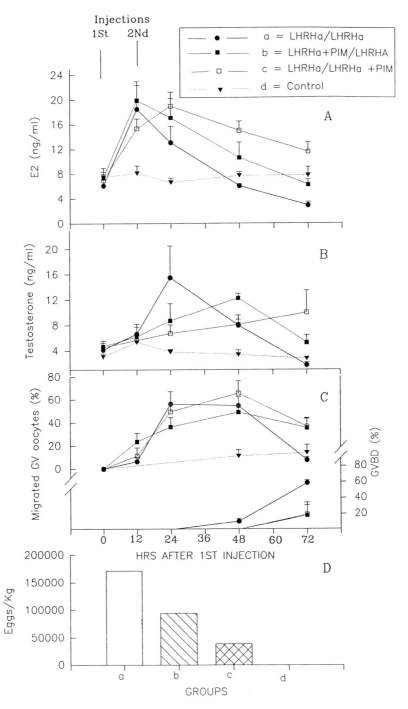

Fig. 7.5 Hormonal profiles of gonadal steroids and relative fecundity in sea bass after two injections (12 h apart) of LHRHa alone (1st injection = $5\,\mu g\,kg^{-1}$, 2nd injection = $20\,\mu g\,kg^{-1}$) or combined with pimozide ($10\,mg\,kg^{-1}$) [D-Ala^6, Pro^9-NEt]-LHRH = LHRHa; GVBD = Germinal Vesicle Breakdown; Relative Fecundity = no. of eggs per kg post-spawning weight. Symbols represent mean of five animals; vertical lines, standard error of mean. (F. Prat, M. Carrillo & S. Zanuy unpublished results.)

148 *Broodstock Management and Egg and Larval Quality*

The percentage of fish responding to the treatments, and their relative fecundities, were also reduced in the pimozide treated group. The worst results were obtained in those fish that received pimozide in the second injection and the best in those that received LHRHa alone (Fig. 7.5D). Levels of 17α20β-P did not show any significant change throughout the experiment (results not shown).

These data indicate that changes in the hormonal profiles induced by pimozide administration produce alterations in the rates of oocyte maturation and ovulation which in turn can affect the quality of the spawnings. It seems that in the sea bass, as in some other marine species (Copeland & Thomas 1989), the dopaminergic inhibition of the gonadotropin release is only poorly developed. Consequently, for commercial practice it is recommended that pimozide or other dopamine inhibitors are not used for spawning induction.

7.3.2 Environmental control of reproduction

Manipulation of natural cycles of day length and temperature

Sea bass inhabiting higher and lower latitudes spawn in response to changing day length. As a consequence, seasonal reproduction of wild stocks has considerable adaptative significance, but is limited to a few months in the year. This is a major disadvantage for intensive farming, where supplies of eggs are required throughout the year. However, the environmental cues can be manipulated to alter spawning times of sea bass broodstock and hence solve the ever-increasing and aseasonal demands for eggs and fry. Manipulations of photoperiod and temperature are the main environmental factors used to alter the spawning time of sea bass. Unfortunately, many of these studies have been performed by varying temperature and photoperiod together. This has made it difficult to define the individual roles these environmental factors might have in reproduction.

The first evidence of an influence of temperature on spawning time came from long-term observations of the reproductive cycle under ambient conditions. In general, in years when it is cold, natural spawnings are advanced, and conversely, in years when it is warmer, spawnings are retarded (Fig. 7.6A). By keeping sea bass broodstocks under natural photoperiod, but heating the water during autumn and winter (Zanuy et al. 1986), spawning is delayed by 1 month with respect to controls (Fig. 7.6B). However, in only 60% of the fish did natural spawning occur. Furthermore, the quality of spawnings was low and some animals became sexually regressed.

Further delays of spawning into May resulted from combining long photoperiods with high temperatures, but here spontaneous spawnings did not occur at all; the fish had to be injected with LHRHa (results not shown). Zanuy et al. (1986) suggested that temperatures above 17°C (i.e. those usually found from

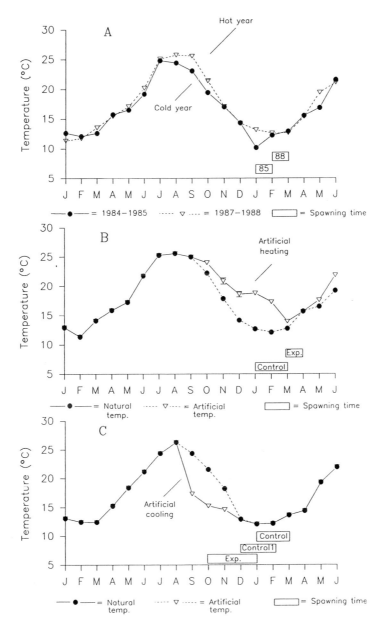

Fig. 7.6 Effects of sea water temperatures on the timing of spawning of sea bass. (A) Indoor natural monthly mean temperatures (NP) and natural photoperiod (NP) (eastern Spain, latitude 40°N and longitude 0°). (B) NP and artificial heated waters (after Zanuy et al., 1986). (C) Artificial constant photoperiod (AP):LD 9:15 (9 h light followed by 15 h dark in each 24 h cycle) until March followed by LD 15:9 until April and then LD 9:15 (not shown) and artificial cooled waters. (After Carrillo et al. 1991b.) Control = NP and NT; Control 2 = AP and NT.

May onwards) impede spontaneous spawning of sea bass despite the presence of mature oocytes in the gonads.

Artificial decreases in temperature (below 17°C) allow spontaneous spawnings of photo-induced mature animals in October. This procedure enables viable eggs to be obtained 3 months in advance of controls (Fig. 7.6C). In fact, the quality of spawnings (percentage of floating eggs), the spawning spread, relative fecundity and spawning index (number of spawnings per fish) of the experimental group were similar to or even higher than the controls (respectively 72.8%, 107 days, 479 000 eggs per kg and 2.09 spawns per fish compared with 77.3%, 45 days, 285 000 eggs per kg and 1.17 spawns per fish, for the controls; Carrillo et al. 1991b). The photo-induced group maintained under natural temperatures started spawning only 1 month before the controls, coinciding with the natural decrease of sea water temperatures to 17°C, which occurs during December (Fig. 7.6C). This confirms previous findings of Zanuy et al. (1986) and Devauchelle & Coves (1988), in which natural ovulations occurred only when water temperatures were in the range of 9–17°C.

Spawning time and rates of maturation of sea bass have been altered by exposing fish to modified seasonally-changing light and/or temperature cycles (Girin & Devauchelle 1978, Barnabé & Paris 1984, Devauchelle 1984, Barnabé & Barnabé-Quet 1985, Coves 1985, Zanuy et al. 1986) with advanced or delayed spawnings following compression or extension of the natural light cycle into respectively shorter or longer time periods than a year. Once the desired out-of-season spawning time is achieved, it can be maintained by a phase-shifted 12 month seasonally-changing photoperiod cycle (Bye 1987).

Use of constant photoperiods

Exposure of fish to 1 month of long days, 15 h light followed by 9 h dark in each 24 h cycle (LD 15:9), in March, April, May, June or July with constant short days (LD 9:15) for the remainder of the year speeds up the rates of maturation, (Carrillo et al. 1989, 1991b, 1993, Bromage et al. 1993a) thus increasing and advancing the proportions of oocytes entering vitellogenesis during October and November; it also brings forward the timing of ovulation and spawning. If the exposure to long days is carried out in August, the timing of vitellogenesis, maturation and spawning is similar to controls. When fish are exposed to long days from September onwards a clear delay in spawning is obtained (Fig. 7.7).

In general exposure to one month of long days during the earlier part of the reproductive cycle advances vitellogenesis, maturation and spawning time of sea bass. The maximal advance occurs when fish are exposed to long days around the time of spawning; the effect diminishes, thereafter, until long days are given at the time of the summer solstice, after which the same photoperiod produces a delay in spawning time (Carrillo et al. 1993).

It is interesting to note that the sequence of hormonal and ovarian changes

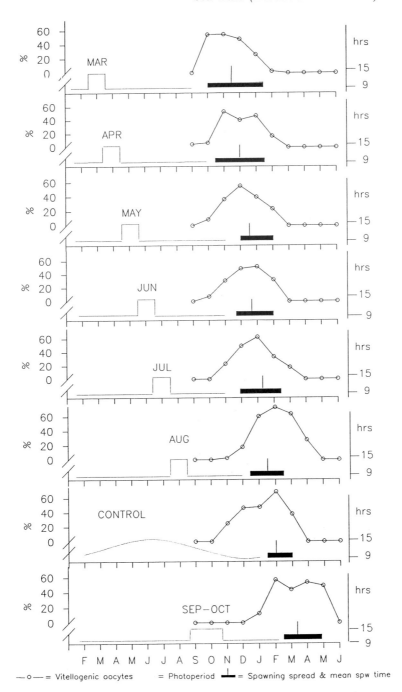

Fig. 7.7 Effects of 1 or 2 months of long photoperiods on the proportion of vitellogenic oocytes and the timing of spawning of sea bass. Vertical lines = mean spawning time. (After Carrillo et al. 1991b.)

induced by the constant photoperiod are similar to those provided by the natural seasonal light cycles. These results parallel those found in salmonids (Bromage *et al.* 1992, 1993a,b).

The profiles of change of the gonadal steroids in plasma of sea bass are also affected by exposure to long days at different times in the annual cycle. Thus, long days in May produce a significant advance in the surge of E2, T and calcium levels when compared with controls. Less pronounced changes are observed in groups exposed to long days in June or July. In each case vitellogenin (Vg) incorporation into the oocytes begins when plasma Ca^{++} levels attain 14–5 mg% (Zanuy *et al.* 1989, Prat 1991). Mañanós *et al.* (1992) confirmed and extended these findings. These authors showed that the displacement of the spawning period was correlated with oocyte development and the plasma profiles of Vg and E2.

An important finding of these studies is that there were no significant reductions in the quality of the eggs and fry and/or the fecundity of the broodstock from fish which had been photoperiodically-advanced (Girin & Devauchelle 1978, Barnabé & Paris 1984, Devauchelle 1984, 1987, Barnabé & Barnabé-Quet 1985, Devauchelle & Coves 1988, Carrillo *et al.* 1989). In contrast, Zanuy *et al.* (1986), Devauchelle & Coves (1988) and Carrillo *et al.* (1991b), all reported poorer fecundities and egg survivals in photoperiodically-delayed fish. Zanuy *et al.* (1986) and Devauchelle & Coves (1988) have suggested that high sea water temperatures were the reason for the poor survivals of progeny. Natural high temperatures can reduce the quality of spawning; this can be overcome by artificial decreases in sea water temperatures (Carrillo *et al.* 1991b). Possibly the mis-match in photoperiod and temperature regimes may significantly affect the neuroendocrine system of the fish to produce poor quality offspring.

The biochemical composition of the eggs may also be affected by unsuitable environmental conditions (Fig. 7.8). Significantly increased levels of total lipids have been reported for eggs spawned outside the normal spawning season; these eggs also showed very poor viability (Devauchelle *et al.* 1982, Serrano *et al.* 1989, Carrillo *et al.* 1991b) and low hatching rates and larval survivals up to first feeding (Serrano *et al.* 1989, Carrillo *et al.* 1991b). This subject clearly requires further investigation.

7.3.3 Endogenous rhythms

An important finding regarding sea bass reproduction is that different light cues are not required for the completion of maturation. Spawning still occurs in fish which are maintained under constant long and short days over successive reproductive cycles (Zanuy *et al.* 1991, Carrillo *et al.* 1992). Under constant short days, the rhythm controlling maturation in sea bass free runs with a periodicity approximating to but significantly different from a year (Fig. 7.9A). These facts strongly suggest that the timing of reproduction of sea bass is being coordinated by an endogenous process.

Sea Bass (Dicentrarchus labrax) 153

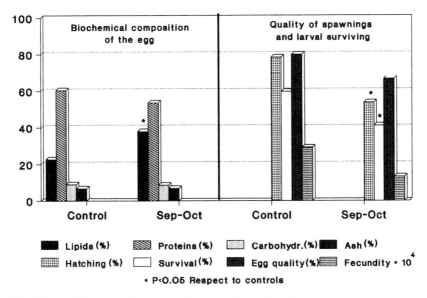

Fig. 7.8 Effects of 2 months of long days (Sep−Oct) in an otherwise constant short day regime on spawning performance and quality of eggs and larvae of sea bass. (After Carrillo et al. 1991b.)

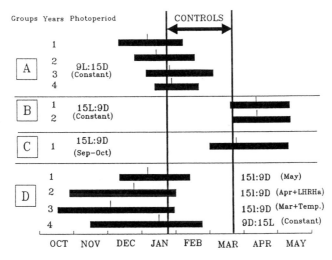

Fig. 7.9 (A) Effects on time of spawning of constant short days (LD 9:15) applied over four consecutive years. (B) Constant long days (LD 15:9) over two consecutive years. (C) Two months of long days (Sep−Oct) in otherwise constant short days over 1 year. (D) One month of constant long days in May (first year); one month of constant long days in April (plus one i.p. injection of 10 µg LHRHa per kg) (second year); 1 month of constant long days in March (plus decrease of sea water temperatures below 16°C) (third year) in an otherwise constant short day regime. Vertical bars represent the mean spawning time. Horizontal bars represent the spawning spread.

Endogenous rhythms of reproduction have also been described for the catfish (Sundararaj et al. 1982), stickleback (Baggerman 1980) and rainbow trout (Duston & Bromage 1986). Different results were obtained when sea bass were exposed to constant long days (LD 15:9) over two successive yearly cycles. In this case delays in spawning time of sea bass were produced (Fig. 7.9B) but again the periodicity of the rhythm over the 2 years of study approximated to, but was significantly different from, a year (Zanuy et al. 1991).

Similar delays in spawning were also observed in fish exposed to 2 months of constant long days from September to October in an otherwise constant short day regime (Fig. 7.9C). An important consideration is that the endogenous rhythms can be entrained by alterations in photoperiod, e.g. 1 month of long days in otherwise constant short days. That is, earlier exposure of long days in the sexual cycle can advance spawning time. However, returning the fish to constant short days leads to a free-running of the reproductive rhythm (Fig. 7.9D).

7.4 Broodstock nutrition and effects on the quality of the progeny

7.4.1 Effects of ration

It is well known that supplies of food are an important determinant of the reproductive potential of fish. However, there are few studies of the effects of ration on fecundity and egg quality in cultured fish and furthermore the results of many of the studies are not very clear (see Springate et al. 1985). This situation is even worse for sea bass, where this aspect has been completely neglected.

Cerdá et al. (1990a) demonstrated that groups of female sea bass (1 kg body weight), fed with a natural diet (trash fish, *Boops boops*) at either full ration adjusted in line with sea water temperatures (from 0.70 to 1.41% of body weight, average 1.05%) or half ration (from 0.17 to 0.70% of body weight, average 0.45%), exhibited significant changes in fecundity (Fig. 7.10). The most significant effects of feeding a low ration to female sea bass were a decrease in egg size and an increase in relative fecundity.

These results parallel findings in other species of teleosts (Springate et al. 1985, Bromage & Cumaranatunga 1988). Full explanation of these results is complicated, however, because fecundity in many fish is also influenced by fish size. The quality of the eggs (percentage floating) was similar in both groups (results not shown) but there was a delay of around 2 weeks and a longer spread of spawning in the fish on half ration. In addition to spawning spread, the number of spawnings per female was elevated in the fish on half ration.

7.4.2 Effects of quality of diet

Studies of the effects of different dietary constituents on egg production and

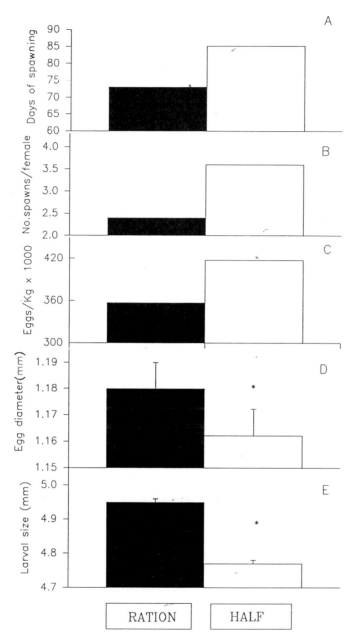

Fig. 7.10 Effects of half and full ration of daily diet on spawning spread, spawning index (No. of spawnings/female), relative fecundity and egg and larval size of sea bass. Vertical bars indicate standard error of mean. * = $P < 0.05$ with respect to full ration. (After Cerdá et al. 1990a.)

quality have prior to this volume only been carried out on a few cultivated fish including a few species of salmonid fish, the carp and Red Sea bream (reviewed by Hardy 1984, Watanabe 1985, Luquet & Watanabe 1986, Watanabe & Kiron this volume Chapter 16). The paucity of information regarding the interaction of nutrition and reproduction in commercially-important marine fish is partially related to the time required for completion of reproduction in captivity. However, the ever-increasing demands of fish farmers for mass production of eggs and the need to know whether nutritional deficiencies may be a significant cause of the poor reproductive performance of a number of cultivated fish means that such studies are now essential.

The effects of natural and commercial diets on growth, metabolism and hormonal regulation of sea bass have also been little studied (Zanuy et al. 1989) and there is no information on the effects of these diets on reproductive performance. Preliminary results of Cerdá et al. (1990b) showed that sea bass broodstock fed with commercial trout diets grew at a slower rate and exhibited significant decreases in relative fecundity, spawning index and spawning spread when compared with control fish which were fed on trash fish. The eggs on the experimental diets were also smaller and produced lower hatch rates and larval survivals. Spawning time was also delayed (Fig. 7.11).

An important effect of the variations in the constituents of the diet was an increase in the levels of atresia in the ovaries of pre-spawning fish. This would explain, in part, the observed alterations in fecundity and egg size. In order to confirm and extend these findings a series of experiments are in progress, changing specifically the constituents of the diets and the times at which they are fed during the year-long cycle. This, together with biochemical and endocrinological approaches applied to broodfish and to eggs and larvae, should help to understand the complex interactions between nutrition and reproduction in fish.

7.5 Egg and larval quality

7.5.1 Assessments of quality

It is thought that good broodstock management produces eggs of good quality and, in turn, good eggs are a guarantee that the maximum number of larvae and fry will survive. However, the term 'egg quality' needs to be defined in order to obtain a reasonable prediction of the larval and fry performances, otherwise much staff time and resources will be wasted in the hatchery.

At present, definitions of quality of the eggs and larvae are rather difficult because there is a lack of standardization of the methods and criteria used in such studies (Kjorsvik et al. 1990, Bromage et al. 1992, 1994; see also N. Bromage this volume Chapter 1). Furthermore there are no clear-cut relationships between the morphological and biochemical characteristics of the eggs and the subsequent survival rates of larvae and fry. Egg and larval quality assessments in sea bass are no exception. In fact a variety of criteria have been used for evaluating egg quality in the sea bass, including viability, bouyancy, fertilization and hatching rates, the number and size of the lipid droplets, egg size, wet and dry weights, incidence of deformities, biochemical composition,

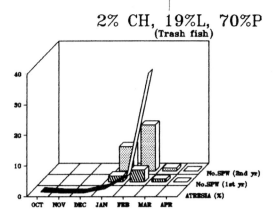
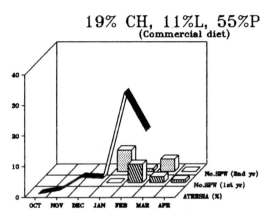
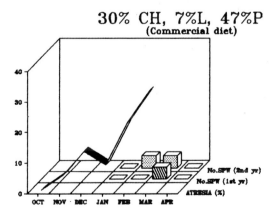

Fig. 7.11 Effects of diet composition administered over two consecutive years on the periodicity of spawnings and on the percentage of atretic oocytes obtained by intraovarian cannulation. Bottom diet on the second year consisted of 19% CH, 11%L and 55% P. (After Cerdá *et al.* 1990b.)

and more recently levels of hormones. Most of these studies have used fertilization rates, egg buoyancy, hatching rate and survival to specific developmental stages as criteria for assessing egg quality.

Although fecundity has been related to egg size (Cerdá *et al.* 1990a,b), egg diameter is not a good indicator of egg and larval quality. More reliable criteria are possibly to be found with the biochemical composition of the eggs. In fact the total lipid content of the eggs has been correlated with egg and larval viability following alterations in spawning time (Devauchelle *et al.* 1982, Serrano *et al.* 1989, Carrillo *et al.* 1991b).

Positive and negative correlations have been found between egg dry weight and larval deformities and hatching rates respectively (Devauchelle & Coves 1988). Marangos *et al.* (1985) observed a positive correlation between the levels of free amino acids found in the larvae of sea bass and salinity. This could be a criterion for assessing larval quality considering the role of amino acids in osmoregulation in this species. Despite these studies there is still a lack of knowledge of the importance of other egg quality characteristics, including pigments, micronutrients, chromosomal aberrations, etc. (Fig. 7.12; see also Fig. 1.7 Chapter 1), all of which may be correlated with larval viability. For many marine fish, including the sea bass, heavy mortalities often occur at the time of yolk sac reabsorption and first feeding.

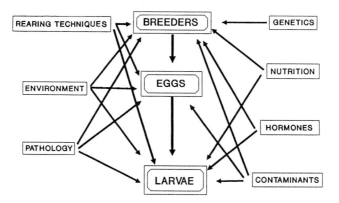

Fig. 7.12 Schematic representation of the factors affecting egg and larval quality in the sea bass.

7.5.2 Factors affecting larval development

In sea bass yolk sac reabsorption occurs 7–10 days after hatching, depending on the temperature. From hatching to yolk sac reabsorption larvae must be kept in complete darkness to inhibit feeding behaviour. In addition, the water surface must be kept clean to allow air to be gulped for swimbladder inflation (Chatain 1991a,b). During the period between yolk sac reabsorption and 30 days of age, larvae must be fed with live food. Weaning generally is performed when larvae are 40 days old (Chatain 1991a,b).

Numerous morphological and behavioural abnormalities have been described during development of larvae from hatching up to 40 days of age (Barahona-Fernández 1982, Johnson & Katavic 1984, Coves 1985). These malformations affect up to 97% of fish stocks in some hatcheries. Recently, rearing techniques aimed to optimize stocking density, the intensity and quality of light, photoperiod, tank colour and water quality have considerably reduced larval abnormalities and produced survivals approaching 50% (Chatain 1991a). Many finfish hatcheries have used the presence of a normal swimbladder and the absence of skeletal abnormalities as reliable criteria to define fry quality.

Although the swimbladder problem has been minimized by improvements in system design and husbandry, lordosis and scoliosis still affect a significant proportion of the population. Scoliosis has usually been associated with the absence of a normal swimbladder induced by muscular pressure due to an erratic swimming behaviour (Chatain 1991b). Nevertheless, other factors, e.g. nutritional deficiencies, can induce morphological abnormalities.

Navarro (1990) reported that larvae of sea bass fed with *Artemia* lacking some essential polyunsaturated fatty acids (PUFA) (e.g. 20:5w3) had developmental malformations, exhibited abnormal swimming behaviour and experienced massive mortalities 10–24 days after feeding with the deficient diets. These problems have been overcome by feeding with nauplii of *Artemia* enriched with PUFA (mainly 22:6 n3) (Van Balaer *et al.* 1985) or by alternately feeding with PUFA-enriched and un-enriched *Artemia* strains (Navarro *et al.* 1988). This last approach is attractive to fish farmers who have only limited supplies of good quality *Artemia*. Notwithstanding these advances, use of live food is rather expensive owing to the extra manpower and equipment requirements and the necessity for strict sanitary control.

In recent years much effort has been expended in trying to find suitable artificial diets, but still with poor results. The main problems encountered have been cannibalism, disease and low survivals. Despite the widespread use of artificial diets (dry and wet pellets) commencing 40 days after hatching, the formulation of these diets has not been shown to meet fully the nutritive requirements of the larvae. Thus a combination of artificial and live foods including copepods (*Eurytemora velox*), lagoonal plankton (*Daphnia pulex*) or saline plankton (*Artemia*) have had to be used. These give survivals of 90% and above (Barnabé 1991).

The development of microcapsules and microparticles for early weaning has made rapid progress in Europe and Japan, and such microdiets have been used with 20 day old (2–3 mg) sea bass larvae. However, on their own they result in survivals of only 40% (Barnabé 1984, 1987, Person-Le Ruyet *et al.* 1993). Despite such advances the weaning of marine fish larvae at first feeding with microdiets remains difficult. Adequate formulation with highly-digestible components is essential because the larval digestive system is not fully developed and functional. At the same time the specific nutritive requirements of larvae are still not fully understood.

7.6 Holding tanks/facilities and management

7.6.1 Broodstock

Sea bass broodfish require holding tanks according to their size and stocking density. There are three characteristic designs in commercial hatcheries, large, medium and small volume systems. The large systems, widely used in Japan and south-east Asia, utilize large outdoor concrete tanks of $50-100\,m^3$ capacity to maintain broodstock. Medium volume systems, widely used in Europe, utilize indoor circular tanks of $15-30\,m^3$ with sophisticated control systems for filtering, heating and cooling. Small volume systems, typical of the Mediterranean area, utilize round tanks of $10-20\,m^3$ with a complete control of environmental condition (Lisac 1988). In general, round dark-coloured tanks, supplied by well-aerated circulating sea water (salinity 35–37‰) provide optimal indoor experimental conditions for sea bass broodstock (Fig. 7.13). Breeding programmes in commercial hatcheries need a mass production of eggs throughout the year. Currently, supplies are restricted to a few months from January to March in most Mediterranean areas. To meet this requirement, light-proof tanks provided with tungsten bulbs or fluorescent lamps, controlled by electronic time clocks, are used to alter the time of spawning of sea bass.

Another important facility is the heat bomb or exchanger which cools seawater

Fig. 7.13 Light-proof tanks for sea bass broodfish at the Instituto de Acuicultura de Torre de la Sal (IATS), Spain.

to temperatures below 16°C for hatchery programmes designed to provide out-of-season eggs, especially during the summer, autumn or spring, when ambient temperatures are elevated. In practice, several groups of broodfish are usually maintained under different photoperiod regimes in order to obtain spawnings every month in such a way that individual groups spawn only once a year. One aspect that has provoked important changes in the quantity and quality of spawnings is the realization that acute and chronic stress interferes with normal maturation (Zanuy et al. 1986, Carrillo et al. 1989, 1991b, 1993, Planas et al. 1990).

As a consequence, it is now considered important to keep broodstocks in isolated units in order to avoid external disturbances during spawning and allow full environmental control.

The absence of a well-formulated diet for sea bass broodstock that will guarantee good egg quality and reasonable larval survival, has meant that aquaculturists have had to depend on natural food (e.g. trash fish) fed alone or in combination with artificial diets. This particular situation has restricted the holding facilities with respect to the use of automatic feeders and requires the broodfish to be hand-fed.

In fish, it is commonly accepted that fecundity and egg quality are always lower at first spawning or puberty than in successive spawning cycles. This is also the case with sea bass. Females first spawn in captivity at the age of 3 years and an average body weight of 700 g. The quality of these initial spawnings are usually poor. Males are more precocious than females and spawn at two years of age at an average body weight of 200 g. The quality of spawning in older females (8+ years) also decreases. Consequently, in programmes of large scale egg production it is advisable to use females between 4 and 8 years of age (second to fifth spawning cycles in the Mediterranean area). This directly affects the size of the holding facilities.

Equal numbers of both sexes are distributed in discrete units. More specifically, ten female fish of 0.8–3.0 kg weight are maintained with 10 male fish of 0.3–1.5 kg. in circular PVC tanks of 2000 l capacity (Fig. 7.11) provided with running, well aerated sea water (37‰ salinity). Photoperiod is controlled by electronic clocks connected to fluorescent lamps giving 480 lux at the water surface. Animals are hand-fed with trash fish.

7.6.2 Egg collection and larval rearing

The outflows of broodstock holding tanks are fitted with fine mesh egg collectors. These are checked daily, and following spawning the contents are transferred to graduated 2 l measuring cylinders. In this way the volume of good quality (floating) and non-viable (sinking) eggs can be determined.

For egg incubation many laboratories utilize PVC containers of 5 l capacity, thus increasing the overall number of containers but spreading the risks of simultaneously losing all eggs. These are fitted with fine mesh bottoms which allow adequate water circulation but at the same time retain eggs and larvae in the units. Each incubator has an independent water and air supply. These incubation units are immersed in large rectangular 500 l tanks filled with filtered, sterilized sea water in a closed system (Fig. 7.14). Sea water is gradually heated from 12°C to 16°C and maintained constant at 16°C throughout larval rearing. Approximately 600–1000 eggs are transferred to each incubation unit to evaluate survivals to hatch and first-feeding. Hatching under these conditions occurs at 4 days post-fertilization and yolk sac reabsorption (first-feeding) 8 days after hatching.

Larval rearing is a very delicate procedure which requires special holding facilities and proper supplies of food according to the size of the larvae. Lissac (1988), has outlined the structural components of hatcheries for sea bass. In brief, large volume hatcheries possess big rectangular rearing tanks (50–100 m^3) with a large area (around 50%) dedicated to live food production such us phytoplankton and rotifers. Larvae are fed with rotifers (the Artemia food step is usually omitted) and subsequently weaned on chopped trash fish and shrimp.

In a medium volume hatchery larval rearing is performed in round 7–20 m^3 capacity tanks provided with filtered and heated sea water. Live food is produced in many small units; rotifers and *Artemia* in tanks of 0.5–1.5 m^3 and phytoplankton in 200–300 l PVC bags occupying 20% of the total hatching area.

Fig. 7.14 Small incubators (5 l capacity) immersed in large tanks provided with an independent water and air supply at the IATS for egg and larval quality assessment.

Larvae are fed with rotifers and *Artemia* before weaning. Finally, in a small volume hatchery larval rearing is performed in circular tanks of 2–4 m^3 with conical bottoms and rotifers and *Artemia* are cultured in tanks of 1–4 m^3, and phytoplankton in small PVC bags of 25–80 l, all with complete environmental control.

This laboratory utilizes San Francisco Bay (USA) or La Mata (Spain) *Artemia* strains for feeding sea bass larvae (Fig. 7.15). Cysts are decapsulated with NaOCl (1 g NaOCl per 2 g cyst). Eclosion takes place in sea water (salinity 38‰) and the nauplii obtained are fed to the larvae at a concentration of 10–200 nauplii per larva.

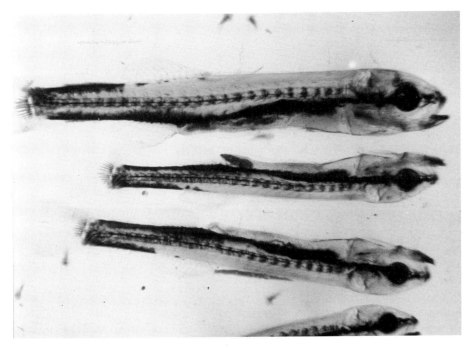

Fig. 7.15 Sea bass larva after yolk sac reabsorption with nauplii of *Artemia salina* provided as live food.

7.7 Future prospects

By using constant light cycles it is possible to reliably induce spawning of sea bass and produce eggs for at least 8 months of the year. For much of the year the eggs and larvae from these altered spawnings are of good quality. However, when water temperatures exceed 17°C egg quality is reduced and hence, for commercial production, some form of temperature control is required.

A further problem that needs to be solved is the elevated proportion of males obtained by hatcheries and their poor growth performance compared with females. A study of sex control is required in order to determine the labile period for artificial sex reversal, the appropriate experimental conditions for ploidy manipulations, e.g. gynogenesis to produce monosex, and finally the environmental and social influences during larval rearing which affect the sex ratio of sea bass. Other fundamental studies, largely neglected, are those relating to puberty and its control in sea bass. These studies are needed to prevent the appearance of premature males and to delay puberty in order to obtain improved growth without interference with gonadal development.

The studies relating to the effects of broodstock nutrition on the quality of spawnings and progeny are also very promising. A direct consequence of these studies could be a net increase in the efficiency of egg and larval production by hatcheries. There is also an urgent need for further genetic studies on sea bass, including the initiation of broodstock selection programmes and procedures for avoiding inbreeding. Finally, the early weaning of sea bass onto microdiets is still possible and rearing depends on supplies of live food. Further studies on the nutritional physiology of larvae are needed for the proper formulation of the diets.

Acknowledgements

This work was supported by the CICYT Project MAR 88-0231, BRIDGE Project BIOT 0188, FAR Project AQ 2.406 and AIR Project AIR 2-CT93-1005 (DG 14 SSMA).

References

Alessio, G., Gandolfi, G. & Schreiber, E.B. (1976) Induction de la ponte, élevage et alimentation des larves et des alevins des poissons euryhalins. *Etudes Revues Conseil Générale Péches Méditerranean, FAO*, **55**, 143–57.

Alvariño, J.M.R., Carrillo, M., Zanuy, S., Prat, F. & Mañanós, E. (1992a) Pattern of sea bass development after ovarian stimulation by LHRHa. *Journal of Fish Biology*, **41**, 965–70.

Alvariño, J.M.R., Zanuy, S., Prat, F., Carrillo, M. & Mañanós, E. (1992b) Stimulation of ovulation and steroid secretion by LHRHa injection in the sea bass (*Dicentrarchus labrax*): effect of time of day. *Aquaculture*, **102**, 177–86.

Baggerman, B. (1980) Photoperiodic and endogenous control of the annual reproductive cycle in teleost fishes. In *Environmental Physiology of Fishes*, (ed. M.A. Ali), pp. 553–68. Plenum Press, New York and London.

Barahona-Fernández, M.-T. (1982) Body deformation in hatchery reared European sea bass *Dicentrarchus labrax* (L.). Types, prevalence and effect on fish survival. *Journal of Fish Biology*, **21**, 239–49.

Barnabé, G. (1973) Contribution à la connaissance de la croissance et de la sexualité du loup (*Dicentrarchus labrax*) de la région de Sète. *Annales Institut Océanographique*, **49**, 49–75.

Barnabé, G. (1980) Exposé synoptique des données biologiques sur le loup ou bar (*Dicentrarchus labrax*). *Synopses FAO Pêches*, **126**.

Barnabé, G. (1984) Aliment composé pour animaux aquatiques et procédés et dispositifs pour élever des larves de poisson. *Brevet USTL* no 84/17155, BNPI, Paris.

Barnabé, G. (1987) L'élevage du loup et de la daurade: Aspects, techniques et réalités économiques. *Océanis*, **13**(1), 59–68.

Barnabé G. (1991) La cría de lubina y de dorada. In *Acuicultura* (Coordinated by G. Barnabé), pp. 573–612. Omega, S.A., Barcelona.

Barnabé, G. & Barnabé-Quet, R. (1985) Avancement et amélioration de laponte induite chez le loup *Dicentrarchus labrax* (L.) à l'aide d'un analogue de LHRH injecté. *Aquaculture*, **49**, 125–32.

Barnabé, G. & Paris, J. (1984) Ponte avancée et ponte normale du loup *Dicentrarchus labrax* (L.) a la Station de Biologie Marine et Lagunaire de Sète. In *L'Aquaculture du Bar et des Sparidés* (eds. G. Barnabé & R. Billard), pp. 63–72. INRA, Paris.

Blázquez, M., Piferrer, F., Zanuy, S. & Carrillo, M. (1992) Identification of the labile period for sex steroid-induced gonadal differentiation in sea bass (*Dicentrarchus labrax* L.). *Abstracts of the Second International Symposium on Fish Endocrinology*, Saint Malo, France.

Bou Ain, A. (1977) *Contribution à l'étude morphologique, anatomique et biologique de* Dicentrarchus labrax *(Linné 1758) et* Dicentrarchus punctatus *(Bloch 1972) des côtes tunisiennes*. Thèse de Doctorat de Spécialité, Faculté des Sciences, Tunis.

Bouget, J.F. & De La Gándara, F. (1987) Seguimiento a escala piloto de la ovogénesis de la lubina (*Dicentrarchus labrax* L.) bajo control de fotoperíodo en diferentes estaciones del año. *Cuad Marisquos Publicacciones Technica*, **12**, 35–40.

Bromage, N. & Cumaranatunga, P.R.T. (1988) Egg production in the rainbow trout. In *Recent Advances in Aquaculture*, Vol. 4 (eds J.F. Muir & R.J. Roberts), pp. 63–138. Croom Helm, Timber Press, London & Sydney Portland, Oregon.

Bromage, N., Carrillo, M. & Zanuy, S. (1988) Light controls spawning in sea bass. *Fish Farming International*, January, 22.

Bromage, N., Jones, J., Randall, C., Thrush, M., Davies, B., Springate, J., Duston, J. & Barker, G. (1992) Broodstock management, fecundity, egg quality and the timing of egg production in the rainbow trout (*Oncorynchus mykiss*). *Aquaculture*, **100**, 141–66.

Bromage, N., Randall, C., Davies, B., Thrush, M., Duston, J., Carrillo, M. & Zanuy, S. (1993a) Photoperiodism and the control of reproduction and development in farmed fish. In *Aquaculture: Fundamental and Applied Research*, (eds Lahlou & P. Vitello), pp. 81–102. American Geophysical Union, Washington.

Bromage, N., Randall, C., Duston, J., Thrush, M. & Jones, J. (1993b) The environmental control of spawning in salmonids. In *Recent Advances in Aquaculture*, Vol. 4 (eds N. Bromage, E.M. Donaldson, M. Carrillo & S. Zanuy, pp. 55–66. Blackwell Science, Oxford.

Bromage, N., Shields, R. Basavaraja, N., Bruce, M., Young, C., Dye, J., Smith, P., Gillespie, M., Gamble, J. & Rana, K. (1994) Egg quality determinants in finfish: the role of overripening with special reference to the timing of stripping in the Atlantic halibut, *Hippoglossus hippoglosus*. *Journal of the World Aquaculture Society*, **25** (in press).

Bruslé, J. & Roblin, C. (1984) Sexualité du loup *Dicentrarchus labrax* en condition d'élevage contrôlé. In *L'Aquaculture du Bar et des Sparidés*, (eds G. Barnabé & R. Billard), pp. 33–43. INRA, Paris.

Bye, V.J. (1987) Environmental management of marine fish reproduction in Europe. *Proceedings of the Third International Symposium on the Reproductive Physiology of Fish*, St. John's, Newfoundland, 289–98.

Cambré, M.L., Verdonck, W., Ollevier, F., Vandesande, F., Batten, T.F.C. & Kühn, E.R. (1986) Immunocytochemical identification and localization of the different cell types in the pituitary of the sea bass (*Dicentrarchus labrax* L.). *General and Comparative Endocrinology*, **61**, 368–75.

Caporiccio, B. & Connes, R. (1977) Etude ultrastructurale des enveloppes périovocytaires et périovulaires de *Dicentrarchus labrax* L. (Poisson Téléostéen). *Annales de Science Naturelle, Paris, Zoologie (12ème Sér.)*, **19**, 351–68.

Carrillo, M., Bromage, N., Zanuy, S., Serrano, R. & Prat, F. (1989) The effect of modifications in photoperiod on spawning time, ovarian development and egg quality in the sea bass (*Dicentrarchus labrax* L.). *Aquaculture*, **81**, 351–65.

Carrillo, M., Zanuy, S. & Kühn, E. (1991a) Seasonal changes in thyroid activity of male sea bass (*Dicentrarchus labrax* Linnaeus 1758) (Perciformes:Serranidae) adapted to different salinities. *Science Marina*, **55**(3), 431–6.

Carrillo, M., Bromage, N., Zanuy, S., Serrano, R. & Ramos, J. (1991b) Egg quality and fecundity in the sea bass (*Dicentrarchus labrax*) and the effects of photoperiodically-induced advances and delays on spawning time. In *Proceedings of the International Symposium on the Repro-*

ductive Physiology of Fish, (eds A.P. Scott, P. Sumpter, D.E. Kime & M.S. Rolfe), pp. 259–61. FishSymp 91, Sheffield.

Carrillo, M., Zanuy, S., Prat, F., Serrano, R. & Bromage, N. (1993) Environmental induction of spawning in sea bass. In *Recent Advances in Aquaculture*, Vol. 4 (eds N. Bromage, E.M. Donaldson, M. Carrillo & S. Zanuy), pp. 43–54. Blackwell Science, Oxford.

Cerdá, J., Ramos, J., Carrillo, M., Zanuy, S. & Serrano, R. (1990a) Efecto de la ración de la dieta suministrada a reproductores de lubina (*Dicentrarchus labrax*, L.) sobre la calindad de la puesta y supervivencia larvaria. *Actas III Congreso Nacionale Acuicultura*, 63–8.

Cerdá, J., Perez, J., Carrillo, M., Zanuy, S. & Serrano, R. (1990b) Efecto de la dieta sobre la calidad de la puesta en reproductores de lubina (*Dicentrarchus labrax* L.). *Actas III Congreso Nacionale Acuicultura*, 57–62.

Chatain, B. (1991a) Current status of the French intensive rearing techniques for sea bass (*Dicentrarchus labrax*) and sea bream (*Sparus auratus*). *Abstracts of 13th Conference of ESCPB on Research for Aquaculture: Fundamental and Applied Aspects.* Antibes 147.

Chatain, B. (1991b) Morpho-anatomical quality of sea bass and sea bream fry cultured in intensive conditions. *Abstracts of 13th Conference of ESCPB on Research for Aquaculture: Fundamental and Applied Aspects.* Antibes, 148.

Copeland, P.A. & Thomas, P. (1989) Control of gonadotropin release in the Atlantic croaker (*Micropogonias undulatus*): evidence for lack of dopaminergic inhibition. *General and Comparative Endocrinology*, **74**, 479–83.

Coves, D. (1985) État actuel de l'élevage du loup en écloserie. *Aqua-Revue*, **3**, 26–30.

Devauchelle, N. (1984) Reproduction decalée au bar (*Dicentrarchus labrax*) et de la daurade (*Sparus aurata*). In *L'Aquaculture du Bar et des Sparidés*, (eds G. Barnabé & R. Billard, pp. 53–62. INRA, Paris.

Devauchelle, N. (1987) Four marine spawners in European hatcheries. *Production controlée en ecloserie. Synthese des Papiers Presentés dans le Cadre du MEDRASP à Roviny-Zadar (Yugoslavia)*, 18–28 January 1986. Rapport FAO, Rome.

Devauchelle, N. & Coves, D. (1988) Sea bass (*Dicentrarchus labrax*) reproduction in captivity: gametogenesis and spawning. *Aquatic Living Resources*, **1**, 215–22.

Devauchelle, N., Brichon, G., Lamour, F. & Stephan, G. (1982) Biochemical composition of ovules and fecond eggs of sea bass (*Dicentrarchus labrax*), sole (*Solea vulgaris*) and turbot (*Scophthalmus maximus*). *Proceedings of the International Symposium on the Reproductive Physiology of Fish*, 155–7.

Duston, J. & Bromage, N. (1986) Photoperiodic mechanisms and rhythms of reproduction in the female rainbow trout. *Fish Physiology and Biochemistry*, **2**, 35–51.

Fernández, J., Gutiérrez, J., Carrillo, M., Zanuy, S. & Planas, J. (1989) Annual cycle of plasma lipids in sea bass, *Dicentrarchus labrax* L.: effects of environmental conditions and reproductive cycle. *Comparative Biochemistry and Physiology*, **93A**(2), 407–12.

Girin, M. & Devauchelle, N. (1978) Décalage de la période de reproduction par raccourcissement des cycles photoperiodiques et thermiques chez les poissons marins. *Annales Biologie Animale Biochimie Biophysique*, **18**, 1059–65.

Gutiérrez, J., Fernández, J., Carrillo, M., Zanuy, S. & Planas, J. (1987) Annual cycle of plasma insulin and glucose of sea bass. *Dicentrarchus labrax. Fish Physiology and Biochemistry*, **4**(3), 137–41.

Hardy, R.W. (1984) Salmonid broodstock nutrition. *Aquaculture*, **43**, 98–108.

Hassin, S., Yaron, Z. & Zohar, Y. (1991) Follicular steroideogenesis, steroid profiles and oogenesis in the European sea bass, *Dicentrarchus labrax*. In *Proceedings of the Fourth International Symposium on the Reproductive Physiology of Fish*, (eds A.P. Sumpter, J.P. Scott, D.E. Kime & M.S. Rolfe, p. 100. FishSymp 91, Sheffield.

Johnson, D.W. & Katavic, I. (1984) Mortality, growth and swimbladder stress syndrome of sea bass (*Dicentrarchus labrax*) larvae under varied environmental conditions. *Aquaculture*, **38**(1), 67–78.

Kah, O., Zanuy, S., Mañanós, E., Anglade, I. & Carrillo, M. (1991) Distribution of salmon gonadotrophin releasing-hormone in the brain and pituitary of the sea bass (*Dicentrarchus labrax* L.): an immunocytochemical and immunoenzymoassay study. *Cell Tissue Research*, **266**, 129–36.

Kennedy, M. & Fitzmaurice, P. (1972) The biology of the bass (*Dicentrarchus labrax*), in Irish waters. *Journal of the Marine Biological Association of the United Kingdom*, **52**, 557–97.

Kjorsvik, E., Mangor-Jensen, A. & Holmefjord, I. (1990) Egg quality in marine fishes. *Advances in Marine Biology*, **26**, 71–113.

Lisac, D. (1988) Marine hatchery technology – Systems Review. In *Aquaculture Engineering Technologies for the Future. IChemE Symposium Series No 111*, pp. 65–76. EFCE Publication Series No 66, Stirling, UK.

Luquet, P. & Watanabe, T. (1986) Interaction 'nutrition-reproduction' in fish. *Fish Physiology and Biochemistry*, **2**, 121–9.

Mañanós, E., Núñez-Rodriguez, J., Zanuy, S., Carrillo, M. & Le Menn, F. (1991a) Sea bass (*Dicentrarchus labrax* L.) vitellogenin: Development of a specific enzyme linked immunosorbent assay (ELISA). *Abstracts of 13th Conference of ESCPB on Research for Aquaculture: Fundamental and Applied Aspects*, Antibes, 191.

Mañanós, E., Zanuy, S., Carrillo, M., Nuñez Rodriguez, J. & Le Menn, F. (1991b) Quantification by ELISA of vitellogenin levels in sea bass (*Dicentrarchus labrax* L.) maintained under different photoperiods. In *Proceedings of the Fourth International Symposium on the Reproductive Physiology of Fish*, (eds A.P. Scott, J.P. Sumpter, D.E. Kim & M.S. Rolfe), p. 325. FishSymp 91, Sheffield.

Mañanós, E., Zanuy, S., Carrillo, M., Nuñez, J. & LeMenn, F. (1992) Vitellogenic events following photoperiod manipulation of the reproductive cycle of female sea bass (*Dicentrarchus labrax*). *Abstracts of the Second International Symposium on Fish Endocrinology*, Saint Malo.

Marangos, C., Yagi, H. & Ceccaldi, J. (1985) Influence de la salinité sur la composition corporelle des acides aminés libres dans les oeufs et les premières larves de *Dicentrarchus labrax* (Linnaeus, 1758) (Pisces, Teleostei, Serranidae). *La Mer*, **23**, 177–83.

Mayer, I., Shackley, S.E. & Ryland, J.S. (1988) Aspects of the reproductive biology of the bass, *Dicentrarchus labrax* L. I. A histological and histochemical study of oocyte development. *Journal of Fish Biology*, **33**, 609–22.

Mayer, I., Shackley, S.E. & Witthames, P.R. (1990) Aspects of the reproductive biology of the bass, *Dicentrarchus labrax* L. II. Fecundity and pattern of oocyte development. *Journal of Fish Biology*, **36**, 141–8.

Navarro, J.C. (1990) *Caracterización de las cepas españolas de Artemia desde el punto de vista de su valor nutritivo y de sus fenotipos electroforéticos. Implicaciones prácticas en acuicultura*. Ph.D. thesis, University of Barcelona.

Navarro, J.C., Hontoria, F., Varó, I. & Amat, F. (1988) Effect of alternate feeding with a poor long chain polyunsaturated fatty acid Artemia strain and a rich one for sea bass (*Dicentrarchus labrax*) and prawn (*Penaeus keraturus*) larvae. *Aquaculture*, **74**, 307–17.

Person-Le Ruyet, J., Ficher, C. & Thébaud, L. (1993) Sea bass (*Dicentrarchus labrax*) weaning and ongrowing onto Serbar. In *Proceedings of the Fourth International Symposium on Fish Nutrition and Feeding, Biarritz, France, June 1991*, (eds S.J. Kaushik & P. Luquet), pp. 623–8. INRA Editions, Versailles, France.

Planas, J., Gutiérrez, J., Fernández, J., Carrillo, M. & Canales, P. (1990) Annual and daily variations of plasma cortisol in sea bass, *Dicentrarchus labrax* L. *Aquaculture*, **91**, 171–8.

Prat, F. (1991) *Control del ciclo reproductor de la lubina* (Dicentrarchus labrax L.) *por manipulación hormonal y ambiental*. Ph.D. thesis, University of Barcelona.

Prat, F., Zanuy, S. & Carrillo, M. (1987) Effects of LHRH analog alone or combined with pimozyde on plasma sex steroids, egg quality and spawning of sea bass (*Dicentrarchus labrax* L.). *General and Comparative Endocrinology*, **66**, 21–2.

Prat, F., Zanuy, S., Carrillo, M., de Mones, A. & Fostier, A. (1990) Seasonal changes in plasma levels of gonadal steroids of sea bass, *Dicentrarchus labrax* L. *General and Comparative Endocrinology*, **78**, 361–73.

Roblin, C. & Bruslé, J. (1983) Ontogenése gonadique et différenciation sexuelle du loup *Dicentrarchus labrax*, en conditions d'élevage. *Reproduction Nutrition et Dévelopment*, **23**, 115–27.

Rodriguez, G.M. (1991) *Analisis genético de poblaciones de lubina* (Dicentrarchus labrax L.) *mediante electroforesis de enzimas*. Ph.D. thesis, University of Malaga.

Scott, A.P., Canario, A.V.M. & Prat, F. (1990) Radioimmunoassay of ovarian steroids in plasmas of ovulating female sea bass (*Dicentrarchus labrax*). *General and Comparative Endocrinology*, **78**, 299–302.

Serrano, R., Zanuy, S. & Carrillo, M. (1989) Determinación de la calidad de huevos fertilizados de lubina (*Dicentrarchus labrax* L.) por medio de parámetros bioquímicos. In *Acuicultura Intermareal*, (ed. M. Yufera), pp. 229–35. Instituta Ciencia Marina Andalucía, Cádiz.

Springate, J.R.C., Bromage, N.R. & Cumaranatunga, R. (1985) The effects of different ration on fecundity and egg quality in the rainbow trout (Salmo gairderi). In *Nutrition and Feeding in Fish*, (eds C.B. Cowey, A.M. Mackie & G.G. Bell), pp. 371–91. Academic Press, London.

Sundararaj, B., Vasal, S. & Halberg, F. (1982) Circannual rhythmic ovarian recrudescence in the catfish. *Advances in Biosciences*, **41**, 319–37.

Tortonese, E. (1986) Moronidae. In *Fishes of the North-Eastern Atlantic and Mediterranean*, Vol. II (eds P.J.P. Whitehead, M.L. Bauchot, J.-C. Hureau, J. Nielsen & E. Tortonese), pp. 793–4. UNESCO, UK.

Van Balaer, E., Amat, F., Hontoria, F., Léger, P. & Sorgelos, P. (1985) Preliminary results on the nutritional evaluation of w3-HUFA-enriched *Artemia* nauplii for larvae of the sea bass *Dicentrarchus labrax*. *Aquaculture*, **49**, 223–9.

Villani, P. (1976) Ponte induite et élevage des larves de poissons marins dans les conditions de laboratoire. *Etudes Revues Conseil Generale Pêches Mediterranean FAO*, **56**, 117–34.

Watanabe, T. (1985) Importance of the study of broodstock nutrition for further development of aquaculture. In *Nutrition and Feeding in Fish*, (eds C.B. Cowey, A.M. Mackie & J.G. Bell), pp. 395–14. Academic Press, New York, NY.

Zanuy S. & Carrillo, M. (1984) La salinité: un moyen pour retarder la ponte du bar. In *L'Aquaculture du Bar et des Sparidés*, (eds G. Barnabé & R. Billard), pp. 73–80. INRA, Paris.

Zanuy, S., Carrillo, M. & Ruiz, F. (1986) Delayed gametogenesis and spawning of sea bass (*Dicentrarchus labrax* L.) kept under different photoperiod and temperature regimes. *Fish Physiology and Biochemistry*, **2**, 1–4.

Zanuy, S., Prat, F., Bromage, N., Carillo, M. & Serrano, R. (1989) Photoperiodic effects on vitellogenesis, steroid hormone levels, and spawning time in the female sea bass (*Dicentrarchus labrax*). *General and Comparative Endocrinology*, **74**, 253.

Zanuy, S., Bromage, N., Carrillo, M., Prat, F. & Serrano, R. (1991) Endogenous circannual rhythms and the control of reproduction in the sea bass (*Dicentrarchus labrax* L.). In *Proceedings of the Fourth International Symposium on the Reproductive Physiology of Fish*, (eds A.P. Scott, J.P. Sumpter, D.E. Kime & M.S. Rolfe). p. 175. FishSymp 91, Sheffield.

Zanuy, S., Carrillo, M., Perez, J., Gutiérrez, J. & Planas, J. (1993) Environmental and nutritional influences on hormonal profiles in cultivated sea bass (*Dicentrarchus labrax* L.). In *Recent Advances in Aquaculture*, Vol. 4 (eds N. Bromage, E.M. Donaldson, M. Carrillo & S. Zanuy), pp. 140–152. Blackwell Science, Oxford.

Zohar, Y., Billard, R. & Weil, C. (1984) La reproduction de la daurade (*Sparus aurata*) et du bar (*Dicentrarchus labrax*): connaissance du cycle sexuel et contrôle de la gamétogenèse et de la ponte. In *L'Aquaculture du Bar et des Sparidès*, (eds G. Barnabé & R. Billard), pp. 3–24. INRA, Paris.

Chapter 8
Atlantic Halibut (*Hippoglossus hippoglossus*) and Cod (*Gadus morhua*)

8.1 Introduction
8.2 Biology and taxonomy
    8.2.1 Cod
    8.2.2 Halibut
8.3 History of marine fish studies
    8.3.1 Cod
    8.3.2 Halibut
8.4 Broodstock management and egg and larval rearing
    8.4.1 Gonad development
    8.4.2 Broodstock feeding and egg production
    8.4.3 Egg production and collection: ovulatory rhythms and egg viability
    8.4.4 Rearing procedures for eggs and yolk-sac larvae
    8.4.5 Larval rearing
8.5 Viability of eggs and yolk-sac larvae
    8.5.1 Environmental factors
    8.5.2 Microbiology and antibiotics
8.6 Assessment of seed quality
8.7 Conclusions
    References

Atlantic halibut (*Hippoglossus hippoglossus* L.) and cod (*Gadus morhua* L.) are considered to be important species for the aquaculture industry in the North Atlantic. Much of the research for bringing these new species into commercial farming has been carried out in Norway, where a substantial input of governmental funding has secured necessary basic and applied research. Emphasis has been placed on problems concerning production of juveniles, and this activity has led to increased knowledge of broodfish husbandry and reproductive biology for both species, as well as important aspects of egg and larval biology.

There is still little knowledge on how broodfish handling and feeding affect offspring quality. Research activities include manipulations of spawning season, studies of seasonal gonadal development and spawning cycles, stripping of eggs vs natural spawning, effects of feed composition and ration, and how the age of the spawning fish may affect viability of the eggs and larvae.

The viability of eggs and yolk-sac larvae are influenced by both parental and environmental factors. Recent results show that the species have different tolerances towards environmental parameters like light and mechanical stress, and the effects are also dependent on the susceptibility of eggs and larvae in various stages of development. Studies of egg and larval viability have used 'quality parameters' like fertilization rate, rate of symmetrical cell cleavages, egg and larval survival, buoyancy, larval behaviour and ability to capture prey, and response to various stimuli.

The quality of larvae at the time of first feeding is variable. This is especially the case in halibut, which has a very long yolk-sac stage compared with most other marine fish larvae, and studies of possible factors affecting the normal development of the halibut larvae are now being carried out.

This paper is a review of results obtained from research activities on broodstock management in relation to egg quality, factors which seem to affect egg and yolk-sac larval viability, and how such variations in quality may be manifested.

8.1 Introduction

The expansion of European aquaculture in recent years has led to a growing interest in the introduction of new marine species, especially turbot (*Scophthalmus maximus* L.), Atlantic halibut (*Hippoglossus hippoglossus* L.) and cod (*Gadus morhua* L.). Of these, the halibut and cod are abundant in the relatively cold waters of the North Atlantic, and they are considered to be potentially important species for the North Atlantic aquaculture industry. The major interest in husbandry of marine cold-water fishes came in the early 1980s. Successful experiments on mass rearing of cod juveniles were carried out in large enclosures (Kvenseth & Øiestad 1984), and the first halibut juveniles were produced from artificially fertilized eggs (Blaxter *et al.* 1983).

The halibut seems especially promising for aquaculture, as the natural stocks are over-exploited (Haug 1990) and prices are high within the distribution area. The growth rate of wild halibut shows a good potential for fast growth (Haug 1990, Haug & Tjemsland 1986, Jakupsstovu & Haug 1988) as do results from growth experiments on juvenile halibut (Haug *et al.* 1989, Berge & Storebakken 1991). Adoff *et al.* (1993) recently demonstrated that three generations of farmed halibut from a commercial farm reached a mean weight of 3 kg after $3\frac{1}{2}$ years (8–12°C). Captive fish do have better growth rates and higher condition factors than wild fish (Jobling 1988, Haug *et al.* 1989, Kjesbu 1989). However, problems of early maturation are now experienced both for cod and halibut reared in captivity. Halibut males as small as 1–2 kg have matured (Lehmann *et al.* 1991, I. Holmefjord, I. Lein, T. Gjedrem, P. Scherrer & S. Bolla unpublished), which may reduce the economic output for farmers. Halibut and cod farming are still in an early phase of development, though, and there is a great potential for improvements in production and growth results.

The research for bringing such new species into commercial farming has been centred upon Norway. The main difficulty in both halibut and cod culture has been a controlled production of juveniles, and the major obstacle is still high mortalities during the larval phase. Emphasis has therefore been put on problems concerning production of juveniles. The strategy has been that development of technology and production methods must be based on knowledge of the requirements of eggs and larvae. Factors like environmental, physical and microbial conditions for eggs and larvae, as well as nutritional requirements through phases of first feeding and weaning, have been regarded as important. This activity has led to increased knowledge of egg and larval biology, broodfish husbandry and reproductive biology for both species. This paper concentrates on results of direct importance to egg and larval viability, and egg quality aspects are considered up to the time when exogenous feeding is necessary.

8.2 Biology and taxonomy

8.2.1 Cod

Atlantic cod, *Gadus morhua*, belongs to the family Gadidae within the order Gadiformes (Fig. 8.1). Cod is the most commercially important species in the family Gadidae. It is distributed in the North Atlantic and adjacent seas, from Bay of Biscay to Greenland, Spitzbergen and Novaya Zemlya (Whitehead *et al.* 1986). Cod may be found south to Cape Hatteras on the west side of the Atlantic (see Harden Jones 1968). There are three sub-species in the Atlantic area, *G.m. morhua* L. around Novaya Zemlya, *G.m. marisalbî* Derjugin in the White Sea, and *G.m. callarias* in the Baltic (Whitehead *et al.* 1986). The largest cod reported was 169 cm (Babayan 1978, Tretyak 1984) and the maximum age of a cod of this length is probably about 30 years, according to the methods of ageing of Tomlinson & Abramson (1961) and Nikolsky (1974). In the eastern/

Fig. 8.1 Four-year-old broodstock cod (50–60 cm in length). The upper more golden-coloured specimen has been fed on sand eel whilst the other has received a white fish, meal-based artificial diet. (Photograph kindly provided by B. Howell, MAFF, Conwy, Wales.)

central area the Arctic-Norwegian cod is the largest and most important stock, with feeding areas in the Barents Sea and around Spitsbergen, and spawning areas off Lofoten and Møre along the Norwegian coast (Bergstad *et al.* 1987). This stock thus undertakes a long spawning migration, while more coastal stocks usually spawn locally.

Growth rate and condition factor seem important for age at first maturity. Wild juvenile cod reared in captivity and fed high rations exhibit a higher growth rate and earlier maturation than fish in nature (Godø & Moksness 1987). Farmed cod reared from the egg stage will mature at 2 years of age, which is much earlier than wild cod from the same area (Braaten 1984, Godø & Moksness 1987). The mean age at first maturity for wild cod will vary between the different stocks, e.g. the age at 50% maturity is 2–3 years for cod from the Irish Sea and 8–9 years for fish from the south-western parts of Greenland (Garrod 1977). This difference also reflects differences in growth rates for the various stocks; the mean weight of 5 years old cod is respectively 6.5 kg and 1.3 kg from these two localities (Garrod 1977).

In Norwegian waters cod spawn from February to May. Peak spawning for the Arcto-Norwegian cod is in March/April at temperatures of 4–6°C (Solemdal 1982, Pedersen 1984, Sundby & Bratland 1987).

8.2.2 Halibut

The Atlantic halibut *Hippoglossus hippoglossus* (L.) is the largest flatfish in the North Atlantic (Fig. 8.2). Within the order Pleuronectiformes (flatfish) the genus *Hippoglossus* belongs to the family Pleuronectidae, sub-family Pleuronectinae (Ahlström *et al.* 1984, Hensley & Ahlström 1984). Females as large as 333 kg have been captured (Mathisen & Olsen 1968), whereas males seldom exceed 50 kg.

The general biology of the Atlantic halibut has been reviewed by Haug (1990). It is distributed in the northern part of the Atlantic Ocean and parts of the Barents Sea and Arctic Ocean (Andriyashev 1954). Atlantic halibut seem very similar to the Pacific halibut *Hippoglossus stenolepis*, but they are now regarded as two separate species (Grant *et al.* 1984).

In wild fish, there is a clear sexual difference in growth rates from the age of 6–7 years, when the males mature (see Haug 1990). There is a considerable variation in age at first maturation in halibut, and maturity seems to be more a function of growth rate and size than age (Haug 1990). Wild Atlantic halibut from Faroese waters, which exhibits the best known growth rates, mature at 55 cm for males and 110–115 cm for females (Jakupsstøvu & Haug 1988), whereas halibut from Newfoundland waters mature at 80 and 115–120 cm respectively (Methven *et al.* 1992).

The Atlantic halibut spawns at great depths (100–700 m) during the period January–April, at temperatures varying between 5 and 8°C (Kohler 1967, Kjørsvik *et al.* 1987, Jakobsstovu & Haugh 1988, Haug & Kjørsvik 1989).

Fig. 8.2 Nine-year-old broodstock Atlantic halibut (25+ kg in weight). (Photograph kindly provided by R. Shields, P. Smith and J. Dye, SFIA, Ardtoe, Argyll, Scotland.)

8.3 History of marine fish studies

8.3.1 Cod

The pelagic nature and first descriptions of egg and larval development in cod were made by the Norwegian fishery biologist Georg Ossian Sars in the 1860s (Sars 1869). This discovery led to the first sea ranching experiments, where large numbers of yolk-sac larvae were reared for release into the sea, with the hope of enhancing the natural fish stocks. The Norwegian production of cod yolk-sac larvae started in 1884, and several countries were soon to follow (review by Solemdal et al. 1984). The Norwegian cod hatchery at Flødevigen produced between 20 million and 400 million yolk-sac larvae annually in the period 1890–1963, whereas American hatcheries produced up to 1.5 billion to 2.5 billion cod larvae annually in the most productive years, which were between 1920 and 1950. As no effect of these releases could be measured over the years the activities gradually ceased, although the last yolk-sac larvae were released as late as in 1971 (Solemdal et al. 1984). In the early period, the first successful attempts to bring cod larvae past metamorphosis were also demonstrated, by releasing yolk-sac larvae into a large enclosure (Rognerud 1887).

Although the release of yolk-sac larvae had no documented effect on recruitment, this activity had a major influence on the development of fisheries research. In the years following this pioneer period, much knowledge was obtained about the ecology, feed preference and survival mechanisms of cod larvae and adult fish, with the aim of understanding the recruitment processes of natural cod stocks. When the first successful mass rearing experiments on cod juveniles occurred in 1983 (Øiestad et al. 1985), it was from an ecological approach and by refining the old concept of a large enclosure and using natural zooplankton as first feed for the larvae.

8.3.2 Halibut

It was not until the 1930s that the first known experiments on halibut eggs and larvae were made by Rollefsen (1934) at Trondheim Biological Station in Norway, where he was able to fertilize and incubate eggs and keep live larvae up to 10 days after hatching. When the next rearing experiments on halibut started in 1974 (Solemdal et al. 1974), almost nothing was known about the early life history of this species. The mesopelagic distribution of halibut eggs was discovered by Haug et al. (1984, 1986) as a result of several years of field work. The egg and larval development has been described in several investigations (Rollefsen 1934, Lønning et al. 1982, Blaxter et al. 1983, Pittman et al. 1987, 1990a, 1990b, Kjørsvik & Reiersen 1992).

Up to 1982 all halibut gametes used in experiments were stripped from fish which were captured by gill nets on the spawning grounds. Thus, a very variable egg performance and viability was obtained, and eggs tended to have a very poor buoyancy. The first halibut broodstocks were established in 1982–4; this was the first real step towards a reliable supply of eggs.

Halibut-rearing is still in a developmental phase, probably because the biological knowledge of halibut was very poor when the first rearing experiments started in 1974, and because halibut larvae have a very long and susceptible yolk-sac phase. The increasing number of surviving fish from Norwegian rearing experiments presented in Table 8.1 is a good example of the improvements in success of the development. Up to 1987 juvenile production of halibut took place only at research stations, but after 1988 commercial production also started. In recent years, there has been a steady increase in the total production of halibut juveniles.

Table 8.1 Total number of live halibut produced from various year-classes (at April 1992) in Norway. The large increase in numbers in 1991 is mostly due to the production of one private company (data from Nortvedt 1992)

	1986–7	1988	1989	1990	1991
Public research stations (4)	53	112	533	1530	2881
Private companies (5)		300	600	7575	23570

8.4 Broodstock management and egg and larval rearing

At present, there is an increasing number of broodstock populations of both cod and halibut on farms. All halibut broodstocks and much of the existing cod broodstocks are fish collected from the wild, but the first recruitment of halibut spawners from juvenile production has now occurred both in Norway (I. Holmefjord et al., unpublished data) and in Scotland (P. Smith, pers. comm.). The halibut are usually kept in large, covered tanks (diameter 8–20 m) with a continuous flow of sea water, whereas cod are mostly kept outdoors in traditional pens or tanks most of the year. Both species are normally held under ambient day length and temperature conditions, 5–14°C, (Lehmann et al. 1991, I. Holmefjord et al. unpublished data).

The thermal preference of cod is 9–17°C (Bøhle 1974), and maximum feed intake occurs at 14–16°C (McKenzie 1934). Temperature tolerance is dependent on the previous temperature at which the fish have been held. However, wild cod are found in areas with sub-zero temperatures (Templeman & Fleming 1965) with an upper temperature limit of 23–24°C (McKenzie 1934). Cod may be fed on soft pelleted feed or commercially available dry pellets (Mangor-Jensen et al. 1991). Optimal environmental parameters for cod farming are further discussed by Jobling (1988).

Young halibut can survive temperatures below 0°C, but the fish were then inactive and feeding stopped below 2°C (Goff & Lall 1989, E. Kjørsvik, unpublished data). The optimum growth and feeding rate for halibut seems to be in the range of 6–12°C, with a higher optimum for young than for larger fish (Smith & Dye 1988, Adoff et al. 1993, I. Holmefjord, unpublished data, A. Folkvord unpublished data). A decreased feeding uptake is experienced in large halibut when temperatures reach 14°C (Nordtvedt 1992), and a rapid shift in temperatures will affect appetite more adversely than slower changes (I. Holmefjord, unpublished data).

Halibut broodstock are usually fed to satiation on a moist pelleted diet or on herring, mackerel, squid and trash white fish supplied with a vitamin mix two to three times a week (Haug et al. 1989, I. Holmefjord et al. unpublished data). Farmed halibut recruited for the broodstock have also been fed dry pellets (Berge & Storebakken 1991, Adoff et al. 1993). No feed is offered to spawning fish.

8.4.1 Gonad development

Attempts to control reproduction and egg quality require basic knowledge of reproductive physiology, and recent work on cod and halibut has aimed at understanding the regulation of gonadal growth and maturation of gametes.

Changes in day length and temperature are the most important environmental parameters affecting the timing of gonad maturation and onset of spawning in temperate species (Woodhead & Woodhead 1965, Lam 1983, Bye 1984, Smith *et al.* 1991). The gonadal growth starts during early autumn for cod (Sivertsen 1935) and halibut (Tjemsland 1960, Haug & Gulliksen 1988a,b). Seasonally-increasing plasma vitellogenin (VTG) and steroid hormone levels in fish are associated with the onset of reproductive activity (Crim & Glebe 1990), and investigations on seasonal sex steroid hormone levels in fish generally show a pattern of peak values during the pre-spawning period and a seasonal minimum after spawning (Wingfield & Grimm 1977, Liu *et al.* 1990).

Plasma levels of VTG, oestradiol-17β and testosterone increase gradually up to the commencement of spawning for cod and halibut (Kjesbu & Norberg 1991, Methven *et al.* 1992). The levels of VTG and hormones then oscillate sharply between batches, indicating a steady inflow of yolk proteins to the eggs during the final maturation (Kjesbu & Norberg, 1991, Norberg and Kjesbu 1991, Methven *et al.* 1992). This pattern is exemplified for halibut in Fig. 8.3. The major influx of protein to the ovary of cod occurs during the period when the first 10% of the eggs are spawned, and it seems that this influx as well as the subsequent influxes between final maturation of the egg batches will determine the final oocyte size (Kjesbu *et al.* 1991, 1992). This pattern thus seems to be similar to turbot, where sex steroids were found to fluctuate on a daily basis during the spawning season (Howell & Scott 1989).

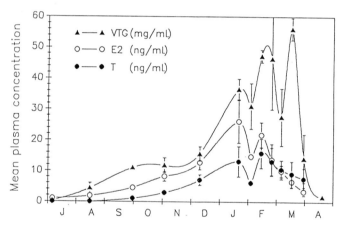

Fig. 8.3 Plasma levels of vitellogenin (VTG), oestradiol-17β (E2) and testosterone (T) in laboratory-held mature female halibut prior to and during the reproductive period. Values are means ± 1 S.E. Missing standard error bars are too small for presentation. n is the number of fish sampled. (Reproduced with permission from Methven *et al.* 1992.)

8.4.2 Broodstock feeding and egg production

Little is known about how broodstock feeding affects egg and larval viability in halibut (McIntyre 1952) and cod. In halibut, research has been hampered by the low availability of fish. In cod, most work has been focused on quantitative aspects of feeding rather than qualitative aspects.

Halibut and cod seem to have a higher protein demand than plaice and salmonids (Lie et al. 1988, Hjertnes & Opstvedt 1989), and additional vitamins in the feed have given improved growth in young halibut (Goff & Lall 1989). Both halibut and cod have a very high potential fecundity, with a positive relationship between relative fecundity and fish size (Oosthuizen & Daan 1974, Haug & Gulliksen 1988a,b, Kjesbu 1988, Kjesbu et al. 1991).

Recent studies on cod have revealed that feeding rations and growth rate have a very significant effect on potential fecundity of the fish, and that cod with high condition factors produced more pre-vitellogenic oocytes than fish of the same size deprived of food (Waiwood 1982, Kjesbu et al. 1991). The proportion of pre-vitellogenic oocytes which undergo maturation and ovulation is also dependent on the nutritional status of the fish. Actual fecundity (i.e. eggs spawned) in cod was 20–80% of the potential fecundity, and the degree of atresia was correlated to the condition factor of the fish (Kjesbu et al. 1991). Kjesbu et al. (1991) observed atresia to be a regulating mechanism during spawning also in overfed cod, probably caused by an overproduction of vitellogenic oocytes. In halibut broodstocks, individual females in very poor condition may omit one spawning season (I. Holmefjord et al., unpublished data).

Feed quality does have an effect on egg and larval viability and performance in other species (for a review, see Kjørsvik et al. 1990), and protein quality, lipid composition and vitamins are also important in this regard (see e.g. Watanabe 1985). One such qualitative study on cod was carried out by Mangor-Jensen et al. (1991), who measured the effects of vitamin C in broodstock diets on egg and larval viability of cod. There seemed to be a low demand for dietary ascorbic acid in the broodstock diet in these experiments.

8.4.3 Egg production and collection: ovulatory rhythms and egg viability

Cod and halibut are batch spawners; individual fish release multiple batches of pelagic eggs according to their ovulatory rhythm during the spawning season. Halibut of 20–60 kg have been reported to give 6–16 egg batches during one spawning season, with mean ovulatory rhythms of 70–90 h (Holmefjord 1991, Norberg et al. 1991, Norberg & Kjesbu 1991). Similarly for cod, Kjesbu (1989) observed 15–20 batches spawned by individual fish in 50–60 days (Table 8.2).

Table 8.2 Spawning characteristics for broodstock halibut and cod. References for spawning data are given in the text; egg size is from Russell (1976)

Species	Spawning interval (h)	Batch size ($\times 10^3$)	No. of batches	Egg size (mm)
Atlantic halibut	70–90	10–200	6–16	3.00–3.80
Cod	60–75	10–350	15–20	1.16–1.89

The egg size will decrease during the spawning season (Kjørsvik et al. 1987, Kjesbu et al. 1991). A decrease in egg size as spawning progresses has been related to exhaustion of reserves for vitellogenesis, and the rate of decrease in egg size is also dependent on the nutritional state of the fish (Kjesbu et al. 1991).

Cod will spawn naturally in captivity. The broodstock is transferred to special spawning tanks or pens before the onset of spawning, and fertilized eggs can be collected daily (Huse & Jensen 1983, Holm & Andersen 1989, Rosenlund et al. 1993). The fertilization rate of eggs obtained by natural spawning is generally high when the fish spawn regularly, and there seem to be few problems with egg over-ripening (Fig. 8.4a). Overripening may occur if the fish are stressed. When spawning cod were separated in pairs in smaller tanks, up to one-third of the fish showed signs of stress, resulting in irregular spawning intervals, low fertilization rates, and increased occurrence of abnormal embryos (Fig. 8.4b, Kjesbu 1989 and pers. comm.). If cod eggs are stripped and stored in ovarian fluid, eggs may achieve a high fertilization rate up to 9 h after stripping (Kjørsvik & Lønning 1983). Older eggs decrease in fertility, and cod larvae hatching from such overripe eggs have poorer viability than larvae from ripe eggs (Annette Thorvik, pers. comm.).

The egg production of halibut is at present mainly based on stripping of males and females. Egg viability after ovulation is time-restricted, and it seems that fertilization should occur within 6–12 h of the estimated time of ovulation (Kjørsvik 1990, Kjørsvik et al. 1990, Holmefjord 1991, Norberg et al. 1991, Bromage et al. 1994), after which egg quality is markedly reduced.

Hatching rates also decreased if unripened halibut eggs were fertilized (Holmefjord 1991, Bromage et al. 1994). Fertilization rates were not reduced, but hatching rates were significantly lowered when eggs were stripped and fertilized 6–9 hours too early (I. Holmefjord, unpublished data). However, these experiments indicated that unripened eggs may at least partially go through some of the final maturation processes if they are kept in ovarian fluid after stripping (Holmefjord 1991, Bromage et al. 1994).

Stripping must therefore be done by carefully surveying the ovulatory rhythm of each female so that effects from overripening processes are minimized. Overripening of eggs after ovulation is a process which seems to occur during a shorter time period for batch spawners than for fish spawning once a year (for review, see Kjørsvik et al. 1990). Stripping time in relation to ovulation has proved to be very important for total egg yield and high fertilization and hatching rates in halibut (Holmefjord 1991, Norberg et al. 1991, Bromage et al. 1994, Holmefjord et al. 1994), as it has for other marine batch spawners (McEvoy 1984, Howell & Scott 1989, McEvoy & McEvoy 1992). Norberg et al. (1991) also experienced that the total yield and fertilization rates for individual halibut females were dramatically improved from one year to the next by such procedures. The ovulatory intervals and egg yield from a small female halibut is shown in Fig. 8.5. Ovulatory rhythms seem to be more or less constant for

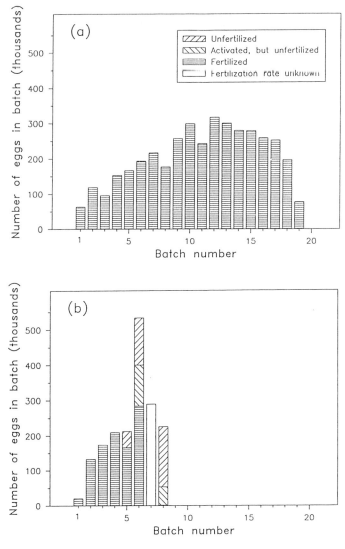

Fig. 8.4 The spawning activity of cod in captivity: (a) natural spawning, (b) from pairs of fish maintained in small tanks. (Reproduced with permission from Kjesbu 1989.)

individual females over the whole spawning period, a finding which has also been reported for turbot by McEvoy (1984) and Howell & Scott (1989), in whose experiments, and in those of Devauchelle et al. (1988), overripening seemed to be temperature-dependent.

Even by considering the ovulatory rhythms for individual fish, fertilization rates and egg viability are still highly variable for the batches that are stripped from halibut (Norberg et al. 1991, Bromage et al. 1994, Holmefjord et al. 1994, Holmefjord, unpublished data). Stripping is very time-consuming and laborious, and probably also quite stressful for the fish, so further studies on halibut

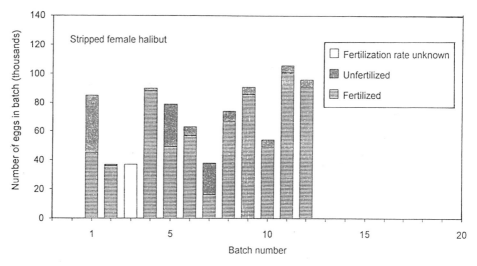

Fig. 8.5 Egg yield from a stripped female halibut (fish weight 30 kg) during one spawning season. The fish gave a total of 16 l of eggs in 12 batches. (Data calculated from Norberg & Kjesbu 1991, Norberg *et al.* 1991.)

reproduction should elucidate the necessary optimal conditions for natural spawning and environmental effects on reproduction. Some broodstock halibut have been observed to give naturally spawned fertilized eggs from 1989 and onwards (I. Holmefjord *et al.*, unpublished data, Jakob Blom, pers. comm.). Fertilization and hatching rates have been highly variable for these eggs, and the factors inducing natural spawning are still unknown. The same problem is also encountered for captive sole, *Solea solea* (L.) and turbot, *Scophthalmus maximus* (L.), where up to half of their annual egg production may be unfertilized (Houghton *et al.* 1985, Bromley *et al.* 1986), although field studies of cod (E. Kjørsvik, unpublished data) and sole (Howell *et al.* 1991) indicate that natural populations have very high fertilization rates.

8.4.4 Rearing procedures for eggs and yolk-sac larvae

One of the major difficulties with the culture of early life history stages of halibut is the long periods of time required for egg incubation and maintenance of yolk-sac fry (Pittman *et al.* 1990b). Hatching occurs 16–19 days post-fertilization at 5°C (Fig. 8.6). During this time in Scotland eggs are maintained in 100 l cylindro-conical tanks supplied with a slow inflow of raised salinity sea water (35.5‰) at densities of 125 eggs per litre (J. Dye & R. Shields, pers. comm., Fig. 8.7). In contrast many of the Norwegian systems use an upwelling current to maintain positive egg buoyancy throughout incubation.

Just before hatching or just after hatching, the eggs or yolk-sac fry are transferred to conical tanks, 700 l or more in volume, supplied with upwelling

Atlantic Halibut and Cod 181

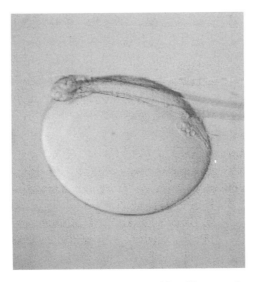

Fig. 8.6 Yolk-sac fry of Atlantic halibut 12 h after hatching. Note very large yolk mass for a marine fish larva. (Photograph kindly provided by R. Shields and J. Dye, SFIA, Ardtoe, Argyll, Scotland.)

Fig. 8.7 Egg incubation tanks (100 l capacity) for eggs of Atlantic halibut (Photograph kindly provided by R. Shields and J. Dye.)

Fig. 8.8 Conical yolk-sac larvae tanks (700 l capacity). (Photograph kindly provided by R. Shields and J. Dye.)

inflows of high salinity waters at 4.5–5.5°C (R. Shields & J. Dye, pers. comm., Fig. 8.8; see also [8.5.1]). The stocking density of the yolk-sac fry is usually about 8–10 per litre. Hatching can be synchronized by manipulating the light. Eggs are maintained in the dark until just before the time hatching is expected to start. The lights are then switched on for 12 h at 20 lux; this inhibits hatching, but when the eggs are returned to darkness hatching occurs in all the eggs a few hours later. The yolk-sac is removed totally by about 330° days but first-feeding begins at 210–250° days.

8.4.5 Larval rearing

Much of the juvenile production is still based on the use of natural zooplankton as organisms for first feeding. However, as enrichment and cultivation techniques for production of live feed organisms suitable for marine fish larvae have developed (Watanabe *et al.* 1983, Bengtson *et al.* 1991, Kanazawa 1991, Devresse *et al.* 1992, Olsen *et al.* 1993a,b, Dhert *et al.* 1993, Reitan *et al.* 1993, Watanabe 1994), an increasing proportion of cod and halibut juveniles has been produced by more intensive methods.

Rearing systems which have been based primarily on natural zooplankton are now using cultured feed organisms to a greater extent, especially during critical periods of development or to ensure a steady supply of feed for the larvae (Lein & Holmefjord, 1992). At present good survivals are being achieved

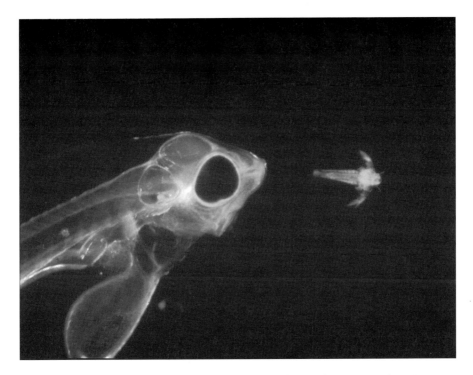

Fig. 8.9 First-feeding halibut larva (230 degree days after hatching) feeding on *Artemia*. (Photograph kindly provided by R. Shields and J. Dye.)

in larvae up to approximately 14 day post first-feeding by using algae, rotifers and enriched *Artemia* (Fig. 8.9). Aside from nutritional value this mixture, particularly the algae, seems to help with the perception of light by the larvae and the consequent development of normal patterns of feeding behaviour and also the improvement of water quality in the tanks (R. Shields & J. Dye, pers. comm.).

The subsequent period of development up to weaning is more problematic. Abnormalities of mouth development, eye migration and pigmentation and scoliosis are commonly encountered. Feeding with copepods gives the best results, but work is proceeding in order to further develop enrichment procedures for *Artemia* as these organisms are at present much easier to maintain in culture throughout the year than copepods. Larvae can be easily weaned onto artificial diets after some 70 days from first feeding, although it is probable that this progression might be carried out much earlier. Larval metamorphosis, in particular the migration of the eyes and the change from a symmetrical to a flat fish, begins some 30 days post-feeding and continues throughout the period of time in which the fish are feeding on live feeds.

8.5 Viability of eggs and yolk-sac larvae

8.5.1 Environmental factors

Some important environmental factors determining egg and larval viability are temperature, salinity, mechanical stress and light, but also microbiology and exposure to antibiotics will influence the survival potential. Larval quality at the time of start-feeding will clearly be affected by its environmental history.

Eggs at different developmental stages have different tolerance limits to stress factors (Westernhagen 1988). Holmefjord & Bolla (1988) found high mortality when a standard mechanical stress was introduced during the first days of embryonic development, while no mortality was observed when the same stress was applied after closure of the blastopore. Similarly, Opstad & Raae (1986) found a linear relationship between increasing mechanical stress (standardized air bubbling) and egg mortality at the gastrula stage for cod eggs, whereas little mortality was observed when the same levels of stress were applied after closure of the blastopore. Transport of eggs should thus be recommended after blastopore closure.

The very long yolk-sac stage of Atlantic halibut larvae make them especially vulnerable to possible sub-lethal stressors. The larvae are currently produced either in small stagnant units or in large incubators with a slow upwelling flow to keep the larvae in the water column (Holmefjord et al. 1994, Opstad & Bergh 1993) (Fig. 8.7). Major considerations are to avoid mechanical stress on the delicate larvae and to provide good hygienic conditions in tanks. The water flow in the large upwelling systems reduces the accumulation of bacteria, but increased flow rates caused increased mortalities (Opstad & Bergh 1993). In small stagnant units, low concentrations of antibiotics have been added to partly control bacterial growth (Holmefjord et al. 1993).

High survival rates and high frequencies of normal development have been obtained for halibut yolk-sac larvae at low temperatures (2–6°C), while 9–10°C caused high frequencies of deformed larvae (Bolla & Holmefjord 1988, Pittman et al. 1990a). The first period after hatching was most sensitive to increased temperatures (Lein & Holmefjord 1993). For cod embryos normal development can occur between −1.5 and 12°C, while yolk-sac larvae seem to tolerate temperatures up to 16°C (Westernhagen 1970, 1988, Laurence & Rogers 1976, Thompson & Riley 1981, Iversen & Danielssen 1984, Makhotin et al. 1984).

Yolk-sac larvae of halibut seem to be more stenohaline than cod larvae. When newly-hatched halibut larvae were transferred to different salinities, the highest yield of functional larvae was found at salinities of 30–34‰. High salinities caused low survival, while low salinities gave high survival rates but low frequencies of normal larvae (Tveite 1992). When different salinities were introduced 30 days after hatching, the larvae tolerated a wider range of salinities (Tveite 1992). Westernhagen (1970) found that eggs from Baltic cod develop normally within a range of 20–33‰. Eggs from more oceanic cod stocks generally have high mortalities in 20‰ but may survive well in salinities up to 41‰ (Laurence & Rogers 1976, Kjørsvik *et al*. 1984). Cod larvae may tolerate salinities down to 2–3‰ (Yin & Blaxter 1987).

Even low light intensities induced abnormal development when newly-hatched larvae of Atlantic halibut were exposed to light (Bolla & Holmefjord 1988). At late larval stages, halibut larvae tolerate more light, and a positive phototaxis was shown for larvae older than 20 days (Naas & Mangor-Jensen 1990). There seem to be no reports on negative effects of normal light regimes on cod larvae, although light might affect the utilization of yolk reserves in the larvae. Growth of yolk-sac larvae of cod seems more affected by light than temperature in the temperature range 3–7°C. Larvae reared under continous light were shorter than larvae reared in total darkness, whereas no significant difference in length was obtained by temperature variations according to Solberg & Tilseth (1987).

8.5.2 Microbiology and antibiotics

Microbial factors have received increasing attention in culture because of reported improvements in egg and larval viability. Mass production of marine fish larvae is characterized by incubation of eggs and larvae in systems with large numbers of individuals, and the organisms may be exposed to dense bacterial populations. It has been recognized for a long time that the surfaces of eggs and larvae provide good substrates for bacterial colonization (Oppenheimer 1955). Cod and halibut eggs can be colonized within a few hours after fertilization (Hansen & Olafsen 1989), and bacteria may cause damage to the outer chorion surface (Hansen & Olafsen 1989, Bergh *et al*. 1992) or unfavourable conditions for the embryos (Barker *et al*. 1989).

Several suspected pathogenic bacterial strains have been isolated from marine eggs and larvae (Bolinches & Egidius 1987, Barker *et al*. 1989, Hansen & Olafsen 1989, Bergh *et al*. 1992), and they may have different effects on larval performance. Exposure of halibut eggs to *Flexibacter* sp. caused high mortalities in late egg stages, at hatching and during the early yolk-sac stage, whereas exposure of eggs to three different *Vibrio* spp. caused low mortality up to hatching followed by a high mortality during the yolk-sac stage (Bergh *et al*. 1992). Bacteria were present in the gill cavity, heart and frontal yolk sac regions in the *Vibrio*-infected larvae, although the *Vibrio* bacteria did not penetrate the egg chorion before hatching. A new species, *Flexibacter ovolyticus*, was recently isolated from halibut eggs (Hansen *et al*. 1992a). This species is an opportunistic pathogen for halibut eggs and larvae, with enzymatic capacity to dissolve the outer layers of the egg shell. It causes disease and high mortality levels when present in high density, or when environmental factors are unfavourable for the larvae.

The ability of the larvae to withstand potentially harmful bacteria is not well understood, nor is the relationship between intestinal microflora and survival rates of fish larvae. Stressed larvae from sub-optimal rearing conditions may be more susceptible to intestinal infections than healthy larvae, owing to an influence on the immune system (Tannock 1984, Angelidis et al. 1987). An intestinal microflora seems to be established in the early larval stage (Yoshimizu et al. 1980, Sugita et al. 1982, 1988, Campell & Buswell 1983, Muroga et al. 1987, Hansen et al. 1992b), probably in a relatively non-selective manner (Cahill 1990). Yolk-sac larvae ingest bacteria during osmoregulatory drinking, which occurs even before the mouth and digestive tract appears functional in cod and halibut (Mangor-Jensen & Adoff 1987, Tytler & Blaxter 1988, Kjørsvik & Reiersen 1992). Bacteria have been observed in the intestinal tract of these larvae before active feeding commences, and gut endocytosis of bacteria was demonstrated in yolk-sac larvae of cod and herring (Olafsen 1984, Hansen & Olafsen 1990, Hansen et al. 1992b, Olafsen & Hansen 1992). Hansen et al. (1992a,b) found that the primary intestinal microflora in cod and halibut larvae developing in intensive systems could originate from the egg epiflora at hatching. Results obtained with turbot larvae by Munro et al. (1993) further indicate that rearing conditions that lead to a slow rate of bacterial colonization of the larval gut may be beneficial.

Problems with variable and unpredictable levels of survival, growth and quality of cod and halibut larvae can to a large extent be explained by microbial conditions and bacterial infections during the egg and yolk-sac stages (Bergh & Jelmert 1990, Vadstein et al. 1993). High mortalities among fish larvae are often associated with bacterial pathogens (Iwata et al. 1978, Kusuda et al. 1986, Masumura et al. 1989, Muroga et al. 1990) although known pathogens are not always identified (Muroga et al. 1987, Nicolas et al. 1989). Shelbourne (1963) experienced a major breakthrough by introducing antibiotics in larval rearing of plaice. The use of antibiotics reduces larval mortalities significantly, and prophylactic treatment with antibiotics in egg and larval incubation systems has been widely used (Blaxter & Hunter 1982, Gatesoupe 1982, 1989, Perez-Benavente & Gatesoupe 1988, Holmefjord et al. 1994).

However, routine use of antibiotics may have several negative consequences in larval rearing, such as the danger of selecting antibiotic-resistant or opportunistic bacteria, or a change in larval microflora which can affect the digestion and general viability of the larvae. Recent findings by Hansen et al. (1992b) demonstrated that the presence of antibiotics resulted in a dramatic change in the intestinal microflora of herring larvae, as bacterial flora of yolk-sac larvae exposed to antibiotics consisted of three or four *Flavobacterium* spp. and control larvae were dominated by *Pseudomonas* or *Alteromonas* strains. Microbial control of the egg and larval environment by other means than antibiotics should therefore be sought, in order to minimize the danger of encouraging opportunistic and potentially harmful bacteria.

Vanbelle et al. (1990) discussed how the use of probiotics can enhance the health and productivity of farmed animals. Possible strategies to obtain microbial control in larval rearing can be based on fundamental principles in microbial ecology, on present knowledge of interactions between larvae and bacteria and on larval immunology. Such a strategy is proposed by Vadstein *et al.* (1993), by combining three elements: non-selective reduction of bacteria, selective enhancement of bacteria and improvement of larval resistance (Table 8.3). Preliminary results along these lines have led to increased viability and growth of larvae, following the development of methods for surface disinfection of eggs (Salvesen *et al.* 1991, T. Harboe, I. Huse & G. Øie, pers. comm.) or selection for desirable bacteria in larval tanks (Gatesoupe 1991).

Better knowledge must be achieved regarding methods for effective surface disinfection of eggs, when a bacterial microflora is established in fish larvae, whether this primary microflora is selective or basically a reflection of the environment, and effects of manipulation of the larval microbial flora.

Table 8.3 Elements of a possible strategy to obtain microbial control in rearing of marine fish eggs and larvae, with examples of possible methods (after Vadstein *et al.* 1993)

Non-selective reduction of bacteria
  Surface disinfection of eggs (Bergh & Jelmert 1990, Salvesen *et al.* 1991, T. Harboe, I. Huse & G. Øie, pers. comm.
  Reduction of input of organic matter
  Removal of organic matter
  Grazer control of bacterial biomass (Maeda & Nogami 1989)

Selective enhancement of bacteria
  Selection for desirable bacteria (Vanbelle *et al.* 1990, Gatesoupe 1991)
  Addition of selected bacteria to tanks
  Incorporation of selected bacteria in feed

Improval of larval resistance against bacteria
  Stimulation of general immune system
  Modulation of maternal immunity

8.6 Assessment of seed quality

Assessment of egg and larval quality should be performed as early in development as possible, and should be easy to perform. Possible indicators of egg quality in general were discussed by Kjørsvik *et al.* (1990), and fertilization rate, morphological characteristics (cell symmetry), rate of normal embryos and hatching rates were most often used as measures of quality. Fertilization rate and cell symmetry at the early blastula stage are considered to be possible direct indicators of viability in cod and halibut (Kjørsvik & Lønning 1983, Kjørsvik *et al.* 1984, Solemdal *et al.* 1992a,b, Bromage *et al.* 1994, Kjørsvik 1994). For cod, there is a highly significant correlation between observed rate of normal embryos and their hatching rate (Westernhagen *et al.* 1988, Kjørsvik, 1994).

Larval quality should be estimated by the end of the period of endogenous energy supply. Larval survival during the yolk sac stage is not a sufficient measure of larval viability. This is clearly demonstrated for halibut larvae, as larval groups often develop high rates of malformations although survival may be high (Pittman et al. 1989, see Pittman 1991). Larval functionality, which may be a more suitable quality parameter of larvae, may be estimated by morphological development, pigmentation pattern, survival rate in stress tests (see Section 15.4.2), or by response to a stimulus. Judged by larval morphology and survival, cod larvae hatching from eggs categorized as abnormal blastulae seemed less viable than larvae hatching from normal embryos (Kjørsvik 1994 and unpublished data). Behaviour may also be a useful tool for assessment of larval quality. Newly-hatched fish larvae have specific behaviour patterns, and behaviour will be irregular or abnormal if larvae are weak or have a poor functionality (Skiftesvik 1992, Solemdal et al. 1992a,b).

8.7 Conclusions

A reproducible high survival and growth of embryos and larvae is necessary for a successful mass rearing of marine juvenile fish. Although much progress has been made recently, this is still the major bottle-neck in halibut farming. Feeding and environmental regimes must be developed according to the biology and reproductive capacity of the broodstock to obtain an optimal production of eggs from batch-spawning fish with pelagic eggs. Many of the basic requirements, especially of halibut are still largely unknown. Stable and optimal conditions during spawning, as well as effective and cautious egg sampling methods should also be obtained. Natural spawning is the standard method for cod. This seems promising also for halibut, and will clearly be less stressful for the broodstock than stripping.

There is a need for standardized reporting procedures for assessment of egg and larval quality, in order to enable better comparison of results from different species and different institutions. The need for predictive measures of egg quality is clearly recognised by the ICES Working Group on Mass Rearing of Juvenile Marine Fish, which recommends that the potential usefulness of cell symmetry at the 8–16 cell stage should be explored (Anon 1993). Further understanding of processes causing varying egg quality should also enable improvements in broodstock performance and survivals of eggs and larvae in the hatchery.

Larval viability will to a large extent depend on the developmental history and environmental influences during egg and larval development. Reliable assessments of larval quality (functionality) should be developed for use at the end of the yolk-sac stage, before exogenous feeding starts. Control of microbial conditions in the early egg and larval phases seems to be an important challenge in obtaining reproducible production of juveniles, and should be further developed.

References

Adoff, G.R., Andersson, T., Engelsen, R. & Kvalsund, R. (1993) Land-based farm for ongrowing of halibut. In *Fish Farming Technology*, (eds H. Reinertsen, L.A. Dahle, L. Jørgensen & K. Tvinnereim), pp. 329–31. Balkema, Rotterdam.

Ahlström, E.H., Amaoka, K., Moser, D.A. & Sumida, B.Y. (1984) Pleuronectiformes: Development. In *Ontogeny and Systematics of Fishes*, (eds H.G. Moser, W.J. Richards, D.M. Cohen, M.P. Fahay, A.W. Kendall Jr & S.L. Richards), pp. 640–70. American Society of Ichthyologists and Herpetologists, Special Publication No. 1.

Andriyashev, A.P. (1954) *'Fishes of the Northern Seas of the USSR'*. Trudy Zoologichskogo Institut Akademiya Nauk SSSR, No. 53 (translated from Russian by the Israel Program for Scientific Translations, IPST Cat. No. 836).

Angelidis, P., Baudin-Laurencin, F. & Youinou, P. (1987) Stress in rainbow trout, *Salmo gairdneri*: effects upon phagocyte chemoluminescence, circulating leucocytes and susceptibility to *Aeromonas salmonicida*. *Journal of Fish Biology*, **31** (supplement A), 113–22.

Anon (1993) Report of the working group on mass rearing of juvenile marine fish to the Mariculture Committee of ICES. ICES CM 1993/F:8.

Babayan, V.K. (1978) Determination of some population parameters and possible catches of cod from the South Barents Sea. *Trudy VNIRO*, **128**, 44–51.

Barker, G.A., Smith, S.N. & Bromage, N.R. (1989) The bacterial flora of rainbow trout, *Salmo gairdneri* Richardson, and brown trout, *Salmo trutta* L., eggs, and its relationship to developmental success. *Journal of Fish Diseases*, **12**, 281–93.

Bengtson, D.A., Léger, P. & Sorgeloos, P. (1991) Use of *Artemia* as a food source for aquaculture. In *Artemia Biology*, (eds R.A. Browne, P. Sorgeloos & C.N.A. Trotman), pp. 255–85. CRC Press, Cleveland, Orio.

Berge, G.M. & Storebakken, T. (1991) Effect of dietary fat level on weight gain, digestibility and fillet composition of Atlantic halibut. *Aquaculture*, **99**, 331–8.

Bergh, Ø. & Jelmert, A. (1990) Antibacterial treatment of eggs of halibut (*Hippoglossus hippoglossus* L.). *ICES CM 1991*/F:39.

Bergh, Ø., Hansen, G.H. & Taxt, R.E. (1992) Experimental infection of eggs and yolk sac larvae of halibut, *Hippoglossus hippoglossus* L. *Journal of Fish Diseases*, **15**, 379–91.

Bergstad, O.A., Jørgensen, T. & Dragesund, O. (1987) Life history and ecology of the gadoid resources of the Barents Sea. *Fisheries Research*, **5**, 119–61.

Blaxter, J.H.S. & Hunter, J.R. (1982) The biology of the clupeoid fishes. *Advances in Marine Biology*, **20**, 1–223.

Blaxter, J.H.S., Danielssen, D., Moksness, E. & Øiestad, V. (1983) Description of the early development of the halibut *Hippoglossus hippoglossus* and attempts to rear the larvae past first feeding. *Marine Biology*, **73**, 99–107.

Bolinches, J. & Egidius, E. (1987) Heterotrophic bacterial communities associated with the rearing of halibut (*Hippoglossus hippoglossus* L.) with special reference to *Vibrio* spp. *Journal of Applied Ichthyology*, **3**, 165–73.

Bøhle, B. (1974) Temperature preference of cod (*Gadus morhua* L.). *Fisken og Havet*, Series B., **20**, 1–20 (in Norwegian).

Bolla, S. & Holmefjord, I. (1988) Effects of temperature and light on development of Atlantic halibut larvae. *Aquaculture*, **74**, 355–8.

Braaten, B. (1984) Growth of cod in relation to fish size and ration level. In *The Propagation of Cod*, Gadus morhua L. Flødevigen Rapportserie, (eds E. Dahl, D.S. Danielssen, E., Moksness & P. Solemdal), **1**, 677–710.

Bromage, N., Shields, R., Young, C., Bruce, M., Basavaraja, N., Dye, J., Smith, P., Gillespie, M., Gamble, J. & Rana, K. (1994) Egg quality determinants in finfish: The role of overripening with special reference to the timing of stripping in the Atlantic halibut *Hippoglossus hippoglossus*. *Journal of the World Aquaculture Society*, **25** (in press).

Bromley, P.J., Sykes, P.A. & Howell, B.R. (1986) Egg production of turbot (*Scophthalmus maximus* L.) spawning in tank conditions. *Aquaculture*, **53**, 287–95.

Bye, V.J. (1984) The role of environmental factors in the timing of reproductive cycles. In *Fish Reproduction. Strategies and Tactics*, (eds G.W. Potts & R.J. Wooton), pp. 187–205. Academic Press Inc., London.

Cahill, M.M. (1990) Bacterial flora of fishes: A review. *Microbial Ecology*, **19**, 21–41.
Campell, A.C. & Buswell, J.A. (1983) The intestinal microflora of farmed Dover sole (*Solea solea*) at different stages of fish development. *Journal of Applied Bacteriology*, **55**, 215–23.
Child, A.R., Howell, B.R. & Houghton, R.G. (1991) Daily periodicity and timing of the spawning of sole, *Solea solea* (L.), in the Thames estuary. *ICES Journal of Marine Science*, **48**, 317–23.
Crim, L.W. & Glebe, B.D. (1990) Reproduction. In *Methods for Fish Biology*, (eds C.B. Schreck & P.B. Moyle), pp. 529–47. American Fisheries Society, Bethesda, MD.
Devauchelle, N., Alexandre, J.C., Corre, N.L. & Letty Y. (1988) Spawning of turbot (*Scophthalmus maximus*) in captivity. *Aquaculture*, **69**, 159–84.
Devresse, B., Léger, P., Sorgeloos, P., Murata, O., Nasu, T., Ikeda, S., Rainuzzo, J.R., Reitan, K.I., Kjørsvik E. & Olsen Y. (1992) Improvement of flatfish pigmentation through the use of DHA enriched rotifers and *Artemia*. *Book of Abstracts, V, International Symposium on Fish Nutrition and Feeding*, Santiago, Chile, September 1992.
Dhert, P., Sorgeloos, P. & Devresse, B. (1993) Contributions towards a specific DHA enrichment in the live food *Brachionus plicatilis* and *Artemia* sp. *Fish Farming Technology. Proceedings of the First International Conference on Fish Farming Technology*, Trondheim, Norway, 9–12 August 1993. (eds H. Reinertsen, L.A. Dahle, L. Jørgensen & K. Tvinnereim), pp. 109–15. A.A. Balkema, Rotterdam.
Garrod, D.J. (1977) The North Atlantic Cod. In *Fish Population Dynamics*, (ed. J.A. Gulland), pp. 216–239. John Wiley and Sons, London.
Gatesoupe, F.-J. (1982) Nutritional and antibacterial treatments of live food organisms: the influence on survival, growth rate and weaning success of turbot (*Scophthalmus maximus*). *Annales de Zootechnie*, **4**, 353–68.
Gatesoupe, F.-J. (1989) Further advances in the nutritional and antibacterial treatments of rotifers as food for turbot larvae, *Scophthalmus maximus* (L.). In *Aquaculture – A Biotechnology in Progress*, Vol. 2 (eds M. De Pauw, E. Jaspers, H. Ackerfors & N. Wilkins), pp. 721–30. European Aquaculture Society, Bredene.
Gatesoupe, F.J. (1991) The use of probiotics in fish hatcheries: Results and prospects. *ICES CM* 1991/F:37.
Godø, O.R. & Moksness, E. (1987) Growth and maturation of Norwegian coastal cod and Northeast Arctic cod under different conditions. *Fisheries Research*, **5**, 235–42.
Goff, G.P. & Lall, S.P. (1989) An initial examination of the nutrition and growth of Atlantic halibut (*Hippoglossus hippoglossus*) fed whole herring with a vitamin supplement. *Proceedings of the Annual Meeting 1989, Aquaculture Association of Canada Symposium, Bulletin of the Aquaculture Association of Canada*, **89**, 53–5.
Grant, W.S., Teel, D.J., Kobayashi, T. & Schmitt, C. (1984) Biochemical population genetics of Pacific halibut (*Hippoglossus stenolepis*) and comparison with Atlantic halibut (*H. hippoglossus*). *Canadian Journal of Fisheries and Aquatic Sciences*, **41**, 1083–8.
Hansen, G.H. & Olafsen, J.A. (1989) Bacterial colonization of cod (*Gadus morhua* L.) and halibut (*Hippoglossus hippoglossus*) eggs in marine aquaculture. *Applied Environmental Microbiology*, **55**, 1435–46.
Hansen, G.H. & Olafsen, J.A. (1990) Endocytosis of bacteria in yolk sac larvae of cod (*Gadus morhua* L.). In *Microbiology in Poikilotherms*, (ed. R. Lésel), pp. 187–91. Elsevier Science Publishers B.V. (Biomedical Division), Amsterdam.
Hansen, G.H., Bergh, Ø., Michaelsen, J. & Knappskog, D. (1992a) *Flexibacter ovolyticus* sp. nov., a pathogen of eggs and larvae of Atlantic halibut, *Hippoglossus hippoglossus* L. *International Journal of Systematic Bacteriology*, **42**, 451–8.
Hansen, G.H., Strøm, E. & Olafsen, J.A. (1992b) Effect of different holding regimens on the intestinal microflora of herring (*Clupea harengus*) larvae. *Applied and Environmental Microbiology*, **58**, 461–70.
Harden Jones, F.R. (1968) *Fish Migration*. Arnold, London.
Haug, T. (1990) Biology of the Atlantic halibut, *Hippoglossus hippoglossus* (L., 1758). *Advances in Marine Biology*, **26**, 1–69.
Haug, T. & Gulliksen, B. (1988a) Fecundity and egg sizes in ovaries of female Atlantic halibut, *Hippoglossus hippoglossus* (L.). *Sarsia*, **73**, 259–61.
Haug, T. & Gulliksen, B. (1988b) Variation in liver and body condition during gonad development

of Atlantic halibut, *Hippoglossus hippoglossus* (L.). *Fiskeridirektoratets Skrifter, Serie Havundersøkelser*, **18**, 351–63.
Haug, T. & Kjørsvik, E. (1989) Comparative studies of Atlantic halibut (*Hippoglossus hippoglossus* L.) spawning in different areas. The Third ICES Symposium – the Early Life History of Fish, Bergen, 3–5 October 1988. *Rapports et Procès Verbaux de la Réunion, CIEM*, **191**, 440.
Haug, T. & Tjemsland, J. (1986) Changes in size- and age-distributions and age at sexual maturity in Atlantic halibut, *Hippoglossus hippoglossus*, caught in North Norwegian waters. *Fisheries Research*, **4**, 145–55.
Haug, T., Kjørsvik, E. & Solemdal, P. (1984) Vertical distribution of Atlantic halibut (*Hippoglossus hippoglossus*) eggs. *Canadian Journal of Fisheries and Aquatic Sciences*, **41**, 798–804.
Haug, T., Kjørsvik, E. & Solemdal, P. (1986) Influence of some physical and biological factors on the density and vertical distribution of Atlantic halibut *Hippoglossus hippoglossus* eggs. *Marine Ecology Progress Series*, **33**, 207–16.
Haug, T., Huse, I., Kjørsvik, E. & Rabben, H. (1989) Observations on the growth of juvenile Atlantic halibut (*Hippoglossus hippoglossus* L.) in captivity. *Aquaculture*, **80**, 79–86.
Hensley, D.A. & Ahlström, E.H. (1984) Pleuronectiformes: relationships. In *Ontogeny and Systematics of Fishes*, (eds H.G. Moser, W.J. Richards, D.M. Cohen, M.P. Fahay, A.W. Kendall Jr & S.L. Richards), pp. 670–87. American Society of Ichthyologists and Herpetologists, Special Publication No. 1.
Hjertnes, T. & Opstvedt, J. (1989) Effects of dietary protein levels on growth in juvenile halibut (*Hippoglossus hippoglossus* L.). *Proceedings of the Third International Symposium on Feeding and Nutrition in Fish, Toba, Japan 28 August – 1 September*, 189–93.
Holm, J.C. & Andersen, E. (1989) Improved spawning pen for Atlantic cod. *World Aquaculture*, **20**, 107.
Holmefjord, I. (1991) Timing of stripping relative to spawning rhythms of individual females of Atlantic halibut (*Hippoglossus hippoglossus* L.). *European Aquaculture Society, Special Publication No. 15*, Ghent, Belgium, 203–4.
Holmefjord, I. & Bolla, S. (1988) Effect of mechanical stress on Atlantic halibut eggs at different times after fertilization. *Aquaculture*, **68**, 369–71.
Holmefjord, I., Gulbrandsen, J., Lein, I., Refstie, T., Léger, Ph., Harboe, T., Huse, I., Sorgeloos, P., Bolla, S., Olsen, Y., Reitan, K.I., Vadstein, O., Øie, G. & Danielsberg, A. (1994) An intensive approach to Atlantic halibut fry production. *Journal of the World Aquaculture Society*, **24**, (in press).
Houghton, R.G., Last, J.M. & Bromley, P.J. (1985) Fecundity and egg size of sole (*Solea solea* L.) spawning in captivity. *Journal du Conseil International pour l'Exploration de la Mer*, **42**, 162–5.
Howell, B.R. & Scott, A.P. (1989) Ovulation cycles and post-ovulatory deterioration of eggs of the turbot (*Scophthalmus maximus* L.). *Rapports et Procès-Verbaum der Réunions, CIEM*, **191**, 21–6.
Howell, B.R., Child, A.R. & Houghton, R.G. (1991) Fertilization rate in a natural population of the common sole (*Solea solea* L.). *ICES Journal of Marine Science*, **48**, 53–9.
Huse, I. & Jensen, P.A. (1983) A simple and inexpensive spawning and egg collection system for fish with pelagic eggs. *Aquaculture Engineering*, **2**, 165–72.
Iversen, S.A. & Danielssen, D.S. (1984) Development and mortality of cod (*Gadus morhua* L.) eggs and larvae in different temperatures. In *The Propagation of Cod Gadus morhua L.*, (eds E. Dahl, D.S. Danielssen, E. Moksness & P. Solemdal), *Flødevigen Rapportserie*, **1**, 49–65.
Iwata, K., Yanahara, Y., & Ishibashi, O. (1978) Studies on factors related to mortality of young red sea bream (*Pagrus major*) in the artificial seed production. *Fish Pathology*, **13**, 97–107.
Jakupsstovu, S.H.I. & Haug, T. (1988) Growth, sexual maturation and spawning season of Atlantic halibut, *Hippoglossus hippoglossus*, in Faroese waters. *Fisheries Research*, **6**, 210–15.
Jobling, M. (1988) A review of the physiological and nutritional energetics of cod, *Gadus morhua* L., with particular reference to growth under farmed conditions. *Aquaculture*, **70**, 1–19.
Kanazawa, A. (1991) Nutritional mechanisms causing abnormal pigmentation in cultured marbled sole larvae, *Limanda yokohamae* (Heterosomata). In *Larvi '91 – Fish and Crustacean Larviculture Symposium, Ghent, Belgium, 27–30 August 1991*, (eds P. Lavens, P. Sorgeloos, E. Jaspers & F. Ollevier), pp. 20–22. *European Aquaculture Society, Special Publication* No. 15.

Kjesbu, O.S. (1988) *Aspects of the reproduction in cod* (Gadus morhua L.): *oogenesis, fecundity, spawning in captivity and stage of spawning*. Dr Scient. thesis, University of Bergen, Norway.

Kjesbu, O.S. (1989) The spawning activity of cod, *Gadus morhua* L. *Journal of Fish Biology*, **34**, 195–206.

Kjesbu, O.S. & Norberg, B. (1991) Oocyte growth in spawning Atlantic cod, *Gadus morhua*. In *Proceedings of the Fourth International Symposium on the Reproductive Physiology of Fish, Sheffield*, (eds A.P. Scott, J.P. Sumpter, D.E. Kime & M. Rolfe), p. 324.

Kjesbu, O.S., Klungsøyr, J. Kryvi, H., Witthames, P.R. & Greer Walker, M. (1991) Fecundity, atresia, and egg size of captive Atlantic cod (*Gadus morhua*) in relation to proximate body composition. *Canadian Journal of Fisheries and Aquatic Sciences*, **48**, 2333–43.

Kjesbu, O.S., Kryvi, H., Sundby, S. & Solemdal, P. (1992) Buoyancy variations in eggs of Atlantic cod (*Gadus morhua* L.) in relation to chorion thickness and egg size: theory and observations. *Journal of Fish Biology*, **41**, 581–99.

Kjørsvik, E. (1990) The effects of different incubation conditions on the eggs of halibut, *Hippoglossus hippoglossus* (L.). *Journal of Fish Biology*, **37**, 655–7.

Kjørsvik, E. (1994) Egg quality in wild and broodstock cod (*Gadus morhua* L.). *Journal of the World Aquaculture Society*, **25** (in press).

Kjørsvik, E. & Lønning, S. (1983) Effects of egg quality on normal fertilization and early development of the cod, *Gadus morhua* L. *Journal of Fish Biology*, **23**, 1–12.

Kjørsvik, E. & Reiersen, A.L. (1992) Histomorphology of the early yolk-sac larvae of the Atlantic halibut (*Hippoglossus hippoglossus* L.) – an indication of the timing of functionality. *Journal of Fish Biology*, **41**, 1–19.

Kjørsvik, E., Stene, A. & Lønning, S. (1984) Morphological, physiological and genetical studies of egg quality in cod (*Gadus morhua* L.). In *The Propagation of Cod* Gadus morhua L. (eds E. Dahl, D.S. Danielssen, E. Moksness & P. Solemdal), *Flødevigen Rapportserie*, **1**, 67–86.

Kjørsvik, E., Haug, T. & Tjemsland J. (1987) Spawning season of the Atlantic halibut (*Hippoglossus hippoglossus*) in northern Norway. *Journal du Conseil International pour l'Exploration de la Mer*, **43**, 285–93.

Kjørsvik, E., Mangor-Jensen, A. & Holmefjord, I. (1990) Egg quality in fishes. *Advances in Marine Biology*, **26**, 71–113.

Kohler, A.C. (1967) Size at maturity, spawning season and food of Atlantic halibut. *Journal of the Fisheries Research Board of Canada*, **24**, 53–66.

Kusuda, R., Yokoyama, J. & Kawaim, K. (1986) Bacteriological study on cause of mass mortalities in cultured black sea bream fry. *Bulletin of the Japanese Society of Scientific Fisheries*, **52**, 1745–51.

Kvenseth, P.G. & Øiestad, V. (1984) Large-scale rearing of cod fry on the natural food production in an enclosed pond. In *The Propagation of Cod* Gadus morhua L. (eds E. Dahl, D.S. Danielssen, E. Moksness & P. Solemdal), *Flødevigen Rapportserie*, **1**, 645–55.

Lam, T.J. (1983) Environmental influences on gonadal activity in fish. In *Fish Physiology*, Vol. IX B (eds W.S. Hoar, D.J. Randall & E.M. Donaldson), pp. 65–116. Academic Press Inc., London.

Laurence, G.C. & Rogers, C.A. (1976) Effects of temperature and salinity on comparative embryo development and mortality of Atlantic cod (*Gadus morhua* L.) and haddock (*Melanogrammus aeglefinus* L.). *Journal du Conseil International pour l'Exploration de la Mer*, **36**, 220–8.

Lehmann, G.B., Karlsen, Ø. & Holm, J.C. (1991) The impact of feeding on growth and sexual maturation in cod. *Institute of Marine Research, Report No. 22*. Bergen, Norway.

Lein, I. & Holmefjord, I. (1992) Age at first feeding of Atlantic halibut larvae. *Aquaculture*, **105**, 157–64.

Lein, I. & Holmefjord, I. (1993) Success factors in rearing of yolk sac larvae of Atlantic halibut. In *Fish Farming Technology*, (eds M. Reinertsen, L.A. Dahle, L. Jørgensen & K. Tvinnereim), pp. 57–8. Balkema, Rotterdam.

Lie, Ø, Lied, E. & Lambertsen, G. (1988) Feed optimization in Atlantic cod (*Gadus morhua*): Fat versus protein content in the feed. *Aquaculture*, **69**, 333–41.

Liu, H., Stickney, R.R. & Dickhoff, W.W. (1990) Changes in plasma concentrations of sex steroids in adult Pacific halibut *Hippoglossus stenolepis*. *Journal of the World Aquaculture Society*, **21**, 62–72.

Lønning, S., Kjørsvik, E., Haug, T. & Gulliksen, B. (1982) The early development of the halibut,

Hippoglossus hippoglossus (L.), compared with other marine teleosts. *Sarsia*, **67**, 85−91.
McEvoy, L.-A. (1984) Ovulatory rhythms and over-ripening of eggs in cultivated turbot, *Scophthalmus maximus* L. *Journal of Fish Biology*, **24**, 437−48.
McEvoy, L.A. & McEvoy, J. (1992) Multiple spawning in several commercial fish species and its consequences for fisheries management, cultivation and experimentation. *Journal of Fish Biology*, **41** (supplement B), 125−36.
McIntyre, A.D. (1952) The food of halibut from North Atlantic fishing grounds. *Marine Research*, 3.
McKenzie, R.A. (1934) The relation of the cod to water temperatures. *Canadian Fisherman*, **21**, 11−14.
Maeda, M. & Nogami, K. (1989) Some aspects of biocontrolling methods in aquaculture. *Proceedings from International Symposium on Marine Biotechnology, Japan 1989*, 395−8.
Makhotin, V.V., Novikov, G.G., Soin, S.G. & Timeiko, V.N. (1984) The peculiarities of the development of White Sea cod. In *The Propagation of Cod Gadus morhua L.* (eds E. Dahl, D.S. Danielssen, E. Moksness & P. Solemdal), *Flødevigen rapportserie*, **1**, 105−20.
Mangor-Jensen, A. & Adoff, G.R. (1987) Drinking activity of the newly hatched larvae of cod *Gadus morhua* L. *Fish Physiology and Biochemistry*, **3**, 99−103.
Mangor-Jensen, A., Sandness, K., Haaland, H. & Rosenlund, G. (1991) Effects of vitamin C in broodstock diets on egg quality of cod (*Gadus morhua* L.). *European Aquaculture Society (Ghent, Belgium) Special Publication No. 15*.
Masumura, K., Yasunobi, H., Okada, N. & Muroga, K. (1989) Isolation of a *Vibrio* species the causative bacterium of intestinal necrosis of Japanese flounder larvae. *Fish Pathology*, **24**, 135−41.
Mathisen, O.A. & Olsen, S. (1968) Yield isopleths of the halibut *Hippoglossus hippoglossus* L., in Northern Norway. *Fiskeridirektoratets Skrifter, Serie Havundersøkelser*, **14**, 129−59.
Methven, D.A., Crim, L.W., Norberg, B., Brown, J.A., Goff, G.P. & Huse, I. (1992) Seasonal reproduction and plasma levels of sex steroids and vitellogenin in Atlantic halibut (*Hippoglossus hippoglossus*). *Canadian Journal of Fisheries and Aquatic Sciences*, **49**, 754−9.
Munro, P.D., Birkbeck, T.H. & Barbour, A. (1993) Influence of rate of bacterial colonization of the gut of turbot larvae on larval survival. In *Fish Farming Technology*, (eds H. Reinertsen, L.A. Dahle, L. Jørgensen & K. Tvinnereim), pp. 85−92. A.A. Balkema, Rotterdam.
Muroga, K., Higashi, M. & Keitoku, H. (1987) The isolation of intestinal microflora of farmed red sea bream (*Pagrus major*) and black sea bream (*Acanthopagrus schlegeli*) at larval and juvenile stages. *Aquaculture*, **65**, 79−88.
Muroga, K., Yasunobu, H., Okada, N. & Masumura, K. (1990) Bacterial enteritis of cultured flounder *Paralichthys olivaceus* larvae. *Diseases of Aquatic Organisms*, **9**, 121−5.
Naas, K.E. & Mangor-Jensen, A. (1990) Positive phototaxis during late yolk-sac stage of Atlantic halibut larvae *Hippoglossus hippoglossus* (L.). *Sarsia*, **75**, 243−6.
Nicolas, J.-L., Robic, E. & Ansquer, D. (1989) Bacterial flora associated with a trophic chain consisting of microalgae, rotifers and turbot larvae: Influence of bacteria on larval survival. *Aquaculture*, **83**, 237−48.
Nikolsky, G.V. (1974) *Theory of the Fish Stock Dynamics*. Pishchevaja Promyshlennost' Press, Moscow.
Norberg, B. & Kjesbu, O. (1991) Reproductive physiology in coldwater marine fish: Applications in aquaculture. In *Reproductive Physiology of Fish. Proceedings of the Fourth International Symposium on the Reproductive Physiology of Fish. University of East Anglia, UK, 7−12 July 1991*, (eds A.P. Scott, J.P. Sumpter, D.E. Kime & M.S. Rolfe), 239−44.
Norberg, B., Valkner, V., Huse, I., Karlsen, I. & Lerøy Grung, G. (1991) Ovulatory rhythms and egg viability in the Atlantic halibut (*Hippoglossus hippoglossus*). *Aquaculture*, **97**, 365−71.
Nortvedt, R. (1992) Halibut farming. Nutrition, feed composition and production optimization. Report to the Norwegian research program 'New Species'. The Norwegian Research Council (in Norwegian).
Øiestad, V., Kvenseth, P.G. & Folkvord, A. (1985) Mass production of cod juveniles (*Gadus morhua*) in a Norwegian saltwater pond. *Transactions of the American Fisheries Society*, **114**, 590−5.
Olafsen, J.A. (1984) Ingestion of bacteria by cod (*Gadus morhua* L.) larvae. In *The Propagation of Cod Gadus morhua L.*, (eds E. Dahl, D.S. Danielssen, E. Moksness & P. Solemdal).

Flødevigen Rapportserie, **1**, 627–43.

Olafsen, J.A. & Hansen, G.H. (1992) Intact antigen uptake in intestinal epithelial cells of marine fish larvae. *Journal of Fish Biology*, **40**, 141–56.

Olsen, Y., Reitan, K.I. & Vadstein, O. (1993a) Dependence of temperature on loss rates of rotifers, lipids, and ω3 fatty acids in starved *Brachionus plicatilis* cultures. *Hydrobiologia*, **255**, 13–20.

Olsen, Y., Rainuzzo, J.R., Reitan, K.I. & Vadstein, O. (1993b) Manipulation of lipids and ω3 fatty acids in *Brachionus plicatilis*. In *Fish Farming Technology. Proceedings of the First International Conference on Fish Farming Technology, Trondheim, Norway, 9–12 August 1993*, (eds H. Reinertsen, L.-A. Dahle, L. Jørgensen & K. Tvinnereim), pp. 101–8. A.A. Balkema, Rotterdam.

Oosthuizen, E. & Daan, N. (1974) Egg fecundity and maturity of North Sea cod, *Gadus morhua*. *Netherlands Journal of Sea Research*, **8**, 378–97.

Oppenheimer, C.H. (1955) The effect of marine bacteria on the development and hatching of pelagic fish eggs and the control of such bacteria by antibiotics. *Copeia*, **1**, 43–9.

Opstad, I. & Bergh, Ø. (1993) Culture parameters, growth and mortality of halibut (*Hippoglossus hippoglossus* L.) yolk sac larvae in upwelling incubators. *Aquaculture*, **109**, 1–11.

Opstad, I. & Raae, A.J. (1986) Physical stress on halibut larvae. *ICES CM 1986/F:18*.

Pedersen, T. (1984) Variation of peak spawning of Arcto-Norwegian cod (*Gadus morhua* L.) during the period 1929–1982 based on indices estimated from fishery statistics. In *The Propagation of Cod Gadus morhua L.*, (eds E. Dahl, D.S. Danielssen, E. Moksness & P. Solemdal). *Flødevigen Rapportserie*, **1**, 301–16.

Perez-Benavente, G. & Gatesoupe, F.-J. (1988) Bacteria associated with cultured rotifers and *Artemia* are detrimental to larval turbot *Scophthalmus maximus* L. *Aquacultural Engineering*, **7**, 289–93.

Pittman, K. (1991) *Aspects of the early life history of the Atlantic halibut* (Hippoglossus hippoglossus *L.): Embryonic and larval development and the effects of temperature*. Doctoral thesis, University of Bergen, Norway.

Pittman, K., Berg, L. & Naas, K. (1987) Morphological and behavioural development of halibut (*Hippoglossus hippoglossus*) larvae with special reference to mouth development and metamorphosis. *ICES CM 1987/F:18*.

Pittman, K., Skiftesvik, A.B. & Harboe, T. (1989) Effect of temperature on growth rates and organogenesis in the larvae of halibut (*Hippoglossus hippoglossus* L.). *Rapports et Procès-verbaux des Réunions, CIEM*, **191**, 421–30.

Pittman, K., Bergh, Ø., Opstad, I., Skiftesvik, A.B., Skjoldal, L. & Strand, H. (1990a) Development of eggs and yolk-sac larvae of halibut (*Hippoglossus hippoglossus* L.). *Journal of Applied Ichthyology*, **6**, 142–60.

Pittman, K., Skiftesvik, A.B. & Bergh, L. (1990b) Morphological and behavioural development of halibut, *Hippoglossus hippoglossus* L. larvae. *Journal of Fish Biology*, **37**, 455–72.

Rabben, H., Nilsen, T.O., Huse, I. & Jelmert, A. (1986) Production experiment of halibut fry in large enclosed water columns. *ICES CM 1986/F:19*.

Reitan, K.I., Rainuzzo, J.R., Øie, G. & Olsen, Y. (1993) Nutritional effects of algal addition in first feeding of turbot (*Scophthalmus maximus* L.) larvae. *Aquaculture*, **118**, 257–75.

Rognerud, C. (1887) Hatching cod in Norway. *Bulletin of the United States Fisheries Commission*, **7**, 113–16.

Rollefsen, G. (1934) The eggs and larvae of the halibut (*Hippoglossus vulgaris*). *Kongelig Norske Videnskabers Selskabs Forhandlinger*, **6**, 20–23.

Rosenlund, G., Meslo, I., Røsjø, R. & Torp, H. (1993) Large scale production of cod. In *Fish Farming Technology. Proceedings of the First International Conference on Fish Farming Technology, Trondheim, Norway, 9–12 August 1993*, (eds H. Reinertsen, L.A. Dahle, L. Jørgensen & K. Tvinnereim), pp. 141–6. A.A. Balkema, Rotterdam.

Russell, F.S. (1976) *The Eggs and Planktonic Stages of British Marine Fishes*. Academic Press, London.

Salvesen, I., Jørgensen, L. & Vadstein, O. (1991) Evaluation of four chemicals for surface-disinfection of marine fish eggs. *European Aquaculture Society, Special Publication*, **15**, 406–8.

Sars, G.O. (1869) Report of practical and scientific investigations of the cod fisheries near Lofotenen Islands, made during the years 1864–1869. Translated from 'Indberetninger til Departementet for det Indre fra Cand. G.O. Sars om de af ham i aarene 1864–69 anstillede praktisk-videnskabelige Undersøgelser angaaende Torskefiskeriet i Lofoten', Christiania 1869. Translated by H. Jacobsen in *Report of the US Commission on Fisheries* 1877 (Pt.IV), 565–705.

Shelbourne, J.E. (1963) A marine fish-rearing experiment using antibiotics. *Nature*, **198**, 74–5.

Sivertsen, F. (1935) The spawning of cod. With special emphasis on the annual cycle in the reproductive organs. *Fiskeridirektoratets Skrifter, Serie Havundersøkelser*, **4**, 1–29.

Skiftesvik, A.B. (1992) Changes in behaviour at onset of exogenous feeding in marine fish larvae. *Canadian Journal of Fisheries and Aquatic Sciences*, **49**, 1570–2.

Smith, P.L. & Dye, J.E. (1988) First glimmers of success with halibut. *Fish Farmer*, **11**, 16–18.

Smith, P., Bromage, N., Shields, R., Ford, L., Gamble, J. Gillespie, M., Dye, J., Young, C. & Bruce, M. (1991) Photoperiod controls spawning in the Atlantic halibut (*Hippoglossus hippoglossus*). In *Reproductive Physiology of Fish*, (eds A. Scott, J. Sumpter, D. Kime & M.S. Rolfe), p. 172. University of East Anglia, Norwich.

Solberg, T.S. & Tilseth, S. (1987) Variations in growth patterns among yolk-sac larvae of cod (*Gadus morhua* L.) due to differences in rearing temperature and light regime. *Sarsia*, **72**, 347–9.

Solemdal, P. (1982) The spawning period of Arcto-Norwegian cod during the years 1976–1981. Report on the working group on larval fish ecology, Lowestoft, England, 3–6 July 1981. *ICES CM/L:3*.

Solemdal, P., Tilseth, S. & Øiestad, V. (1974) Rearing of halibut. I. Incubation and the early larval stages. *ICES CM 1974/F:41*.

Solemdal, P., Dahl, E., Danielssen, D.S. & Moksness, E. (1984) The cod hatchery in Flødevigen – background and realities. In *The Propagation of Cod Gadus morhua L.*, (eds E. Dahl, D.S. Danielssen, E. Moksness & P. Solemdal). *Flødevigen Rapportserie*, **1**, 17–45.

Solemdal, P., Kjesbu, O.S. & Kjørsvik, E. (1992a) The effects of maternal status of Arcto-Norwegian cod on egg quality and vitality of early larvae. I. The collection and characteristics of the cod females, a pilot study. *ICES CM 1992/G:78*.

Solemdal, P., Bergh, Ø., Finn, R.N., Fyhn, H.J., Grahl-Nielsen, O., Homme, O., Kjesbu, O.S., Kjørsvik, E., Opstad, I. & Skiftesvik, A.B. (1992b) The effects of maternal status of Arcto-Norwegian cod on egg quality and vitality of early larvae. II. Preliminary results of the experiment in 1992. *ICES CM 1992/G:79*.

Sugita, H., Enomoto, A. & Deguchi, Y. (1982) Intestinal microflora in the fry of *Tilapia mossambica*. *Bulletin of the Japanese Society of Scientific Fisheries*, **48**, 875.

Sugita, H., Tsunohara, M., Onishi, T. & Deguchi, Y. (1988) The establishment of an intestinal microflora in developing goldfish (*Carassius auratus*) of culture ponds. *Microbial Ecology*, **15**, 333–44.

Sundby, S. & Bratland, P. (1987) Spatial distribution and production of eggs from Northeast-Arctic cod at the coast of Northern Norway 1983–1985. *Fisken og Havet*, **1987**(1), 1–58.

Tannock, G.W. (1984) Control of gastrointestinal pathogens by normal flora. In *Current Perspectives in Microbial Ecology*, (eds M.J. Klug & C.A. Reddy), pp. 374–82. American Society for Microbiology, Washington, D.C.

Templeman, W. & Fleming, A.M. (1965) Cod and low temperatures in St Mary's Bay, Newfoundland. *ICNAF Special Publications*, **6**, 131–6.

Thompson, B.M. & Riley, J.D. (1981) Egg and larval development studies in the North Sea cod (*gadus morhua* L.). *Rapports et Procès-Verbaux de la Réunion, CIEM*, **178**, 553–9.

Tjemsland, J. (1960) *The halibut in Northern Norway*. Cand. Real. thesis, University of Bergen (in Norwegian).

Tomlinson, J.K. & Abramson, V.J. (1961) Fitting a von Bertalanffy growth curve by least squares. *California Department of Fish and Game*, 116.

Tretyak, V.L. (1984) A method of estimating the natural mortality rates of fish at different ages (exemplified by the Arcto-Norwegian cod stock). In *The Proceedings of the Soviet-Norwegian Symposium on Reproduction and Recruitment of Arctic Cod, Leningrad, 26–30 September 1983*. (eds O.R. Godø & S. Tilseth), pp. 241–74. Institute of Marine Research, Bergen.

Tveite, S. (1992) *Salinity effects on yolk-sac larvae of Atlantic halibut* (Hippoglossus hippoglossus L.). Cand. Scient., thesis, AKVAFORSK, NLH-Ås, Norway (in Norwegian).

Tytler, P. & Blaxter, J.H.S. (1988) Drinking in yolk-sac stage larvae of the halibut, *Hippoglossus hippoglossus* (L.). *Journal of Fish Biology*, **32**, 493–4.

Vadstein, O., Øie, G., Olsen, Y., Salvesen, I., Skjermo, J. & Skjåk-Bræk, G. (1993) A strategy to obtain microbial control during larval development of marine fish. In *Fish Farming Technology*, (eds H. Reinertsen, L.A. Dahle, L. Jørgensen & K. Tvinnereim), pp. 69–75. A.A. Balkema, Rotterdam.

Vanbelle, M., Teller, E. & Focant, M. (1990) Probiotics in animal nutrition: a review. *Archives of Animal Nutrition*, **40**, 543–67.

Waiwood, K.G. (1982) Growth history and reproduction in Atlantic cod (*Gadus morhua*). In *Proceedings of the International Symposium on Reproductive Physiology of Fish, Wageningen, The Netherlands, 2–6 August 1982*, (eds C.J.J. Richter & H.J.T. Goos), pp. 206–8. Pudoc, Wageningen.

Watanabe, T. (1985) Importance of the study of broodstock nutrition for further development of aquaculture. In *Nutrition and Feeding in Fish*, (eds C.B. Cowey, A.M. Mackie & S.G. Bell), pp. 395–414. Academic Press, London.

Watanabe, T. (1994) Importance of docosahexaenoic acid in marine larval fish. *Journal of the World Aquaculture Society*, **24** (in press).

Watanabe, T., Tamiya, T., Oka, A., Hirata, M., Kitajima, C. & Fujita, S. (1983) Improvement of dietary value of live foods for fish larvae by feeding them on ω3 highly unsaturated fatty acids and fat-soluble vitamins. *Bulletin of the Japanese Society of Scientific Fisheries*, **49**, 471–9.

Westernhagen, H. von (1970) Erbrütung der Eier von Dorsch (*Gadus morhua* L.), Flunder (*Pleuronectes flesus* L.) und Scholle (*Pleuronectes platessa* L.) unter kombinierten temperatur- und salzgehaltsbedingungen. *Helgoländer Wissenschaftliche Meeresuntersuchungen*, **21**, 21–102.

Westernhagen, H., von (1988) Sublethal effects of pollutants on fish eggs and larvae. In *Fish Physiology*, Vol. XI, part A (eds W.S. Hoar & D.J. Randall), pp. 253–346. Academic Press, London.

Westernhagen, H. von, Dethlefsen, V., Cameron, P., Berg, J., Fürstenberg, G. (1988) Developmental defects in pelagic fish embryos from the western Baltic. *Helgoländer Wissenschatliche Meeresuntersuchungen*, **42**, 13–36.

Whitehead, P.J.P., Bauchot, M.-L., Hureau, J.-C., Nielsen, J. & Tortonese, E. (eds) (1986) *Fishes of the North-eastern Atlantic*, Vol. II. UNESCO/Scientific and Cultural Organization, UK.

Wingfield, J.C. & Grimm, A.S. (1977) Seasonal changes in plasma cortisol, testosterone and oestradiol-17β in the plaice, *Pleuronectes platessa* L. *General and Comparative Endocrinology*, **31**, 1–11.

Woodhead, A.D. & Woodhead, P.M.J. (1965) Seasonal changes in the physiology of the Barents Sea cod, *Gadus morhua* L., in relation to its environment. I. Endocrine changes particularly affecting migration and maturation. *Special Publications of the International Commission of North-West Atlantic Fisheries*, **6**, 1–11.

Yin, M.C. Blaxter, J.H.S. (1987) Temperature, salinity tolerance and buoyancy during early development and starvation of Clyde and North sea herring, cod and flounder larvae. *Journal of Experimental Marine Biology and Ecology*, **107**, 279–90.

Yoshimizu, M., Kimura, T. & Sakai, M. (1980) Microflora of the embryo and the fry of salmonids. *Bulletin of the Japanese Society of Scientific Fisheries*, **46**, 967–75.

Chapter 9
Pacific Salmon (*Oncorhynchus* spp.)

9.1 Introduction
9.2 Taxonomy and life history
9.3 Culture of broodfish
    9.3.1 Holding facilities
    9.3.2 Stress and disease prevention and treatment
    9.3.3 Judging ripeness
    9.3.4 Spawning
    9.3.5 Acceleration of spawning
    9.3.6 Genetic considerations
    9.3.7 Supplementation hatcheries
    9.3.8 Commodity production
9.4 Hatcheries, conservation and rescue programmes
    9.4.1 Broodstock selection
    9.4.2 Genetic considerations
    9.4.3 Synchronization of spawning
    9.4.4 Genetic monitoring
Acknowledgements
References

The semelparous, anadromous life history of the Pacific salmon imposes important considerations for management of their broodstock. Because most cultured salmon are released, selection of donor stocks and breeding designs that preserve locally adapted genotypes are critical. Meeting the objectives of the hatchery programme is a major consideration for broodstock selection and breeding. Hatchery strategies range from preservation of the wild type, as in endangered species recovery programmes, to developing hatchery stocks that differ from endemic wild fish to minimize the effects of liberated hatchery fish on wild stocks. Other strategies concern totally captive stocks where the fish are never liberated. The degree of wildness of the hatchery populations varies and is determined by the number of generations that the broodstock is removed from the wild. Furthermore, because these fish are released into the natural environment, the aquaculturist has no control of factors affecting the well being of the fish for the majority of their lives; prevention and treatment of disease are thus absolutely essential. Because broodstock may be in very limited supply, maximizing the taking and viability of eggs is important.

    A major difficulty for the majority of Pacific salmon hatcheries is the need to establish production strategies and goals that harmonize with policies for management of wild stocks. Hatchery production in most cases can no longer be assessed by the number of juveniles released; it is the yield of adults that is important. However, given the current priority of managing river systems and fisheries for wild stocks, it is now critical that hatchery strategies employ concepts that either totally eliminate or at least minimize potential impacts on wild salmonids. Such strategies and practices must include genetic considerations that minimize potential effects should the hatchery fish interbreed with wild spawners. Quality of the hatchery product consequently needs to be defined in terms of genetic similarity of the hatchery gene pool to that of geographically proximate wild populations.

A second major difficulty in the culture of Pacific salmon is that spawning is accompanied by death. Pre-spawning mortality or poor gamete quality is a non-recoverable loss of reproductive potential from those individuals. A third problem is a general shortage of adequate adult broodfish. The lack of hatchery adults can be attributed to a poor return rate by the hatchery stock, unavailability of wild donor stocks, presence of the diseases necessitating culling, or genetic factors limiting the effective population size.

9.1 Introduction

The Pacific salmon of the genus *Oncorhynchus* have culture requirements that are similar to those of other members of the family Salmonidae. The semelparous nature and polymorphic life history characteristics of the various species of Pacific salmon impose some specialized broodstock management strategies and facilities that may differ within the group and from other fish. Furthermore, the goal for management of these broodstocks can be for vastly different purposes, including commercial aquaculture, enhancement of depleted wild stocks for the recreational and commercial fishery, and rescue and conservation of rare and endangered species.

9.2 Taxonomy and life history

Facilities and management of Pacific salmon broodstock are species and race dependent because life history patterns of the taxa are quite diverse. Salmon of the genus *Oncorhynchus* inhabit coastal streams of the northern Pacific rim countries from Japan in the west to the United States in the east. Seven species of salmon are currently recognized in this genus: the chinook (*O. tshawytscha*), coho (*O. kisutch*), sockeye (*O. nerka*), chum (*O. keta*), pink (*O. gorbuscha*), masu (*O. masou*), and amago (*O. rhodurus*) salmon. Masu salmon are found only in Asia in countries bordering the Sea of Japan and the Sea of Okhotsk; the amago salmon is restricted to Japan. The taxonomy of these species has been recently reviewed by Stearley & Smith (1993).

An excellent review of the life histories of these species has recently been edited by Groot & Margolis (1991). Whereas the timing of spawning by a species may be quite broad, the spawning period tends to be relatively short and may range from a few weeks in one stock to a few months in others (Table 9.1). Typically, spawning is in the autumn (fall), the exact timing often dependent on temperature. Some species like the pink and chum salmon typically make short spawning migrations in fresh water, entering the mouths

Table 9.1 Life history information for normal adult forms of anadromous Pacific salmon, including time spent in fresh water before migrating to the ocean (Juv FW), length of rearing in the ocean (Ocean), length of time as adults in fresh water prior to spawning (Adult FW), range of spawning date for the species (Dates of spawning), range of average eggs per female for stocks of the species (Fecundity), maximum number of eggs (Max), and the source (Reference)

Species	Time			Dates of spawning	Fecundity (Max)	Reference
	Juv FW	Ocean	Adult FW			
Chinook	3 months–1+ year	1–4 years	up to 9 months	Apr–Jan	4300–9400 (17 000)	Healey 1991
Coho	1.5 years	1.5 years	Few months	Aug–Apr	2000–5000 (7600)	Sandercock 1991
Sockeye	1+–2+ years	1–4 years	Several months	Jun–Feb	2000–2400 (5000)	Burgner 1991
Pink	weeks–months	2 years	weeks–months	Jun–Oct	1200–1900 (2800)	Heard 1991
Chum	few months	3–5 years	weeks–months	May–Jan	2000–4000 (8000)	Salo 1991
Masu	2–3+ years	few months	months	Jun–Oct	1900–3800	Kato 1991
Amago	1–2 years	6+ months	months	Oct–Nov	95 815	Kato 1991

of rivers in a relatively ripe condition. Other species may travel thousands of km upstream; some, such as the chinook salmon, reside for more than 6–9 months in fresh water prior to spawning. Salmon are categorized by spring, summer, fall (autumn), and winter runs, depending on when they return to fresh water from the ocean.

Polymorphism in the age at which adults ripen is considerable, even within a species. Males may mature precociously as smolt-sized individuals or as jacks after only a few months at sea. The age of maturity of typical adults is rather fixed in some species and quite variable in others (Table 9.1). The length of time in fresh water prior to smolting, the ocean rearing period, and the time in fresh water prior to spawning when they do not feed differs between and within the species (Table 9.1). Some species evolved forms that are not anadromous but spend their entire lives in fresh water (Healey 1991, Heard 1991, Salo, 1991, Sandercock 1991). These species are also cultured to some extent and include the kokanee, a form of sockeye, the yamame (or yamabe), a form of masu, and most populations of amago salmon (Burgner 1991, Kato 1991). In fact, the anadromous amago salmon spend very little time in the ocean and in open-ocean migration; non-migratory masu salmon males can spawn more than once (Kato 1991).

Because Pacific salmon can be spawned only once, the taking of eggs from each female must be maximized. Although these fishes are relatively fecund (Table 9.1), the male to female ratio may differ from 1:1 and thereby present problems in terms of maximizing the effective population size. Frequently, males compose the early part of each run and the sex ratio becomes more equal during the peak of spawning; however, it is difficult to generalize about the sex ratio of stocks of a species (see chapters in Groot & Margolis 1991). General references on the culture of these species are available (Leitritz & Lewis 1980, Laird & Needham 1988).

9.3 Culture of broodfish

9.3.1 Holding facilities

Broodstock can ideally be obtained by designing hatcheries that allow the fish to swim directly into the holding pond. Broodstock can also be captured with:

(1) weirs, the slats of which must be spaced to avoid gilling;
(2) rock or electric fences;
(3) gill or drift nets, which are not very desirable because of the injury they may cause;
(4) electroshocking, which can also cause injury; and
(5) gaffs or snag hooks that should be used only as a last resort.

Nets or hooks should be used only when the fish are near spawning and at low water temperatures (Anon., 1976, Senn *et al.* 1984). Adults may be steered to a certain location by using imprinting chemicals (Hasler et al. 1983), but they may not be particularly helpful in attracting adults back to the site where they were released as smolts (Rehnberg et al. 1985).

Holding facilities for broodstock must have good water quality and be constructed with smooth sides to avoid injury to the fish. Adults can be held in tanks at low densities, in double-stop weirs in rivers where water velocity is low, or in ponds, either as a side channel to a river or on land. If the holding facility is close to salt water intrusion, care must be taken to avoid stagnation at high tide. The fish cannot be overcrowded because of the potential for physical injury and oxygen depletion. As a general rule, for each kg of fish of about 8 kg in size, $35\,m^3$ of space is needed if the water flow rate is $3.8\,l\,min^{-1}$. Dissolved oxygen must remain above $6-7\,mg\,l^{-1}$. Holding adults in water above $13-14°C$ may lead to low egg quality. An upwelling water supply is optimum (Senn et al. 1984).

Holding facilities for the adults should allow for separation of the sexes. The fish need to be sorted by sex, and the males should be held downstream of the females. Ripe and unripe fish must also be sorted. Sick or dying fish must be removed as quickly as possible (Anon. 1976).

Fish should be allowed to mature in the river before they are taken into the hatchery. This is particularly advantageous for species like chinook and sockeye salmon that react unfavourably to holding conditions and must frequently be held for many months during the summer when water temperature exacerbates reactions to stress and pathogens (McNeil & Bailey 1975). The salmon may be held in sea water, but at warm temperatures diseases such as vibriosis can be a major problem. Allowing salmon to mature in full-strength sea water may be detrimental. Several species tend to do better if held in brackish water or sea water with a freshwater lens on top, but fresh water appears to be best (Wertheimer 1984, Wertheimer & Martin 1981).

The number of broodfish needed to meet a production goal can easily be calculated if one knows the fecundity of the fish and the egg-to-adult survival (Senn et al. 1984).

9.3.2 Stress and disease prevention and treatment

Broodstock must frequently be held for many months in the hatchery prior to spawning, often under stressful conditions of overcrowding and warm summer temperatures that render them particularly susceptible to stress and pathogens. In addition, maturation in these salmon is accompanied by senescence, a physiological development that includes significantly elevated corticosteriod hormone levels in these fish (Robertson & Wexler 1957, Donaldson & Fagerlund 1968).

Stress places a load on fish, making them less capable of coping with other challenges (Schreck & Li 1991). Corticosteroids also suppress the immune system and make resistance to pathogens more difficult (Maule et al. 1989).

Adult salmon face a double risk because holding and handling are stressful, thereby elevating corticosteroid concentrations (Schreck 1981), and their immune systems become naturally suppressed as they mature. Stress and pathology may even be caused by human shadows, vehicles and noise (Anon. 1976). Agitation of the fish can also increase oxygen demand, and transportation can be quite stressful for adults.

An integrated fish health management plan needs to be developed for each facility to minimize stress and maintain health of the fish. Stress can be reduced by holding the fish in cool water, having low densities, providing cover over the raceways or ponds (either physical or by spraying water), and by minimizing handling of the fish. We have observed that with particularly sensitive species, crowding and handling in direct sunlight should be avoided. If stress cannot be avoided, fish can tolerate it better in the morning before sunrise or on a cloudy day. Multiple stressors seem to leave cumulative effects (Barton et al. 1986); adequate recovery thus needs to be allowed between individual stressful practices. We have noted that species such as coho and chinook salmon undergoing final maturation in sea water are particularly sensitive to stress.

Adult salmon harbour a variety of bacterial and viral pathogens. They are also particularly susceptible to fungal infections. A fish health protection programme should be established that includes:

(1) monitoring for pathogens to check disease status, particularly if there is pre-spawning mortality (Bauer et al. 1988);
(2) treatment of diseases;
(3) methods for disinfection; and
(4) disposal of infected carcasses.

The Pacific Northwest Fish Health Protection Committee in 1989 prepared an excellent 'Model Comprehensive Fish Health Protection Program'. Adults can be injected with antibiotics (DeCew 1972). For example, erythromycin is used to fight bacterial kidney disease (*Renibacterium salmoninarum*) (BKD); it may also help to reduce the infection of progeny. Oxytetracycline may be the drug of choice if several pathogens are present and if the water is warm, and oxolinic acid may control furunculosis (*Aeromonas salmonicida*) (Dr R. Holt, pers. comm.).

Broodstock should be managed to avoid subsequent disease in progeny. BKD is vertically transmitted (Elliott et al. 1989), as perhaps is infectious hematopoietic necrosis virus, but this is open to question (Mulcahy & Pascho 1985, LaPatra et al. 1991). While isolation of broodfish may not be critical, eggs from individual spawners must be quarantined after fertilization so that the parents can be screened for these pathogens (Pascho et al. 1991). Eggs from parents with high rates of infection should be destroyed.

9.3.3 Judging ripeness

Pacific salmon do not typically complete spawning in captivity (except in spawning channels), and usually reproduction must be achieved through artificial fertilization. The quality of gametes peaks during the maturation period. Therefore, to optimize gamete quality, aquaculturists must discriminate between immaturity, full maturity and over-maturity. Furthermore, the race between death and maturation in Pacific salmon magnifies the importance of having a means for judging the ripeness of animals (especially females). However, the need for judging ripeness must be balanced against the need to minimize handling animals to reduce stress-related mortality. Usually, the fish are anaesthetized during inspection for maturity, facilitating the examination while keeping the fish in water. Numerous anaesthetics are available for use with salmon (Senn *et al.* 1984).

Many of the techniques developed for determining maturity in salmon are similar to those in other fish and include:

(1) Abdominal palpation; in females, the individual is gently raised with the tail higher than the head (and sometimes turned over), which reveals swelling and softening of the abdomen beneath the pectoral fins and collapse of the abdominal wall by the vent in ripe individuals (Anon. 1976); in males, gentle pressure applied to the abdomen in an anteroposterior direction results in expression of milt from the vent in ripe fish.
(2) Swollen, protruding genital papilla (females).
(3) Colouration and kipe formation (especially males): useful for relative measure of ripeness, but not for precise timing of maturity.
(4) Calendar date; same as colouration.
(5) Ultrasound; has been used to distinguish between males and females within 1 month of spawning (Martin *et al.* 1983), but not for precise timing of maturity.
(6) Plasma hormone levels; steroids and gonadotropins have been correlated with ovulation and spermiation in salmon (Fitzpatrick *et al.* 1986, Swanson 1991); however, the delay between sampling and assay results as well as the expense render these techniques experimental.

Examination of fish for ripeness should be performed with cotton, nylon or wool gloves and both hands should be used to support the weight of the fish to avoid damage to the eggs. Expulsion of the eggs from the vent by gentle abdominal pressure is generally not a good indicator of ripeness because eggs mature first in the posterior part of the skein. In addition, pressure on the abdomen can kill eggs. Premature or late spawning results in reduced fertility (Anon. 1976), although in coho salmon high quality gametes were produced even 20 days after ovulation (Fitzpatrick *et al.* 1987a).

9.3.4 Spawning

Once determined as ripe, the fish are killed by a blow to the head, anaesthetics or electric shock. Carbon dioxide, not a true anaesthetic, has been used with the advantage that the resulting carcasses are not contaminated with a chemical. Fish should be carried head downwards and should not be piled on top of one another, thus preventing damage to the eggs.

In females, the head, gills or tail are cut to allow the blood to drain, thus minimizing the chance for contamination of the eggs. The eggs may remain in a female for up to 1 h without loss of viability if the fish are kept cool. After draining the blood, a spawning knife is placed in the vent to prevent premature loss of eggs, the female is lifted by the gill cavity over a clean dry bucket, and an incision is made from the vent up (around the two pelvic fins) to about the pectoral fins. Most of the eggs should flow freely from the female into the bucket; the remaining ripe eggs can be dislodged by gentle shaking of the ovarian tissue and should be discarded.

In males, the ventral side of the fish is dried and then the fish is held firmly by the caudal peduncle and the head. By bending the tail and head upwards, milt is expressed from the vent. More milt can be expressed by gently moving the hand along the ventral part of the fish toward the vent. The gametes can be mixed together immediately or stored separately (Leitritz & Lewis 1980). Stored eggs remain viable longer than sperm (Withler & Humphreys 1967) and duration of gamete viability is inversely related to temperature (Withler & Morley 1968). Fertilization takes place when the sperm is activated by water in the presence of eggs.

Although the quality of egg and sperm should be of primary importance, few measurements exist for quantifying *a priori* such quality. Instead, the following characteristics are used; in females clear ovarian fluid, no blood in fluid, good (as judged from experience) colour in eggs, good shape and no broken eggs; and in males sperm disperses without adhesion or coagulation, good colour of milt, no blood or faeces in fluid. These judgments of quality are best left to experienced aquaculturists.

The one quantitative exception to these qualitative measurements is sperm motility. Because spermatocrits do not necessarily correspond to successful fertilization, microscopic measurements of sperm motility such as percentage of sperm that are active, speed of movement, and duration of activity following activation by water are used (Crim & Glebe 1990; see also Chapters 2 and 3). The overripening process in rainbow trout eggs is accompanied by an increase in water content, free lipid and calcium and a loss of precipitable protein and bound lipid (Craik & Harvey 1984). All measurements are terminal, thus requiring sub-sampling of individual egg lots, and none can be done rapidly to facilitate fertilization in a production hatchery.

Other factors that can affect egg quality include temperature, gamete storage time, and mechanical shock (Jensen 1981a). However, these parameters tend to be defined only in negative terms. Eggs that are not uniform in colour within a female, are from a female that has excessive water in her body cavity, or remain tightly bound in the skein, should not be used. Gametes should not be exposed to direct sunlight or rapid temperature changes (Anon. 1976).

Transportation of fertilized gametes must be done with care to avoid excessive mechanical shock or light (Jensen 1981a). In coho salmon, three periods of sensitivity to mechanical shock occurred after fertilization. These periods corresponded to between 5 and 10 min after water activation, between 45 min and 2 h, and between 4 and 14 days (Jensen 1981b). Eggs should, therefore, be moved to their incubators as soon as possible after water activation. Eyed eggs are relatively resistant to mechanical shock and can be transported with good success as long as temperatures are kept cool (but not too cold) and the eggs are kept moist (Anon. 1976).

9.3.5 Acceleration of spawning

Maturation of Pacific salmon in captivity can be problematic for a variety of reasons, including:

(1) some animals do not mature;
(2) some animals die prior to full maturity;
(3) production goals require eggs or progeny outside the time that fish can produce offspring; and
(4) unsynchronized maturation in small populations may result in females ripening when no ripe males are available or result in limited potential crosses available to maintain genetic diversity in broodstock.

To circumvent these problems, several techniques for manipulating spawning time have been developed. Most research has concentrated on altering the timing of reproduction in female salmon, although many of the techniques are transferable to males.

The reproductive development of female salmonids can be divided into distinct, yet continuous phases, previtellogenesis, vitellogenesis (production and incorporation of yolk proteins into the oocytes), final maturation, and ovulation (van Bohemen *et al.* 1981). The endocrine system directs final ovarian maturation in fish through the hypothalamo-pituitary-gonadal (H-P-G) axis, a hierarchical system of hypothalamic releasing hormones that affect gonadotropin secretion from the pituitary (Peter 1983), that in turn stimulates gonadal development directly or through stimulation of gonadal steroid production (Nagahama 1990).

Technologies to control the reproductive cycle of salmon can be divided into two categories, hormone treatments and photoperiod manipulations. Techniques for hormone-induced acceleration of reproduction manipulate the H-P-G axis to compress the time required to complete final maturation. Hormone treatments that induce final maturation and ovulation in females evolved from injections of gonadotropins (Clemens & Sneed 1962) to the use of gonadotropin-releasing hormone analogues (GnRHa) (Donaldson et al. 1981, Sower et al. 1982, Fitzpatrick et al. 1984).

GnRHa can be administered to the fish through intraperitoneal injection of hormone dissolved in sterile saline (Fitzpatrick et al. 1984) or through implantation of a pellet containing hormone (Crim & Glebe 1984), a technique not widely reported for use in Pacific salmon. GnRHa is ineffective if administered too long before the time of natural spawning (Crim & Glebe 1984, Fitzpatrick et al. 1987b). Therefore, the determination of the period of sensitivity to GnRHa is critical to successful acceleration of maturation. Plasma testosterone levels were predictive of ovulatory response to GnRHa in coho salmon (Fitzpatrick et al. 1987b). In general, GnRHa is effective when administered about 6 weeks before the time of natural ovulation and a successful response to GnRHa is ovulation within 2 weeks of treatment.

Photoperiod manipulations successfully accelerated maturation in rainbow trout (Bromage et al. 1982, Duston & Bromage 1987). However, the technique has not been widely used for Pacific salmon. Zaugg et al. (1986) reported maturation accelerated by about 4 weeks in spring chinook salmon exposed to decreasing photoperiods, whereas Johnson (1984) delayed maturation in the same species by 4–5 weeks using long day-length photoperiod during naturally decreasing day length. Maintaining coho and pink salmon under artificial photoperiods for their entire life cycles resulted in off-season spawning (MacQuarrie et al. 1981). Maturation and spawning were accelerated 2–3 months in masu salmon under an initial 1–2 months of long photoperiod during early maturation followed by short photoperiod during the latter stages of maturation (Takashima & Yamada 1984).

At the Little White Salmon National Fish Hatchery, spring chinook salmon arrive at the hatchery in mid-May from their up-river migration. About 1300 fish are held in raceways enclosed in a building equipped with sodium vapour lamps. The photoperiod begins at L:D 12:12 and is then reduced by 4 min per day until spawning is complete. Spawning takes place in mid-July under the accelerated photoperiod, rather than in mid-August under natural photoperiod. Although the eggs are smaller from accelerated fish, no problems have arisen with egg viability. The spawning season lasts the same length of time under accelerated and natural photoperiods (Jack Bodle, pers. comm.).

The use of photoperiod manipulation and GnRHa treatment together has not been documented for Pacific salmon. Uncertainty as to when sensitivity to GnRHa occurs under accelerated photoperiod may have discouraged attempts to combine these manipulations. However, the use of physiological monitoring may facilitate future attempts. The correlation between egg size and sensitivity to GnRHa in other species (Tamura & Lee 1987) and the small egg size in accelerated chinook salmon noted above suggests that acceleration of final maturation with both photoperiod manipulation and GnRHa treatment may be problematic. Administration of other hormones that increase the secretion and uptake of egg proteins may be needed if photoperiod and GnRHa are to be combined.

9.3.6 Genetic considerations

The purpose of broodstock choice, mating designs, rearing strategies, and genetic monitoring is to exploit efficiently genetic variation of broodstock to achieve aquacultural goals. Genetic variation is the biological potential of these programmes and Pacific salmon are genetically highly variable species (Grant *et al.* 1980, Beacham *et al.* 1985, Beacham *et al.* 1987, Wehrhahn & Powell 1987, Utter *et al.* 1989). Production strategies to achieve these goals are constrained primarily by two variables, the extent of control breeders have over their broodstock throughout its life and the degree to which genetic variation may be altered (Fig. 9.1). When broodstock are collected from the wild and progeny are to be released in the wild, maintenance of genetic diversity must be exploited to allow populations to respond to unpredictable environments. Alternatively, when salmon are captive for their entire life-cycle, genetic variation may be altered or exploited to produce desired yields.

9.3.7 Supplementation hatcheries

The goal of supplementation programmes is ultimately to increase natural production by restoring or increasing self-sustaining populations in streams

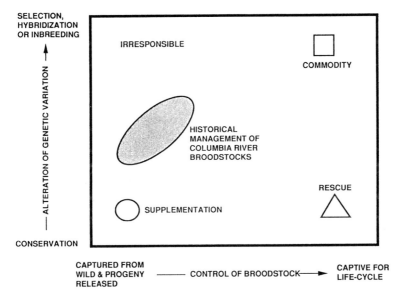

Fig. 9.1 Ideal production strategies for supplementation, commodity, and rescue hatcheries for Pacific salmon.

where they have been lost or are less abundant than desired. Historically, supplementation has had unpredictable and often detrimental genetic consequences for wild populations (Hindar *et al.* 1991) and may not succeed in increasing production (Reisenbichler & McIntyre 1986, Steward & Bjornn 1990). Consequently, protection of genetic variation of wild populations, as well as of hatchery broodstock, has become necessary.

Choice of broodstock is critically important. Broodstock taken from local populations generally perform better in native environments than exogenous hatchery strains or hybrids of the two (Bams 1976, Altukhov & Salmenkova 1987, Reisenbichler 1988, Hindar *et al.* 1991), presumably because Pacific salmon evolved genetically differentiated, ecologically specialized populations. First choice for donor stock is the wild population that is to be supplemented (see Tave 1986, Allendorf & Ryman 1987, Kapuscinski *et al.* 1994, for discussion of the appropriate numbers of founding fish.) Removal of large numbers of natural spawners from a population that may already be depressed may, however, accelerate decline of the wild population if the supplementation does not produce more returning spawners than are produced naturally. If large numbers of natural spawners are not available, rescue considerations (discussed later in this chapter) may be more appropriate.

Second choice (first choice for restoration programmes) is a neighbouring population or group of closely-related neighbouring populations of the same genetic lineage, with similar life-history traits, that occupy similar environments. Unique genetic variation is lost when even closely-related populations are combined and the performance of hybrid strains may be unpredictable because of outbreeding depression or heterosis. Neither outbreeding depression nor heterosis have been well documented in anadromous Pacific salmon. However, the certain loss of unique genetic variation is a necessary cost of supplementation in these situations. Although individual costs may be small, collectively the replacement of many genetically differentiated, ecologically specialized populations with fewer homogenous populations limits the potential of supplementation. The least desirable choice is an existing hatchery strain that has been derived from such a neighbouring population or the extirpated population that is to be restored.

Loss of genetic variation in each generation of hatchery population by genetic drift and inbreeding must be minimized in supplementation programmes to prevent the divergence of wild and hatchery populations and the increases in abnormal or less viable progeny. This may be accomplished by using as many individuals as possible, maintaining a 1:1 sex ratio during mating, and minimizing the variance in family size (the number of offspring surviving to reproduce). When large numbers of broodstock are used (>200), these conditions will be met by single pair matings in which sperm from a single male is used to fertilize eggs of a single female (Allendorf & Ryman 1987, Withler 1988).

The extent to which genetic variation has been lost by genetic drift and inbreeding in Pacific salmon because of less-than-ideal mating practices is uncertain. Heterozygosities of chinook salmon were not significantly lower in hatchery strains than in wild populations from the same geographical area (Utter et al. 1989). However, substantial allelic frequency variation and gametic disequilibrium in some chinook salmon populations were interpreted as evidence of low effective numbers of breeders in these programmes (Waples & Smouse 1990, Waples & Teel 1990). Evidence of low effective numbers of broodstock was documented in a coho salmon hatchery that consistently produced thousands of returning adults (Simon et al. 1986). Because one male may fertilize several females, production practices for Pacific salmon have often optimized smolt production in terms of the number of broodstock that can be held by using one male to several females (Senn et al. 1984) and by pooling milt; collectively these practices effectively lower the number of broodstock contributing to following generations (Withler 1988).

Intentional or unintentional artificial selection in the hatchery must also be minimized to prevent genetic differences, which might be unfavourable, from evolving between hatchery fish and the wild donor population. Effects of artificial selection will be minimized by choosing spawners randomly from the entire spawning period, maintaining phenotypic variation of offspring under uniform rearing conditions, and where the wild donor population can withstand the loss of adults, using wild broodstock every second or third generation (Allendorf & Ryman 1987).

Evidence of unintentional artificial selection in Pacific salmon is equivocal. Non-random choice of broodstock for physiological or behavioural traits, such as time of final maturation, has genetically altered some hatchery strains of Pacific salmon (reviewed in Steward & Bjornn 1990). The hatchery environment may also impose selection differentials on allele frequencies of juvenile traits that result in divergence of hatchery and wild populations (or domestication selection). This consequence could potentially be reversed by natural selection differentials during migration and ocean rearing or by non-random choice of returning adults that are used for spawning.

Rigorous evidence of domestication selection or the necessary conditions for the occurrence of domestication selection is lacking for Pacific salmon. However, inadvertent genetic change must still be considered likely, although the direction of the change may be unpredictable, precisely because the goal of supplementation programmes is to reduce the high mortality that occurs in wild populations from the egg to the smolt stage. This will lead to genetic divergence, unless one of several unlikely conditions occurs: all mortality is random with respect to genotype, or selection differentials in the smolt-to-spawner stage oppose natural selection in the egg-to-smolt stage.

9.3.8 Commodity production

Commodity production includes ocean ranching, in which broodstock may be taken from returning hatchery adults, as well as production programmes that rely on captive broodstock, such as some net-pen or cage culture of Pacific salmon. Genetic considerations in ocean ranching should be similar to supplementation programmes, except that ocean ranching programmes may not have similar access to wild local populations for broodstock, and concern about genetic divergence of hatchery and wild populations may be enforced by regulatory requirements rather than economic self-interest or conservation goals. In contrast, captive broodstock culture that prevents escape or release of salmon to the wild may use many of the techniques developed for genetic improvement of livestock. The theory, potential and application of these techniques are readily available elsewhere (e.g. Falconer 1981, Schultz 1986, Tave 1986, Refstie 1990).

Obtaining information on genetic performance necessary to choose broodstock may be the single aspect of genetic broodstock-management that most enhances the success of a programme for Pacific salmon. Obtaining such information is much more difficult than for domesticated livestock because of the recent wild origins and uses of most hatchery strains. For example, three kinds of information are useful for choosing broodstock:

(1) history of donor population;
(2) descriptions of performance of different strains in different environments; and
(3) estimates of the amount of additive genetic variation that can be exploited by artificial selection.

Examination of genealogies of Pacific salmon hatchery strains in the Columbia River Basin (Howell *et al.* 1985) revealed that of 26 major strains designated as local strains, 22 have undergone uncontrolled hybridization with exogenous wild or hatchery strains (Fig. 9.1) Although neither the magnitude nor often the exact identity of exogenous populations could be determined, allozyme studies can suggest genetic relationships (Schreck *et al.* 1986, Utter *et al.* 1989). With genealogies and hatchery transfer information as guidelines, estimates of the effective population sizes might be calculated from hatchery records of the numbers of spawners used, sex ratios and fertilization techniques. Disease resistance, disease history, age and size structure, time of migration, and spawning period were usually available, but estimates of growth, food conversion efficiency, or resistance to stress by different strains in different environments were lacking. Heritabilities for production traits of specific hatchery strains were also lacking.

9.4 Hatcheries, conservation and rescue programmes

As wild Pacific salmon populations decline in North America (Nehlser *et al.* 1991), rescue of endangered populations will increasingly replace supplementation. Rescue differs from supplementation in one fundamental way: sampling of the endangered, donor population for enough broodstock of each sex to

Pacific Salmon (Oncorhynchus *spp.*) 211

avoid significant loss of genetic variation will reduce the wild population to levels from which it is unlikely to recover by itself. Consequently, many rescue programmes must maintain the population *ex situ* until causes of decline have been corrected and the population can be restored to the wild. Lack of adequate numbers of broodstock means that special efforts in mating designs, determining sex before final maturation, synchronizing of spawning, preservation of gametes, and rearing strategies for juveniles must be made to balance losses of genetic variation from low effective population numbers, outbreeding and domestication selection.

9.4.1 Broodstock selection

When the decision is made to use hatcheries for the rescue of endangered populations, two basic strategies for collecting broodstock must be considered:

(1) collect all reproductively capable fish; and
(2) take limited numbers of fish from a number of closely-related populations.

The choice depends on the population structure of the target populations (Meffe & Vrijenhoek 1988). If fish are restricted to a single location, the first option should be considered. If fish are restricted to several locations, but historical patterns of gene flow have been limited between locations and interpopulation genetic diversity is great (for that species), then the first option should be considered, but groups from each area should be kept separate if possible. However, if endangered fish occur in several locations but interpopulation genetic diversity is low (for that species), then the second option may be more appropriate (Fig. 9.2).

Because sex ratios may be skewed in wild, declining populations, collecting equal numbers of males and females for broodstock will skew ratios of remaining wild fish even further and increase the likelihood of extinction. In such cases, it may be more appropriate to combine the sampling strategies by taking all reproductively capable fish from one location and just a few individuals from other locations. For species with overlapping generations, collections may be made in consecutive years and gametes preserved. However, continual use of small, wild populations increases the likelihood of divergence of hatchery and wild populations through genetic drift.

9.4.2 Genetic considerations

Rescue programmes are subject to similar genetic concerns as supplementation programmes. Two possible mating designs should be considered. In first generation matings of wild fish, factorial or diallel crosses allow unequal numbers of males and females to be used while increasing genotypic diversity in progeny that are to be returned to the wild (Fig. 9.3). In subsequent matings, pedigreed mating can be used to nearly double effective population size (Tave 1986). In

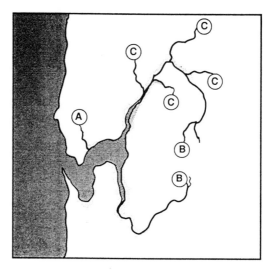

Fig. 9.2 Three situations requiring different sampling strategies for rescue of declining Pacific salmon populations. (A) All fish restricted to a single location. (B) Fish restricted to different locations, but historical patterns of geneflow have been limited, and interpopulation genetic diversity is large (for that species). (C) Fish restricted to different locations, but geneflow has occurred regularly, and interpopulation genetic diversity is small (for that species).

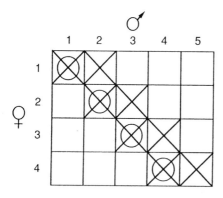

Fig. 9.3 Three hypothetical mating designs. Circles are single pair matings, diagonal crosses are factorial mating and squares are diallel matings.

pedigreed mating, an equal number of males and females are chosen from each family for broodstock, eliminating variation in family size and increasing genetic variance. This requires that families of the previous matings be kept separate until fish can be marked with long-lasting identification. Pedigreed matings are especially important in captive situations where large increases in numbers of broodstock cannot be maintained. Use of cryopreserved sperm may also increase genetic variance and effective population size (Wheeler & Thorgaard 1991).

Because endangered populations may be maintained for several generations before being released, broodstock and progeny must be reared in environments that minimize unnatural variation that could result in differential survival of different families or that impair performance in the wild (Kapuscinski *et al.* 1994). Considerable evidence suggests that hatchery fish raised in environments that simulate natural substrate, light, cover, food availability, density and flows for exercise may perform better than those raised under more typical hatchery conditions (Vincent 1960, Leon 1986, Murray & Beacham 1986, Gray & Cameron 1987, Fuss & Johnson 1988, Brannas 1989).

Although this is logistically more feasible in small rescue efforts than in large supplementation programmes, two problems have to be resolved. First, most existing Pacific salmon hatcheries would require extensive modification. Second, some domestication selection may be an unavoidable cost of rescue, but research is desperately needed to compare benefits of using more natural, heterogeneous environments to produce better performing fish with the theoretical benefits of using uniform environments to monitor changes in phenotypic variation that may indicate harmful selection differentials.

9.4.3 Synchronization of spawning

The techniques for accelerating maturation may have special application in the operation of conservation hatcheries. Conservation hatcheries will probably be expected to attain their production goals with limited population sizes. Therefore, to maintain genetic diversity it is critical that males and females mature synchronously so that the number of crosses between parents can be maximized.

For example, in 1991 the staff at the Coleman National Fish Hatchery was able to obtain only ten individuals of the threatened winter run of chinook salmon for the purposes of artificial propagation. The sex or maturity status of the fish could not be discerned from external characteristics, and plasma samples were therefore analysed for steroid content, which revealed that the population contained only one male (which was close to being ripe) and that four of the remaining females were maturing and probably sensitive to GnRHa treatment (M.S. Fitzpatrick, C.B. Schreck, J.S. Foott & D. Farnsworth, unpublished data). Such information facilitated decisions to capture more fish to increase the number of males and to initiate GnRHa treatment with those animals deemed most likely to respond. Despite the success of identifying the sex and maturity of these fish by measuring plasma hormone levels, the technique required invasive sampling techniques that may lead to excessive stress in some animals. A number of non- or minimally-invasive techniques have been reported which, with further development, may provide a less stressful means of gaining some of the same information.

Gordon *et al.* (1984) reported the use of an assay for vitellogenin in skin mucus from coho salmon to discriminate between males and females within 4 to 5 months of spawning. Ultrasound was also reported to be an effective discriminator of sex and maturity (Martin *et al.* 1983, Reimers *et al.* 1987).

Finally, Devlin *et al.* (1991) reported the ability to discriminate between males and females with a DNA probe. Once the sex of individual fish has been identified and the maturity judged, photoperiod or hormone treatments can be used to induce maturation. The use of hormone therapy may be more effective than photoperiod manipulation for synchronizing maturation because hormones can override endogenous rhythms that lead to natural variation in spawning time; photoperiod tends to accelerate the period of spawning but not compress it.

9.4.4 Genetic monitoring

Monitoring allows detection of genetic changes for supplementation and rescue programmes and evaluation of broodstock performance. Both allelic and quantitative genetic data are necessary for monitoring (Lande & Barrowclough 1987). However, measures that can detect a statistically significant, direct or indirect response of the population should be used. Captive broodstock improvement programmes normally include monitoring survival, growth, food conversion, yield, resistance to disease and stress, incidence and types of abnormalities, fecundity and viability, smoltification indices, as well as other specific traits of interest (Tave 1986). Controls can be used to partition the effects of environmental variation and genetic changes. Effective population sizes may be estimated and controlled to avoid unintentional genetic changes, but loss of heterozygosity and rare alleles may be monitored with allozyme data (Allendorf & Ryman 1987).

Because natural environmental variation cannot be controlled, monitoring in supplementation and rescue programmes must rely primarily on genetic measures or on quantitative traits that are determined very early in development, such as allozyme data or fluctuating morphological asymmetry (Leary *et al.* 1984, 1985, Allendorf & Ryman 1987) and must be conducted on both hatchery and wild populations. Demographic, physiological, and life-history data complement genetic data (Vrijenhoek 1989, Quatro & Vrijenhoek 1989), although few adequate models have been developed for monitoring Pacific salmon. For example, Kapuscinski & Lannan (1986) provided methods for estimating effective population size in Pacific salmon using demographic and life-history variables, some of which might be estimated with allozyme data, whereas Waples (1990a,b) derived an alternative method that relies primarily on allele frequency changes and average age of reproduction.

Acknowledgements

We appreciate the assistance of W. McNeil, Department of Fisheries and Wildlife, and J. Rohovec, Department of Microbiology, Oregon State University; R. Holt and T. Walters, Oregon Department of Fish and Wildlife; and J. Bodle, US Fish and Wildlife Service, for technical information. Cooperators on the Oregon Cooperative Fishery Research Unit are the US Fish and Wildlife

Service, Oregon State University, and the Oregon Department of Fish and Wildlife.

References

Allendorf, F.W. & Ryman, N. (1987) Genetic management of hatchery stocks. In *Population Genetics and Fishery Management*, (eds N. Ryman & F. Utter), pp. 1141–60. Washington Sea Grant, University of Washington, Seattle, WA.

Altukhov, Y.P. & Salmenkova, E.A. (1987) Stock transfer relative to natural organization, management, and conservation of fish populations. In *Population Genetics and Fishery Management*, (eds N. Ryman & F. Utter), pp. 333–43. Washington Sea Grant, University of Washington, Seattle, WA.

Anon. (1976) An outline of the management of salmon hatching and rearing. *Hokkaido Salmon Hatchery, Fishery Agency of Japan*, **9**, 1–43. Translated in 1978 by the Canadian Fisheries Marine Service Translation Series.

Bams, R.A. (1976) Survival and propensity for homing as affected by presence or absence of locally adapted paternal genes in two transplanted populations of pink salmon (*Oncorhynchus gorbuscha*). *Journal of the Fisheries Research Board of Canada*, **33**, 2716–25.

Barton, B.A., Schreck, C.B. & Sigismondi, L.A. (1986) Multiple acute disturbances evoke cumulative physiological stress responses in juvenile chinook salmon. *Transactions of the American Fisheries Society*, **115**, 245–51.

Bauer, J., Holt, R., Groberg, W., Onjukka, S., Amandi, A., Smith, L., Cummings, E., Banner, C., Kreps, T., Vendshus, S., LaPatra, L. & Kaufman, J. (1988) Augmented fish health monitoring in Oregon. *Bonneville Power Admin. Portland, Oregon, Annual Report 1987*. DE-A179-87BP33823.

Beacham, T.D., Withler, R.E. & Gould, A.P. (1985) Biochemical genetic stock identification of pink salmon (*Oncorhynchus gorbuscha*) in southern British Columbia and Puget Sound. *Canadian Journal of Fisheries and Aquatic Sciences*, **42**, 1474–38.

Beacham, T.D., Gould, A.P., Withler, R.E., Murrary, C.B. & Barner, L.W. (1987) Biochemical genetic survey and stock identification of chum salmon (*Oncorhynchus keta*) in British Columbia. *Canadian Journal of Fisheries and Aquatic Sciences*, **44**, 1702–13.

van Bohemen, C.G., Lambert, J.G.D. & Peute, J. (1981) Annual changes in plasma and liver in relation to vitellogenesis in the female rainbow trout. *General and Comparative Endocrinology*, **44**, 94–107.

Brannas, E. (1989) The use of a simulated redd for incubating Baltic salmon (*Salmo salar*) and trout (*Salmo trutta*) alevins. *Aquaculture*, **83**, 261–7.

Bromage, N.R., Whitehead, C. & Breton, B. (1982) Relationships between serum levels of gonadotropin, oestradiol-17β, and vitellogenin in the control of ovarian development in the rainbow trout. II. The effects of alterations in environmental photoperiod. *General and Comparative Endocrinology*, **47**, 366–76.

Burgner, R.L. (1991) Life history of sockeye salmon (*Oncorhynchus nerka*). In *Pacific Salmon Life Histories*, (eds C.G. Groot & L. Margolis), pp. 2–117. UBC Press, Vancouver.

Clemens, H.P. & Sneed, K.E. (1962) Bioassay and use of pituitary materials to spawn warm-water fishes. *US Department of the Interior, Fish and Wildlife Service Research Report*, **61**.

Craik, J.C.A. & Harvey, S.M. (1984) Biochemical changes associated with overripening of the eggs of rainbow trout *Salmo gairdneri* Richardson. *Aquaculture*, **37**, 347–57.

Crim, L.W. & Glebe B.D. (1984) Advancement and synchrony of ovulation in Atlantic salmon with pelleted LHRH analog. *Aquaculture*, **43**, 47–56.

Crim, L.W. & Glebe, B.D. (1990) Reproduction. In *Methods for Fish Biology*, (eds C.B. Schreck & P.B. Moyle), pp. 529–53. American Fisheries Society, Bethesda, Maryland.

DeCew, M.G. (1972) Antibiotic toxicity, efficacy, and teratogenicity in adult spring chinook salmon (*Oncorhynchus tshawytscha*). *Journal of the Fisheries Research Board of Canada*, **29**, 1513–17.

Devlin, R.H., McNeil, B.K., Groves, T.D.D. & Donaldson, E.M. (1991) Isolation of a Y-chromosonal DNA probe capable of determining genetic sex in chinook salmon (*Oncorhynchus tshawytscha*). *Canadian Journal of Fisheries and Aquatic Sciences*, **48**, 1606–12.

Donaldson, E.M. & Fagerlund, U.H.M. (1968) Changes in the cortisol dynamics of sockeye salmon (*Oncorhynchus nerka*) resulting from sexual maturation. *General and Comparative. Endocrinology*, **11**, 552–61.

Donaldson, E.M., Hunter, G.A. & Dye, H.M. (1981) Induced ovulation in coho salmon (*Oncorhynchus kisutch*). II. Preliminary study of the use of LH-RH and two high potency analogues. *Aquaculture*, **26**, 129–41.

Duston, J. & Bromage, N. (1987) Constant photoperiod regimes and the entrainment of the annual cycle of reproduction in the female rainbow trout (*Salmo gairdneri*). *General Comparative Endocrinology*, **65**, 373–84.

Elliott, D.G., Pascho, R.J. & Bullock, G.L. (1989) Developments in the control of bacterial kidney disease of salmonid fishes. *Diseases of Aquatic Organisms*, **6**, 201–15.

Falconer, D.S. (1981) *Introduction to Quantitative Genetics*. Longman, New York.

Fitzpatrick, M.S., Suzumoto, B.K., Schreck, C.B. & Oberbillig, D. (1984) Luteinizing hormone-releasing hormone analogue induces precocious ovulation in adult coho salmon (*Oncorhynchus kisutch*). *Aquaculture*, **43**, 67–3.

Fitzpatrick, M.S., Van Der Kraak, G. & Schreck, C.B. (1986) Plasma profiles of sex steroids and gonadotropin in coho salmon (*Oncorhynchus kisutch*) during final maturation. *General and Comparative Endocrinology*, **62**, 437–51.

Fitzpatrick, M.S., Schreck, C.B., Ratti, F.D. & Chitwood, R. (1987a) Viabilities of eggs stripped from coho salmon at various times after ovulation. *Progressive Fish Culturist*, **49**, 177–80.

Fitzpatrick, M.S., Redding, J.M., Ratti, F.D. & Schreck, C.B. (1987b) Plasma testosterone concentration predicts the ovulatory response of coho salmon (Oncorhynchus kisutch) to gonadotropin-releasing hormone analog. *Canadian Journal of Fisheries and Aquatic Science*, **44**, 1351–7.

Fuss, H.J. & Johnson, C. (1988) Effects of artificial substrate and covering on growth and survival of hatchery-reared coho salmon. *Progressive Fish Culturist*, **50**, 232–7.

Gordon, M.R., Owen, T.G., Ternan, T.A. & Hilderbrand, L.D. (1984) Measurement of a sex-specific protein in skin mucus of premature coho salmon (*Oncorhynchus kisutch*). *Aquaculture*, **43**, 333–9.

Grant, W.S., Milner, G.B., Krasnowski, P. & Utter, F.M. (1980) Use of biochemical genetic variants for identification of sockeye salmon *Oncorhynchus nerka* stocks in Cook Inlet, Alaska. *Canadian Journal of Fisheries and Aquatic Sciences*, **38**, 1665–71.

Gray, R.W. & Cameron, J.D. (1987) A deep-substrate streamside incubation box for Atlantic salmon eggs. *Progressive Fish Culturist*, **49**, 245–56.

Groot, C.G. & Margolis, L. (eds) (1991) *Pacific Salmon Life Histories*. UBC Press, Vancouver.

Hasler, A.D., Scholz, A.T. & Goy, R.W. (1983) *Olfactory Imprinting and Homing in Salmon: Investigations into the Mechanism of the Imprinting Process*. Springer-Verlag, New York.

Healey, M.C. (1991) Life history of chinook salmon (*Oncorhynchus tshawytscha*). In *Pacific Salmon Life Histories*, (eds C.G. Groot & L. Margolis), pp. 311–93. UBC Press, Vancouver.

Heard, W.R. (1991) Life history of pink salmon (*Oncorhynchus gorbuscha*). In *Pacific Salmon Life Histories*, (eds C.G. Groot & L. Margolis), pp. 119–230. UBC Press, Vancouver.

Hindar, K., Ryman, N. & Utter, F. (1991) Genetic effects of aquaculture on natural fish populations. *Canadian Journal of Fisheries and Aquatic Sciences*, **48**, 945–57.

Howell, P., Jones, K., Scarnecchia, D., LaVoy, L., Kendra, W. & Ortmann, D. (1985) *Stock Assessment of Columbia River Anadromous Salmonids*. Bonneville Power Administration, Portland, Oregon, Final Report, 1985. DE-A179-84BP12737.

Jensen, J.O.T. (1981a) Some spawning and incubation conditions that affect salmonid gamete viability and embryo development. In *Salmonid Broodstock Maturation*, (ed. T. Nosho), pp. 50–3. Washington Sea Grant, University of Washington Seattle, WA.

Jensen, J.O.T. (1981b) Mechanical shock sensitivity of coho eggs. In *Salmonid Broodstock Maturation* (ed. T. Nosho), p. 84. Washington Sea Grant, University of Washington Seattle, WA.

Johnson, W.S. (1984) Photoperiod induced delayed maturation of freshwater reared chinook salmon. *Aquaculture*, **43**, 279–87.

Kapuscinski, A.R. & Lannan, J.E. (1986) A conceptual genetic fitness model for fisheries management. *Canadian Journal of Fisheries and Aquatic Sciences*, **43**, 1606–16.

Kapuscinski, A.R., Steward, C.R., Goodman, M.L., Krueger, C.C., Williamson, J.H., Bowles, E.

& Carmichael, R. (1994) Genetic conservation guidelines for salmon and steelhead supplementation. NOAA Technics Report NMFS (in press).

Kato, R. (1991) Life histories of masu and amago salmon (*Oncorhynchus masou* and *Oncorhynchus rhodurus*). In *Pacific Salmon Life Histories*, (eds C.G. Groot & L. Margolis), pp. 447–520. UBC Press, Vancouver.

Laird, L. & Needham, T. (1988) *Salmon and Trout Farming*. Ellis Horwood Ltd, Chichester, UK.

Lande, R. & Barrowclough, G.F. (1987) Effective population size, genetic variation, and their use in population management. In *Viable Populations for Conservation*, (ed. M.E. Soule), pp. 87–123. Cambridge University Press, New York.

LaPatra, S.E., Groberg, W.J., Rohevec, J.S. & Fryer, J.L. (1991) Delayed fertilization of steelhead (*Oncorhynchus mykiss*) ova to investigate vertical transmission of infectious hematopoietic necrosis virus. In *Second International Symposium on Viruses of Lower Vertebrates*, (ed. J.L. Fryer), pp. 261–8. Oregon State University Printing Department, Corvallis.

Leary, R.F., Allendorf, F.W. & Knudsen, K.L. (1984) Superior developmental stability of enzyme heterozygotes in salmonid fishes. *American Naturalist*, 124, 540–51.

Leary, R.F., Allendorf, F.W. & Knudsen, K.L. (1985) Developmental instability as an indicator of the loss of genetic variation in hatchery trout. *Transactions of the American Fisheries Society*, 114, 230–5.

Leitritz, E. & Lewis, R.C. (1980) Trout and salmon culture. *California Department of Fisheries and Game, Fishery Bulletin*, 164.

Leon, K.A. (1986) Effect of exercise on feed consumption, growth, food conversion, and stamina of brook trout. *Progressive Fish Culturist*, 48, 43–6.

McNeil, W.J. & Bailey, J.E. (1975) *Salmon Ranchers Manual*. National Marine Fisheries Service, Settle.

McQuarrie, D.W., Markert, J.R. & Vanstone, W.E. (1981) Photoperiod induced off-season spawning of coho and pink salmon. In *Salmonid Broodstock Maturation*, (ed. T. Nosho), pp. 41–2. Washington Sea Grant, University of Washington, Seattle, WA.

Martin, R.W., Myers, J., Sower, S.A., Phillips, D.J. & McAuley, C. (1983) Ultrasonic imaging, a potential tool for sex determination of live fish. *North American Journal of Fishery Management*, 3, 258–64.

Maule, A.G., Tripp, R.A., Kaattari, S.L. & Schreck, C.B. (1989) Stress alters immune function and disease resistance in chinook salmon (*Oncorhynchus tshawytscha*). *Journal of Endocrinology*, 120, 135–42.

Meffe, G.K. & Vrijenhoek, R.C. (1988) Conservation genetics in the management of desert fishes. *Conservation Biology*, 2, 157–69.

Mulcahy, D.M. & Pascho, R.J. (1985) Vertical transmission of infectious hematopoietic necrosis virus in sockeye salmon (*Oncorhynchus nerka*): Isolation of virus from dead eggs and fry. *Journal Fish Disease*, 8, 393–6.

Murray, C.B. & Beacham, T.D. (1986) Effect of incubation density and substrate on the development of chum salmon eggs and alevins. *Progressive Fish Culturist*, 48, 242–9.

Nagahama, Y. (1990) Endocrine control of oocyte maturation in teleosts. In *Progress in Comparative Endocrinology*, (eds A. Epple, C. Scanes & M. Stetson), pp. 385–92. Wiley-Liss Inc., New York.

Nehlsen, W., Williams, J.E. & Lichatowich, J.A. (1991) Pacific salmon at the crossroads: stocks at risk from California, Oregon, Idaho and Washington. *Fisheries*, 16(2), 4–21.

Pascho, R.J., Elliott, D.G. & Streufert, J.M. (1991) Brood stock segregation of spring chinook salmon *Oncorhynchus tshawytscha* by use of the enzyme-linked immunosorbent assay (ELISA) and the fluorescent antibody technique (FAT) affects the prevalence and levels of *Renibacterium salmoninarum* infection in progeny. *Diseases of Aquatic Organisms*, 12, 25–40.

Peter, R.E. (1983) The brain and neurohormones in teleost reproduction. In *Fish Physiology*, Vol. 9A (eds W.S. Hoar, D.J. Randall & E.M. Donaldson), pp. 97–135. Academic Press, New York.

Quatro, J.M. & Vrijenhoek, R.C. (1989) Fitness differences among remnant populations of the endangered Sonoran topminnow. *Science*, 245, 976–8.

Refstie, T. (1990) Application of breeding schemes. *Aquaculture*, 85, 169–9.

Rehnberg, B.G., Curtis, L. & Schreck, C.B. (1985) Homing of coho salmon (*Oncorhynchus kisutch*) exposed to morpholine. *Aquaculture*, 44, 253–5.

Reimers, E., Landmark, P., Sorsdal, T., Bohmer, E. & Solum, T. (1987) Determination of salmonids' sex, maturation and size: an ultrasound and photocell approach. *Aquaculture Magazine*, **13**(6), 41–4.

Reisenbichler, R.R. (1988) Relation between distance transferred from natal stream and recovery rate for hatchery coho salmon. *North American Journal of Fishery Management*, **8**, 172–4.

Reisenbichler, R.R. & McIntyre, J.D. (1986) Requirements for integrating natural and artificial production of anadromous salmonids in the Pacific Northwest. In *Fish Culture in Fisheries Management*, (ed. R.H. Stroud), pp. 365–74. American Fisheries Society, Bethesda, MD.

Robertson, O.H. & Wexler, B.C. (1957) Pituitary degeneration and adrenal tissue hyperplasia in spawning Pacific salmon. *Science*, **125**, 1295–6.

Salo, E.O. (1991) Life history of chum salmon (*Oncorhynchus keta*) In *Pacific Salmon Life Histories* (eds C.G. Groot & L. Margolis), pp. 231–309. UBC Press, Vancouver.

Sandercock, F.K. (1991) Life history of coho salmon (*Oncorhynchus kisutch*). In *Pacific Salmon Life Histories*, (eds C.G. Groot & L. Margolis), pp. 395–445. UBC Press, Vancouver.

Schreck, C.B. (1981) Stress response in teleostean fishes: response to social and physical factors. In *Stress and Fish*, (ed. A.D. Pickering), pp. 295–321. Academic Press, London.

Schreck, C.B. & Li, H.W. (1991) Performance capacity of fish: Stress and water quality. In *Aquaculture and Water Quality*, (eds D.E. Brune & J.R. Tomasso). *Advances in World Aquaculture*, **3**, 21–29.

Schreck, C.B., Li, H.W., Hjort, R.C. & Sharpe, C.S. (1986) Stock identification of Columbia River chinook salmon and steelhead trout. *Bonneville Power Administration*. DE-A179-83BP13499, Portland, OR.

Schultz, F.T. (1986) Developing a commercial breeding program. *Aquaculture*, **57**, 65–76.

Senn, H., Mack, J. & Rothfus, L. (1984) Compendium of low-cost Pacific salmon and steelhead trout production facilities and practices in the Pacific Northwest. *Bonneville Power Administration*. DE-AC79-83BP12745 No. 83–353.

Simon, R.C., McIntyre, J.D. & Hemingsen, A.R. (1986) Family size and effective population size in a hatchery stock of coho salmon (*Oncorhynchus kisutch*). *Canadian Journal of Fisheries and Aquatic Sciences*, **43**, 2434–42.

Sower, S.A., Schreck, C.B. & Donaldson, E.M. (1982) Hormone-induced ovulation of coho salmon (Oncorhynchus kisutch) held in seawater and fresh water. *Canadian Journal of Fisheries and Aquatic Sciences*, **39**, 627–32.

Stearley, R.F. & Smith, G.R. (1993) Phylogeny of the Pacific trouts and salmons (*Oncorhynchus*) and genera of the family Salmonidae. *Transactions of the American Fisheries Society*, **122**, 1–33.

Steward, C.R. & Bjornn, T.C. (1990) Supplementation of salmon and steelhead stocks with hatchery fish: a synthesis of published literature. *US Fish and Wildlife Service Technical Report*, 90–1.

Swanson, P. (1991) Salmon gonadotropins: reconciling old and new ideas. In *Proceedings of the Fourth International Symposium on the Reproductive Physiology of Fish*, (eds A.P. Scott, J.P. Sumpter, D.E. Kime & M.S. Rolfe), pp. 2–7. Fish Symp 91, Sheffield.

Takashima, F. & Yamada, Y. (1984) Control of maturation in masu salmon by manipulation of photoperiod. *Aquaculture*, **43**, 243–57.

Tamura, C.S. & Lee, C.-S. (1987) Testosterone and estradiol-17β profiles of female milkfish, *Chanos chanos* undergoing chronic LHRH-A and 17α-methyltestosterone. In *Proceedings of the Third International Symposium on Reproductive Physiology of Fish*, (eds D.R. Idler, L.W. Crim & J.M. Walsh), p. 218. Memorial University of Newfoundland, St John's.

Tave, D. (1986) *Genetics for Fish Hatchery Managers*. Avi, Westport, Connecticut.

Utter, F., Milner, G., Stahl, G. & Teel, D. (1989) Genetic population structure of chinook salmon in the Pacific Northwest. *Fisheries Bulletin US*, **85**, 13–23.

Vincent, R.C. (1960) Some influences of domestication upon three stocks of brook trout (*Salvelinus fontinalis* Mitchell). *Transactions of the American Fisheries Society*, **89**, 35–52.

Vrijenhoek, R.C. (1989) Population genetics and conservation. In *Conservation for the Twenty-First Century*, (eds D. Western & M. Pearl), pp. 89–98. Oxford University Press, New York.

Waples, R.S. (1990a) Conservation genetics of Pacific salmon. II. Effective population size and the rate of loss of genetic variability. *Journal of Heredity*, **81**, 267–76.

Waples, R.S. (1990b) Conservation genetics of Pacific salmon. III. Estimating effective population size. *Journal of Heredity*, **81**, 277–89.

Waples, R.S. & Smouse, P.E. (1990) Gametic disequilibrium analysis as a means of identifying mixtures of salmon populations. *American Fisheries Society Symposium*, **7**, 439–58.

Waples, R.S. & Teel, D.J. (1990) Conservation genetics of Pacific salmon. I. Temporal changes in allele frequency. *Conservation Biology*, **41**, 144–56.

Wehrhahn, C.F. & Powell, R. (1987) Electrophoretic variation, regional differences, and gene flow in the coho salmon (*Oncorhynchus kisutch*) of southern British Columbia. *Canadian Journal of Fisheries and Aquatic Sciences*, **44**, 822–31.

Wertheimer, A.C. (1984) Maturation success of pink salmon (*Oncorhynchus gorbuscha*) and coho salmon (*O. kisutch*) held under three salinity regimes. *Aquaculture*, **43**, 195–212.

Wertheimer, A.C. & Martin, R.M. (1981) Viability of gemetes from adult anadromous coho salmon ripened in an estuarine pen. *Progressive Fish Culture*, **43**, 40–2.

Wheeler, P.A. & Thorgaard, G.H. (1991) Cryopreservation of rainbow trout semen in large straws. *Aquaculture*, **93**, 95–100.

Withler, R.R. (1988) Genetic consequences of fertilizing chinook salmon (*Oncorhynchus tshawytscha*) eggs with pooled milt. *Aquaculture*, **68**, 15–25.

Withler, F.C. & Humphreys, R.M. (1967) Duration of fertility of ova and sperm of sockeye (*Oncorhynchus nerka*) and pink (*O. gorbuscha*) salmon. *Journal of the Fisheries Research Board of Canada*, **24**, 1573–8.

Withler, F.C. & Morley, R.B. (1968) Effects of chilled storage on viability of stored ova and sperm of sockeye and pink salmon. *Journal of the Fisheries Research Board of Canada*, **25**, 2695–9.

Zaugg, W.S., Bodle, J.E., Manning, J.E. & Wold, E. (1986) Smolt transformation and seaward migration in 0-age progeny of adult spring chinook salmon (*Oncorhynchus tshawytscha*) matured early with photoperiod control. *Canadian Journal of Fisheries and Aquatic Sciences*, **43**, 885–8.

Chapter 10
Channel Catfish (*Ictalurus punctatus*)

10.1 Introduction
10.2 Broodstock management
    10.2.1 Sources of broodfish
    10.2.2 Selection of broodfish
    10.2.3 Pond culture systems for broodfish
    10.2.4 Nutrition of broodfish
    10.2.5 Parasites and diseases
    10.2.6 Predators and competition
10.3 Spawning
    10.3.1 Pond method
    10.3.2 Pen method
    10.3.3 Aquarium method
10.4 Egg incubation and hatching
    10.4.1 Natural incubation (open pond)
    10.4.2 Mechanical incubation
    10.4.3 Care of eggs
    10.4.4 Care of yolk-sac fry
10.5 Production of fry and small fish
    10.5.1 Trough and raceway culture
    10.5.2 Pond culture
    References

The channel catfish, *Ictalurus punctatus*, is the most important of farmed fish in the United States. Annual production of channel catfish is over 260 thousand tonnes and is projected to double by the turn of the century. In the 1960s and 1970s, research on broodstock selection and management, spawning and care and culture of larvae resulted in techniques that could provide the necessary numbers of quality seedstock as this warm-water aquaculture industry grew. Continued research in these areas and in water quality management, feeds development and disease control allowed production levels to increase from less than $1\,t\,ha^{-1}$ 25 years ago to over $4\,t\,ha^{-1}$ today, and has made catfish farming a success story in United States aquaculture.

However, there remain areas for improvement in disease prevention and control, nutrition and feed development, and in techniques for water quality management for both larvae and broodfish. There is no cure for channel catfish virus disease, that can devastate a whole population of small fish, and losses from enteric septicemia caused by the bacterium, *Edwardsiella ictaluri*, make this infectious disease the most important in catfish farming. Control of parasites remains a pressing need especially to producers of larvae. Of immediate concern in the United States and gaining in importance worldwide is the lack of enough approved chemicals and drugs for the treatment of diseases and control of parasites, for water quality management, and for elimination of other husbandry problems.

Great advances have been made in nutrition and feed development; however, price constraints and the limited availability of nutrients such as fish meal will require more research to fortify feeds properly for larvae and broodfish channel catfish. Vitamin and mineral supplementation for these specialty feeds will require more research as production intensifies.

10.1 Introduction

Channel catfish, *Ictalurus punctatus*, is the most important of the farmed fish in the United States. It is in the family Ictaluridae, order Siluriformes. Other species in the family that have been cultured and have some potential for commercial production, in order of priority, are blue catfish, *I. furcatus*; white catfish, *I. catus*; flathead catfish, *Pylodictus olivaris*; brown bullhead, *I. nebulosus*; yellow bullhead, *I. natalis*; and black bullhead, *I. melas*.

Most of these fishes are similar in appearance, with cylindrical bodies and skin without scales. The fins of the ictalurids are soft-rayed, but the pectoral and dorsal fins have sharp spines along the leading edges. A most noticeable characteristic in this family is the presence of barbels arranged around the mouth in a definite pattern of two above the jaws, four below the jaws, and one on each tip of the maxilla. The characteristics of the channel catfish that distinguish it from the other ictalurids are the deeply forked tail, anal fins with 24–29 rays, and irregular spots on the skin. The spots, however, may be absent in individuals larger than 0.5 kg.

The channel catfish, compared in numerous studies with other catfishes in the 1950s and 1960s, was shown to possess the best combination of characteristics of commercial importance. Seedstocks can be readily produced in large numbers at competitive prices. The species tolerates crowding in production systems of many different types, including ponds, cages and raceways, and the accompanying water quality conditions characteristic of intensive culture. Channel catfish efficiently convert to flesh economical and readily available feed ingredients in a variety of formulations, and they are resistant to many of the common fish diseases. At harvest, the processor finds that the channel catfish retains its quality in a variety of products in storage, and the consumer is presented with excellent-textured and mild-flavoured table fare.

The development of channel catfish is a major success story and serves as the model for new programmes in private aquaculture. The first known spawning of the species in captivity was reported in 1892 (Leary 1908), and the state fish hatchery in Pratt, Kansas, began propagation in 1910. However, production levels remained at less than 500 kg ha^{-1} until 1958, when the late Dr H.S. Swingle introduced pelleted feeds and produced what was then an astounding 2.4 t ha^{-1} (Swingle 1959). Research intensified, and in 1960 the first farmed fish were produced using the results of studies up to that time.

Subsequent research in feeds development, disease control and water quality management, which was quickly utilized by the producers, resulted in production levels increasing from $1.3\,t\,ha^{-1}$ in the 1960s to $4\,t\,ha^{-1}$ today, with some producers exceeding $6\,t\,ha^{-1}$ and total annual production now exceeds 200 000 t. Even larger production intensities and overall industry expansion are projected, as the results of research in disease prevention, water quality management, feeds development and other areas are made available to the private culturist (US Department of Agriculture 1990). Improvements in pond, cage and raceway systems, and possibly the development of affordable recirculation systems, will allow for even greater production intensities and profitability.

Late in coming, the ongoing research on genetics and selective breeding is expected to provide the catfish culturist with the same improved qualities of commercial interest, such as disease resistance, feed utilization efficiency and processing conversions, as have been demonstrated for poultry, swine, and other domestic animals. Aquaculture, as in other enterprises, requires the availability of good quality seedstock at affordable prices, which was once thought to be the major limiting factor in the development of channel catfish culture. Researchers and private fish farmers cooperated to develop the techniques for broodfish management and the production of eggs and larvae, which are now easily sufficient to meet the needs of the industry.

Basically, the programme involves the stocking of broodfish into an earthen pond, offering the fish formulated feeds and natural foods, and managing the pond to ensure good water quality. At the time of spawning, the culturist provides suitable vessels in which the female can deposit the eggs for normal hatching by the male, or the culturist removes the eggs for artificial hatching. Larvae obtained by either technique are stocked into earthen ponds specially prepared for that purpose, or they are retained in troughs and cultured in a more controlled environment to the juvenile stage before being transferred to the final grow-out facility. The techniques of broodstock management and egg and larval production commonly used by culturists are described below.

10.2 Broodstock management

The success of any culture programmes begins with broodstock management. Without broodfish capable of producing sufficient numbers of eggs and larvae that can be reared at affordable prices to sizes required by the culturist, the industry cannot grow sufficiently large to support the feed mills, processing plants and other service industries, or to encourage market development for the species. Successful management of broodfish begins with their selection for the qualities desired by the culturists and ends with the successful production of larvae.

10.2.1 Sources of broodfish

Only domesticated stocks of fish that have been subjected to several generations of intensive culture should be considered for potential broodfish. For the beginning producer, it may be more prudent to purchase 3 year old fish or preferably 4 year old fish and place them in a good management programme. While channel catfish are generally 4 years of age before becoming dependable spawners, well-cared-for-fish can be productive at 3 years of age. When mature fish capable of immediate spawning are required, purchase them from a reputable dealer, even if you must pay a premium price.

10.2.2 Selection of broodfish

Selective breeding of channel catfish has been slow in developing, and it will probably be many years before improved catfish breeds are available to producers. However, a number of techniques can be practised by the producer at this time to improve catfish performance.

(1) Select only fish without deformities, since some deformities are genetic. Avoid fish with a history of channel catfish virus (CCV) disease. The stock should be inspected by a trained fish pathologist and prophylactic treatments applied if appropriate. Fish should be free of sores and bruises and should not be thin and underfed.

(2) Select fish that appear to have good dress-out weight, since net weight is the true value of a fish's worth. Thus, selecting a male fish with a large head and slim body, or a female fish with a short body and full abdomen, should be avoided, since these characteristics do not appear essential for reproduction.

(3) Select the larger males and females from the first seining of the production pond not only for seineability, but also to ensure that the fish from subsequent harvests would not result in the saving of the slower-growing fish. Since the fish would be about 2 years old, avoid retaining those individuals that appear to be nearing sexual maturity. Since male fish are generally larger than female fish of the same age, individually inspect the fish to ensure that sufficient female fish are retained to have a 1:1 or 3:2 female:male ratio.

(4) Maintain a breeding population that contains several hundred breeding pairs to reduce in-breeding depression. Pairing female fish from one pond with males from another pond should be considered when dealing with small numbers of breeding pairs.

10.2.3 Pond culture systems for broodfish

Construction

Adequate facilities to care for the channel catfish broodfish and to permit the culturist to take the spawns or propagate the larvae and small fish are necessary for well-managed and profitable production. Facilities most commonly used are levee ponds built on level land, which require a source of surface or ground water, and watershed ponds built on sloping land, which use run-off water from the surrounding area as the water source. Tanks and raceways have been used successfully to maintain broodfish, but the programme requires continuous attention by experienced aquaculturists. At today's level of technology, this procedure probably should be attempted only if conventional facilities are not available.

Levee ponds, the most commonly used facility, require for their construction the presence of at least 1 m depth of clay soil that is free of pesticides and will retain water. A ground water supply of at least a minimal $500 \, l \, min^{-1} \, ha^{-1}$ is essential. The pond bottoms must be smooth and free of stumps and snags to permit seining. Water depths are generally 2 m at the deep end and slope upward to about 0.5 m at the opposite end and around the margin to accommodate the spawning containers. Levee tops should be a minimum of 3 m in width and covered with gravel to permit all-weather access for feeding, management and egg-taking.

Watershed ponds are generally used in areas where the terrain and groundwater supplies are not conducive to the construction of levee ponds. Owing to the lie of the land, the levee at the deep end of the pond is usually deeper than 2 m, but the culturist should construct the facility to ensure accessibility, as described for the levee ponds. Because water depths and water quality management are much more difficult in watershed ponds, the culturist should consider devoting the most easily managed production facilities to broodfish, larvae and small fish.

Pond preparation

The next year's harvest of eggs and larvae depends on this year's preparation of the broodfish pond, stocking and management. Preparation of the ponds to receive the broodfish should take place prior to the spawning season, so that the pond is ready to receive the broodfish as soon as spent or at the conclusion of the spawning period, normally between April and July. Many culturists rotate their ponds between food fish production, larvae and small fish production, and broodfish maintenance. The transfer of broodfish into ponds previously used for food fish production affords the opportunity to dry out the pond to eliminate fish diseases and unwanted wild fish, to eliminate fish-made depressions and to restore the levees.

After performing the required maintenance and repairs, the culturist applies a pre-emergent herbicide such as Aquazine[1] (Simazine) at $5-25\,kg\,ha^{-1}$ (Schnick et al. 1989) to prevent the growth of rooted aquatic vegetation, and then fills the pond with surface water or, preferably, well water. Surface water should be free of pesticides and filtered to prevent contamination with fish and other predators and competitors. Forage fish such as the fathead minnow (*Pimephales promelas*) may be stocked at the rate of $20-25\,kg\,ha^{-1}$.

Sexing of fish and stocking

The task of identifying and separating male and female fish is relatively easy when the fish are sexually mature and in good condition (Fig. 10.1). Character-

Fig. 10.1 Channel catfish. Females (right) in good condition display an egg-enlarged abdomen and have a head narrower than their bodies. Males (left) have wide mouths, larger head (sometimes with large muscles), and the body is narrower than the head.

---

[1] Use of trade names does not imply endorsement by the US Fish and Wildlife Service. Mention of chemical uses is not an endorsement of the chemical or the practice. Only chemicals specifically approved for use on food fish should be used and then only in accordance with product labels.

istics of the male include a broad, muscular head that is wider than the body, and often a mottled gray pigmentation on the underside of the jaw and abdomen. The female has a head that is narrower than the body, and usually has no distinct muscular pads or pigmentation on the underside of the jaw and abdomen. When the secondary sexual characteristics are not well defined, one must examine the genital areas. The urogenital pore of the male is located on the genital papilla, which looks like a small fleshy nipple behind the anus (Fig. 10.2, Norton *et al.* 1976). In contrast the female has an anterior genital pore and a posterior urinary pore for separate expulsion of eggs and urinary products (Fig. 10.2).

Broodfish used in the current year's programme and still suitable as broodfish, based on size, condition, and, most importantly, being proven spawners, are usually retained for the following year's programme. Generally, broodfish 5 years or older and weighing in excess of 4 kg are not restocked but sold for food. The proven broodfish are sexed and stocked at a rate of 500–1000 kg ha^{-1} and at a ratio of 1:1 to 5:3 females to males.

Replacement broodfish, preferably 3 years of age, are selected as described previously and are stocked at a rate of 400–700 kg ha^{-1} and at a ratio of 1:1 to 3:2 females:males. In other ponds, the lower stocking density and more equal stocking ratio are to allow for more intensive feeding and to offset the anticipated higher mortality of the males due to spawning-related injuries.

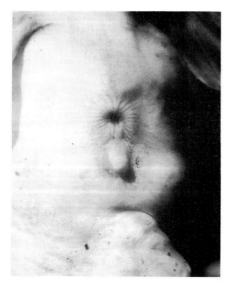 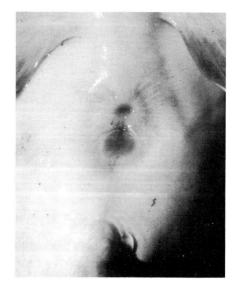

Fig. 10.2 The genital papilla posterior to the anus of the male channel catfish (left) appears as a fleshy nipple in contrast with the slit for the female (right).

Feeding techniques and water quality management

The culturist must regulate the biological, chemical and physical aquatic environment so that the conditions favour survival of the broodfish and production of large numbers of quality eggs and viable larvae. Because fish farming is a business, any action of the producer must be cost effective to ensure a profit.

The fish are offered these feeds at the rate of 2–4% of the body weight daily during the summer months when dissolved oxygen levels are above 4 ppm and when the water temperature exceeds 20°C, and they are fed *ad libitum* when the water is cooler, when they are still accepting feed. Feed should be offered around the entire pond. Forage should not be considered as a part of the daily ration, but if consumption of formulated feed decreases, apparently owing to the availability of live forage, decrease the feed allowance so as not to accelerate water quality deterioration.

The culturist may have initially filled the broodfish ponds with water of adequate quality. However, feeding the fish initiates deterioration in water quality. Continued reduction of water quality, characterized by low dissolved oxygen levels, high levels of ammonia and nitrites, carbon dioxide and sometimes hydrogen sulfide, stresses the fish and opens the way for diseases and parasites. Also excessive phytoplankton may be produced because of the nitrogen and phosphate in the feed waste, resulting in even greater demands on the existing dissolved oxygen at night and supersaturation during the day.

Poor water quality, regardless of the cause, must be corrected before the fish become stressed and diseased or die. Options for improving water quality include temporary reduction in the feed allowance, aeration to increase the level of dissolved oxygen, chemical oxidation to reduce the Biological Oxygen Demand (potassium permanganate), and pond flushing (Lay 1971, Boyd 1979 Huner & Dupree 1984a, Boyd & Ahmad 1987). Adverse water quality is correctable, but the techniques are expensive.

Emergency aeration is an essential tool to the commercial culturist, since catastrophic loss of dissolved oxygen due to phytoplankton die-off, water overturn or excessive biological oxygen demand (BOD) from decaying feed and organic matter can cause loss of an entire production pond (Boyd & Ahmad 1987). Emergency aeration includes use of large pumps that spray water over the pond surface or transfer well-oxygenated water from an adjacent pond. Large paddle-wheel aerators that splash the water into the air and cause strong water currents are the most popular in catfish farming. An alternate, but expensive, method of emergency aeration is the addition of large quantities of well water to lower the pond water temperature and to dilute the waste metabolites. However, well water is usually devoid of oxygen, and aeration by splashing or spraying is still required.

10.2.4 Nutrition of broodfish

Channel catfish production levels of 200–400 kg ha^{-1} were increased to over 2000 kg ha^{-1} by the use of pelleted feeds, the breakthrough that initiated catfish farming (Swingle 1959). Nutritional research over the years has resulted in the availability of suitable formulations at reasonable costs for the different life stages produced in ponds, raceways, and cages.

Channel catfish feeds contain from 25 to 36% protein derived from animal sources (fish meal and animal by-product meals) and plant sources (soybean and other oilseed meals, and cereals). Energy is derived primarily from 3 to 8% fats (vegetable oils and animal by-product fats) and starches. Essential vitamins, minerals and other nutrients are supplied in pre-mixes and fatty acids in fish oil. Nutrient requirements and deficiency signs for the different sizes of fish and the nutrient content, digestibility, and other values required in formulating suitable feeds are contained in National Research Council (1983) and Robinson (1989).

10.2.5 Parasites and diseases

Parasites and diseases are believed to cause up to US$30 million in lost production and mortalities annually. The incidence of disease is increased by stress (inherent to intensive culture), to poor water quality, inadequate nutrition and the presence of the infective organisms. Reducing these losses is dependent on continued research on improved management and the development of treatment techniques, including the registration of effective chemicals and drugs (MacMillian 1985).

The bacterial diseases of greatest concern in small fish are enteric septicemia (ESC) and those caused by *Pseudomonas* and *Aeromonas*; Romet and oxytetracycline are the drugs of choice. Ectoparasites, which are common to all warmwater fish, affect broodfish, larvae and small channel catfish. Chemical treatments of choice are formalin and copper sulphate. Channel catfish virus disease (CCV) is specific to larvae and small catfish; no cure is known (Schnick *et al.* 1989).

10.2.6 Predators and competition

Predators can cause major fish losses in brood ponds, and competitors can cause significant reduction in egg quality and numbers. Predators of principal concern to broodfish are the great blue heron and the alligator, but the most devastating predator is the human poacher.

The great blue heron is a large wading bird that can cause major losses of even large broodfish by spearing them with its sharp beak. Scaring devices such as firecrackers, nets and overhead wires have limited effect; the birds usually return to the fish farm. In areas where alligators are present in large numbers, broodfish could be held in ponds near the owner's residence, since alligators are often shy and avoid human contact. Game biologists and state enforcement personnel will usually assist in removing the alligators if requested. Both species are federally protected, and special permits are required for their control.

Human poachers generally have the most devastating effect on the farmer's broodfish population. The fish are easily captured by hook and line, and because of their excellent condition they bring a premium price. Many farmers have reported almost total fish loss in ponds frequented by night-time poachers over a 1–2 month period. Effective law enforcement is one way to control this problem, but fish loss might be better controlled by locating the broodfish ponds at sites away from intruder access and near the manager's residence.

10.3 Spawning

Channel catfish were spawned in ponds as early as 1916 (Shira 1917). However, as late as 1958 the predictability of spawning was low, and the most limiting factor to the establishment of a catfish industry was perceived to be the production of sufficient numbers of larvae and small fish at an affordable price (Swingle 1959). Research in the 1960s and 1970s in the areas of feeds development, water quality management and disease control resulted in techniques that are used today to produce the 400–500 million small fish used annually in the catfish industry. Most broodfish are spawned in ponds. Broodfish have been spawned commercially in pens and aquaria, but these types of facilities are now used primarily in research.

10.3.1 Pond method

Some producers prefer to let the broodfish remain in the brood pond, but most culturists transfer the broodfish to specially prepared spawning ponds in spring, 2–3 weeks before the start of the spawning season. The spawning pond should be free of sunfish and other predators that will eat the eggs or prey on the larvae and small fish. Good water quality is essential to spawning success, so generally the culturist fills the ponds with well water, applies an appropriate herbicide (Schnick et al. 1989) to prevent the growth of rooted vegetation, and offers the fish a commercial feed at a rate that will satisfy their nutritional needs but not unduly deteriorate the water. A pond of 2–5 ha, with levees allowing easy access and water depth of 0.5–1 m around most of the margin, is ideal.

Transferring the broodfish to spawning ponds provides the culturist with the opportunity to inspect the fish and to discard those that are in a nonspawnable condition. Since a male can usually spawn several females during the season, many researchers recommend a female:male stocking ratio of 2:1 or 3:2 (Jensen et al. 1983) or even as high as 4:1 (Bondari 1983).

Several weeks prior to the spawning season, spawning containers are placed in the ponds. Spawning containers may include bulk milk cans (about 20–40 l), wooden nail kegs, clay milk churns and small metal barrels (Busch 1983). However, any container that is obtainable and affordable, is of sufficient size to hold a pair of spawning fish, can be lifted and handled by the culturist, and does not contain pesticides or other toxic compounds would be satisfactory. The containers are usually placed on their sides in a line along the pond shore, 2–10 m apart in 0.3–0.7 m of water (Fig. 10.3). The number of containers needed is between one-third and the total number of males present. Marking each container with a stake or a float makes it easier to locate the container.

When spawning begins, the culturist should inspect each container every third to fourth day to determine whether eggs are present. Late morning is generally preferred, since it is less disruptive to the overall work programme and is prior to the hottest time of day (Jensen *et al.* 1983). The inexperienced culturist inserts a hand into the container opening to feel for the eggs and often finds a large, irate, possessive male guarding the egg mass. The experienced culturist, however, carefully lifts the container to near the water surface and tilts the container opening slightly downward to encourage the male, if present, to slide out. The interior of the container can then be visually inspected and

Fig. 10.3 Spawning containers located in a row along the side of the pond provide good spawning sites for the broodfish and easy access for the fish farmer.

the spawn, if present, located. A carpenter's putty knife is a suitable tool to separate the adhered egg mass from the floor of the spawning container. The egg mass is placed in a tub or insulated box, and the spawning container is replaced to be reused.

Several containers can be inspected before transferring the egg masses to a larger transport unit or carrying the eggs directly to the hatchery building. The culturist must ensure that the eggs are not exposed to the direct sunlight, that the temperature of the egg-transport water does not exceed 30°C, and that the dissolved oxygen in the transport water does not decline below 5 ppm. The period from recovery of the eggs to their placement in the hatching trough should not exceed 30 min unless provisions can be made to aerate the water and control the water temperature.

Early research demonstrated that spawning can be accelerated or delayed (Meyer *et al.* 1973, Sneed 1975), but this seems to be of little practical value to the modern catfish farmer. However, techniques to bring about the resumption of spawning activity that has ceased owing to declining water quality or unseasonably hot weather are extremely important to the larvae and small fish producer, if the larvae numbers obtained are insufficient to meet projected stocking needs and sales. Producers have found that draining one-fourth to one-half of the pond volume and replacing it with high-quality, cool ground water will often bring about a resumption of spawning.

10.3.2 Pen method

Pen spawning was practised by governmental fish producers and a few commercial farmers in the 1960s, but today it is largely limited to the research community. It has the advantage that specific pairings can be made, and, should hormone inducement be incorporated, large numbers of spawns can be obtained in a short time interval from a minimum of broodfish. It is not the method of choice for the large-scale producer of larvae and small fish.

Designs of pen systems are many. However, most pens are about 4 m long × 2 m wide, most commonly constructed of a wooden frame covered with wire. The wire mesh is of the size that will permit water exchange but not allow fish to escape. Generally, the three sides of the pen are embedded into the bottom mud, and the fourth side is the pond bank to allow ease of access. Water depth at the deep end is usually about 0.7 m and at the bank side is 0.3 m. The wire usually projects about 0.3 m above the water surface to prevent escape.

Spawning containers are as described earlier. The containers are usually located in the middle of the pen, with the opening away from the bank. Recovery of the eggs is as previously described, after which the spent female may be removed and replaced with another female of about the same size as the male, or another pair of fish may be used.

10.3.3 Aquarium method

Aquarium spawning was practised by researchers and producers in the late 1950s and early 1960s, when hormone inducement was believed to be needed to obtain the eggs and larvae essential for the industry's needs (Sneed & Clemens 1959, 1960). In this technique, modified by Huner & Dupree (1984b), the gravid and ready-to-spawn female fish was injected intraperitoneally with 1100–1800 i.u. human chorionic gonadotropin or 5 mg dried carp pituitary per kg body weight of the female fish and stocked into a $0.7 \times 0.7 \times 0.3$ m glass aquarium. A comparable-sized, but non-injected, male was then placed into the aquarium. Water was continuously flushed through the aquarium to maintain adequate oxygen levels. Spawning often occurred in about 17 h, after which the pair of fish was immediately removed and the eggs recovered for artificial hatching.

10.4 Egg incubation and hatching

After spawns have been observed in the pond containers, pens, or aquaria, the fish producer can elect to have the male fish incubate the eggs, or the eggs can be removed for mechanical incubation.

10.4.1 Natural incubation (open pond)

Natural incubation is the most primitive cultural method, whereby the male fish incubates the eggs, and the spawning pond also serves as the nursery pond for the larvae. Although primitive, the technique may be the most justifiable for the small producer who needs only a few thousand larvae, since no special equipment is required and the producer has little involvement or expense. The major disadvantages are that it is almost impossible to monitor production, to control insects and wild fish, to prevent the spread of diseases and to control predation from the adult fish.

As an alternative, the fish producer can inspect the spawning containers for the presence of larvae. Larvae usually stay in the spawning containers for 4–7 days after hatching, before scattering. Recovery and transfer of the larvae to a specially prepared pond permits the culturist to inventory the fish and conduct the other hatching management tasks similar to those practised with mechanical incubation.

10.4.2 Mechanical incubation

Mechanical incubation is the method of choice for most of the catfish farming industry. It has the advantage that the eggs and early larvae can be observed and treated for diseases of larvae. Electrical services for the hatching trough motors, lights, and other equipment may best be located overhead to protect the hatchery workers.

An auxiliary generator is essential in a hatchery unit to prevent the loss of eggs and larvae when electrical service is disrupted. Loss of water flow and aeration for even less than 5 min can cause almost total loss of eggs near the time of hatching and loss of larvae at all stages. To save money in the purchase of an auxiliary generator, many hatchery operators separate the electrical services into those essential for life support, including hatching trough motors, water pumps, air blowers, and nonessential lights and other devices. The life-support systems are connected to the auxiliary generator system. Qualified electricians and equipment vendors can assist in designing the auxiliary power system and specifying suitable equipment. Fully automatic equipment is available for switching to auxiliary power, or an alarm system can be installed that sounds when power has been disrupted.

Mechanical hatching troughs of various designs have been in use for over 50 years. The troughs are constructed of a variety of materials, including aluminum, steel, fibreglass and rigid plastic material, and are typically 3.5–7 m long, 0.4 to 0.8 m wide, and 0.3 m deep. Most troughs are flat-bottomed, but many producers are using 50 cm diameter polyvinyl chloride (PVC) pipe, split lengthwise, to construct roundbottomed units. A 1.3–2 cm aluminum or steel shaft is supported by pillow bearings attached to the top of the trough and running the length of the trough. Flexible metal paddles about 4 cm in width and of the length that will extend about 10 cm into the water when the paddle is in the down position are located about 0.5–0.8 m along the shaft, the exact distance depending on the width of the egg-holding baskets suspended between the paddles in the trough. A gear-reduced or belt-driven pulley reduction motor of about 30 rev/min turns the shaft and paddles (Fig. 10.4). The water supply is piped into one end of the trough and out through a standpipe drain at the other end.

Temperature of the water for the egg-hatching troughs and larvae-holding units in the hatchery should range from 25 to 28°C. When no heated water is available, some producers delay spawning until the season has progressed and the water has warmed. Temperatures below 25°C retard the development of the eggs and accelerate the growth of fungus that can smoother the eggs. Water temperatures above 30°C can cause the eggs to develop abnormally. Flow rates can vary depending on the number of egg masses in the trough, but generally sufficient water should be flushed through the trough to maintain dissolved oxygen levels above 5 ppm, and to achieve a water exchange every 15–30 min to dilute the ammonia and other waste metabolites.

Egg baskets are constructed to fit between the paddles and inside the sides of the trough. For example, when the distance between the paddles is 0.8 m and the trough is 0.6 m wide, the overall basket would be 0.4 m long × 0.5 m wide × 10 cm deep. Cross partitions are usually installed that, in this example, would result in six sections, each 23 × 25 cm. Flexible tie wire is usually employed to hold the basket at a depth of about 8 cm. Most baskets are constructed of galvanized woven wire with a mesh size near 0.6 cm.

Fig. 10.4 Mechanical incubators, such as this egg hatching trough, are typically used by most of the catfish industry.

Larvae-holding troughs of sizes and construction similar to the egg-hatching units are usually a component of the hatchery. Usually, two troughs are employed to hold the newly hatched larvae recovered from each hatching unit until the yolk has been absorbed and the fish are ready to be placed in specially prepared ponds or other culture facilities. A water supply with a minimum capacity of $20 \, l \, min^{-1}$ is necessary to provide sufficient water exchange to maintain the dissolved oxygen level above 5 ppm and to flush out the waste metabolites. Most producers supply supplemental oxygen by suspending an agitator in each trough. The agitator paddle screen is covered with fine wire mesh to prevent the larvae from being drawn in and killed. A removable screen is positioned at the foot of the trough to prevent the fish from being lost through the standpipe drain.

10.4.3 Care of eggs

Spawns recovered from the open ponds, pens or aquaria are placed into the baskets in the freshly cleaned and disinfected hatching trough. Channel catfish eggs are sensitive to sudden temperature changes, so the eggs should be tempered to the trough water temperature. Large spawns should be separated into smaller portions about the size of a person's double fists. Several small spawns or several pieces from a large spawn can be placed in each basket section, provided that there is sufficient space for the water to move freely around and through the egg masses. Daily maintenance is essential to extract dead eggs from the spawns and to remove the organic films and debris from the the tank sides and bottoms. Many of the later problems with disease have their beginning with dirty and unsanitary hatching facilities. Dead eggs on which fungus has started to grow are best removed, even though some good eggs may be lost.

Chemical control of fungus on the spawns is often required in the hatchery. No chemical suitable for this need is currently registered by the US Goverment's Food and Drug Administration. Research has shown that Diquat, which is registered as an algicide, and Betadine, which is registered as a disinfectant, will control fungus if applied for 15 min at the rate of 25–50 ppm static application (Schnick et al. 1989).

Channel catfish eggs require 3–8 days to hatch, depending on the age of the spawn at recovery and the incubating temperature. The eggs change from yellow to pink to red (or brown) as the embryo's circulatory system develops. At hatching, the yolk-sac larvae fall through the wire mesh to the trough bottom and then collect along the trough sides and in the corners.

10.4.4 Care of yolk-sac fry

Once to several times daily the hatched larvae should be removed from the hatching trough and transferred to the holding troughs in the hatchery or, less desirably, stocked directly into the rearing ponds. Larvae can be captured with a 10 cm × 10 cm small-mesh net of the type commonly available from stores that sell aquarium fish, or siphoned using a 1.5–2 cm flexible hose. The larvae can then be placed into a water-filled container (tare weight known), their weight determined, and then immediately transferred to the trough. Subsamples by weight and number will provide a good estimate of the number of larvae transferred. Studies have shown that the number of larvae can vary from 16 to 25 per g (Sneed 1975), and experience has demonstrated that holding troughs 3.5 m long, 0.4 m wide, and 0.3 m deep can be stocked with 10 000–50 000 larvae. A 5–20 l min^{-1} supply of well-aerated water will usually meet the dissolved oxygen requirements and flush out the waste metabolites. A fine-mesh-screen-covered agitator is usually provided to supply supplemental aeration.

Larvae require 5–8 days to absorb the yolk at the 25–28°C water temperatures. During the absorption of the yolk, the larvae develop mouthparts, turn from grey to black and swim to the surface of the water. These larvae, commonly called swim-up fry, are capable of seeking out and consuming natural food organisms or manufactured feeds specifically formulated for larval channel catfish.

10.5 Production of fry and small fish

Depending on circumstances and needs, the larvae produced in the hatchery can be immediately transferred to ponds prepared specifically for their culture, or they can be retained in troughs for various lengths of time.

10.5.1 Trough and raceway culture

Larvae and various life stages of channel catfish can be cultured in troughs and raceways throughout their entire life. The advantage is that the fish are completely acclimatized to living in a controlled environment and have not been exposed to many of the diseases and parasites commonly found in the less controllable ponds. The major disadvantage is that more stringent management is required, thus increasing the cost of the fish.

The programme requires a variety of facilities of increasing sizes. In the initial larvae holding trough of about $3.5\,m$ long $\times$ $0.4\,m$ wide $\times$ and $0.3\,m$ deep, about 10 000 larvae can be stocked and reared to about $5\,cm$. These fish, as they grow, can be transferred to larger troughs or concrete raceways and ultimately into the final grow-out facility.

Sufficient water of temperature $22-30°C$ to effect a complete water exchange each $\frac{1}{4}-\frac{1}{2}\,h$, and to maintain a dissolved oxygen level about $5\,ppm$, is needed. Supplemental aeration may be required. Expanded (floating) feeds must be of graduated sizes manufactured and formulated for the different life stages and must meet the complete nutritional requirements.

10.5.2 Pond culture

Most of the 400–500 million larvae and small fish required each year by the aquaculture industry are cultured in ponds. Pond culture is less labour-intensive and less expensive than trough and raceway culture, and it can be readily incorporated into most farm programmes. Technology for pond culture of larvae and small fish is advanced, having been practised at government fish hatcheries and in some private facilities since before the turn of century. Since the 1950s the techniques of production of numerous warm-water species in ponds have advanced significantly, and many of these practices have been adopted by catfish lavae and small-fish producers. Discussion of techniques for the pond method of larvae and small-fish production in levee ponds and run-off ponds follows.

Pond design

Levee or diked ponds constructed on relatively flat land suitable for larvae and small-fish production are generally $0.5-4\,ha$ in surface area, with water depths of about $1.5\,m$ in the deep end, sloping upward to $0.6\,m$ in the shallow end. A suitable drain must be installed to permit complete removal of the water and drying of the pond bottom before the larvae are stocked. Smoothing of the pond bottom to fill in the fish-made depressions and to eliminate seining obstacles prior to refilling and stocking is often necessary. A groundwater supply is preferred, but surface water can be used if free of pesticides and filtered to prevent contamination of the pond with parasites and wild fish.

Watershed ponds suitable for larvae and small-fish production are usually 0.5–3 ha in surface area and are filled with run-off water. Knowledge of the agricultural activities taking place on the land in the watershed is essential to ensure that pesticides or other compounds are not flushed into the pond. If wild fish are present in the streams supplying the pond, provision may be made to filter the water before it enters the pond.

Preparation for stocking

Ponds to be stocked with larvae or small fish should be drained and dried, and any pools of water or damp areas should be treated with hydrated lime at the rate of $250\,kg\,ha^{-1}$ to eliminate wild fish and predatory insects. If it is not possible to dry the pond completely, Wellborn et al. (1984) recommended the application of 1 ppm emulsifiable rotenone to eliminate the sunfish. Detoxification of the roten one usually occurs within a few days, but the process can be accelerated by the addition of potassium permanganate (Lay 1971).

After adding water to a depth of 0.5 m or more, organic and chemical fertilizers are added to develop a bloom consisting of various microscopic plants and animals that serve as the food source for the larvae and shade to retard the growth of rooted aquatic plants. Several fertilizing regimes are commonly used with good results, including $400-1000\,kg\,ha^{-1}$ animal or poultry manures, 200–400 kg Bermuda hay, 100 kg oilseed protein meals (soybean meal or cottonseed meal), and 100 kg of an inorganic fertilizer such as 16–20–0 or 10–20–0 (N-P-K). When inorganic fertilizers alone are used, about $200\,kg\,ha^{-1}$ of 16–20–0 applied every 2–3 weeks will usually produce a suitable bloom.

Just prior to stocking the larvae, the predaceous aquatic insects must be eliminated, especially in those ponds that have contained water for several weeks. The air-breathing insects can be controlled by applying 23–28 l of diesel fuel or kerosene mixed with 1 l of engine motor oil per hectare, repeated several times at intervals of 4 days, until the larvae are sufficiently large enough to evade the predatory insects (Chappell 1981). Gill-breathing insects cannot be controlled in this manner. Research has shown that Baytex, methyl Parathion, or Dylox will control these insects, but the compounds have not been registered with the US Government's Food and Drug Administration for this purpose (McGinty 1980, Piper et al. 1982).

Prevention of the growth of rooted aquatic vegetation is essential in good catfish management. Rooted aquatics compete with the larvae for nutrients, but more importantly, for access to the food organisms on the pond bottom. Fish can be entangled and lost, especially when the pond is drained. Application of a pre-emergent herbicide such as Aquazine (Simazine) at the rate of 5–25 kg ha^{-1} on the pond bottom prior to filling is a recommended treatment (Schnick *et al.* 1989).

Stocking

Yolk-sac larvae may be recovered from the hatching trough and immediately placed into the nursery ponds, or stocking could consist of the later life stages previously described. It is desirable to stock the numbers that will result in the maximum number of the desired sizes of small fish for later grow-out programmes (Table 10.1). Some producers stock larger numbers and harvest at intervals, thus making available different sizes for sale before settling on the final number of fish to be grown to the desired size.

For good survival the larvae must not be stressed at stocking. The fish are very sensitive to temperature changes and low oxygen levels. Many producers prefer to stock the larvae at about daybreak, when the hatchery and pond water temperatures are nearly equal, and when the air is cool. Stocking the larvae where there is shelter and where formulated feed can be made available will enhance survival (Snow 1962, Jensen *et al.* 1983).

Feeding

Offering formulated feeds to larvae and small fish to supplement the natural food blooms is practised by most fish farmers. Formulated feeds are usually offered in excess of the amount that the fish can consume, but the large

Table 10.1 Projected small fish sizes under ideal management in 120–150 days at various stocking densities[a]

Larvae stocking density (fish per ha)	Average length (cm)
25 000	17–25
75 000	15–20
132 000	13–17
183 000	10–15
238 000	8–13
300 000	8–13
350 000	8–10
500 000	5–8
750 000	2–5
1 250 000	2

[a] Modified from Jensen *et al.* (1983)

allowance is necessary to ensure that the feed is available to all the fish. Uneaten or wasted feed, if not excessive, poses little problem to water quality management, and in practice acts as an organic fertilizer.

Feeds are initially offered near the fish shelters or near the site of stocking. However, within a few days to about a week after stocking the fish will disperse, thus necessitating the distribution of feed around the entire shoreline. Best survival and growth are obtained when the feed is offered three to six times daily initially; within a few weeks the frequency of feeding can be reduced to once or twice daily.

In intensive production, formulated feed is initially offered the larvae at the rate of $10-30 \, kg \, ha^{-1}$ daily, but this allowance is gradually increased to $50-100 \, kg \, ha^{-1}$ as the fish increase in size. The feeding allowance is governed by management decisions based on achieving good growth but also maintaining suitable water quality.

Water quality

The primary concerns of the fish producer are maintaining dissolved oxygen above 5 ppm and minimizing the level of ammonia and nitrites, carbon dioxide, and hydrogen sulphide. Techniques previously discussed for the management of the broodfish are applicable here.

Parasites and disease

Prevention of parasites and diseases is extremely important to the larvae and small-fish producer, since these fish are the essential component to the later grow-out programme. Fish that die cannot be replaced until the following spawning season, and fish that are parasitized and diseased do not grow as well as healthy fish. Diseases and parasites that affect channel catfish and methods for their control and treatment were discussed previously. Culture methods that reduce stress on the fish, such as good water quality management and adequate nutrition, are of first importance, since stress in the presence of the disease organism will in all probability result in disease.

Predators and competition

Some fish and animals compete with larval catfish and small catfish for space and food, and others prey on these fish. Not controlling these competitors and predators often results in reduction in growth rate and significant mortality, possibly as high as 100%. Bullfrogs prey on small fish, and tadpoles compete for space and food. Sunfish and black bass are competitors when small and often grow large enough to become predators. Water snakes of all sizes are efficient predators. Losses to the migratory water birds, such as the herons and cormorants, are now a major problem for the catfish producer.

Control measures vary with the problem. Snakes and frogs can be reduced by frequent mowing to eliminate high grass. Infestation by wild fish can be prevented by filtering the incoming water through Sarar cloth, a fine-mesh screen or gravel bed. Sometimes, only scare tactics or actual killing of the unwanted animals may be effective. In some cases, such as the control of migratory water birds, special permits are required.

Harvesting and grading

Catfish larvae stocked in large numbers are usually partially harvested as small fish several times during the year to reduce the standing crop and to meet the market demand for the different sizes of fish. In practice the full or nearly full pond is seined to remove up to 80% of the total crop before total draining. Special care must be taken during the summer to reduce stress from high water temperatures and low dissolved oxygen. Seining is usually done in the early morning hours. Harvested fish can be graded in the pond, with the small sizes returned to the pond, or they may be transported to the holding shed, put into water that is 5–10°C cooler, graded, and held for several days with no damage. The addition of 1% table salt, antibiotics or bacteriostatics and an anaesthetizing agent to the transport water is often beneficial (Huner *et al.* 1984).

References

Bondari, K. (1983) Efficiency of male reproduction in channel catfish. *Aquaculture*, **35**, 79–82.

Boyd, C.E. (1979) *Water Quality in Warmwater Fish Ponds*. Alabama Agricultural Experiment Station, Auburn University. Craftmaster Printers, Inc., Opelika, Alabama.

Boyd, C.E. & Ahmad T. (1987) *Evaluation of aerators for channel catfish farming*. Alabama Agricultural Experiment Station, Auburn University, Bulletin, 584.

Busch, R.L. (1983) Evaluation of three spawning containers for channel catfish. *Progressive Fish Culturist*, **45**(2), 97–9.

Chappell, J. (1981) Management and selection of brood fish. *Aquaculture Magazine*, **7**(2), 24–29.

Huner, J.V. & Dupree H.K. (1984a) Pond management. In *Third Report to the Fish Farmers: the Status of Warmwater Fish Farming and Progress in Fish Farming Research*, (eds H.K. Dupree & J.V. Huner), pp. 17–43. US Fish and Wildlife Service, Washington, DC.

Huner, J.V. & Dupree H.K. (1984b) Methods and economics of channel catfishes production, and techniques for the culture of flathead catfish and other catfishes. In *Third Report to the Fish Farmers: the Status of Warmwater Fish Farming and Progress in Fish Farming Research*, (eds H.K. Dupree & J.V. Huner), pp. 44–82. US Fish and Wildlife Service, Washington, DC.

Huner, J.V., Dupree, H.K. & Greenland, D.C. (1984) Harvesting, grading, and holding fish. In *Third Report to the Fish Farmers: the Status of Warmwater Fish Farming and Progress in Fish Farming Research*, (eds H.K. Dupree & J.V. Huner), pp. 158–64. US Fish and Wildlife Service, Washington, DC.

Jensen, J., Dunham, R.A. & Flynn, J. (1983) *Producing channel catfish fingerlings*. Cooperative Extension Service, Auburn University, Alabama, Circular ANR-327.

Lay, B.A. (1971) Applications for potassium permanganate in fish culture. *Transactions of the American Fisheries Society*, **100**(4), 813–5.

Leary, J.L. (1908) Description of San Marcos Station with some of the methods of propagation in use at that station. *Transactions of the American Fisheries Society*, **37**, 75–78.

MacMillian, J.R. (1985) Infectious diseases. In *Channel Catfish Culture*, (ed. C.S. Tucker), pp. 405–96. Elsevier, New York.

McGinty, A.S. (1980) *Survival, growth and variations in growth of channel catfish fingerlings*. PhD dissertation, Auburn University, Alabama.

Meyer, F.P., Sneed, K.E. & Eschmeyer, P.T. (eds) (1973) *Second Report to the Fish Farmers: Status of Warmwater Fish Farming and Progress in Fish Farming Research*. Resource Publication 113, US Fish and Wildlife Service, Washington, DC.

National Research Council (1983) *Nutrient Requirements of Warmwater Fishes and Shellfishes. Nutrient Requirement of Domestic Animals*. National Academy Press, Washington, DC.

Norton, V.M., Nishimura, H. & Davis, K.B. (1976) A technique for sexing channel catfish. *Transactions of the American Fisheries Society*, **105**(3), 460–62.

Piper, R.G., McElwain, I.B., Orme, L.E., McCaren, J.O., Fowler, L.G. & Leonard, J.L. (1982) *Fish Hatchery Management*. US Fish and Wildlife Service, Washington, DC.

Robinson, E.H. (1989) Channel catfish nutrition. *Reviews in Aquatic Sciences*, **1**(3), 365–91.

Schnick, R.A., Meyer, F.P. & Gray, D.L. (1989) *A Guide to Approved Chemicals in Fish Production and Fishery Resource Management*. Available from: Publications Distribution Offices, University of Arkansas Cooperative Extension Service, PO Box 391, Little Rock, Arkansas 72203.

Shira, A.F. (1917) Fish-cultural activities of the Fairport Biological Station. *Transactions of the American Fisheries Society*, **47**(1), 39–44.

Sneed, K.E. (1975) Channel catfish culture methods. In *European Inland Fisheries Advisory Commission Workshop on Controlled Reproduction of Cultivated Fishes*, pp. 166–73. EIFAC Technical Paper 25, FAO, Rome.

Sneed, K.E. & Clemens, H.P. (1959) The use of human chorionic gonadotropin to spawn warmwater fishes. *Progressive Fish Culturist*, **21**, 117–20.

Sneed, K.E. & Clemens, H.P. (1960) *Use of fish pituitaries to induce spawning in channel catfish*. Special Scientific Report – Fisheries, 329, US Fish and Wildlife Service, Washington DC.

Snow, J.R. (1962) A comparison of rearing methods for channel catfish fingerlings. *Progressive Fish Culturist*, **24**(3) 112–18.

Swingle, H.S. (1959) Experiments on growing fingerling channel catfish to marketable size in ponds. *Proceedings of the Twelfth Annual Conference Southeastern Association of Game and Fish Commissioners, Louisville, KY*, 63–72.

US Department of Agriculture (1990) *Aquaculture: Situation and Outlook Report*. USDA Commodity Economics Division, Economic Research Service, AQUA-4.

Wellborn, T.L., Herring, A.J. & Callahan, R. (1991) *Managing Mississippi Farm Ponds*. Publication 1428. Mississippi Cooperative Extension Service, Mississippi State University, Starkville.

Chapter 11
African Catfish (*Clarias gariepinus*)

11.1 Introduction
11.2 Taxonomy and systematics
11.3 Annual rhythms of reproduction
11.4 Maintenance of pubertal broodfish
11.5 Selection, maintenance and breeding of broodfish
11.6 Triploidy and gynogenesis
11.7 Survival of larvae
    11.7.1 Larval rearing systems
    11.7.2 Effects of egg size
    11.7.3 Yolk supplies and nutrient utilization
References

The seasonality of spawning is a major problem in broodstock management of African catfish, *Clarias gariepinus*. Their circannual endogenous rhythms of gonadal recrudescence and regression occurring in nature can be avoided in captivity by raising the broodfish from egg to maturity under a constant high temperature.

Maintenance conditions of broodfish can profoundly affect the quantity and quality of gametes. Temperatures above or below the optimum of 25°C disturb reproduction. Atresia of post-vitellogenic follicles and regression of testes tubules occur at 30°C. Retarded development of both ovaries and testes occurs at 20°C.

The feeding level, irrespective of temperature, determines the total number of oocytes, being highest in better-fed females. Poorly-fed females have the highest gonadosomatic index (GSI), indicating that food reduction leads to investment of energy in the ovaries at the expense of somatic tissue.

Intersexual contact is also an important requirement in the provision of maintenance conditions. Males stimulate ovarian growth and development by olfactory and tactile cues. Females tend to hamper male gonadal growth. This is most pronounced in the testes but absent in seminal vesicles.

Egg size has an important implication for viability of larvae. This is shown by the eggs of African and Asian catfish (*Clarias batrachus*), the former species having a larger egg size than the latter. After absorption of the yolk sac, the larvae of *C. gariepinus* survive starvation much longer than *C. batrachus*. Since egg size is directly proportional to fish size larger females produce larger eggs.

Commercial producers of catfish fingerlings use large females of 3–10 kg. This broodstock size represents a compromise between producing large larvae at the onset of exogenous feeding and the economic goal of reducing the costs of broodfish maintenance.

11.1 Introduction

The African catfish, *Clarias gariepinus*, was introduced into Netherlands in 1977. Artificially-induced breeding, larval rearing and the energy balance and nutritional requirements of ongrowing fish were the research topics of the Department of Fish Culture & Fisheries, Wageningen Agricultural University. Commercial interest in the Netherlands for intensive culture of this species was raised in 1985 after gourmet restaurants had paid high prices for the experimentally-reared fish. Within 1 year, about 50 catfish farms realised a production of about 300 t (Table 11.1). The Dutch market, however, could not absorb this production and half of the farms went out of business in 1986. The production soon resumed a slow and steady growth owing to successful efforts to create an export market. Small production units in Belgium, Germany, Hungary, Poland and Russia were also established. In 1991 a limited number of hatcheries in the Netherlands provided about 2 000 000 fry for the production of 1000 t of African catfish. Since 1970 the African catfish has been raised successfully in Africa. The development of a reliable method for the production of fry is a priority for research.

The number and quality of gametes of the species can be profoundly affected by the conditions under which the broodstock are maintained. In ponds circannual rhythms of gonadal recrudescence and regression occur regularly. In the present chapter the effects of ration, temperature, and the interaction of sexes on the reproductive performance of broodfish will be reviewed. Recent results on growth and metabolism of eggs and larvae are also given.

Table 11.1 Production of African catfish in the Netherlands

Year	Number of farms	Production (tonnes per year)
1985	25	80
1986	50	300
1987	30	300
1988	30	400
1992	25	1200

11.2 Taxonomy and systematics

By the early 1970s, the first publications on the culture of African *Clarias* species appeared (De Kimpe & Micha 1974, Micha 1974). Confusion existed regarding the identification of economically important species such as *C. lazera, C. mossambicus, C. gariepinus, C. anguillaris* and *C. senegalensis*. In his systematic revision of these five African species of the genus Clarias, Teugels (1986) distinguished two valid species *Clarias gariepinus* and *Clarias anguillaris*. In his key to the species the following characteristics were used:

- The number of gill rakers on the first branchial arch varies between 24, in a specimen of 27.7 mm Standard Length (SL), up to 110, in a specimen of 600.0 mm SL; according to the geographic locality of the specimen, the regression coefficient (b) in the regression equation for the number of gill rakers (y) as function of the standard length (x) ranges between 0.122 and 0.267...*Clarias gariepinus* (Burchell 1822).
- The number of gill rakers on the first branchial arch varies between 16 (in a specimen of 31.5 mm SL) and 50, in a specimen of 650.0 mm SL; according to the geographic origin of the specimen, the regression coefficient (b) in the regression equation for the number of gill-rakers (y) as function of the standard length (x) ranges between 0.013 and 0.060...*Clarias anguillaris* (Linnaeus 1758).

The following affinities between the two species were also given by Teugels (1986). *Clarias anguillaris* is closely related to *C. gariepinus* (Burchell 1822) described from the Orange river in South Africa. Apart from the Orange river, *C. gariepinus* is reported from nearly all over Africa and even from Asia Minor. *Clarias anguillaris*, however, has a more restricted distribution and occurs from the Nile to west Africa. In most of the river systems both species are sympatric. Identification in these areas of sympatricity can be achieved only by counting the numbers of gill rakers on the first gill arch. Although this number varies with the standard length and with the geographic origin, both species can be easily discerned: in specimens of *Clarias anguillaris* we counted 16–40 gill rakers, while in equal sized specimens of *C. gariepinus* 24–110 gill rakers were found. The regression coefficient (b) in the regression equation for the number of gill rakers related to the standard length was calculated for both. According to the geographic origin of the specimen, the b values for both ranged as follows:

Clarias anguillaris : $0.013 \leq b \leq 0.060$
Clarias gariepinus : $0.122 \leq b \leq 0.267$

Using the F-test, this difference proved to be statistically significant.

It is remarkable that the two species can be distinguished only on the basis of the number of gill rakers. Munro (1967) observed that *Clarias* species have a rather euryphagous niche. Zooplankton becomes more important with increasing size and predominates in the diet of the largest fishes. The number of gill rakers increases, presumably resulting in more efficient filter feeding (Jubb 1961). Variability in numbers of gill rakers might thus be caused by the ecological conditions in which the species/population is living. Karyograms and genetic biochemical markers of both species are needed to sustain their taxonomic position within the genus *Clarias*.

The present systematic status of African catfish species, based on Teugels' (1986) revision and currently used in aquaculture, is summarized in Table 11.2. It is nevertheless interesting to examine a morphometric characteristic of the

Table 11.2 Systematic status of African catfish species used in aquaculture (after Teugels 1986)

Genus	Species/hybrid	Junior synonyms
Clarias	*gariepinus*	*lazera, mossambicus*
Clarias	*anguillaris*	*senegalensis*
Heterobranchus	*longifilis*	
	C. gariepinus × *H. longifilis*	

junior synonyms, which was described by Boulenger (1915) in his catalogue of the freshwater species of Africa in the British Museum. In the Northern and Central part of Africa the African catfish or sharptooth barbel (a Southern African name) has been described as *C. lazera*, in the Eastern part as *C. mossambicus* and in the Southern part as *C. gariepinus*. The variability in the ratio of depth of body/total length is very large:

C. lazera   5–9
C. mossambicus 6–8
C. gariepinus 6–7

Fish culturists have also noticed this variability and want to exploit this (genetic?) characteristic in their breeding/production programme, since the filleted *C. gariepinus* is much bigger than the filleted *C. lazera* (Fig. 11.1).

Karyological analysis of three strains of '*C. lazera*' originating from Israel (Hula swamp), Central African Republic (Unbangui river) and Ivory Coast (Fish culture station, Bouaké) revealed an identical karyotype (Ozouf-Costaz *et al.* 1990). The synonymy between *C. gariepinus*, *C. mossambicus* and *C. lazera* should, however, still be confirmed on the genetic level. The origin of strains currently used in European intensive aquaculture is summarised in Table 11.3.

An intergeneric hybridization between the species *Clarias gariepinus* and *Heterobranchus longifilis* (Table 11.2) appeared to be successful (Hecht & Lublinkhof 1985). In Zimbabwe and Nigeria the hybrid raised under pond

Table 11.3 'Strains' of *Clarias gariepinus* used in European intensive aquaculture

Strains	Geographic origin			
	Central African Republic	Ivory coast	Israel	South Africa/ Zimbabwe
Wageningen	+			
Utrecht	+		+	
Belgium	+	+	?	
Germany				+

Fig. 11.1 African catfish from Southern Africa (upper specimen) and Central Africa (lower specimen). Note the difference in depth of body/total length. This difference is also apparent in their fillets (lower picture).

conditions showed a better growth rate than that of the parental species. The introduction of the African catfish in Asian fish culture (Thailand and Indonesia) raises the question whether (natural) hybridization with the Asian catfish *Clarias batrachus* is also possible. The hatching curves of both species have been assessed in our broodstock (Fig. 11.2). The mean oocyte diameter of *C. gariepinus* and *C. batrachus* varied between 1.41 and 1.51 and between 1.32

*African Catfish (*Clarias gariepinus*)*

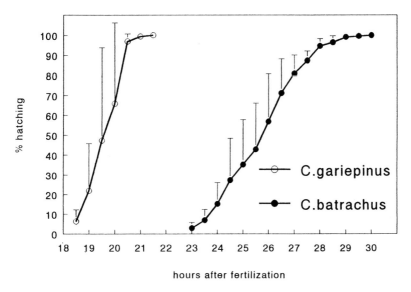

Fig. 11.2 Cumulative mean percentage (± S.D.) hatching curves of *Clarias gariepinus* and *C. batrachus* at 30°C incubation temperature.

and 1.37 respectively. At a water temperature of 30°C the larvae of *C. gariepinus* hatched earlier than those of *C. batrachus*. Hatching started in the former species at 18.30 h and was completed at 21.30 h after fertilization of eggs; respective time intervals were 23 h and 30 h for *C. batrachus*.

The fertilization rate and the embryo mortality in intra- and interspecific crosses of *C. gariepinus* and *C. batrachus* was also studied in the Department of Fish Culture & Fisheries, Wageningen Agricultural University, the Netherlands (Table 11.4). The cross *C. batrachus* female × *C. gariepinus* male showed a very low fertilization rate, which is probably due to the relatively

Table 11.4 Embryo mortality in intra- and interspecific crosses* of *Clarias gariepinus* (G) and *C. batrachus* (B)

Cross ♀ ♂	Fertilization (%) 0 h	Embryo mortality (%)		
		10 h	15 h	20 h
G × G	100	7.4	9.5	9.5
G × B	100	18.8	56.5	100
B × G	3.6	4.4	91.7	100
B × B	100	1.7	3.0	5.2

* Two broodfish per species
  Four egg samples per cross

small size of the micropyle of the former species in relation to the large size of the sperm of the latter species. The cross *C. gariepinus* female × *C. batrachus* male as well as both intraspecific crosses showed high fertilization rates. The embryonic development of the interspecific crosses followed the pattern of the maternal species. Mass mortality occurred, however, during the blastula stage when the blastopore was formed. All hybrids died within 20 h after fertilization. It is surprising that this interspecific cross was not viable.

11.3 Annual rhythms of reproduction

In nature the annual reproductive cycle of *C. gariepinus* comprises a resting period, a period of gametogenesis and a breeding period (Fig. 11.3). In the swamps of the Hula Nature Reserve (Israel) the resting period lasts from August until March (van Oordt *et al.* 1987). At that time the ovaries contain mainly oogonia, which are located in the ovarian villi (Richter & van den Hurk 1982) (Fig. 11.4). The testes have only spermatogonia located in the cysts. During this period the GSI is very low (Fig. 11.3). Gametogenesis begins in March and ends in May. Two oocyte stages, i.e. vesicle formation and exogenous yolk formation, can be distinguished in the ovaries. The former differs from the latter by the presence of numerous vacuoles containing chromophobic material. These vacuoles play a role in the formation of cortical alveoli, which release their acidic mucopolysaccharide into the perivitelline space after fertilization (Fig. 11.16). Exogenous yolk formation is observed in larger oocytes (300–1200 μm) and is characterized by the presence of yolk granules.

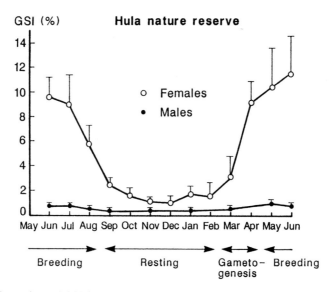

Fig. 11.3 Fluctuations of GSI (gonadal weight × 100/body weight) of female and male *Clarias gariepinus* during one annual cycle in the swamp of the Hula Nature Reserve (Israel). (After van Oordt *et al.* 1987.)

African Catfish *(Clarias gariepinus)*   249

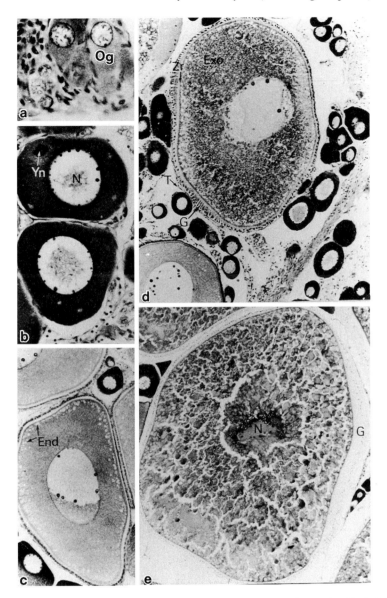

Fig. 11.4 Follicle development in *Clarias gariepinus*. (After Richter & van den Hurk 1982.) (a) Oogonium (× 560): (b) Previtellogenic oocyte (× 224): (c) Follicle with chromophobic endogenous vacuoles (× 140): (d) Vitellogenic follicle (× 88): (e) Postvitellogenic follicle (× 88).

Oogonium (Og), Yolk nucleus (Yn), Nucleus (N), Endogenous Cortical alveoli/vesicles (E), Exogenous Yolk (Exo), Theca (T), Granulosa (G), Zone Radiata (ZR).

The testes also start recrudescence, which is characterized by a strong spermatogenic activity in the cysts. In the breeding phase they contain spermatids and spermatozoa. The GSI of females increases rapidly owing to a strong vitellogenic activity of oocytes. During the breeding season from May until

August, ovaries are in the post-vitellogenic or post-ovulation stage. Post-vitellogenic follicles have oocytes with diameters of 1000–1200 µm and centrally located nuclei (Fig. 11.4). The stimulus to spawn appears to be associated with a rise in water level and inundation of marginal areas. The continuing decrease in GSI in August and September, however, shows that after ovulation vitellogenesis is limited and does not lead to a restoration of the original number of post-vitellogenic follicles. A full grown ovary can also enter a phase of regression when environmental conditions are not favourable. This happens occasionally in August and September in the Hula Nature Reserve.

Females from this population kept in nearby ponds (Ginosar) also follow the same annual pattern of ovarian development (Richter et al. 1987a) (Fig. 11.5). Confinement, including optimum feeding, seems to affect the relative weight of the ovaries, being 12–13% in ponds and 10–11% in nature. In both populations the changes in GSI followed the annual changes in daily photoperiod and water temperature (Fig. 11.6). The cyclical changes, both of increasing and decreasing GSI in pondfish, however, started about 1 month later than in nature.

The spawning seasons of African catfish in countries north and south of the equator last from June till September and from November till February respectively (Richter 1976). The seasonality of spawning imposes a considerable problem in extensive African catfish farming. The primary concern of any fish hatchery in the tropics is to produce the maximum number of the highest

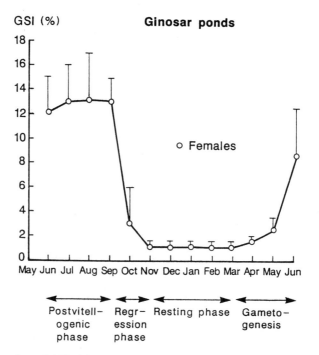

Fig. 11.5 Fluctuations of GSI of female *Clarias gariepinus* during one annual cycle in a pond in Ginosar, Israel. (After Richter et al. 1987a.)

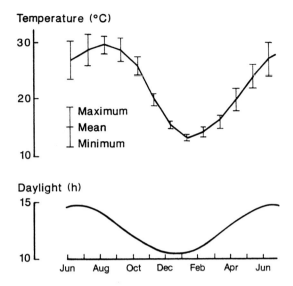

Fig. 11.6 Fluctuations of water temperature and daylight at the fish culture station in Ginosar (Israel). (After Richter *et al.* 1987a.)

quality eggs and fry at any time of the year. This objective was the subject of a research programme at the Central African Republic Fish Culture Station in Bangui. Female broodfish originating from the Unbangui river were transferred from ponds to indoor tanks with the aim of prolonging their reproductive cycle. They were maintained with optimum feeding at a constant high temperature and under local light conditions. Unfortunately, these broodfish continued their annual interrupted reproductive cycle for 1 year. In the subsequent years, however, they were artificially spawned at intervals of about 6 weeks. This experiment indicated that the annual rhythm of ovarian recrudescence and regression of *C. gariepinus* can be erased by a constant temperature regime (Janssen 1985).

In a complementary series of experiments the influence of photoperiod and origin of a hatchery-raised African broodfish population was examined in Wageningen (Richter *et al.* 1987a). Female broodfish kept in tanks at a constant high temperature and with a local semi-natural photoperiod showed uninterrupted ovarian activity (Fig. 11.7). The broodfish could be stripped monthly for more than 1 year and the fertilized eggs developed into normal fry (Fig. 11.8). Another set of broodfish were stocked without stripping for 1 year. These fish remained postvitellogenic until the end of the experiment and were able to be spawned artificially with carp pituitary suspension. These results indicate that annual rhythms of ovarian recrudescence and regression are not present in broodfish, which were raised from egg to maturity under constant high temperature. The cooperative research effort of scientists working in the Central African Republic, Israel and the Netherlands has resulted in a practical manual

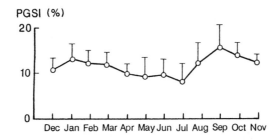

Fig. 11.7 Pseudogonadosomatic index [weight of egg mass collected by stripping × 100/(body weight before injection − weight of stripped eggs)] of female *Clarias gariepinus* during one annual cycle at the hatchery of Wageningen (The Netherlands). (After Richter *et al.* 1987a.) Standard deviations refer to 10 broodfish.

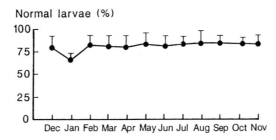

Fig. 11.8 Percentages of normal larvae hatching from ovulated eggs (see also Fig. 11.7).

for the culture of the African catfish (*Clarias gariepinus*) in the tropics (Viveen *et al.* 1985).

11.4 Maintenance of pubertal broodfish

Ambient factors have been shown to influence the reproductive rhythm of adult *Clarias gariepinus*. The effects of ration, temperature and the presence of conspecifics on the reproductive performance of pubertal fish have been studied in our laboratory (Richter *et al.* 1982). Catfish were raised from egg to fingerling in troughs with recirculating water at 30°C. Under optimum feeding conditions a mean weight of 40 g was reached in about 60 days. The fingerlings were subsequently distributed into fibreglass tanks of 600 l and fed on trout pellets. The effects of feeding level and temperature (20, 25 and 30°C) were studied under North European light conditions (March−January). Feeding level 1 was defined as that giving the most efficient feed conversion and the other two levels were derived from it being a third as much and three times as much as the optimum. Gonadal characteristics including the GSI, the sperm index (an index based on the proportion of testis tubules containing ripe spermatozoa) and the hatching rate of eggs (determined after incubating of fertilized eggs at 30°C) were studied after 240 days of treatment (Figs 11.9, 11.10).

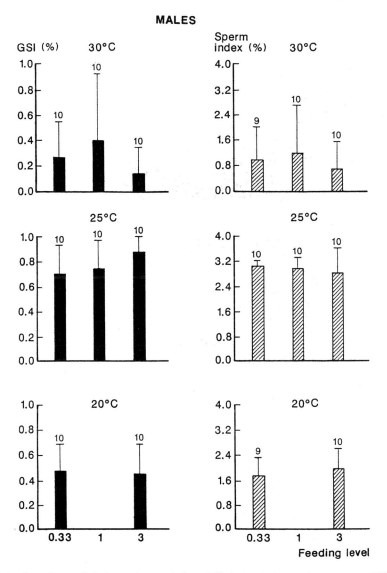

Fig. 11.9 Gonadosomatic index and sperm index of *Clarias gariepinus* males at an age of 300 days in relation to various feeding levels and temperatures. (After Richter *et al.* 1982.) Number of fish on top of S.D. bars.

In males (Fig. 11.9) the feeding level had no clear effect on GSI nor sperm index. Temperature, however, strongly affected testis development. Both GSI and sperm index reached the highest values at 25°C. In females (Fig. 11.10) the feeding level, irrespective of temperature, determined the total number of oocytes, being highest in the better-fed fish. The poorly-fed females, however, had the highest GSI, indicating that food reduction leads to the investment of energy in the ovaries at the expense of somatic tissue. Hatching rates of eggs

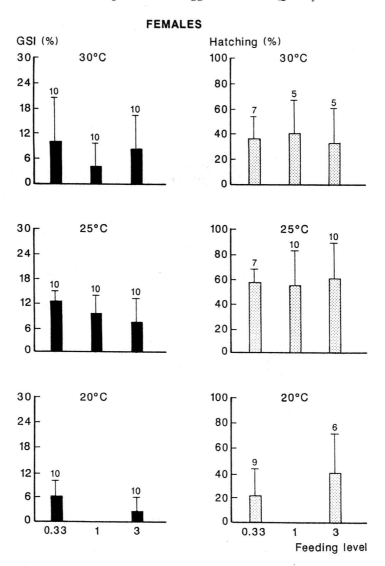

Fig. 11.10 Gonadosomatic index and hatching rate of eggs of *Clarias gariepinus* females at an age of 300 days (see Fig. 11.9).

were clearly affected by the temperature at which the broodfish had been maintained. Eggs originating from females raised at 25°C showed the highest rates of hatching. The present experiments with pubertal African catfish were in the temperature range of its natural habitat. Remarkably, a constant temperature above or below the apparent optimum temperature of 25°C disturbed reproduction. Atresia of post-vitellogenic follicles and regression of testis tubules occurred at 30°C, whilst retarded development of both ovaries and testes was observed at 20°C.

Stimulation and inhibition of gonadal development by intersexual contact was also studied in pubertal catfish. Juvenile fish, 137 days old, were sexed on the basis of morphological features and fed Trouvit trout pellets at a rate decreasing from 1.3 to 1.1% of their body weight per day. The set-up of the experiment was such that females were exposed to various combinations of male stimuli (Van Weerd et al. 1988, 1990). The experiment was conducted in a two-layered aquarium system at a water temperature of 24–25°C. Over the 100-day experimental period mean female body weight increased from 112 g to 297 g. It appeared that in pubertal C. gariepinus, male stimulation of ovarian development is brought about by chemical (pheromonal) and tactile cues but not by visual, auditory or electric ones. A basic GSI level of 1.9–2.7% was found in the control group (no contact) and in the treatments in which visual and auditory contact was possible (Fig. 11.11). An intermediate GSI level of about 4% was reached in treatments where olfactory but not tactile stimuli were involved. A maximum GSI level of 7.3% was reached in the unlimited contact treatment, in which females received all possible male stimuli (Fig. 11.11).

In summary it can be concluded that males stimulate ovarian development by both olfactory and tactile cues. A similar experiment was conducted with pubertal males. Juvenile males (150 days old) were exposed over a 90 day period to various combinations of female and male stimuli (Van Weerd et al. 1991). It appeared that females did not stimulate but rather tended to hamper male gonadal growth. These effects were most pronounced with regards to testicular development and not to the seminal vesicles. Emission of inhibiting (pheromonal) cues released by females as a result of interaction with males is assumed to play a role in the depression of testis development. The GSI of inhibited males was 0.38% versus 0.51% in the controls. Stimulation of testis development by olfactory stimuli from holding water of male populations is another surprising finding in this study. These males reached a GSI of 0.51% versus 0.36% in the controls. A similar phenomenon has been reported earlier (Stacey & Hourston 1982) for pheromonal stimulation of male and female spawning behaviour through milk, in *Clupea harengus pallasi*. Stimulation of testicular development of male *Sarotherodon mossambicus* by conspecific stimuli has also been described (Silverman 1978), regardless of the sex of the stimulatory animals.

11.5 Selection, maintenance and breeding of broodfish

Selection of broodfish in the Netherlands is carried out at production farms by the Catfish Hatchery Fleuren at Someren. Stocks of fingerlings are monitored during the 5 months rearing period.
The following criteria are used for selection:

- quiet behaviour and constant feed intake;
- feed conversion of 0.8–0.9;

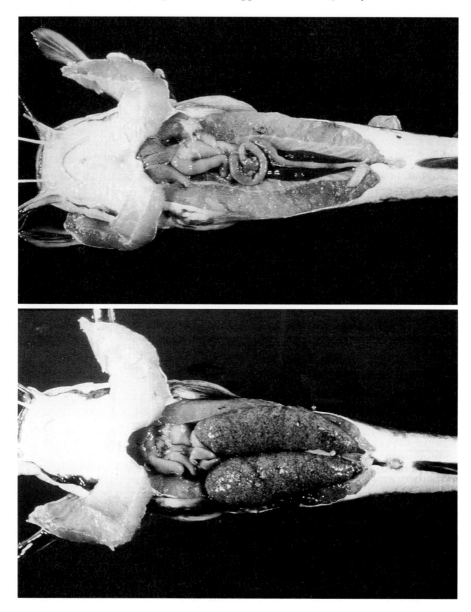

Fig. 11.11 Ovarian development in 8 months old broodfish of *Clarias gariepinus*. Above: female raised without male contact. Below: female raised under conditions of unlimited contact with males.

- mean increase of bodyweight from 10–900 g;
- mortality less than 15%.

Large specimens of about 1.5 kg are selected from the production units and transferred to the catfish hatchery. They are kept in flow-through tanks at high densities of $100 \, \text{kg m}^{-3}$. The oxygen level is maintained above $3 \, \text{mg l}^{-1}$, ammonia-nitrogen below $2\,mg\,l^{-1}$, nitrite-nitrogen below $1\,mg\,l^{-1}$ and nitrate-nitrogen below $150\,mg\,l^{-1}$. The stock are kept indoors on a constant light:dark cycle (12L:12D) or in continuous semi-darkness. The water temperature is maintained close to 25°C. The fish are fed Trouvit No. 4 or Biocatfish (Trouw) at a rate of 0.5% of their wet body weight per day.

Slow growers and fish which mature precociously are eliminated from the broodstock. The broodstock are raised for 6–12 months in the hatchery and have reached a weight of 3–5 kg at the time that they are used for induced breeding. In the research facilities of the Department of Fish Culture & Fisheries in Wageningen (Fig. 11.12) 1 year old female and male broodfish of 300–1000 g are used for reproduction. They are kept together in 140 l aquaria (Fig. 11.13). A unit of 48 aquaria is connected to a water-treatment unit consisting of a trickle filter and a lamellar separator (Fig. 11.14). For the purpose of hormonally-induced breeding, broodfish are stocked individually without food for about 36 h so that the alimentary system will be empty at the time of stripping. Determination of the developmental stage can be done macroscopically (thickness and softness of the belly region) or microscopically (size of post-vitellogenic oocytes). The first method is not reliable. Slender females with distended bellies often have regressed watery oocytes and are not suitable for induced breeding (Fig. 11.15). The post-vitellogenic stage of the

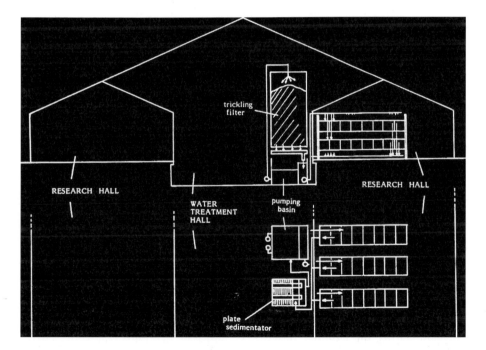

Fig. 11.12 Cross section and upper view of the fish hatchery of the Department of Fish Culture and Fisheries, Agricultural University, Wageningen. A maintenance unit for broodstock of the African catfish is indicated (see also Figs 11.13, 11.14).

Fig. 11.13 Detail of a trickling filter, pumping unit and lamellar plate separator in the water treatment hall (see Fig. 11.12).

gonad can be confirmed by canulation of the ovaries and by measurement of the oocytes. The mean diameter of the oocytes should be more than 1.05 mm. Suitable broodfish are marked by PIT tags (Fig. 11.15).

A variety of natural and synthetic hormones have been used for artificially induced breeding of female broodfish (Table 11.5). 17α-hydroxy-progesterone is not often applied because two injections are required with a time interval of 4 h (Richter *et al.* 1985, Richter *et al.* 1987b). Desoxycorticosterone acetate (DOCA) is no longer used because of its harmful side effects such as the infections under the skin which follow intramuscular injections of the hormone. Spawning induction of *C. gariepinus* using homoplastic pituitaries has become standard practice on commercial catfish farms in Southern Africa. For donor and recipient fish of equivalent weight, a single homogenised pituitary gland, collected during the summer months, is sufficient to induce spawning (Britz 1991). The latency time between hypophysation and spawning is temperature-dependent (Hogendoorn & Vismans 1980) (Table 11.6). Chorulon (human chorionic gonadotropin = hCG) is a very reliable hormone for artificially

Fig. 11.14 A two layer aquarium system with male and female broodstock of the African catfish in the research hall (see Fig. 11.12).

induced breeding. The dose response/latency time relationships have been described (Eding *et al.* 1982).

The possible negative effects of the repeated use of hCG as spawning inducing hormone have also been investigated (Richter *et al.* 1987c). The presence of hormone residues and the long-term effects on absolute fecundity and hatching rate of eggs were investigated in ten broodfish. As a control the

Table 11.5 Agents used for hormonally-induced breeding of *Clarias gariepinus* at 25°C

Agent	Dosage ($\mu g\,g^{-1}$)	Latency time (h)
Carp pituitary	4	11
Pimozide and LHRH$_a$	5 and 0.05	12,5
17α-Hydroxy-progesterone	8[1]	12,5
Desoxycorticosterone acetate	50[2]	14
Human chorionic gonadotropin	4[3]	16

[1] Two injections; [2] no ovulation; [3] IU per g

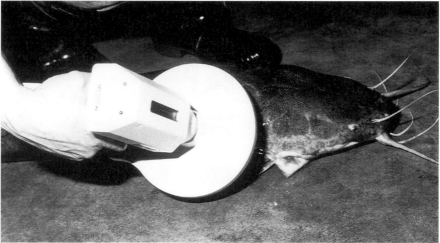

Fig. 11.15 Female broodfish of *Clarias gariepinus*. Above: females with distended belly and ovaries in regression. Below: PIT-tagged broodfish.

Table 11.6 Induced breeding in females of *Clarias gariepinus* with carp pituitary suspension

Water temperature (°C)	Latency time (h)	Incubation of eggs (h)
20	21	57
22	15,5	38
24	12	29
26	10	25
28	8	22
30	7	20

same procedure was followed for estimating egg quality and egg quantity parameters with another set of broodfish using carp pituitary as spawning inducing hormone. Four weeks after the first and second spawnings low residues of 17.8 and 19.6 mIU hCG per ml were found. No immune response against hCG could be detected. The mean absolute fecundity of three subsequent hCG spawnings were 56 000 ± 12 000, 80 500 ± 13 400 and 81 500 ± 21 500. The corresponding hatching rates (%) were 81.3 ± 10.8, 86.3 ± 1.7 and 69.7 ± 7.2. Comparisons with carp pituitary injected controls showed no significant differences.

The histological changes occurring in the ovary after hormonal administration are illustrated in Fig. 11.16 (Richter & van den Hurk 1982). After about 4 h post-injection the germinal vesicle of the oocyte migrates towards the animal pole. At that time the micropyle is formed by penetration of the theca, the granulosa and the zona radiata. The granulosa cells begin to secrete a sticky attachment disc around the micropyle. After about 8 h the germinal vesicle breaks down and the karyoplasm is interspersed in the oocyte plasm. The oocyte gradually hydrates and finally ovulates from the follicular envelope towards the lumen of the ovary. During maturation the first polar body is extruded just prior to ovulation. DOCA is an exceptional hormone in that it induces only maturation and not ovulation. Oviposition of eggs seldom occurs in confinement and therefore the eggs are extruded by hand-stripping. Water absorption through the chorion towards the perivitelline space takes place when the eggs reach water. They remain in the metaphase of the second meiotic division until the sperm penetrates the egg.

In the commercial Catfish Hatchery Fleuren at Someren females are hormonally induced and each is stripped at time intervals of 6–7 weeks in order to maintain egg quality. In contrast to females it is not possible to judge externally whether male catfish have developed mature testes. In a mixed broodstock with post-vitellogenic females, mature males should be present. They are operated on twice per year to serve as sperm donors (Fig. 11.17). The operations are performed under anaesthesia with tricaine methane sulfonate. The testes are only partially removed in order to facilitate regeneration. Cryopreservation of milt has been successfully accomplished in Southern Africa. A cryodiluent of 11% glycerine combined with an extender displayed the best sperm survival. Cryopreserved milt, which had been stored in liquid nitrogen for 16 months, resulted in hatching rates of 41%. This relatively low hatching success did not differ from the results obtained with fresh milt. A shorter time lapse between stripping and insemination of ova will probably improve the results (Steyn & Van Vuren 1987).

11.6 Triploidy and gynogenesis

There is a considerable interest in the application of induced triploidy in fish culture. Triploid fish are often sterile, and therefore depression of growth rate

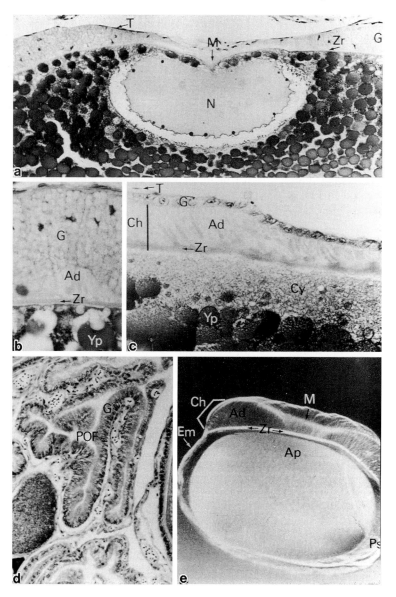

Fig. 11.16 Maturation and ovulation of oocytes in *Clarias gariepinus*. (After Richter and van den Hurk 1982.)
(a) Migrating nucleus in maturing follicle (× 224): (b) Secretion of an attachment disc by the granulosa (× 500): (c) Germinal vesicle breakdown (× 224): (d) Postovulatory follicle (× 224): (e) Ovulated ovum with chorion (× 60);

Nucleus (N), Granulosa (G), Zone Radiata (Zr), Micropyle (M), Theca (T), Attachment Disc (AD), Yolk plate (YP), Oocyte cytoplasm (Cy), Post Ovulatory Follicle (POF), Animal pole (Ap), Chorion (Ch), Perivitellogenic space (Ps).

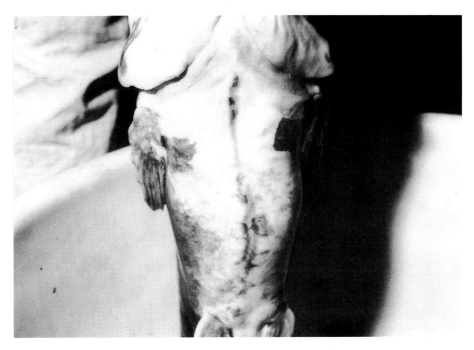

Fig. 11.17 Ventral view of an adult male *Clarias gariepinus*, showing the scars of a testes operation. Selected males are operated on twice per year. They function as sperm donor for the breeding programme of the Catfish Hatchery Fleuren, Someren.

and feed utilization related to sexual maturity does not occur (Purdom 1976).

Genetically sterile fish can also be produced in catfish by artificial retention of the second polar body (Wolters *et al.* 1981). The time at which the second meiotic division takes place varies from species to species. Triploidy was successfully induced in African catfish by cold-shocking eggs at 5°C for 40 min starting 2–4 min after fertilization at 27°C (Richter *et al.* 1987d). Levels of ploidy in larvae and adult fish were determined by chromosome counts in larval tissue and adult kidney tissue (Fig. 11.18). Larval tissues showed 80–100% triploidy in cold shocked groups and 100% diploidy in controls. Hatching of cold-shocked eggs, expressed as a percentage of the survival rate of controls, ranged from 9 to 40%. Triploid African catfish have poorly developed sterile gonads in comparison with untreated fish (Fig. 11.19). The ovaries of triploid fish contain mainly oogonia, so it appears that meiosis is blocked when oogonia are transformed into primary oocytes.

The testes of cold-treated triploid fish (Fig. 11.20) contain cysts with spermatogonia and primary spermatocytes, which are blocked in prophase 1 of meiosis. In young males aged about 6 months, primary spermatocytes predominate in histological preparations of the testes. At later stages, when the males have reached an age of 8 months, the primary spermatocytes start to degenerate and the cysts fuse, giving rise to degenerating tubules.

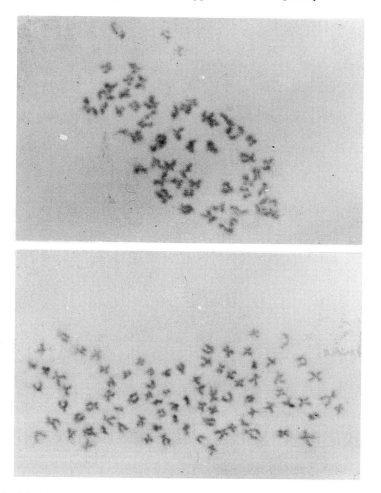

Fig. 11.18 Metaphase chromosome spreads (× 2000) from 7 days old larval tissue of *Clarias gariepinus*. Chromosome number 54. Above: diploid. Below: triploid.

In African catfish, growth rate at either high or low feeding levels was not significantly affected by triploidy (Henken *et al.* 1987). At both feeding levels less protein and more fat per unit fresh body weight gain were deposited in triploids. Decisions in favour of triploid culture of *C. gariepinus* should be based on expected advantages in body composition and gutted weight rather than on expected increases in growth rate.

In fish such as salmonids (Donaldson & Benfey 1987) and cyprinids (Nagy 1987), where the female is homogametic (XX), all-female fish can be obtained by artificial gynogenesis. The term gynogenesis implies that genomic inheritance of the embryo is entirely female. In practice, it means that the paternal chromosomes of the fertilizing spermatozoon need to be inactivated, without affecting its functional ability to initiate development of the embryo. Radiation is often used for the genetic inactivation of spermatozoa. Although initially

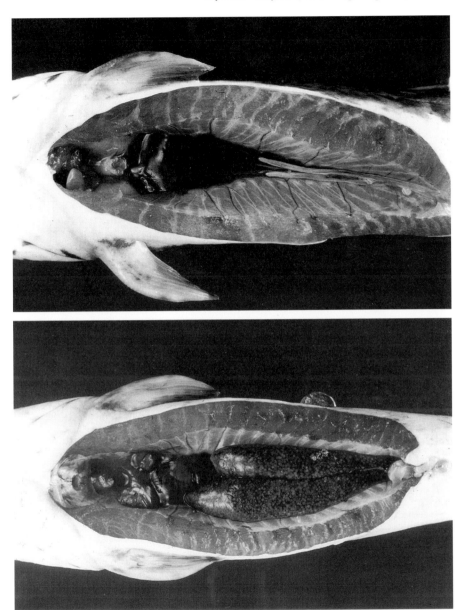

Fig. 11.19 Ovarian development of 8 months old *Clarias gariepinus*. Above: triploid female with reduced ovaries. Below: diploid female with ovaries, containing post-vitellogenic oocytes.

viable, gynogenetic haploid fish do not survive beyond yolk absorption. Diploidy can be restored to gynogenetic haploids by interfering either with meiosis, by retaining the second polar body with its haploid set of chromosomes, or with mitosis, by preventing first cell division. The degree of homozygosity in offspring of the former depends on the rate of crossing over between non-sister chromatids

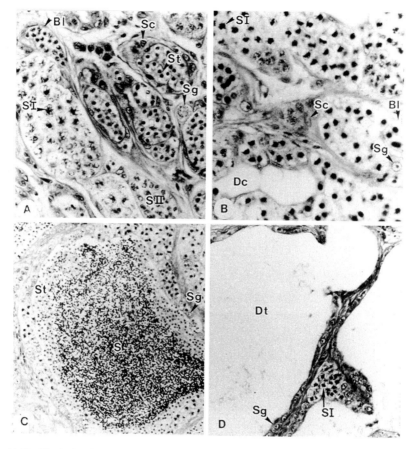

Fig. 11.20 Testis development of diploid 6 months old (A and C) and triploid 8 months old (B and D) *Clarias gariepinus*.
Basal lamina (Bl), Sertoli cells (Sc), Spermatogonia (Sg), Primary spermatocytes (SI), Secondary spermatocytes (SII), Spermatids (St), Spermatozoa (S), Degenerating cysts (Dc), Degenerating tubules (Dt). A and B ($\times$ 280), C and D ($\times$ 56).

during the first meiotic division. Completely homozygous diploid offspring are produced in the latter case.

In a preliminary experiment with *C. gariepinus* heterozygous gynogenetic zygotes have been produced by cold shocking eggs 2–4 min after fertilization. Genetical inactivation of sperm was achieved by UV irradiation. Some gynogenetic offspring of *C. gariepinus* developed into a white-greyish phenotype (Fig. 11.21). This colour variant is possibly determined by the absence of melanophores (Komen *et al.* 1991). The reproductive characteristics as well as the growth performance of gynogenetic fish is followed in the practice of African catfish culture. As in common carp, crossbreeding experiments involving inbred gynogenetic strains will possibly allow for selection of superior hybrids in terms of late maturation and high growth rate (Komen *et al.* 1992).

African Catfish (Clarias gariepinus) 267

Fig. 11.21 Female broodstock of the Catfish Hatchery Fleuren. Note the difference between the white–greyish gynogenetic female and the black normal female.

11.7 Survival of larvae

11.7.1 Larval rearing systems

In Europe 75% of the *Clarias* fingerling demand is covered by two producers (Verreth & Eding 1993). The following rearing procedure is followed for individual egg batches. About 50 000 fertilized eggs are allowed to adhere to sieves with a mesh size of 1 mm and suspended in a 300 l aquarium at a temperature of 25°C. (Fig. 11.22). It takes about 27 h for the eggs to hatch. Mean hatching rates of 50% are obtained. After hatching, the larvae (Fig. 11.22) pass through the mesh but the egg remnants remain. Within 3 days of hatching the yolk-sac will be absorbed. The larvae are subsequently fed Artemia for 7 days, after which gradual weaning to a commercial dry trout starter begins.

The larvae are fully weaned to this diet 14–21 days after the start of exogenous feeding. When the larvae attain a weight of about 100 mg (about 2 weeks after starting feeding) they are transferred from the aquaria to 1000 l fry rearing tanks. Fry rearing takes about 3 weeks, by which time a weight of 0.5 g is reached. About 20 000 hatchlings reach that mean weight. Mortality is supposed to be strongly related with cannibalism in the tanks. To reach the market fingerling size of 5–10 g, a third rearing step (4–5 weeks) is included, during which the fish are raised in a recirculation system, at a density of 75 kg m^{-3} of rearing water. Owing to the wide variation in individual growth

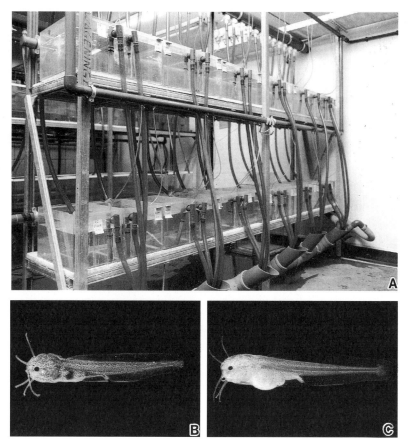

Fig. 11.22 (A) Experimental rearing unit for larvae of *Clarias gariepinus*. The unit consists of 24 aquaria (volume = 17 l) in which 1000 larvae are stocked at the onset of exogenous feeding, and a water purification unit (not visible) consisting of a UV-filter, a gravel filter and a heating unit. Waterflow is maintained at 0.5–1 l minute per aquarium.
(B) Larva of *Clarias gariepinus* before exogenous feeding
(C) Larva with a filled gut (30 minutes after first exogenous feeding).
At the start of exogenous feeding, the larvae measure 6.5 mm and weigh 2.5 mg.

rates, the 0.5 g fry have to be graded before this last rearing step starts. At the Department of Fish Culture & Fisheries in Wageningen the same rearing principles are followed. The hatching and survival rates, however, are usually higher than on the commercial farms. It is supposed that the size of the rearing units (Fig. 11.22A) and the improved levels of control are major factors behind these output differences.

The final goal of hatchery production is to provide fish farmers with sufficient fingerlings of a good quality. The meaning of the term 'quality' is, however, not very well defined. Mostly it is translated into terms of health, strength and disease resistance. From a management point of view, it would be interesting to know the factors which control the so-called quality of the offspring.

11.7.2 Effects of egg size

The previous interspecific comparison between *Clarias gariepinus* and *Clarias batrachus* shows another difference which is directly related to the mentioned quality of the larvae. If after absorption of the yolk sac no food is given to larvae of both species, *C. gariepinus* larvae survive starvation much longer than those of *C. batrachus* (Figure 11.23). Consequently, these differences in survival rate are probably due to species-specific factors. One of the most striking differences between the species is the size of the eggs, with *C. gariepinus* having much larger eggs than *C. batrachus*. In his review of fish development Blaxter (1988) emphasizes the role of egg size on the subsequent viability of the larvae. Fish species which produce larger eggs endure starvation better and reach the point of no return later than species with small eggs. This agrees very well with the results of the present experiment, and it may, therefore, be concluded that species of the genus *Clarias* with larger eggs also have a higher viability and endurance to starvation than those with smaller eggs.

In a similar manner to other fish species, where egg size is directly proportional to fish size, larger female catfish produce larger eggs. At the Department of Fish Culture & Fisheries, Wageningen Agricultural University, average broodstock size of *C. gariepinus* varies between 0.5 and 1 kg. However, in

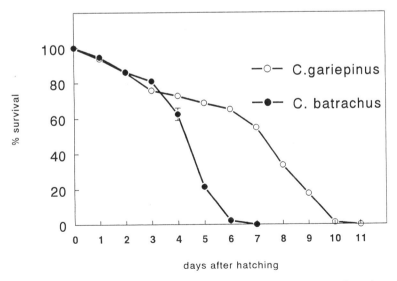

Fig. 11.23 Mean percentage ± S.E. survival rates of *Clarias gariepinus* and *Clarias batrachus* larvae during starvation at 30°C.

270   Broodstock Management and Egg and Larval Quality

the Netherlands, commercial producers of catfish fingerlings use females of 3–10 kg. According to their information, this broodstock size represents a compromise between producing large eggs, and hence larger larvae at the onset of exogenous feeding, and the economic goal of reducing the costs of broodstock maintenance.

11.7.3 Yolk supplies and nutrient utilization

Several studies have been carried out recently which reveal some interesting points regarding the qualitative aspects of larval survival. These include investigations of the metabolism of embryos and starving larvae of *Clarias gariepinus* and the conversion of yolk material into body tissue (Polat et al. 1994). Samples were taken from fertilization onwards, through the yolk sac period and during the larval period of fish which were not provided with any food after completion of yolk absorption. The embryonic period was defined as the period between fertilization and complete yolk absorption, after which the

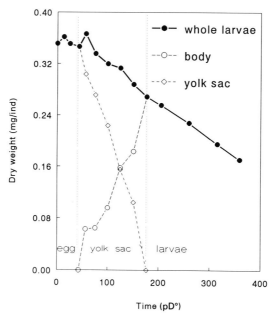

Fig. 11.24 The changes in individual dry weight of embryos and of starving larvae of *Clarias gariepinus*. Samples were taken from fertilization (0 pD°) through the egg phase (hatching around 55 pD°), the eleuthero-embryonal phase (55–170 pD°), and until 50% mortality in the larval phase in fish that did not receive any food (170–360 pD°). During the eleuthero-embryonal phase, samples were taken from whole larvae and body tissue separately. Data on the yolk sac were estimated by the difference between whole animal and body tissue. All data were obtained from individuals raised at 27.5°C. Development time was expressed in terms of physiological day degrees (pD°), which is the number of day degrees (the product of temperature and hours divided by 24 h per day) divided by the Winberg coefficient q to standardize development time at a theoretical standard temperature of 20°C.

larval period started. The experiment was terminated at 50% mortality of the starving larvae. The samples were analysed for biochemical composition, and from studies on oxygen consumption and nitrogen excretion, energy and nitrogen budgets for yolk sac larvae of *Clarias gariepinus* were calculated (Verreth *et al.* 1994a).

Figure 11.24 shows the changes in individual dry weight in whole larvae, yolk-sac and body tissue of embryos and starving larvae of *Clarias gariepinus*. At 28°C, the temperature at which these data were collected, development is very fast, with hatching occurring 24 h after fertilization, and complete yolk absorption 50 h later. During embryonal development, yolk is converted into body tissue, and some material is used for maintenance and tissue growth, resulting in a continuously decreasing individual dry weight of the whole larvae (yolk plus body) (Fig. 11.24). All these data were integrated and used for the development of a simulation model. According to this model, larger and stronger larvae may be produced by an increased yolk conversion efficiency (growth in body tissue/amount of yolk absorbed) during the yolk sac period (Conceição *et al.* 1993). According to the same study, yolk conversion efficiency is dependent upon environmental factors such as temperature and the composition of the yolk.

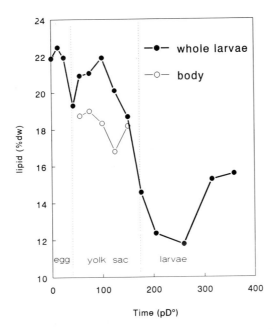

Fig. 11.25 The lipid content (percentage of the dry matter) in embryos and starving larvae of *Clarias gariepinus*. Lipid was extracted by a modified Bligh & Dyer extraction procedure and determined gravimetrically. All data refer to samples from the same study as in Fig. 11.24. For definition of pD° see Fig. 11.24 caption.

dw = dry weight

As shown in Fig. 11.25, the concentration of total lipid in the dry weight of developing yolk sac embryos and larvae of *Clarias gariepinus* decreases steeply from about 22% in newly fertilized eggs to about 14.5% at the moment of complete yolk absorption, after which it decreases further below 12% when no food is given, increasing again to about 15% just before complete starvation (50% mortality). Obviously, during the yolk sac period and immediately after complete yolk absorption, lipids are used for energy, whereas the later increase can be explained only by a preferential use of other compounds, like proteins.

Further fractionation of the total lipid into the different lipid classes provides insight into the differential use of different lipids (Verreth *et al.* 1994b). Figure 11.26 shows the concentration of the different classes of total lipid during the development of yolk sac embryos and starving larvae of *Clarias gariepinus*. Phosphatidylcholine is clearly the first most abundant lipid in the early life stages of this catfish, with a stable concentration of 71–75.8% of the total lipid during the whole period of study. The minor changes in concentration of this phospholipid indicate that it is used in amounts proportional to total lipid.

The second most abundant lipid in the eggs and early yolk sac embryos consists of triglycerides, reaching levels of 12–15% of the total lipid. However, before the end of the yolk sac period is reached, triglycerides are entirely

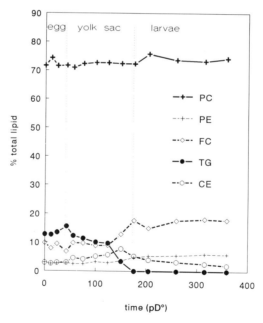

Fig. 11.26 The content of different lipid classes (as a percentage of the total lipid) in embryos and starving larvae of *Clarias gariepinus*. All data were derived from samples taken in the same study as in Fig. 11.19. Lipid classes were determined by TLC-separation, followed by densitometric quantification. The mentioned lipid classes are: Phosphatidyl choline (PC), Phosphatidyl ethanolamine (PE), Free cholesterol (FC), Triglycerides (TG), Cholesteryl esters (CE). For definition of pD° see Fig. 11.24 caption.

eliminated from the system and presumably combusted. The latter hypothesis is substantiated by Fig. 11.27, where the triglyceride content of whole larvae and body tissue is shown. Obviously, only minor amounts of the yolk triglycerides are converted into body tissue lipid.

At this point of the discussion, it is tempting to put forward the following speculative hypothesis. Optimization of the conversion of yolk material into body tissue is partly dependent upon the composition of the yolk. It would be interesting to verify whether a similar phenomenon as the so-called 'protein sparing action' occurs in these early life stages as well. If protein sparing could be induced, it would save protein for growth and result in larger larvae. Further, if it would be biologically feasible to enrich the yolk with lipid classes which serve as preferred energy supplies, this protein sparing action could possibly be induced during the conversion of yolk to body tissue. To date, few data are available to confirm this hypothesis or to prove the technical feasibility of the approach. More research should be dedicated to the energy and protein metabolism of early life stages of fish to test the potential of shifts in energy resource partitioning. In parallel to this, for the African catfish *Clarias gariepinus*, it would be interesting to confirm the feasibility of increasing the triglyceride content of eggs through a combination of selection programmes and broodstock nutrition studies.

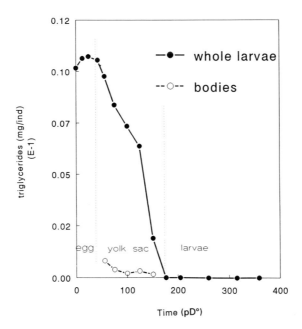

Fig. 11.27 The triglyceride level (mg per individual) in whole larvae and body tissue of *Clarias gariepinus* embryos (eggs and eleuthero-embryos) and starving larvae. The triglycerides were determined after TLC-separation and densitometric quantification of lipids extracted from samples, described in Fig. 11.24.

References

Boulenger, G.A. (1915) *Catalogue of the Freshwater Fishes of Africa in the British Museum (Natural History)*, 4 volumes. British Museum (Natural History), London.

Blaxter, J.H.S. (1988) Pattern and variety in Development. In *Fish Physiology, Volume XI A, The Physiology of Developing Fish: Eggs and Larvae*, (eds W.S. Hoar & D.J. Randall), pp. 1–58. Academic Press Inc., London, New York.

Britz, P.J. (1991) The utility of homoplastic pituitary glands for spawning induction of the African catfish (*Clarias gariepinus*) in commercial aquaculture in Africa. Short communication, Department of Ichthyology and Fisheries Science, Rhodes University. *Water South Africa*, **17**(3), July 1991.

Conceição, L., Verreth, J., Scheltema, T. & Machiels, M. (1993) A simulation model for the metabolism of yolk-sac larvae of the African catfish, *Clarias gariepinus* (Burchell). *Aquaculture and Fisheries Management*, **24**, 431–43.

De Kimpe, P. & Micha, J.C. (1974) First guidelines for the culture of *Clarias lazera* in Central Africa. *Aquaculture*, **4**, 227–48.

Donaldson, E.M. & Benfey, T.J. (1987) Current status of induced sex manipulation. In *Reproductive Physiology of Fish*, (eds D.R. Idler, L.W. Crim & J.M. Walsh), pp. 108–19. Memorial University of Newfoundland, St John's.

Eding, E.H., Janssen, J.A.L., Kleine Staarman, G.H.J. & Richter, C.J.J. (1982) Effects of human chorionic gonadotropin (HCG) on maturation and ovulation of oocytes in the ovary of the African catfish Clarias lazera (C. & V.). In *Proceedings of the International Symposium on Reproductive Physiology of Fish, Wageningen, The Netherlands, 2–6 August 1982*, (eds C.J.J. Richter & H.J.Th. Goos), p. 195. Pudoc, Wageningen.

Hecht, T. & Lublinkhof, W. (1985) *Clarias gariepinus* × *Heterobranchus longifilis* (Clariidae:Pisces): a new hybrid for aquaculture? *South African Science*, **81**, 620–1.

Henken, A.M., Brunink, A.M. & Richter, C.J.J. (1987) Differences in growth rate and feed utilization between diploid and triploid African catfish, *Clarias gariepinus* (Burchell 1822). *Aquaculture*, **63**, 233–42.

Hogendoorn, H. & Vismans, M.M. (1980) Controlled propagation of the African catfish. *Clarias lazera* (C. & V.). II. Artificial reproduction. *Aquaculture*, **21**, 39–53.

Janssen, J.A.L. (1985) *L'élevage du poisson-chat Africain, Clarias gariepinus, en République Centrafricaine*. FAO, Rome.

Jubb, R.A. (1961) *An Illustrated Guide to the Freshwater Fishes of the Zambezi River, Lake Kariba, Pungwe, Sabie, Lundi and Limpopo Rivers*. Stuart Manning, Bulawayo.

Komen, J., Bongers, A.B.J., Richter, C.J.J., Van Muiswinkel, W.B. & Huisman, E.A. (1991) Gynogenesis in common carp (*Cyprinus carpio* L.) II. The production of homozygous gynogenetic clones in F_1 hybrids. *Aquaculture*, **92**, 127–42.

Komen, J., Wiegertjes, G.F., Van Ginneken, V.J.T., Eding, E.H. & Richter, C.J.J. (1992) Gynogenesis in common carp (*Cyprinus carpio* L.). III. The effects of inbreeding on gonadal development of heterozygous and homozygous gynogenetic offspring. *Aquaculture*, **104**, 51–66.

Micha, J.C. (1974) Fish populations study of Ubangui river: trying local wild species for fish culture. *Aquaculture*, **4**, 85–7.

Munro, J.L. (1967) The food of a community of East African freshwater fishes. *Proceedings of the Zoological Society of London*, **151**, 389–415.

Nagy, A. (1987) Genetic manipulations performed on warm water fish. In *Selection, Hybridization and Genetic Engineering in Aquaculture*, (ed K. Tiews), pp. 175–94. Heenemann Verlag, Berlin.

van Oordt, P.G.W.J., Peute, J., van den Hurk, R. & Viveen, W.J.A.R. (1987) Annual correlative changes in gonads and pituitary gonadotropes of feral African catfish, *Clarias gariepinus*. *Aquaculture*, **63**, 27–41.

Ozouf-Costaz, C., Teugels, G.G. & Legendre, M. (1990) Karyological analysis of three strains of the African catfish, *Clarias gariepinus* (Clariidae), used in Aquaculture. *Aquaculture*, **87**, 271–7.

Polat, A., Conceição, L., Sarihan & Verreth, J. (1994) The protein, lipid and energy metabolism in eleuthero-embryos and starving larvae of the African catfish *Clarias gariepinus* (Burchell).

ICES Marine Sciences Symposium, (in press).
Purdom, C.E. (1976) Genetic techniques in flatfish culture. *Journal of the Fisheries Research Board of Canada*, **33**, 1088–93.
Richter, C.J.J. (1976) The African catfish, *Clarias lazera*, (C. & V.), a new possibility for fish culture in tropical regions? In *Aspects of Fish Culture and Breeding*. Miscellaneous Paper, 13 (ed. E.A. Huisman), pp. 51–71. Agricultural University, Wageningen.
Richter, C.J.J. & van den Hurk, R. (1982) Effects of 11-desoxycorticosterone-acetate and carp pituitary suspension on follicle maturation in the ovaries of the African catfish, *Clarias lazera* (C. & V.). *Aquaculture*, **29**, 53–66.
Richter, C.J.J., Eding, E.H., Leuven, S.E.W. & Van der Wijst, J.G.M. (1982) Effects of feeding level and temperature on the development of the gonad in the African catfish, *Clarias lazera*, (C. & V.). In *Proceedings of the International Symposium on the Reproductive Physiology of Fish, Wageningen, The Netherlands, 2–6 August 1982*, (eds C.J.J. Richter & H.J.Th. Goos), pp. 147–50. PUDOC, Wageningen.
Richter, C.J.J., Eding, E.H. & Roem, A.J. (1985) 17α-hydroxy-progesterone-induced breeding of the African catfish, *Clarias gariepinus* (Burchell), without priming with gonadotropin. *Aquaculture*, **44**, 285–93.
Richter, C.J.J., Viveen, W.J.A.R., Eding, E.H., Sukkel, M., Rothuis, A.J., Van Hoof, M.F.P.M., Van Den Berg, F.G.J. & van Oordt, P.G.W.J. (1987a) The significance of photoperiodicity, water temperature and an inherent endogenous rhythm for the production of viable eggs by the African catfish, *Clarias gariepinus*, kept in subtropical ponds in Israel and under Israeli and Dutch hatchery conditions. *Aquaculture*, **63**, 169–85.
Richter, C.J.J., Eding, E.H., Goos, H.J.Th., De Leeuw, R., Scott, A.P. & Van Oordt, P.G.W.J. (1987b) The effect of pimozide-LHRHa and 17α-hydroxy-progesterone on plasma steroid levels and ovulation in the African catfish, *Clarias gariepinus*. *Aquaculture*, **63**, 157–68.
Richter, C.J.J., Sukkel, M. & Blom, J.H. (1987c) Repeated Human Chorionic gonadotropin (Chorulon) induced spawning in female African catfish *Clarias gariepinus*. In *Reproductive Physiology of Fish*, (eds D.R. Idler, L.W. Crim & J.M. Walsh), p. 99. Memorial University of Newfoundland, St John's.
Richter, C.J.J., Henken, A.M., Eding, E.H., Van Doesum, J.H. & De Boer, P. (1987d) Induction of triploidy by cold-shocking eggs and performance of triploids in the African catfish, *Clarias gariepinus* (Burchell 1822). *Proceedings of the EIFAC/FAO Symposium on Selection, Hybridization and Genetic Engineering in Aquaculture of Fish and Shellfish for Consumption and Stocking, Bordeaux, France, 27–30 May 1986*, Vol. **2**, 225–37.
Silverman, H.I. (1978) Effects of different levels of sensory contact upon reproductive activity of adult male and female *Sarotherodon (Tilapia) mossambicus* (Peters). Pisces: Chichlidae. *Animal Behaviour*, **26**, 1081–90.
Stacey, N.E. & Hourston, A.S. (1982) Spawning and feeding behaviour of captive Pacific herring, *Clupea harengus pallasi*. *Canadian Journal of Fisheries and Aquatic Science*, **39**, 489–98.
Steyn, G.J. & Van Vuren, J.H.J. (1987) The fertilizing capacity of cryopreserved sharptooth catfish (*Clarias gariepinus*) sperm. *Aquaculture*, **63**, 187–93.
Teugels, G.G. (1986) A systematic revision of the African species of the genus *Clarias* (Pisces, Clariidae). *Annales Musée Royale de l'Afrique Centrale*, **247**, 1–199.
Van Weerd, J.H., Sukkel, M. & Richter, C.J.J. (1988) An analysis of sex stimuli enhancing ovarian growth in pubertal African catfish. *Clarias gariepinus*. *Aquaculture*, **75**, 181–91.
Van Weerd, J.H., Sukkel, M., Bin Awang Kechik, I., Bongers, A.B.J. & Richter, C.J.J. (1990) Pheromonal stimulation of ovarian recrudescence in hatchery-raised adult African catfish, *Clarias gariepinus*. *Aquaculture*, **90**, 369–87.
Van Weerd, J.H., Sukkel, M., Bongers, A.B.J., Van der Does, H.M., Steynis, E. & Richter, C.J.J. (1991) Stimulation of gonadal development by sexual interaction of pubertal African catfish, *Clarias gariepinus*. *Physiological Behaviour*, **49**, 217–23.
Verreth, J. & Eding, E.(1993) European farming industry of African catfish (*Clarias gariepinus*): facts and figures. *Aquaculture Europe*, **18**(2), 6–13.
Verreth, J., Polat, A., van Herwaarden, H. & Conceição, L. (1994a) Methods to study energy resource partitioning in early life stages of freshwater fish, with special reference to the African catfish *Clarias gariepinus* (Burchell). *ICES Marine Sciences Symposium* (in press).
Verreth, J., Custers, G. & Melger, W. (1994b) The metabolism of neutral and polar lipids in

eleuthero-embryos and starving larvae of the African catfish *Clarias gariepinus* (Burchell). *Journal of Fish Biology* (in press).

Viveen, W.J.A.R., Richter, C.J.J., van Oordt, P.G.W.J., Janssen, J.A.L. & Huisman, E.A. (1985) *Practical Manual for the Culture of the African Catfish (Clarias gariepinus)*. Directorate General International Cooperation of the Ministry of Foreign Affairs, The Hague, The Netherlands.

Wolters, W.R., Liberg, G.S. & Chrisman, C.L. (1981) Induction of triploidy in channel catfish. *Transactions of the American Fisheries Society*, **110**, 310–12.

Chapter 12
Nile Tilapia (*Oreochromis niloticus*)

12.1 Introduction
    12.1.1 Taxonomy
    12.1.2 Characteristics, status and potential
    12.1.3 Subject area reviewed
12.2 Reproductive biology of *Oreochromis niloticus*
    12.2.1 Origin and genetic background
    12.2.2 Reproductive cycle
    12.2.3 Egg characteristics
    12.2.4 Reproductive capacity
    12.2.5 Fry characteristics
12.3 Traditional fry production methods
    12.3.1 Natural recruitment
    12.3.2 Fry production in breeding ponds
    12.3.3 Feeding and nutrition of fry
12.4 Improved management of tilapia broodstock
    12.4.1 Current trends
    12.4.2 Broodfish selection
    12.4.3 Conditioning and nutrition
    12.4.4 Management of broodstock in tanks or hapas
    12.4.5 Conclusions
12.5 Artificial incubation of tilapia eggs and hatchlings
    12.5.1 Advantages of artificial incubation
    12.5.2 Design and operation of egg incubation units
    12.5.3 Hatchling rearing in trays
    12.5.4 System design
    12.5.5 Performance of fry trays
    12.5.6 Conclusions
12.6 Production of all-male fry populations
    12.6.1 Advantages of hormonal sex reversal
    12.6.2 Hormone preparation and administration
12.7 Future developments in broodstock management
    Acknowledgements
    References

Despite their world-wide distribution and prominence in aquaculture, tilapias still contribute little more than 5% of the total farmed freshwater fish supply. The Nile tilapia, *Oreochromis niloticus*, is the most important species on account of its fast growth rate, adaptability to a wide range of culture conditions and high consumer acceptability.

Many of the problems associated with tilapia farming stem from the exceptional mode of reproduction of the *Oreochromis* species. In addition to being mouthbrooders, these fish mature precociously under certain conditions and energy is diverted from growth into reproduction. While this is a natural phenomenon, it can become acute within cultured populations, rapidly leading to overcrowding and stunting. To overcome these problems, various means of controlling reproduction in tilapia culture systems have been considered, the usual commercial practice being to eliminate breeding by rearing only male fish. The methods adopted include hand-sexing, hybridization and male hormone treatment to produce all-male stocks. Hormonal sex reversal is generally regarded as the most viable technique, but failure to apply the treatment properly commonly results in problems of incomplete masculinization and/or poor survival of fry during treatment.

Because the natural breeding cycles of tilapia females are asynchronous within a given population, relatively small quantities of fry are produced simultaneously and cannibalism by older (larger) recruits becomes an increasing source of fry mortality unless fry harvesting is very frequent. The main challenge for tilapia broodstock management is therefore to develop conditioning and breeding methods which can generate a high output of fry suitable for hormonal sex reversal on a sustainable basis.

Although very frequent removal of swim-up fry from breeding units can reduce the incidence of cannibalism, it is labour-intensive and never 100% efficient. To produce high yields of young fry which can be hormonally sex reversed, it is more efficient to rob eggs or yolk-sac fry from mouth-brooding females and incubate the seed artificially. Various systems of broodstock management for *O. niloticus*, including conditioning of broodfish to improve spawning synchrony combined with robbing of seed, are described. Details are also given of low cost egg incubation and hatchling rearing units designed and tested for artificial intensive rearing of seed to produce *O. niloticus* fry suitable for hormone treatment.

12.1 Introduction

12.1.1 Taxonomy

Almost 80 species of fish are referred to by the common name tilapia, but only eight or nine species feature significantly in aquaculture (Schoenen 1982, Pullin 1983). The tilapias belong to the Tribe Tilapiini, an exclusively African group of fish within the Family Cichlidae. Previously regarded as members of a single genus, *Tilapia* (Trewavas 1966), three main genera are now generally recognised (Table 12.1) based on the taxonomic revision published by Trewavas (1983). In addition to anatomical characteristics, her criteria for generic distinction includes the following differences in their reproductive biology: *Tilapia* (substrate spawners), *Sarotherodon* (paternal or biparental mouthbrooders) and *Oreochromis* (maternal mouth-brooders).

While these biological characteristics are crucially important to aquaculturists, their significance with regard to the precise taxonomic relationships between the three main tilapiine groups is still argued (Chen 1990). It is thought

Table 12.1 Reproductive characteristics of the tilapiine genera and the main species of importance in aquaculture

Genus	Mode of reproduction	Important species in aquaculture
Tilapia	Substrate-spawners (guarded nests)	*T. zillii* *T. rendalli*
Sarotherodon	Paternal or bi-parental mouthbrooders	*S. galilaeus*
Oreochromis	Maternal mouthbrooders	*O. niloticus* *O. mossambicus* *O. aureus* *O. urolepis-hornorum* *O. andersoni* *O. macrochir* *O. spilurus*

probable that tilapias were originally substrate spawners and that the mouth-brooding *Sarotherodon* and *Oreochromis* forms arose from substrate-spawning *Tilapia* fish, but of different species (Trewavas 1983).

Although substrate spawners, *Tilapia* species also care for their eggs and young and will even transfer their eggs by mouth from one territory to another; indeed it is presumed that oral incubation evolved from this very practice (Myers 1937). Particularly in older publications on tilapia culture, both mouth-brooding and substrate-spawning forms are denoted as *Tilapia*.

Nearly every commercial strain of tilapia needs to be assessed on its own merit. Although hybridization is not normal in undisturbed natural populations, it has been used widely in aquaculture. The movement of fish species around Africa and the disturbance of water bodies by man (river diversions, dams, etc.) also means that many wild populations may no longer be pure. The wide range of environmental tolerances and growth characteristics seen in tilapias makes it important to know the exact genetic make-up of any commercial strain. As morphological characteristics are unreliable, various biochemical techniques such as protein electrophoresis are necessary for accurate genetic identification (McAndrew & Majumdar 1983).

12.1.2 Characteristics, status and potential

Nearly all the large tilapias most suitable for culture belong to the *Oreochromis* group. These maternal mouth-brooders engage in communal breeding based on a lek system. Males build and defend territories within a spawning area, (the 'lek' or 'arena'), which is visited by receptive females; there is no pair bond and instead polygamy is usual. Under favourable culture conditions, breeding in a tilapia population is continuous but fluctuating; the reproductive cycles of females within a breeding group are not synchronized, although each fish may breed up to 12 times in a year (Macintosh 1985).

There is usually a significant degree of sexual dimorphism and dichromatism within the *Oreochromis* species, males being larger than females of equivalent age and more highly coloured when breeding. It is this male size advantage, coupled with the capacity of tilapias to quickly over-populate ponds and most other culture systems, which has stimulated strong interest in the use of hormone treatments or hybridization to produce all-male fry populations for aquaculture (Lovshin 1982, Mires 1982, 1983, Macintosh et al. 1985, Mair et al. 1991a,b, McAndrew 1993).

Despite the early spread of tilapias worldwide to over 100 countries (Balarin & Hatton 1979), and the widespread awareness of the advantages of using all-male tilapia stocks, world production of tilapias from aquaculture was only 391 000 t in 1990 according to the Food and Agriculture Organization (FAO) (Table 12.2). This represents only about 5% of the total freshwater fish production, compared with 65% contributed by carps (FAO 1992). Moreover, the bulk of the tilapia production comes from relatively few countries, chiefly China, Taiwan, Indonesia, Philippines, Thailand and Israel.

Small-scale fisheries production of tilapias from lakes and reservoirs (see Petr 1985) probably adds considerably to the total supply of tilapia (Luquet 1991), but clearly there is tremendous scope to expand production to help to meet the rising demand for fish. Globally, by the year 2000 aquaculture and inland fisheries will have to rise to 20 million t to meet a projected shortfall between demand (120 million t) and an expected marine fisheries production of about 100 million t annually (Manzi 1989, Haight 1992).

Of the ten or so tilapia species with recognized aquaculture potential (Table 12.1), the Nile tilapia, *Oreochromis niloticus*, is by far the most important. Over the past 30 years, it has been transferred throughout the world to become the mainstay of tilapia farming in many different environments and at all levels from subsistence production to highly intensive farming (Fig. 12.1).

Many examples of the Nile tilapia's versatility in aquaculture can be cited. In Africa, community participation in integrated rural aquaculture is being promoted by international agencies, such as ALCOM (Aquaculture for Local Community Development: a regional programme of FAO). ALCOM's activities in Zambia and several other southern African countries are focussed on small-

Table 12.2 World aquaculture production, 1990

World total production	15 317 000 t
All fin fish	8 409 000 t
Freshwater finfish	7 666 000 t
Carps and carp-like fish	4 981 000 t
Tilapias	391 000 t
Miscellaneous freshwater fish	1 597 000 t
River eels	104 000 t
Salmonids (including smolts)	593 000 t

Source: FAO (1992)

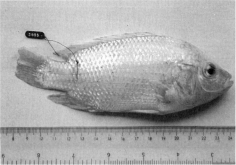

Fig. 12.1 The Nile tilapia, *Oreochromis niloticus*. *Top left*: normal and *top right* 'red' forms of *O. niloticus*. The red tilapia has a numbered tag for broodstock identification. *Above*: commercial intensive fattening of hormonally sex-reversed (all-male) *O. niloticus*; in aerated, concrete tanks, at Rawang, Malaysia. The fish are a 'red' strain, but of mixed colour varieties (white, pink, orange, crimson). (Photos: D.J. Macintosh.)

scale rural fish farming (de Kartzow *et al.* 1992). In contrast, intensive tilapia production in tanks and raceways is operating commercially in Zimbabwe to supply fish to urban supermarkets (Madhu 1992).

In Asia, integrated aquaculture using *O. niloticus* reared in small ponds enriched with agricultural wastes has been promoted vigorously in poor regions, such as north-east Thailand and the Philippines (Edwards 1983, Guerrero 1987, Edwards *et al.* 1988). Some entrepreneurs have also successfully expanded into commercial semi-intensive tilapia production, often in polyculture with carps and catfish, using livestock and agro-industrial wastes. Such advances represent a modern development from the traditional waste-water-fed fish farming methods practised in many Asian countries. However, the latter still continue to be important economically, as for instance the use of domestic waste water to rear carps and tilapias in Vietnam, and their integration with rice and/or vegetable production (Pham & Vo 1990).

Intensive pond, raceway and tank culture systems to produce *O. niloticus* or its hybrids are now operated commercially in many countries (reviewed by Balarin & Haller 1982). Examples recently described include Taiwan (Chen 1990), Florida (Sipe 1992), Zimbabwe (Madhu 1992), Malaysia and Costa Rica (Macintosh 1993). There also have been attempts at similar forms of commercial culture in much more marginal environments for aquaculture using highly managed systems, e.g. Kenya (Haller & Parker 1981), Rwanda (Hanson *et al.* 1988), Belgium (Mélard & Philippart 1980), Israel (Pruginin *et al.* 1988, Mires & Amit 1992), and Colorado, USA (Lauenstein 1978).

The remarkable salt-water tolerance of the tilapias (Chervinski 1982, Perschbacher & McGeachin 1988) means that it is also feasible to grow *O. niloticus* over a wide range of salinities (reviewed by Suresh & Lin 1992). Although *O. niloticus* and its hybrids are not as salt tolerant as some of the other tilapias, fry and adult fish can certainly accommodate 15 and 30–35‰, respectively (Watanabe *et al.* 1984, 1985, Hopkins *et al.* 1989), and there are individual reports of *O. niloticus* being farmed in ponds reaching 50‰, but with much reduced growth (Fineman–Kalio 1988).

It is this diversity of application in aquaculture, coupled with a high growth rate and wide consumer acceptability (Sipe 1992) which makes *O. niloticus* virtually unique as a potential global foodfish and justifies its popular accolade as an 'aquatic chicken'.

12.1.3 Subject area reviewed

The tilapias are one of the best studied groups in aquaculture and have been the subject of three significant symposium volumes in the last decade (Pullin & Lowe-McConnell 1982, Fishelson & Yaron 1983, Pullin *et al.* 1988), plus many reviews and manuals (e.g. Balarin & Hatton 1979, Jauncey & Ross 1982, Balarin & Haller 1982, Guerrero 1987, Tave 1988, Popma & Green 1990, Perschbacher 1992, Suresh & Lin 1992).

Biological and evolutionary aspects of tilapia reproduction have been described in a number of publications, notably by Lowe-McConnell (1959), Fryer & Iles (1972), Noakes & Balon (1982), Trewavas (1982) and Peters (1983). Several recent studies and reviews describe our current knowledge of tilapia reproductive processes, particularly maturation, egg and hatchling development and sex determining mechanisms (Mair 1988, Rana 1990a,b, Mair *et al.* 1991a,b, McAndrew 1993).

Practical aspects of tilapia broodstock management on a commercial scale are now receiving more attention as the introduction of more intensive culture systems increases the demand for reliable supplies of high quality fingerlings.

There is a strong preference for all-male stocks in order that tilapia producers can benefit from the superior growth potential of the male fish.

Several studies have already reported on the practical problems of tilapia broodstock management and fry production (Macintosh 1985, Guerrero & Guerrero 1988, Little 1989, Popma & Green 1990). These and other reviews also describe the advantages of hormonal sex reversal as a means of producing all-male fingerlings in order to prevent breeding in tilapia production units.

It is the intention here to describe the practical application of advances in broodstock management for *O. niloticus* and in techniques for on-rearing tilapia seed for the routine mass production of hormonally-treated, all-male fingerlings.

12.2 Reproductive biology of *Oreochromis niloticus*

12.2.1 Origin and genetic background

The natural distribution of *O. niloticus* has spread from an origin in the Nile Valley to central and western Africa (via the Chad and Niger basins) and south to the Ethiopian lakes and Lake Turkana (Philippart & Ruwet 1982). Seven subspecies of *O. niloticus*, are recognized within this range (Trewavas 1983). Artificial introductions of tilapias within Africa began as early as 1924, including releases into Lake Victoria in the 1950s. By the 1960s *O. niloticus* had been introduced to the USA and, via Japan, to many Asian countries (Trewavas 1983), where it has become a major species in aquaculture.

The normal coloration of *O. niloticus* is silver-grey with dark grey to black markings (Fig. 12.1). In recent years the consumer popularity of *O. niloticus* has been boosted by the emergence of red forms of this fish (e.g. Pullin 1983, McAndrew *et al.* 1988). The availability of these red varieties (see Fig. 12.1) has transformed consumer acceptability in markets not previously disposed to tilapias, especially in the USA (Sipe 1992) and Europe, but also in the more developed Asian countries, notably Taiwan, Singapore and Malaysia (Chen 1990, Macintosh 1993).

The red forms of *O. niloticus* include a red strain from pure Egyptian stock (McAndrew *et al.* 1988); red *O. mossambicus* backcrossed with *O. niloticus*; and the progeny of this red hybrid crossed with either *O. aureus* or *O. urolepishornorum* (El Gamal *et al.* 1988, Kuo 1988).

Because of many generations of selective breeding for the red colour within strains, importations of red tilapias from other countries and back crossing of red tilapia hybrids with pure species, the genetic composition of red tilapia populations is now rather confused and variable from country to country. In Malaysia, for example, red tilapia hybrids from Taiwan have been imported for several years and have no doubt interbred with the local *O. niloticus* (normal colour) and with *O. mossambicus* introduced much earlier (Hickling 1962). Local tilapia producers now breed their own red tilapia, and small quantities of red hybrids from Israel (via Singapore) and Thailand have also been introduced (Macintosh 1993). In neighbouring Thailand, in contrast, both the normal and red strains of Nile tilapia appear to be relatively pure (Pullin & Capili 1988).

12.2.2 Reproductive cycle

The Nile tilapia can mature sexually within 6 months and breed when still very small (i.e. below 40 g). This diversion of energy from growth into reproduction is a natural phenomenon shown by wild tilapias under certain conditions (Iles 1973), but it becomes acute within cultured populations (Mair & Little 1991). Once mature, tilapia females produce multiple batches of eggs, the oral incubation of each egg batch being followed by only a short period of recovery before she is ready to breed again.

A mature female visits the nesting area or 'lek', where adult males are actively displaying. Courtship and ovulation are rapid, with the female picking up the eggs in her mouth as soon as they have been released at ovulation and fertilized. On average, *O. niloticus* produces from several hundred to about 2000 eggs per batch. There is then an incubation period of about 10 days, when the female leaves the nesting area to quietly carry her egg batch until the fry are ready for release. There may be a short nursing period thereafter, usually 1–4 days, during which the female continues to protect the fry even though they have left her mouth and are feeding for themselves. Because the female does not feed during the period of incubation, a phase of intense feeding and recovery follows, usually lasting about 2–4 weeks, before she is ready to spawn again.

The reproductive characteristics described above and illustrated in Fig. 12.2 mean that approximately 1 month, on average, is required for a female tilapia to complete each reproductive cycle. Breeding may not occur every month, but under favourable environmental conditions a female *Oreochromis* will normally produce several batches of young in a year. Macintosh (1985) found that egg production by groups of *O. mossambicus* and *O. niloticus* females maintained under identical and constant conditions in glass aquaria was equivalent to 11 and 6 egg batches per female per year respectively.

A key problem for the tilapia hatchery producer is that spawning by *Oreochromis* tilapias is not synchronized among females, so that a population of broodfish will usually produce continuously, but at a low rate (Little *et al.* 1993). Moreover, in a relatively short period of time, fry output in conventional breeding ponds will decrease as they become overcrowded with recruits from the early spawnings, it being impossible to net out all the fry as they are released.

Even in a more controllable spawning environment, such as a tank, natural fry production will decrease rapidly after an initial peak in spawning (Fig. 12.3), showing that continuous high output cannot be maintained because of the development of asynchronous spawning activity.

Nile Tilapia (Oreochromis niloticus) 285

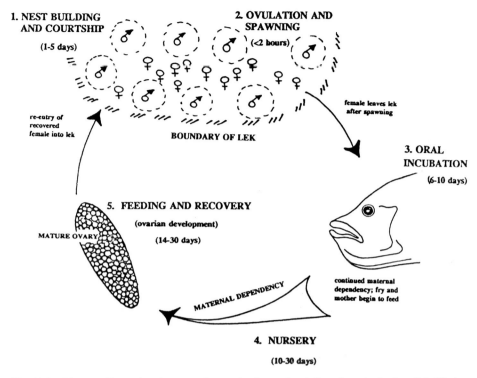

Fig. 12.2 Diagram illustrating the natural reproductive cycle of *Oreochromis* tilapias. (Modified from Little 1989.)

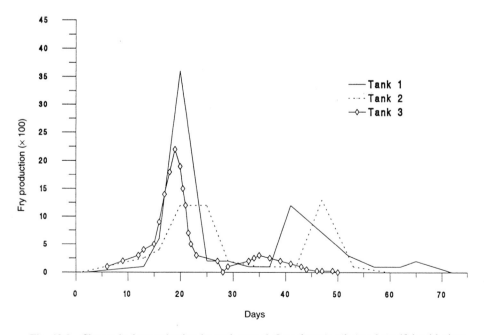

Fig. 12.3 Change in fry production by tank-reared *Oreochromis niloticus* broodfish with time. (From Guerrero & Guerrero 1985.)

Cannibalism of young fry by the older recruits remaining in the pond is another major cause of declining fry output in broodstock ponds, and usually the only effective solution is to drain the pond and start again with new broodfish. Gregory (1987) showed that an age difference between *O. niloticus* fry of 4 days was sufficient to induce cannibalism and the rate of mortalities increased dramatically if the age difference exceeded 24 days. However, cannibalism can even develop significantly among sibling fry from the same brood (Macintosh & De Silva 1984).

12.2.3 Egg characteristics

As an adaptation to their highly specialized form of parental care, the eggs of *Oreochromis tilapias* are large, yolky and without an adhesive layer; egg volume in *O. niloticus* ranges from 2.85 to 11.15 mm^3 (Rana 1986a). Several significant relationships between broodstock traits and egg and fry quality have been demonstrated which are relevant to broodstock selection and management.

In *Oreochromis niloticus* and *O. mossambicus*, older females produce larger eggs, but within a group of females of the same age class there is no significant relationship between body size and egg size (Rana 1986a). Because eggs from older fish contain more yolk reserves the fry from older broodstock can tolerate starvation for a longer period. Also, larger eggs result in larger fry, and this size effect is still detectable in fry 60 days after hatching (Rana & Macintosh 1988). Relevant data on the egg to parent size relationship for *O. niloticus* are summarised in Table 12.3.

Table 12.3 Broodstock and egg relationships in *O. niloticus* (from Rana & Macintosh 1988)

Parameter	Year class of female broodfish		
	0+	1+	2+
Age (months)	8–10	12–14	23–25
Mean weight (g)	53	192	316
Range	31–90	160–220	218–486
Mean egg wt (mg)	1.93	2.79	3.68
Weight of fry after 60 days (g)	1.96	2.50	2.98
Survival time if fry starved (ST_{50})	13.5	16.0	17.5

12.2.4 Reproductive capacity

For economic reasons, commercial hatchery operators are usually interested in gaining maximum benefit from the reproductive potential of tilapia females. There is still much to learn about the best ways of managing *O. niloticus* broodstock to achieve optimum egg output over the productive period of their reproductive phase, but some useful information has emerged from tilapia breeding trials.

On average, larger females produce more eggs per batch than smaller females, although there is high variability associated with the number of eggs per spawning (Macintosh 1985). While egg number increases with female length (L), it does so at a significantly lower rate ($L^{1.74}$ for hatchery-reared *O. niloticus*; Rana 1986a) than in other commercial fish, and this is a presumed consequence of the mouth brooding adaptation (Babiker & Ibrahim 1979). Relative fecundity is not significantly correlated to body length (Rana 1986a).

Fecundity is apparently influenced by genetic factors, as well as by environmental conditions, especially those influencing the nutritional status of the fish. However the reproductive capacity of tilapias is not well expressed by fecundity values because of their multiple spawning habit; females are capable of producing as many as 10–12 egg batches annually (Philippart & Ruwet 1982, Macintosh 1985). Consequently, those factors influencing spawning frequency are at least as significant as size-related fecundity in determining reproductive capacity.

It has also been noticed that larger females are better at incubating their egg batches than smaller fish (Lee 1979, Siraj *et al.* 1983). Since size is usually also indicative of age, this observation can be explained largely by the fact that younger (and smaller) females are less efficient at incubation because of their relative inexperience (Rana 1986a).

Under natural tropical conditions in Thailand, Little (1989) compared fry output from *O. niloticus* broodfish of two different size groups stocked in earthen ponds. Fry were removed six times daily to reduce the potential effect of fry cannibalism and wild fish predation on the results. Table 12.4 shows that, over a significant experimental period (105 or 116 days), swim-up fry yield was more than double from the pond containing small broodfish (average female weight 207 g) compared with output from an equivalent pond with large fish

Table 12.4 Average yields of *Oreochromis niloticus* fry every five days from 1740 m² earthen ponds in relation to broodfish size and frequency of fry harvesting (from Little 1989)

Pond no. (duration)	Broodstock size (Management system)	Mean fry yield every 5 days (SE)	
		Small fry[a]	Large fry[a]
1 (116 days)	Mixed size class (six times daily harvest)	21 403 (3763)	132 (14)
2 (116 days)	Mixed size class (three times daily harvest)	13 369 (2062)	106 (10)
3 (105 days)	Large size class[b] (six times daily harvest)	19 704 (4516)	130 (13)
4 (105 days)	Small size class[b] (six times daily harvest)	48 131 (7914)	207 (18)

[a] Small fry = suitable for hormonal sex reversal (<15 mg)
   Large fry = too large (old) for sex reversal (>15 mg)
[b] Large broodfish mean weight 262 g, small broodfish 207 g

(average female weight 262 g). There were no age-related effects on reproductive output because the fish used in this study originated from a single age group.

The higher productivity achieved from smaller broodstock can probably be explained by their more rapid recovery after fry release, i.e. shorter inter-spawning interval. This study also demonstrated the importance of monitoring reproductive output over a meaningful period of time (preferably several months) before reaching conclusions about the practical importance of different broodstock traits on reproductive output.

Faced with the situation that smaller broodfish may collectively yield more eggs and fry per culture unit, but that larger females individually produce more eggs and fry than smaller ones, the hatchery operator must decide on an optimal management system regarding the size of broodfish selected and the length of time they are used for breeding. It should be noted too that tilapia females continue to grow significantly (1 g per day approximately; Table 12.5) even when reproducing at a high frequency. This raises the question of how frequently broodfish should be replaced in order to optimize productivity in relation to the additional costs that replacement involves (see [12.4]).

Table 12.5 Numbers of *Oreochromis niloticus* broodstock stocked in spawning and conditioning tanks, mean weight gain (g per fish) and feeding rate under different management regimes during an experimental period of 155 days (from Little 1989)

Treatment	Number of Fish			Daily weight gain (g)		Feeding rate (% body weight per day)
	males[a]	females[a]	females[b]	Female	Male	
1	50	52	—	1.54	2.97	2.6
2	50	52	—	1.21	1.91	1.3
3	50	52	19	1.10	2.41	1.3
4	50	52	52	0.65	2.41	1.4

[a] Number of fish in spawning tanks during each inter-harvest interval
[b] Maximum number of fish held in conditioning tank

12.2.5 Fry characteristics

Nile tilapia eggs take about 4 days to hatch at 28°C. Development time is strongly temperature dependent and may vary from 6 days at 20°C to 3 days at 30°C (see [12.5.2]). The stages of egg and hatchling development in *O. niloticus* are described in detail by Rana (1990a). After hatching, the yolk-sac fry show weak movement as they would normally remain protected within the mother's mouth. Early yolk-sac fry, if incubated artificially, are easily damaged (e.g. if the water flow is too strong). The yolk-sac is gradually absorbed over about 4–6 days at 28°C, the gut opens and the swim bladder inflates. This transformation to the swim-up fry stage is accompanied by a change to very active swimming and feeding. This is the point at which they would normally leave the mother's mouth, although she may continue to protect them by readmitting them to her mouth, usually for up to 4 more days. By 10−12 days after hatching the fry are fully independent and clearly tilapia-like.

A key aspect of tilapia fry biology which affects their performance in culture systems is the development of size variation between fry and the subsequent emergence of cannibalism. This is a major problem in tilapia broodstock ponds used for fry production, because older fry which evade capture consume many young fry as soon as the latter are produced. Gregory's (1987) experimental observations on cannibalism between *O. niloticus* of different age groups indicate that there would be significant fry losses from cannibalism by older recruits within 1 month of operating a breeding pond. He concluded that *O. niloticus* as small as 15 mg (17 mm standard length) were capable of preying on swim-up fry of their own species.

Cannibalism is even a significant problem among tilapia fry of the same age because of the appearance of individuals or 'shooters' which are much larger than the others (Sampson 1986). Macintosh & De Silva (1984) showed that cannibalism among fry of the same age group was most serious between 10 and 30 days after first feeding and was inversely related to the amount of food given, i.e. well fed fry were less cannibalistic. Cannibalism accounted for between 10 and 35% of the total fry mortality they observed.

Tilapia fry produced by traditional pond methods are also highly susceptible to predation from wild fish. In Asian fishponds, the typical carnivorous or omnivorous species which consume tilapia fry include climbing perch (*Anabas testudineus*), snakeheads (e.g. *Channa striata*), walking catfish (*Clarias batrachus*) and various gouramies. Little (1989) found *Anabas* and *Channa* always present in a freshwater tilapia broodstock pond studied near Bangkok.

Parasites are another potential biological problem associated with tilapia fry production. The ciliate *Trichodina* is particularly associated with mouth-brooders (Fryer & Iles 1972) because these parasites enter into the buccal cavity of the broodfish and infect the developing fry. 'Flashing' by tilapia fry, usually followed by high mortalities, are the classic symptoms of *Trichodina* infections. Protozoan parasites like *Trichodina* and *Chilodonella* are ubiquitous in pond fish in low numbers, but they can flare into serious infections if the fish are stressed by poor handling, poor nutrition and especially when there is a fall in temperature (Sarig 1971). There is at least one record indicating high mortality from parasitic infestation during high density hormone treatment of hatchery-reared tilapia fry (du Feu 1987).

12.3 Traditional fry production methods

12.3.1 Natural recruitment

Tilapias are unique among cultivated fish in breeding freely under normal conditions without the need for any form of artificial inducement. Natural recruitment in production ponds is still welcomed by many small-scale farmers, who view it as a free source of seed to replace the fish they harvest. However, the associated problems of self-recruitment are crowding and stunting (Iles 1973), a phenomenon probably associated with a shift in the balance between somatic growth and gametogenesis increasingly in favour of the latter as their environment becomes more limiting or stressful (Pullin 1982).

Faced with this habitual problem of over-recruitment and stunting in cultured populations (Cross 1976), methods for controlling reproduction and recruitment are needed for all but subsistence levels of tilapia farming (Mair & Little 1991). Ideally, this should involve a stock management system which separates the production and reproductive phases of the life cycle, coupled with genetic or hormonal manipulation of the cultured stock to eliminate normal reproductive activity during the grow-out phase.

12.3.2 Fry production in breeding ponds

Netting fry released by tilapia broodfish stocked in specially established breeding ponds is still the most widely used method of obtaining tilapia seed. Many farmers simply rely on seine netting their ponds to collect the larger recruits (fry and fingerlings), but of course these are already sexually differentiated. Figure 12.4 illustrates one such system followed in Eastern Thailand. Broodfish are stocked in small earthen ponds at low densities ($0.1-0.3$ females per m^2) and with a sex ratio of between two and five females per male. The ponds are seine netted every 10 days. This yields tilapia fry and fingerlings of mixed sex, the smallest and most abundant being in the $2-3$ cm size range. If fry suitable for hormonal sex reversal are required, the best alternative is to harvest at least daily using hand nets to catch the young swim-up fry as they shoal at the pond edge (Fig. 12.4).

As already noted, the main limitation of breeding ponds is the lack of sustainability, as harvestable fry production usually declines rapidly with time owing to a combination of factors. These include predation by other fish, cannibalism by older recruits, asynchronous spawning by broodstock and a reduction in spawning frequency as the ponds become overcrowded. Some of these limitations, however, especially those connected with predation and cannibalism of young fry, can be overcome with improved management. These include the maintenance of fertile water conditions and a low sediment concentration, precautions against the entry of wild predatory fish and regular pond harvesting. The yield of swim-up fry obtained by edge netting can be enhanced significantly by increasing the frequency of fry harvesting to several times daily.

In the study by Little (1989, and Table 12.4), fry were collected either six times per day (high frequency collection) or three times daily (low frequency) from identical 1740 m^2 earthen ponds containing *O. niloticus* broodfish (317 fish per pond, sex ratio 1:1). One circuit of the pond was made per collection time, using a small hand net to scoop shoaling fry. The time taken to complete each

*Nile Tilapia (*Oreochromis niloticus*)* 291

Fig. 12.4 Netting of Nile tilapia fry from earthen broodstock ponds in Thailand. *Top*: commercial harvesting of fry of mixed age and sex by seine netting; ponds average $1200\,m^2$ and are harvested every 10 days, yielding fish $2-3\,cm$ and above. *Bottom*: hand netting newly released fry suitable for hormonal sex reversal (pond surface area $1740\,m^2$); yields of young fry are greatly improved by frequent daily netting. (Photos: *Top* D.C. Little, *bottom* D.J. Macintosh.)

collection circuit was timed to provide an index of fishing effort. After 105 or 116 days operation, the ponds were drained and the broodstock and other fish present were collected, counted and weighed.

Table 12.4 shows the average fry production in each 5 day period, divided into two categories: small fry suitable for sex reversal (less than 0.015 g) and larger fry too old to respond to hormone treatment. By harvesting at such frequent intervals the problems of predation and cannibalism were greatly reduced and nearly all the fry obtained were young enough for hormone treatment. Production was significantly higher (by 60%) for the high frequency harvesting system, which indicates how rapidly newly released tilapia fry were preyed upon by other fish in the pond. However, fry production was highly variable over each 5 day period, ranging from less than 5000 fry per pond to 60 000 fry per pond. Average yields from three of the four ponds studied were equivalent to 280 000–450 000 fry approximately, or about 2500–4000 fry per pond per day (= 2 fry per m^2 per day approximately). The fourth pond, which had the same biomass of broodstock but consisting of smaller fish (female mean weight 207 g compared with 262 g), gave 866 000 fry, or about double the output (equivalent to 4 fry per m^2 per day).

On draining the ponds for harvest, the biomass of broodfish and offspring not collected during fry harvesting was calculated. Broodfish increased in weight during the trial by almost 1 g per day (0.84 and 0.95 g per day in the high and low frequency treatments respectively). But the largest source of net production in the ponds came from their offspring. These recruits accounted for more than 61% of net production in the low frequency pond and almost 50% in the high frequency pond.

Fry output from tilapia broodstock ponds is generally highly variable. Not only does productivity tend to decline with time as spawning becomes more asynchronous and cannibalism increases, there is also a strong temperature effect on spawning activity. Natural periods of low temperature in the ponds are correlated with declines in fry production, and in countries with significant cool seasons such as Taiwan and Israel fry production in open ponds is curtailed. However, a short phase of low temperature can promote spawning synchrony. Srisakultiew & Wee (1988) found that the gonads of broodfish in Thailand regressed if they were held at low temperature (22°C) for long periods, but a few hours' exposure to cold followed by return to ambient temperature (30°C) improved breeding synchrony.

Whatever effect short-term environmental conditions have on spawning, there will be an eventual decline in fry output correlated with the build up of recruits. In Israel broodfish cease to spawn in ponds which become overcrowded with fry (Mires 1982). Frequent drainage of broodstock ponds to harvest all fish is the most effective way to eliminate the negative impact of recruits, but involves considerable time and effort. In the Philippines, farmers drainage harvest their ponds monthly to remove recruits; the broodfish are then restocked (Guerrero 1987).

12.3.3 Feeding and nutrition of fry

Swim-up fry shoal together initially and will naturally seek warm, shallow water. For this reason they do well in shallow rearing units with a large surface area to volume ratio and grow best at relatively high temperatures (30–32°C). They are primarily plankton feeders, although they seem attracted to a wide variety of particulate feeds. In culture, the growth performance of tilapia fry is much better if they are fed frequently so that they can browse on the food over a prolonged period. It was found that *O. spilurus* fry grew significantly better if fed five times daily compared with three times daily, but continuous feeding was not more efficient than giving five feeds (New *et al.* 1984).

High feeding rates and high quality feeds significantly improve the survival and growth of fry. Santiago *et al.* (1987) fed *O. niloticus* fry (initial weight 12 mg) for 5 weeks on crumbled pellets containing 35% crude protein. Fish receiving 15, 30, 45 and 60% of body weight reached mean weights of 63, 198, 232 and 228 mg respectively, with survival rates of 53, 85, 87 and 84% respectively. When different feed protein levels were compared (20, 30, 40 and 50%), Siddiqui *et al.* (1988) found that *O. niloticus* fry performed best on 40% dietary protein.

12.4 Improved management of tilapia broodstock

12.4.1 Current trends

Given the many limitations associated with natural fry production in ponds, the introduction of more advanced forms of broodstock management is a key to improving fry yields and overall system efficiency. Moreover, better management systems for tilapia broodstock can provide the additional control necessary to achieve large-scale production of fry suitable for hormonal treatment. They also offer the potential to introduce other technical advances in hatchery production, such as manipulation of environmental conditions to improve breeding activity, and broodstock genetic selection to improve fry quality.

Recent improvements in Nile tilapia broodstock management have focussed on conditioning broodfish before spawning; on removing eggs for artificial incubation (i.e. robbing eggs from the mouths of incubating females); and on regular replacement of broodstock (or spawned females only) with rested and conditioned fish to improve spawning synchrony.

To achieve this degree of management it is necessary to confine broodfish in tanks or fine-net cages (hapas) in order to facilitate handling (Figs 12.5, 12.6). Which system is chosen will depend on economic factors, management capabilities and individual choice.

A general comparison of the technical and economic aspects of commercial pond-based, tank-based and cage-based tilapia hatcheries in the Philippines is provided by Guerrero (1987), while Little's (1989) study offers a more analytical

Fig. 12.5 Harvesting seed from Nile tilapia broodfish stocked in 40 m^2 hapas, Asian Institute of Technology, Thailand. (Photo: D.J. Macintosh.)

approach showing quantitatively how improved broodstock management can increase fry output from each of these three main hatchery systems.

12.4.2 Broodfish selection

Since *O. niloticus* has a wide natural range in Africa, with seven sub-species recognised (Trewavas 1983), the genetic base for broodstock selection is potentially very good. In practice, the primary choice for commercial producers is currently between culturing the normal (grey) *niloticus* or one of the red *niloticus* or *niloticus* hybrid strains. The red varieties actually consist of several main colour types, crimson, orange, gold, pink and albino and many intermediate forms (Galman *et al.* 1988). These red variants are now present in many countries as a result of introductions and selective breeding programmes (Lester, *et al.* 1988). In addition, recognized strains of *O. niloticus*, as well as hybrid varieties, are sold commercially by live fish exporters operating in several countries, particularly Israel, the USA and Taiwan.

Under various rearing conditions (ponds, tanks, cages), growth and production rates of normal and red *O. niloticus* have not been significantly different (Rifai

Fig. 12.6 Hatchery facility for broodstock management of Nile tilapia in tanks, involving artificial seed collection and broodstock exchange; top right, 4 m diameter spawning tanks; bottom left, 1.5 m diameter tanks for conditioning broodfish/rearing fry; see text for operational details. (Photo: D.J. Macintosh.)

1988, Macintosh *et al.* 1988, Abella *et al.* 1990). Results from selection trials for growth within strains have been variable (Abella *et al.* 1990, Hulata *et al.* 1988), but both colour forms have shown positive growth responses. Recent work on red *niloticus* in Thailand showed a body weight gain of more than 15% over five generations in a selected line, representing a heritability factor of 0.32 (Jarimopas 1990).

There is clearly also scope to improve tilapia hatchery production by studying the heritability of parameters such as fecundity, egg quality and inter-spawning interval, but genetic selection for desirable reproductive traits remains one of the most undeveloped aspects of tilapia research. There has been little objective comparison of the reproductive performance of different strains of *O. niloticus*. Some differences in fecundity, expressed as average number of incubated eggs per batch, have been shown between different strains of *O. mossambicus* (Macintosh 1985), and other available evidence indicates that the red varieties of *O. niloticus* are less vigorous than the normal colour form. A particular difficulty with this line of research is the strong influence of environmental conditions and phenotypic characteristics, such as size, age, nutritional status, on reproductive performance.

12.4.3 Conditioning and nutrition

In comparison with other fish species in aquaculture, the nutritional requirements of female tilapias are greatly affected by their unique mode of reproduction. Mouth-brooders deprive themselves of food throughout each period of oral incubation. Since female *Oreochromis* can produce several broods in succession, they may ingest food for only 4–5 days between non-feeding incubation periods each lasting 10–13 days (Macintosh 1985). In the often short feeding periods between broods, female tilapias have to feed voraciously to regain body condition lost during incubation and to obtain energy to support further reproductive activity. Studies with tagged fish have shown that body weight loss is well correlated to the duration of incubation (Little 1989). D.J. Macintosh (unpublished) found that female *O. mossambicus* consumed commercial pelleted trout feed up to the equivalent of 40% of their body weight in the 48 h immediately after releasing a batch of fry.

While commercial pelleted diets are now routinely manufactured for tilapias, little is known about the underlying nutrition of tilapia broodstock. This is especially true with respect to their lipid, essential fatty acid, vitamin and mineral requirements (reviewed by Luquet 1991). The optimum level of dietary protein for growth of fry and young adult tilapias seems to be about 27–35% and 25% respectively (Jauncey & Ross 1982, Wee & Tuan 1988, Luquet 1991).

A dietary level of 35% crude protein resulted in optimum growth and spawning by *O. niloticus* broodstock held in clear-water tanks (Wee & Tuan 1988). In the same study, higher protein levels in the diet (42 or 50% crude protein) stimulated earlier maturation and resulted in larger eggs and slightly higher hatching rates (Table 12.6), but had a negative effect on spawning frequency and fecundity. The latter were explained by behavioural factors, with the richer diets stimulating a stronger hierarchy and greater aggression among the broodfish (Wee & Tuan 1988). De Silva & Radampola (1990) also

Table 12.6 Effect of dietary protein intake on egg quality in *Oreochromis niloticus* (values shown are means (SD)); (from Wee & Tuan 1988)

Parameter	Protein content of diet (%)				
	20.2	27.5	35.0	42.6	50.1
Egg wet weight (mg)	4.6 (0.7)[a]	4.8 (0.6)[ab]	4.9 (0.6)[abc]	5.1 (0.6)[bc]	5.3 (0.9)[c]
Egg dry weight (mg)	2.0 (0.5)[a]	2.0 (0.2)[a]	2.0 (0.3)[a]	2.1 (0.3)[a]	2.1 (0.3)[a]
Volume (mm^3)	5.01 (0.84)[a]	5.38 (0.92)[b]	5.43 (0.65)[b]	5.77 (0.97)[b]	5.72 (0.85)[b]
Moisture (%)	58.6 (3.1)[a]	59.2 (1.8)[a]	59.5 (1.6)[a]	59.2 (1.9)[a]	59.6 (1.6)[a]
Hatching rate (%)	64.9 (13.3)[a]	69.8 (12.9)[a]	68.5 (13.5)[a]	73.0 (13.7)[a]	71.3 (12.6)[a]

Values in each row with the same superscript are not significantly different ($P = 0.05$)

found that spawning frequency decreased with protein level, whereas Santiago et al. (1982) reported that higher dietary protein levels (40 and 50%) increased spawning frequency in *O. niloticus*. In conclusion, the interaction between the nutritional regime of broodfish, both before and after the onset of maturation, and their spawning environment seems important in determining the strength of hierarchies between spawners and consequently their spawning performance.

Uchida & King (1962) identified 35–40% dietary protein as optimal for fry production by tank-reared *O. mossambicus* broodfish. Good breeding results in tanks have also been obtained at Stirling University with this species and *O. niloticus* using commercial trout pellets containing 43–46% protein (Macintosh 1985).

There is a strong argument in favour of conditioning broodfish in 'greenwater' ponds because of the presumed nutritional benefits provided by the natural food available and the fact that the fish can graze continuously. It is believed that phytoplankton provides tilapias with a significant source of supply of B vitamins and, moreover, intestinal synthesis of vitamin B12 has been shown in *O. niloticus* (Luquet 1991). In fertilized ponds, *O. niloticus* feeds almost continuously on a mixed diet of green algae, blue-green algae, diatoms and detritus (Dewan & Saha 1979) and so is clearly adapted nutritionally to a diet of mixed phytoplankton.

In actual fact, well-fertilized ponds have proved adequate for tilapia breeding without the use of supplementary feeding and this has been a normal commercial practice in the Philippines (Yater & Smith 1985). However, additional feeding is advisable within economic limits. Supplementation with a pelleted catfish diet (crude protein content 30%) fed three times daily to appetite has produced excellent breeding results with *O. niloticus* in fertilized ponds in Thailand (Little 1989).

In nature, *Oreochromis* species graze almost continuously during daylight hours. They consume phytoplankton, zooplankton, benthic algae and detritus, the proportions varying with species and stage in their life cycle. The tilapia gut is adapted to receive food in small amounts and to digest it slowly. More than 18 h is required for evacuation of a meal (Ross & Jauncey 1981). This mode of feeding suggests that it is advantageous to feed cultured stock several times daily. In clear-water culture, frequent feeding also reduces the risk of uneaten food fouling the water. Balarin & Haller (1982) found that when tank reared tilapias were fed three to six times daily both food consumption and water quality were improved.

For *O. niloticus* broodstock, it is concluded that fish should be fed 2–3% body weight daily, or to appetite, giving the ration over at least three feeds in daylight hours. However in systems in which fish were spawning synchronously, Little (1989) found that it was impossible to feed in excess of the equivalent of 1.5% body weight per day because much of the time the fish were incubating eggs. Indeed a reduction in the feeding response of tilapia broodstock is a good indicator of the onset of spawning synchrony. The apparent feeding ration required by broodfish is also dependent on the type of feed given. Practical experience suggests that there is less food wastage when a floating diet is used (unpublished observation).

Diets containing 25–30% crude protein are satisfactory for broodstock reared in green-water systems, while 30–40% crude protein is probably optimal for fish spawning in clear-water units. The effects on reproduction of vitamin and mineral supplementation have hardly been studied and this would be a fruitful area for further research. The addition of dietary ascorbic acid was found to improve hatchability and fry quality in O. mossambicus (Soliman et al. 1986), but the broodfish used were reared exclusively in clear water.

12.4.4 Management of broodstock in tanks or hapas

Systems comparison

Further results from the comparative study of O. niloticus breeding systems in Thailand by Little (1989) are provided in Table 12.7. The data show seed production (eggs and fry) by broodstock in fertilized earth ponds and hapas in fertilized earth ponds (i.e. green-water systems) and also in concrete tanks supplied with recirculated water (clear-water system). This study also examined the benefits to seed production of good conditioning and of frequently exchanging broodfish in the breeding units (every 5 or 10 days).

On a unit area productivity basis (fry per m^2 per day) tank-held broodfish performed best because they could be held at higher densities. Output per unit weight of female tilapia stocked was variable and did not show any consistent trend in relation to the different broodstock rearing systems used. In terms of relative productivity, output from a 4 m diameter broodstock tank was equivalent to that from broodstock held in a $40\,m^2$ hapa (average output 1700–2800 fry per system per day).

Provided tilapia fry are harvested frequently, earthen ponds can be just as productive as tanks or hapas, or even more so, but of course a much greater water area is required (fry production from $1740\,m^2$ earth ponds reached 2610–10788 per day: Table 12.7). However, these productivity data do not take into account the other negative aspects of pond management, namely the lower degree of control over environmental conditions and the high labour requirement, especially when draining and preparing the ponds between breeding periods.

Effect of broodfish exchange on reproductive performance

Fry output can be increased considerably from tanks (or hapas) by exchanging some or all of the broodfish every 5 or 10 days. The tank systems used in the series of studies at the Asian Institute of Technology (Little 1989) comprised 4 m diameter circular spawning tanks and 1.5 m diameter circular conditioning

Table 12.7 Comparison of swim-up fry production by *Oreochromis niloticus* maintained in different breeding systems (tanks, hapas and ponds) under equivalent conditions in Thailand (from Little 1989)

Treatment no.	Female stocking density (kg m^{-2})	Female productivity (fry per kg per month)	System productivity (fry per day)	Unit area productivity (fry per m^2 per day)
T-11*	0.64–1.27	3344	1332	106
T-12*	0.86–0.96	1941	1483	118
T-13*	0.84–1.48	361	176	14
T-14*	0.68–1.06	3034	1659	132
T-21	0.66–0.77	2152	1790	142
T-22	0.57–0.87	2589	2322	185
T-23	0.56–0.80	2524	2096	167
T-24	0.55–0.89	2412	2160	172
H-11	0.20	3322	2318	58
H-12	0.18–0.24	3171	2835	71
H-13	0.21	3108	2560	64
H-14	0.18–0.25	3021	2364	59
T-31	0.19–0.60	4523	2106	168
T-32	0.21–0.69	4079	2329	185
T-33	0.27–0.80	4138	2818	224
T-34	0.31–0.96	2398	1963	156
H-21	0.04–0.20	3815	1783	45
H-22	0.06–0.27	3330	2133	53
H-23	0.06–0.32	3430	2513	63
H-24	0.08–0.37	2906	2593	65
H-31	0.05–0.12	5215	1834	46
H-32	0.05–0.12	4145	1993	50
H-33	0.05–0.13	2636	1595	40
H-34	0.05–0.13	8463	2592	65
H-35	0.06–0.12	7039	2421	61
P-12	0.02–0.03	3351	4524	2.6
P-13	0.02–0.03	2022	2610	1.5
P-22	0.03	2328	4002	2.3
P-23	0.03–0.04	5077	10788	6.2

T = tank unit, area 12.57 m^2; H = hapa unit, area 40 m^2; P = pond unit, area 1740 m^2;
* Data for 10-day and 5-day harvest intervals combined
Tank and hapa treatments harvested as eggs and yolk-sac fry and swim-up fry output estimated from mean survival after artificial incubation

tanks (Fig. 12.6). Male and female Nile tilapia broodfish in equal numbers were stocked in the spawning tanks at eight fish per m^2, while additional broodfish were maintained in the smaller conditioning tanks. A summary of egg and fry production achieved in these tanks is provided in Table 12.8.

Fry production from the spawning tank in which the broodfish were left undisturbed (i.e. the females were left to incubate their eggs naturally) averaged only 31 fry per kg per day. This increased to 106 seed per kg per day if eggs and fry were robbed from the females every 10 days, thereafter returning them

Table 12.8 Yield of seed (eggs and fry) from *Oreochromis niloticus* females under different management regimes held in spawning tanks before and after a reduction in inter-harvest interval from 10 to 5 days (from Little et al. 1993)

Treatment	Broodfish management system		Productivity (mean seed harvested per kg female per day)	
	Fish exchange	Seed harvest	Inter-harvest interval (days)	
			10	5
1	none	five times daily		
		tank	31.1	7.9
		mouth	—	—
2	none	tank	23.3	—
		mouth	82.6	278.5
3	spawned females only	tank	37.2	—
		mouth	122.6	231.6
4	all females	tank	48.8	—
		mouth	225.6	321.7

to the spawning tank. Harvesting was performed by draining the tank to a water depth of 20–30 cm, carefully netting each female, and forcing her to spit out her seed batch into a small basin. In the other treatments the same procedure was followed, but either the females which had spawned were replaced with rested females from the conditioning tanks, or all the females were replaced, every 10 days. These exchanges of broodfish increased seed output to 160 and 274 seed per kg per day respectively.

Reducing the time interval between seed collection and broodfish exchange from 10 to 5 days further increased daily average seed production, to 322 seed per kg per day, but also increased the degree of management and labour effort required. An important biological consequence of reducing the inter-harvest interval to 5 days is that all the collectable seed are at an early stage of development.

By using a combination of spawning and conditioning tanks for broodfish management, seed output of more than 350 seed per kg female per day can be achieved over a period of 200 days if females are conditioned for 10 days, followed by a spawning opportunity period of 5 days (Little 1989). However, this system of exchange requires management of three groups of females, i.e. two batches under conditioning while a third group is in the spawning tank.

Broodfish exchange also has consequences for hatchery operating costs as much larger numbers of broodfish and holding units need to be maintained. Holding tilapia females for periods of time longer than 10 days, while slightly increasing the number of eggs per individual female, was found to sharply depress the output per female unit in large spawning hapas suspended in ponds (Little et al. 1994). A 15 day period of conditioning gave optimal seed output per unit of spawning arena. However the best broodfish productivity (8463 fry per kg female per month) was attained using selective exchange of females based on either their spawning during the previous 5 day period, or their being in poor condition.

The replacements were fish kept in a separate conditioning hapa. Conditioning broodfish also uses more hapa and pond space which may be limiting, but costs may be reduced by holding the fish at high density. Stocking rates as high as 2.5 kg m^{-2} in hapas can be used provided an adequate feeding regime and water quality are maintained (Amballi & Little 1994). The exchange of male broodfish, in addition to female exchange, has also been shown to increase breeding output by at least one third (Little et al. 1994).

12.4.5 Conclusions

The above examples show how fry production can be boosted by managing tilapia broodstock to reduce the interval between spawnings, thereby increasing spawning synchrony and reproductive output. The highest seed production can be achieved by replacing spawned fish with rested females from conditioning tanks or hapas every 5–10 days. However, these production gains can be achieved only through a high degree of broodstock management, so the hatchery operator must decide whether the returns compensate for the additional labour and supervision required.

12.5 Artificial incubation of tilapia eggs and hatchlings

12.5.1 Advantages of artificial incubation

Because the natural breeding cycles of tilapia females are asynchronous within a given population, relatively small quantities of fry are produced simultaneously and cannibalism by older (larger) recruits becomes an increasing source of fry mortality. Although very frequent removal of swim-up fry from breeding units can reduce the incidence of cannibalism (see Table 12.4), it is labour intensive and never 100% efficient. To eliminate fry cannibalism and produce sufficient yields of similar sized fry suitable for sex reversal it is far better to rob eggs and yolk-sac fry from incubating female broodfish and incubate the seed artificially. As the studies reported in [12.4] have shown, this also has the advantage of increasing spawning synchrony and decreasing the average inter-spawning time interval.

In addition to its direct commercial application, artificial incubation of seed has a further significant advantage for tilapia research. Individual batches of eggs can be reared and the resulting fry analysed precisely to give accurate data on genetic traits (e.g. colour variation or sex ratios among offspring of a particular cross). Coupled with the techniques of stripping eggs and milt from tilapia broodstock and fertilizing tilapia artificially (Rana *et al.* 1990), offspring from any chosen mating can be reared, thereby offering a powerful research tool for the genetic improvement of tilapia stocks.

12.5.2 Design and operation of egg incubation units

In theory, eggs from *Oreochromis* tilapias can be incubated in any container which allows the eggs to rotate gently through the water column. As tilapia eggs are large and negatively buoyant, they sink rapidly to the bottom of a container and clump together. Egg clumping must be avoided by using some form of mild water circulation (Fig. 12.7). The aim is to re-create the gentle rolling movement which the eggs experience when incubated naturally in the female tilapia's mouth.

There has been exhaustive experimental work conducted on the artificial incubation of *Oreochromis* eggs and the environmental factors affecting their hatchability and subsequent development from embryos to fry (Watanabe *et al.* 1984, Subasinghe & Sommerville 1985, 1986, 1988, 1992, Rana 1986b, 1990a,b, Subasinghe *et al.* 1990). Rana (1990c) has reviewed the many early attempts to incubate tilapia eggs routinely using shaking tables or conical upwelling vessels modelled on the Zuger jars developed for carp eggs (Rothbard & Hulata 1980). High and unpredictable egg mortalities were frequently experienced, ranging from 40 to 100% (Shaw & Aronson 1954, Rothbard & Pruginin 1975).

Fig. 12.7 Simple round-bottomed container (made from a soft drinks bottle) for artificially incubating tilapia eggs; a gentle downward waterflow (1 l min^{-1}) maintains the eggs in the water column. (Photo: D.J. Macintosh.)

The principal causes of poor survival are probably attributable to physical damage caused to the egg chorion and subsequent stress due to osmotic imbalance and invasion by microorganisms, initially bacteria, followed by fungi (Subasinghe & Sommerville 1986, 1988, Rana 1990c).

Awareness of these sources of egg mortality has stimulated efforts to reduce mechanical abrasion of tilapia eggs by using round-bottomed containers and a downward water flow. Round-bottomed incubation containers for tilapia eggs were compared with conical vessels under equivalent operating conditions by Rana (1986b). He found that when *O. niloticus* and *O. mossambicus* eggs were hatched experimentally, survival rates were higher in the round-bottomed containers (average 85% compared with 60%), but the time to hatching was faster in the conical ones (48−72 h compared with 90−102 h, at 27−28°C). In both cases, artificial incubation resulted in faster hatching times than equivalent naturally incubated eggs (96−120 h). Under equivalent incubation conditions, hatching times and survival rates were similar for these two tilapia species.

Many studies have now shown that round-bottomed incubators can give good results provided that water flow rates and water quality are carefully regulated (Little 1989, Rana 1990a, Subasinghe & Sommerville 1992). The simplest (and cheapest) containers available are plastic soft drinks bottles (Fig. 12.7). These are manufactured with a rounded base which facilitates water circulation. The upper conical part of the bottle is cut off and, if required, an overflow pipe can be fitted in the side of the bottle about 2 cm below the top. Two-litre soft drink bottles give a working volume of about 1.5 l, sufficient for incubating up to 2000 eggs, but even smaller vessels can be employed for incubating individual egg batches, or replicate samples of eggs for studies of how environmental conditions affect egg survival and development (Subasinghe & Sommerville 1986, 1992).

Because the operation of round-bottom incubators is so straightforward, commercial-scale units can easily be assembled using other inexpensive plastic containers, e.g. recycled pharmacy bottles or water carboys. Figure 12.8 shows a set of 20 l plastic carboys modified for this purpose. A down-welling water flow system is used to promote slow circulation of the eggs in the bottom of each incubator. The incubation vessels are situated above fry trays (as illustrated) so that any hatchlings carried out with the overflow water can be collected. It is important to have a water valve to control the rate of water flow to each incubator as the flow rates must be regulated to suit the number and stage of eggs being hatched.

The required flow rate of water through the incubator is dependent on design and size. The volume and stage of development of eggs also affects the required flow rate and incubation success. Larger 20 l incubators can be used to successfully incubate very large numbers of eggs (in excess of 80 000) with much greater efficiency of water use (10 000 eggs require $1 l s^{-1}$, compared with about $1 l min^{-1}$ for batches of 1000−2000 eggs reared in modified soft drinks vessels).

Fig. 12.8 Close-up view of 20 l incubation vessels (diameter 27 cm, depth 34.5 cm) forming part of the recirculation system for artificially rearing tilapia seed shown diagrammatically in Fig. 12.9. The water spill tray below the incubators has one hatching collecting tray *in situ*. (Photo: D.J. Macintosh.)

As the quality of water is also important for successful hatching of artificially incubated tilapia eggs, the system illustrated in Fig. 12.9 uses an inexpensive slow sand-filter to clean the water, which is then recirculated (Little 1989). A flow through water supply could be used provided the source water was clean, free of chemicals and had a stable temperature. Recirculation systems have the advantage that temperature can be controlled and the system can operate as a fully independent unit in the hatchery, thereby reducing the risk of disease transmission. As an alternative to a sand filter or gravel filter (Rothbard & Hulata 1980), a UV sterilizer can be used. The latter is more expensive but effective for controlling bacteria (Subasinghe & Sommerville 1985).

The development time for tilapia eggs is temperature-dependent (Rana 1990a, Subasinghe & Sommerville 1992). Thus, by regulating the temperature of water flowing through the egg incubator, the time to hatching can be controlled precisely.

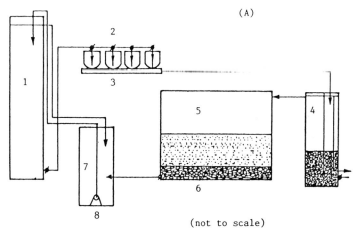

1 Head/storage tank
2 Incubators
3 Collecting tray
4 Prefilter
5 Slow-sand filter
6 Gravel bottom layer
7 Sump/reservoir
8 Pump

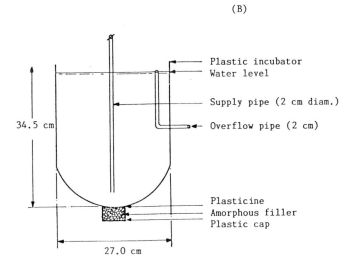

Fig. 12.9 Design of a low-cost artificial incubation system for tilapia seed using 20 l incubation vessels connected to a recirculated water supply operated through a slow sand-filter. (From Little 1989.) (A): Layout of system, (B): Detail of one incubation vessel constructed from a plastic carboy and illustrated in Fig. 12.8.

The stages of development of *O. niloticus* eggs to hatching have been described by Rana (1990a). The rate of hatching varies from about 6 days at 20°C to 2.3 days at 34.5°C (Fig. 12.10). The early stages of egg development are more tolerant of temperature than are the later stages, with the hatching

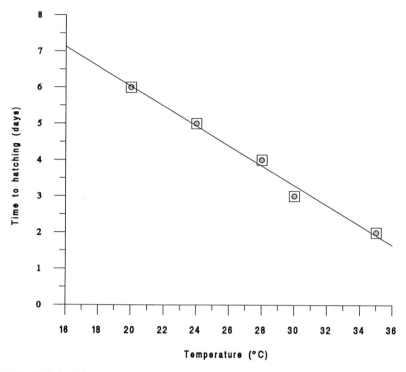

Fig. 12.10 Relationship between incubation temperature and time to hatching of artificially incubated Oreochromis niloticus eggs. (From Rana 1990a.)

phase being the most critical. A high survival rate throughout egg development (>80%) can be obtained in *O. niloticus*, by maintaining water temperature within the range 24–32°C. At the optimum temperature of 28–29°C, time to hatching for *O. niloticus* eggs is about 4 days (Fig. 12.10).

12.5.3 Hatchling rearing in trays

The yolk-sac fry which emerge after hatching are rather delicate and it is best to rear them carefully in specially designed trays until at least the swim-up stage (Fig. 12.11).

In practice, shallow trays may be used for intensive rearing of hatchlings for up to 20 days after hatching. Within this period they should be swimming and feeding actively and can be transferred to larger fry rearing units as required. The trays are ideal for pooling batches of hatchlings of similar age for intensive commercial production, as a single tray can accommodate 10 000–12 000 fry (see below). Hormone feeding for all-male fry production can be started during this hatchling rearing stage as this should commence within the first 10 days after swim-up.

Fig. 12.11 Close-up view of 2.4 l capacity tilapia fry rearing trays in operation (forming part of the system shown diagrammatically in Fig. 12.12).

12.5.4 System design

Inexpensive aluminium or plastic trays with approximate dimensions 40 cm by 25–30 cm by 8–10 cm deep are used, the sort sold everywhere as kitchenware. Two rows of holes are drilled along each side of the tray as shown (Fig. 12.11). Fine plastic or nylon mesh is fixed along the inside of the tray using silicone sealer. This allows water to flow through the sides of the tray, but prevents the hatchlings from escaping. The holes are positioned to regulate the water depth in the tray to 3 cm, thus giving the hatchlings the very shallow water conditions they prefer. The water volume in the tray is about 2.5 l.

Rows of trays can be operated as part of a recirculated water system or the water supply can be a simple flow through one. A 64 tray recirculated water system for tilapia fry developed in Thailand by Little (1989) is illustrated in Fig. 12.12. A suitable filter is required to maintain adequate water quality, but the system illustrated has the advantage over a flow-through water supply that optimum water temperatures for early fry development can be maintained more easily.

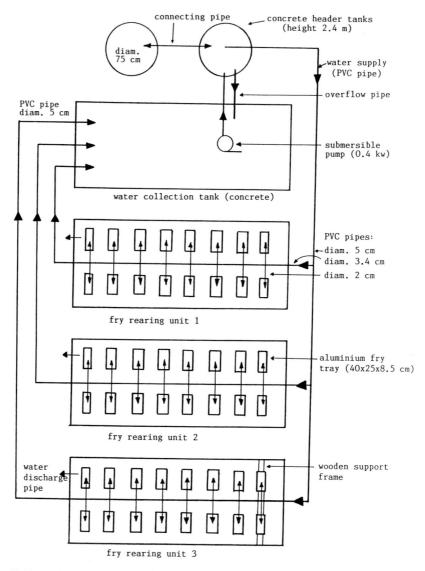

Fig. 12.12 A tilapia hatchling rearing unit consisting of 64 aluminium trays (water capacity 2.4 l) operated as a recirculated water system with physical and biological filtration. (From Little 1989.)

The main risks during early fry rearing are probably from diseases and parasites, especially infection with *Trichodina*. This skin and gill parasite can kill 70–80% of fry within the first 10 days rearing period (du Feu 1987). Artificially incubated seed are much less at risk than naturally incubated ones, but treatment of all broodfish against *Trichodina* is probably a worthwhile step to reduce the potential transmission of this or other parasites to the fry.

12.5.5 Performance of fry trays

The tray system has been evaluated in terms of the effects of water flow rate and hatchling stocking density on fry performance. Yolk-sac hatchlings from several batches can be pooled and reared together provided that they are within the same 24 h period of development. Hatchling densities of 5000, 8000, 12 000, 14 000 and 16 000 per tray (tray water volume 2.4 l) were tested at water flow rates of 2, 3, 4, 5 or 6 l min^{-1} (Rab 1989). The time taken to completely absorb the yolk-sac (i.e. to reach the swim-up stage) ranged from 4.0 to 5.5 days and decreased with flow rate (Fig. 12.13).

The fry survived better at higher flow rates, whereas their specific growth rate (SGR) was inversely related to water flow (Fig. 12.14). Survival was close to 90% on average for trays with 5000–12 000 fry and flow rates of 3 or 4 l min^{-1} and about 70–75% for the other densities and flow rates (Table 12.9). The best SGR of about 10% per day on average occurred in the lowest flow rate of 2 l min^{-1}.

An additional and important finding from the above study (Rab, 1989) was that hatchlings produced by incubating eggs artificially grew better than

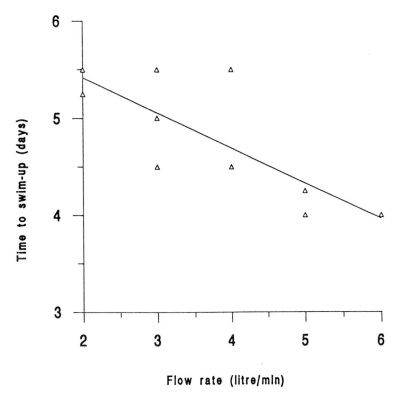

Fig. 12.13 Relationship between water-flow rate and development time for *Oreochromis niloticus* reared in aluminium trays from the yolk-sac fry stage to swim-up. (From Rab 1989.)

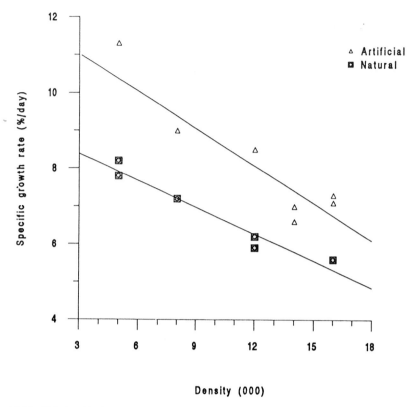

Fig. 12.14 Relationship between specific growth rate and hatchling stocking density for *Oreochromis niloticus* fry reared in aluminium trays with a water-flow rate of $3 l min^{-1}$. Performance of fry produced by natural and artificial incubation is compared. (From Rab, 1989.)

equivalent fry produced naturally (Fig. 12.14) and also showed a higher survival rate.

12.5.6 Conclusions

Artificial egg and hatchling rearing systems are remarkably effective for *O. niloticus* and enable large numbers of fry to be reared in a very limited space. The most important requirements concern maintenance of good water quality and a steady water flow rate through the incubation/rearing units, plus frequent feeding of the developing fry after swim-up (preferably 4–8 times per day). Based on a stocking rate of 10 000 hatchlings per tray, a rearing period of 10 days and an average survival of 80%, it follows that a small unit consisting of 64 trays (see Figs 12.11 and 12.12) can be operated to produce in excess of one million fry per month.

Table 12.9 Effect of hatchling stocking density and water-flow rate on percentage survival (± SE) of *Oreochromis niloticus* fry reared in 2.4 l capacity trays (from Rab 1989)

Stocking density (no. of fry per tray)	Water-flow rate (l min^{-1})				
	2	3	4	5	6
5 000	77.5 (2.0)	95.0 (1.3)	83.4 (1.2)	72.0 (0.5)	66.8 (1.3)
8 000	72.1 (1.8)	92.5 (1.8)	88.5 (1.7)	74.6 (1.2)	72.2 (0.7)
12 000	66.7 (2.2)	90.1 (1.1)	90.6 (1.3)	76.5 (2.0)	75.2 (0.3)
14 000	88.8 (0.5)				
16 000	72.7 (2.8)				

12.6 Production of all-male fry populations

12.6.1 Advantages of hormonal sex reversal

Over the past 30 years various means of controlling reproduction in tilapia production systems have been attempted (reviewed by Guerrero 1982, Mair & Little 1991). While some methods are designed to reduce breeding (e.g. high density stocking, use of cages as culture units), or recruitment (e.g. use of predatory fish to control fry, or intermittent harvesting), techniques to eliminate the female sex altogether have generally been favoured (Popma & Green 1990).

Hand-sexing, hybridization or hormone treatment are the most common methods of producing all-male stocks in commercial practice. Hand-sexing is regarded as being extremely labour intensive with a high risk of human error leading to failure (Guerrero 1982, Popma & Green, 1990). Several interspecific crosses are known to produce only male offspring (reviewed by McAndrew 1993) and all-male tilapia hybrids have become popular in commercial production in Israel, the USA and Taiwan. However, this method is not reliable unless the genetic purity of the parental stocks is carefully maintained. There is also a lower reproductive compatibility between different *Oreochromis* species which adversely affects hybrid fry production. Overall, hormonal sex reversal is widely recognized as having significant advantages over both hand-sexing or hybridization (Guerrero 1987, McAndrew 1993).

Many different hormones have been experimented with in order to manipulate sexual development in the tilapias (reviewed by McAndrew 1993). The synthetic androgen, 17-α-methyl testosterone (MT), has proved both an effective and relatively inexpensive means of masculinizing fry of various tilapia species, including *O. mossambicus*, *O. niloticus*, *O. aureus* and *Tilapia zilli* (Guerrero 1975, Woiwode 1977, Owusu-Frimpong & Nijjhar 1981, Macintosh *et al.* 1985,

Macintosh *et al.* 1988). The incorporation of MT in feed and its administration to young fry has generally proved at least 95% effective in sex reversing genetically female *O. niloticus*, i.e. in most cases 92–100% of the total stock are phenotypically male after treatment (Guerrero 1987, Macintosh *et al.* 1988, Popma & Green 1990, Phelps *et al.* 1992).

12.6.2 Hormone preparation and administration

The method of incorporating MT into feed for sex reversing tilapia fry is described in detail by Popma & Green (1990). Almost any feed can be used as the hormone carrier provided it has a sufficient nutritional value for tilapia fry and is readily consumed. Commercial trout or catfish pellets are suitable, as are simple feeds made locally, provided that they have a high enough protein content (30–40%).

Various hormone concentrations and feeding periods have been used to successfully produce all-male tilapia fry (reviewed by McAndrew 1993). However, it is not the concentration in the feed that is the determining factor, but rather the actual quantity of hormone consumed; the latter is dependent on the feeding rate and how well the fry consume the hormone-feed mixture. The feed should be given generously in at least two, but preferably up to five or six rations daily (Macintosh *et al.* 1988; Popma & Green 1990). On a body weight basis, feeding should be equivalent to about 20% body weight per day initially, falling gradually to about 10% from day 25 onwards (Macintosh 1985, Guerrero 1987, Popma & Green 1990).

In general, hormone dosages within the range 30–60 mg MT per kg feed, given for 25–60 days from first-feeding, have worked effectively on *O. niloticus* fry. Hormone feeding must cover the period of development when sexual differentiation is initiated, which in tilapias seems to occur typically around 16–20 days after hatching (Macintosh 1985). While this suggests that it should be possible to achieve sex reversal after only about 20 days hormone-feeding, in practice such a short treatment period has sometimes been only 95% successful (Popma & Green 1990). It is possible to delay hormone administration for up to 10 days after first-feeding, but with an increasing degree of risk that it will fail to be 100% effective (unpublished data). If the age of the fry is not known, then it can be assumed that those fry less than 15 mg (or 12–15 mm total length, approximately), will be responsive to hormone treatment (Little 1989). Consistently good results have been achieved with MT treatments of 40 mg kg^{-1} for 40 days from first-feeding (Macintosh *et al.* 1988) or 60 mg kg^{-1} given for 21 days (D.C. Little unpublished data) or 25–28 days (Popma & Green 1990).

Problems associated with incomplete sex reversal and/or poor survival of tilapia fry during hormonal treatment have been described by Mair & Little (1991). Any conditions which lead to poor intake of the treated food, such as low temperature, disease, poor feed attractiveness/quality, low feeding frequency, can reduce the effectiveness of the treatment. Under green-water conditions, high fry rearing densities are necessary to ensure that the proportion of natural food to given food consumed is low, and the procedure is economically viable, but this may lead to reduced survival. If adequate water quality is maintained, mean survival of about 70% is typical for tilapia stocked at 12 fry per litre and hormone-treated for 21 days in hapas suspended in fertilized earthen ponds (Asian Institute of Technology, unpublished data).

12.7 Future developments in broodstock management

It is beyond the scope of this review to deal with the complexities of tilapia genetics, but selection for desirable traits such as colour, body shape, growth rate, fecundity and environmental tolerance are obviously an important aspect of broodstock improvement. Research findings are still limited, but sufficient to indicate that genetic inheritance for at least some of these qualities is measurable and may be selected for among different intraspecific stocks and between pure species and hybrids (Hulata et al. 1988, Pullin & Capili 1988, Tave 1988). It is widely recognized, however, that heritability in the tilapias is complex and often is obscured by their highly developed environmental adaptability (e.g. Behrends et al. 1988).

Moreover, a great anomaly with the tilapias is that their genetic resource is almost entirely within Africa, whereas their greatest aquaculture production is within Asia (Pullin & Capili 1988). The gene pool in Asia is probably still extremely limited because of the way introductions have been made (usually involving very small numbers of fish each time). There is clearly scope to widen the gene pool in aquaculture, particularly in Asia, via further careful introductions of African *O. niloticus*.

By breeding *O. niloticus* under controlled conditions in tanks or hapas, coupled with artificial incubation of eggs and fry, selection for desirable traits can proceed rapidly, aided further by their short generation time (about 4 months) and capacity to breed virtually the year round if a suitable environment is maintained.

In short, the potential for genetic improvement in tilapia culture is enormous (McAndrew 1993) and *O. niloticus* is an excellent candidate for further applied fish genetics research. Exciting developments can be expected from combining new research findings on genetic selection and sex manipulation in *O. niloticus* (e.g. Scott et al. 1989, Hussain et al. 1991, Mair et al. 1992) with the practical advances in broodstock management and fry production described here.

Acknowledgements

A major part of this review is based on research funded by the Science and Technology for Developing Countries Programme (STD) of the Commission of the European Communities, through contract no. TSD-090 (STD-1, 1983–86) awarded to Dr Donald J. Macintosh, University of Stirling, Scotland. We are also grateful to Prof. Peter Edwards and the Asian Institute of Technology, Bangkok, Thailand for providing practical facilities for the work, plus technical support. Dr David C. Little is currently seconded to the Aquaculture Programme of the Asian Institute of Technology by the Overseas Development Administration, UK. The ASEAN-EC Aquaculture Development and Coordination Programme (AADCP) and Department of Fisheries, Malaysia, kindly provided additional support to Dr Macintosh to prepare this information for a regional workshop in Malaysia on Tilapia Hatchery Technology. Finally, we wish to thank Dr Brendan J. McAndrew, University of Stirling, for critically reading a draft version of this paper and Miss Julia L. Overton for assistance in preparing the final manuscript.

References

Abella, T.A., Palada, M.S. & Newkirk, G.F. (1990) Within family selection for growth rate with rotation mating in *Oreochromis niloticus*. In *The Second Asian Fisheries Forum*, (eds R. Hirano & I. Hanyu), pp. 515–8. Asian Fisheries Society, Manila, Philippines.

Amballi, A.J.D. & Little, D.C. (1994) Studies on the effect of manipulating hapa size on broodstock conditioning of *Oreochromis niloticus* in fertilised earthen ponds. In *Proceedings of the Third International Symposium on Tilapia in Aquaculture, 11–16 November 1991, Abidjan, Côte d'Ivoire*.

Babiker, M.M. & Ibrahim, H. (1979) Studies on the biology of reproduction in the cichlid *Tilapia nilotica* (L.): gonadal maturation and fecundity. *Journal of Fish Biology*, **14**, 437–48.

Balarin, J.D. & Haller, R.D. (1982) The intensive culture of tilapia in tanks, raceways and cages. In *Recent Advances in Aquaculture*, (eds J.F. Muir & R.J. Roberts), pp. 265–355. Croom Helm Ltd, London & Canberra.

Balarin, J.D. & Hatton, J.P. (1979) *Tilapia: A Guide to their Biology and Culture in Africa*. University of Stirling, Stirling, Scotland.

Behrends, L.L., Kingsley, J.B. & Price III, A.H. (1988) Bidirectional-backcross selection for body weight in a red tilapia. In *The Second International Symposium on Tilapia in Aquaculture, ICLARM Conference Proceedings*, **15** (eds R.S.V. Pullin, T. Bhukaswan, K., Tonguthai & J.L. Maclean), pp. 125–34. Department of Fisheries, Bangkok, Thailand, and International Center for Living Aquatic Resources Management, Manila, Philippines.

Chen, L.-C. (1990) *Aquaculture in Taiwan*. Fishing News Books, Oxford.

Chervinski, J. (1982) Environmental physiology of tilapias. In *The Biology and Culture of Tilapias, ICLARM Conference Proceedings*, **7** (eds R.S.V. Pullin & R.H. Lowe-McConnell), pp. 119–28. International Center for Living Aquatic Resources Management, Manila, Philippines.

Cross, D.W. (1976) Methods to control over-breeding of farmed tilapias. *Fish Farming International*, **3**, 27–9.

De Silva, S.S. & Radampola, K. (1990) Effect of dietary protein level on the reproductive performance of *Oreochromis niloticus*. In *The Second Asian Fisheries Forum*, (eds R. Hirano & I. Hanyu), pp. 559–63. Asian Fisheries Society, Manila, Philippines.

Dewan, S. & Saha, S.N. (1979) Food and feeding habits of *Tilapia nilotica* (L.) (Perciformes: Cichlidae). II. Diel and seasonal patterns of feeding. *Bangladesh Journal of Zoology*, **7**, 75–80.

Edwards, P. (1983) *Final Report on Small-Scale Fishery Project in Pathumthani Province, Central Thailand: A Socio-Economic and Technological Assessment of Status and Potential*. AIT Research Report No. 158. Asian Institute of Technology, Bangkok, Thailand.

Edwards, P., Pullin, R.S.V. & Gartber, J.A. (1988) *Research and Education for the Development of Integrated Crop–Livestock–Fish Farming Systems in the Tropics, ICLARM Studies and Reviews*, **16**. International Center for Living Aquatic Resources Management, Manila, Philippines.

El Gamal, A.A., Smitherman, R.O. & Behrends, L.L. (1988) Viability of red and normal-coloured *Oreochromis aureus* and *O. niloticus* hybrids. In *The Second International Symposium on Tilapia in Aquaculture, ICLARM Conference Proceedings*, **15** (eds R.S.V. Pullin, T. Bhukaswan, K. Tonguthai & J.L. Maclean), pp. 153–8. Department of Fisheries, Bangkok, Thailand, and International Center for Living Aquatic Resources Management, Manila, Philippines.

FAO (1992) *Aquaculture Production (1984–1990)*. FAO Fisheries Circular No. 815, Rev. 4, Fishery Information, Data and Statistics Service, Fisheries Department, FAO, Rome.

du Feu, T.A. (1987) *An evaluation of a recirculated hatchery in Thailand for the sex reversal of tilapia fry*. M.Sc. thesis, Institute of Aquaculture, University of Stirling.

Fineman-Kalio, A.S. (1988) Preliminary observations on the effect of salinity on the reproduction and growth of freshwater Nile tilapia *Oreochromis niloticus* (L.), cultured in brackishwater ponds. *Aquaculture & Fisheries Management*, **19**, 313–20.

Fishelson, L. & Yaron, Z. (eds), (1983) *International Symposium on Tilapia in Aquaculture*. Tel Aviv University Press, Tel Aviv.

Fryer, G. & Iles, T.D. (1972) *The Cichlid Fishes of the Great Lakes of Africa*. Oliver & Boyd, London.

Galman, O.R., Moreau, J. & Avtalion, R.R. (1988) Breeding characteristics and growth performance of Philippine red tilapia. In *The Second International Symposium on Tilapia in Aquaculture, ICLARM Conference Proceedings*, **15**, (eds R.S.V. Pullin, T. Bhukaswan, K. Tonguthai & J.L. Maclean), pp. 169–76. Department of Fisheries, Bangkok, Thailand, and International Center for Living Aquatic Resources Management, Manila, Philippines.

Gregory, R.G. (1987) *Con-specific predation in* Oreochromis niloticus *juveniles in the commercial production of fingerlings*. M.Sc. thesis, Institute of Aquaculture, University of Stirling.

Guerrero, R.D. (1975) Use of androgens for the production of all-male *Tilapia aurea* (Steindachner). *Transactions of the American Fisheries Society*, **104**, 342–8.

Guerrero, R.D. (1982) Control of Tilapia reproduction. In *The Biology and Culture of Tilapias, ICLARM Conference Proceedings*, **7** (eds R.S.V. Pullin & R.H. Lowe-McConnell), pp. 15–60. International Center for Living Aquatic Resources Management, Manila, Philippines.

Guerrero, R.D. (1987) *Tilapia Farming in the Philippines*. Technology and Livelihood Resource Center, Manila, Philippines.

Guerrero, R.D. & Guerrero, L.A. (1985) Effect of breeder size on fry production of Nile tilapia in concrete pools. *Transactions of the National Academy of Science and Technology* (Philippines), **7**, 63–6.

Guerrero R.D. & Guerrero, L.A. (1988) Feasibility of commercial production of sex-reversed Nile tilapia fingerlings in the Philippines. In *The Second International Symposium on Tilapia in Aquaculture, ICLARM Conference Proceedings*, **15** (eds R.S.V. Pullin, T. Bhukaswan, K. Tonguthai & J.L. Maclean), pp. 183–6. Department of Fisheries, Bangkok, Thailand, and International Center for Living Aquatic Resources Management, Manila, Philippines.

Haight, B.A. (1992) Growing towards the 21st Century: World Aquaculture Conference in the USA. *ALCOM News*, **7**, 4–6. Aquaculture for Local Community Development Programme, FAO, Harare, Zimbabwe.

Haller, R.D. & Parker, I.S.C. (1981) New tilapia breeding system tested on Kenya farm. *Fish Farming International*, **8**, 14–18.

Hanson, B.J., Moehl, J.F., Veverica, K.L., Rwangano, F. & Van Speybroaeck, M. (1988) Pond culture of tilapia in Rwanda, a high altitude equatorial African country. In *The Second International Symposium on Tilapia in Aquaculture, ICLARM Conference Proceedings*, **15** (eds R.S.V. Pullin, T. Bhukaswan, K. Tonguthai & J.L. Maclean), pp. 553–60. Department of Fisheries, Bangkok, Thailand, and International Center for Living Aquatic Resources Management, Manila, Philippines.

Hickling, C.F. (1962) *Fish Culture*. Faber & Faber, London.

Hopkins, K., Ridha, M., Leclercq, D., Al-Ameeri, A.A. & Al-Ahmad, T. (1989) Screening tilapias for sea water culture in Kuwait. *Aquaculture & Fisheries Management*, **20**, 389–97.

Hulata, G., Wohlfarth, G.W. & Halevy, A. (1988) Comparative growth tests of *Oreochromis niloticus* × *O. aureus* hybrids derived from different farms in Israel, in polyculture. In *The Second International Symposium on Tilapia in Aquaculture, ICLARM Conference Proceedings*, **15** (eds R.S.V. Pullin, T. Bhukaswan, K. Tonguthai & J.L. Maclean), pp. 191–6. Department of Fisheries, Bangkok, Thailand, and International Center for Living Aquatic Resources Management, Manila, Philippines.

Hussain, M.G., Chatterji, A., McAndrew, B.J. & Johnstone, R. (1991) Triploidy induction in Nile tilapia. *Oreochromis niloticus* L. using pressure, heat and cold shocks. *Theoretical and Applied Genetics*, **81**, 6–12.

Iles, T.D. (1973) Dwarfing or stunting in the genus *Tilapia* (Cichlidae) a possible unique recruitment mechanism. In *Fish Stocks and Recruitment*, (ed. B.B. Parrish). *Rapports et Procès-Verbaux de la Réunion, CIEM*, **164**, 246–54.

Jarimopas, P. (1990) Realized response of Thai red tilapia to five generations of size-specific selection for growth. In: *The Second Asian Fisheries Forum*, (eds R. Hirano & I. Hanyu), pp. 519–22. Asian Fisheries Society, Manila, Philippines.

Jauncey, K. & Ross, B. (1982) *A Guide to Tilapia Feeds and Feeding*. University of Stirling.

de Kartzow, A., van der Heijden, P. & van der Schoot, J. (1992) *Integration of Fish Farming into the Farm-Household System in Luapula Province, Zambia. ALCOM Field Document No. 16*. Aquaculture for Local Community Development Programme, FAO, Harare, Zimbabwe.

Kuo, H. (1988) Progress in genetic improvement of red hybrid tilapia in Taiwan. In *The Second International Symposium on Tilapia in Aquaculture, ICLARM Conference Proceedings*, **15** (eds R.S.V. Pullin, T. Bhukaswan, K. Tonguthai & J.L. Maclean), pp. 219–21. Department of Fisheries, Bangkok, Thailand, and International Center for Living Aquatic Resources Management, Manila, Philippines.

Lauenstein, P.C. (1978) Intensive culture of Tilapia with geothermally heated water. In *Culture of Exotic Fishes*, (eds R.O. Smitherman, W.L. Shelton & J.H. Grover), pp. 82–5. Auburn University, Alabama.

Lee, J.C. (1979) *Reproduction and hybridization of three cichlid fishes*, Tilapia aurea *(Steindachner)*, T. hornorum *(Trewavas) and* T. nilotica *(Linnaeus) in aquaria and in plastic pools*. PhD thesis, Auburn University, Alabama.

Lester, L.J., Abella, T.A., Palada, M.S. & Keus, H.J. (1988) Genetic variation in size and sexual maturation of *Oreochromis niloticus* under hapa and cage culture conditions. In *The Second International Symposium on Tilapia in Aquaculture, ICLARM Conference Proceedings*, **15** (eds R.S.V. Pullin, T. Bhukaswan, K. Tonguthai & J.L. Maclean), pp. 223–30. Department of Fisheries, Bangkok, Thailand, and International Center for Living Aquatic Resources Management, Manila, Philippines.

Little, D.C. (1989) *An evaluation of strategies for production of Nile tilapia* (Oreochromis niloticus L.) *fry suitable for hormonal treatment*. PhD thesis, Institute of Aquaculture, University of Stirling.

Little, D.C., Macintosh, D.J. & Edwards, P. (1993) Improving spawning synchrony in the Nile tilapia (*Oreochromis niloticus*). *Aquaculture and Fisheries Management*, **24**, 319–25.

Little, D.C., Macintosh, D.J. & Edwards, P. (1994) Selective broodfish exchange of *Oreochromis niloticus* in large breeding hapas suspended in earthen ponds. In *The Third International Symposium on Tilapia in Aquaculture, ICLARM Conference Proceedings*, (eds R.S.V. Pullin, J. Lazard, M. Legendre, J.B. Amon Kothias & D. Pauly).

Lovshin, L.L. (1982) Tilapia hybridization. In *The Biology and Culture of Tilapias, ICLARM Conference Proceedings*, **7** (eds R.S.V. Pullin & R.H. Lowe-McConnell), pp. 279–308. International Center for Living Aquatic Resources Management, Manila, Philippines.

Lowe-McConnell, R.H. (1959) Breeding behaviour patterns and ecological differences between Tilapia species and their significance for evolution within the genus Tilapia (Pisces: Cichlidae). *Proceedings of the Zoological Society of London*, **132**, 1–30.

Luquet, P. (1991) Tilapia, *Oreochromis* spp. In *Handbook of Nutrient Requirements of Finfish*, (ed R.P. Wilson), pp. 169–79. CRC Press, New York.

McAndrew, B.J. (1993) Sex control in tilapiines. In *Recent Advances in Aquaculture*, IV (eds J.F. Muir & R.J. Roberts), pp. 87–98. Blackwell Science, Oxford.

McAndrew, B.J. & Majumdar, K.C. (1983) Tilapia stock identification using electrophoretic markers. *Aquaculture*, **30**, 249–61.

McAndrew, B.J., Roubal, F.R., Roberts, R.J., Bullock, A.M. & McEwen, I.M. (1988) The genetics and histology of red, blond and associated colour variants in *Oreochromis niloticus*. *Genetica*, **76**, 127–37.

Macintosh, D.J. (1985) *Tilapia Culture: Hatchery Methods for* Oreochromis mossambicus *and*

O. niloticus, with Special Reference to All-Male Fry production. Institute of Aquaculture, University of Stirling, Stirling.

Macintosh, D.J. (1993) *Tilapia hatchery technology for the ASEAN Region*. Working paper of the ASEAN-EEC Aquaculture Development Programme (AADCP), AADCP Coordination Office, Bangkok, Thailand.

Macintosh, D.J. & De Silva, S.S. (1984) The influence of stocking density and food ration on fry survival and growth in *Oreochromis mossambicus* and *O. niloticus* female × *O. aureus* male hybrids reared in a closed circulated system. *Aquaculture*, **41**, 345–58.

Macintosh, D.J., Varghese, T.J. & Rao, G.P.S. (1985) Hormonal sex reversal of wild-spawned tilapia in India. *Journal of Fish Biology*, **26**, 87–94.

Macintosh, D.J., Singh, T.B., Little, D.C. & Edwards, P. (1988) Growth and sexual development of 17α-methyltestosterone- and progesterone-treated Nile tilapia (*Oreochromis niloticus*) reared in earthen ponds. In *The Second International Symposium on Tilapia in Aquaculture, ICLARM Conference Proceedings*, **15** (eds R.S.V. Pullin, T. Bhukaswan, K. Tonguthai & J.L. Maclean), pp. 457–63. Department of Fisheries, Bangkok, Thailand, and International Center for Living Aquatic Resources Management, Manila, Philippines.

Madhu, S.R. (1992) Commercial fish farming in Zimbabwe. *ALCOM News*, **7** 14–18. Aquaculture for Local Community Development Programme, FAO, Harare, Zimbabwe.

Mair, G.C. (1988) *Studies on sex determining mechanisms in* Oreochromis *species*. PhD thesis, University College of Swansea, Wales.

Mair, G.C. & Little, D.C. (1991) Population control in farmed tilapias. *NAGA, The ICLARM Quarterly*, **4**(2), 8–13. International Center for Living Aquatic Resources Management, Manila, Philippines.

Mair, G.C., Scott, A., Penman, D.J., Beardmore, J.A. & Skibinsi, D.O.F. (1991a) Sex determination in the genus *Oreochromis*. 1. Sex-reversal, gynogenesis and triploidy in *O. niloticus* (L.). *Theoretical and Applied Genetics*, **82**, 144–52.

Mair, G.C., Scott, A., Penman, D.J., Beardmore, J.A. & Skibinsi, D.O.F. (1991b) Sex determination in the genus *Oreochromis*. 2. Sex-reversal, hybridisation, gynogenesis and triploidy in *O. aureus* (Steindachner). *Theoretical and Applied Genetics*, **82**, 153–60.

Mair, G.C., Capili, J.B., Skibinski, D.O.F. & Beardmore, J.A. (1992) The YY-male technology for production of monosex male tilapia. In *Proceedings of the International Workshop on Genetics in Aquaculture and Fisheries Management, University of Stirling, 31 August – 4 September 1992*, pp. 93–6. ASEAN-EEC Aquaculture Development and Coordination Programme, AADCP/PROC/3. AADCP Coordination Office, Bangkok, Thailand.

Manzi, J.J. (1989) Aquaculture research priorities for the 1990s. *World Aquaculture*, **20**, 29–32.

Mélard, C. & Philippart, J.C. (1980) *Pisciculture intensive de* Sarotherodon niloticus *dans les effluents thermiques d'une centrale nucléaire en Belgique*. Paper in FAO/EIFAC Symposium on New Developments in the Utilization of Heated Effluents and of Recirculation Systems for Intensive Aquaculture. EIFAC/80/Symp/Document E/11.

Mires, D. (1982) A study of the problems of the mass production of hybrid tilapia fry. In *The Biology and Culture of Tilapias, ICLARM Conference Proceedings*, **7** (eds R.S.V. Pullin & R.H. Lowe-McConnell), pp. 317–30. International Center for Living Aquatic Resources Management, Manila, Philippines.

Mires, D. (1983) Current techniques for the mass production of tilapia hybrids as practised at Ein Hamifrats hatchery. *Bamidgeh*, **35**, 3–8.

Mires, D. & Amit, Y. (1992) Intensive culture of tilapia in quasi-closed water-cycled flow-through ponds – the Dekel aquaculture system. *Bamidgeh*, **44**, 82–6.

Myers, G.S. (1937) A possible method of evolution of oral brooding habits in cichlid fishes. *The Aquarium Journal, San Francisco*, **10**, 4–6.

New, M.B., Hopkins, K.D. & El Dakour, S. (1984) Effects of feeding frequency on survival and growth of tilapia fry. *Kuwait Institute for Scientific Research*, Safat, KISR Publication 1287.

Noakes, D.L.G. & Balon, E.K. (1982) Life histories of tilapias: an evolutionary perspective. In *The Biology and Culture of Tilapias, ICLARM Conference Proceedings*, **7** (eds R.S.V. Pullin & R.H. Lowe-McConnell), pp. 61–82. International Center for Living Aquatic Resources Management, Manila, Philippines.

Owusu-Frimpong, M. & Nijjhar, B. (1981) Induced sex reversal in *Tilapia nilotica* (Cichlidae) with methyl testosterone. *Hydrobiologia*, **78**, 157–60.

Perschbacher, P.W. (1992) A review of seawater acclimation procedures for commercially important euryhaline tilapias. *Asian Fisheries Science*, **5**, 241–8.

Perschbacher, P.W. & McGeachin, R.B. (1988) Salinity tolerances of red hybrid tilapia fry, juveniles and adults. In *The Second International Symposium on Tilapia in Aquaculture, ICLARM Conference Proceedings*, **15** (eds R.S.V. Pullin, T. Bhukaswan, K. Tonguthai & J.L. Maclean), pp. 415–19. Department of Fisheries, Bangkok, Thailand, and International Center for Living Aquatic Resources Management, Manila, Philippines.

Peters, H.M. (1983) Fecundity, egg weight and oocyte development in tilapias (Cichlidae, Teleostei). *ICLARM Translations* 2, International Center for Living Aquatic Resources Management, Manila, Philippines.

Petr, T. (ed.) (1985) Inland fisheries in multi-purpose river basin planning and development in tropical Asian countries: three case studies. *FAO Fisheries Technical Paper* No. 265. (FIRI/T 265).

Pham, A.T. & Vo, V.T. (1990) Reuse of wastewater for fish culture in Hanoi, Vietnam. In *Wastewater-Fed Aquaculture. Proceedings of the International Seminar on Wastewater Reclamation and Reuse for Aquaculture, Calcutta, India, 6–9 December 1988*, (eds P. Edwards & R.S.V. Pullin), pp. 69–72. Environmental Sanitation Information Center, Asian Institute of Technology, Bangkok, Thailand.

Phelps, R.P., Cole, W. & Katz, T. (1992) Effect of fluoxymesterone on sex ratio and growth of Nile tilapia, *Oreochromis niloticus* (L.). *Aquaculture and Fisheries Management*, **23**, 405–10.

Philippart, J.C. & Ruwet, J.C. (1982) Ecology and distribution of tilapias. In *The Biology and Culture of Tilapias, ICLARM Conference Proceedings*, **7** (eds R.S.V. Pullin & R.H. Lowe-McConnell), pp. 15–60. International Center for Living Aquatic Resources Management, Manila, Philippines.

Popma, T.J. & Green, B.W. (1990) Aquaculture Production Manual – Sex Reversal of Tilapia in Earthen Ponds. *Research and Development Series* No. 35. Auburn University, Alabama, USA.

Pruginin, Y., Fishelson, L. & Koren, A. (1988) Intensive tilapia farming in brackish water from an Israeli desert aquifer. In *The Second International Symposium on Tilapia in Aquaculture, ICLARM Conference Proceedings*, **15** (eds R.S.V. Pullin, T. Bhukaswan, K. Tonguthai & J.L. Maclean), pp. 75–82. Department of Fisheries, Bangkok, Thailand, and International Center for Living Aquatic Resources Management, Manila, Philippines.

Pullin, R.S.V. (1982) General discussion on the biology and culture of tilapias. In *The Biology and Culture of Tilapias, ICLARM Conference Proceedings*, **7** (eds R.S.V. Pullin & R.H. Lowe-McConnell), pp. 331–51. International Center for Living Aquatic Resources Management, Manila, Philippines.

Pullin, R.S.V. (1983) Choice of tilapia species for aquaculture. In *International Symposium on Tilapia in Aquaculture*, (eds L. Fishelson & Z. Yaron), pp. 64–76. Tel Aviv University Press, Tel Aviv.

Pullin, R.S.V. & Lowe-McConnell, R.H. (1982) The Biology and Culture of Tilapias. *ICLARM Conference Proceedings*, **7**. International Center for Living Aquatic Resources Management, Manila, Philippines.

Pullin, R.S.V. & Capili, J.B. (1988) Genetic improvement of tilapias: problems and prospects. In *The Second International Symposium on Tilapia in Aquaculture, ICLARM Conference Proceedings*, **15** (eds R.S.V. Pullin, T. Bhukaswan, K. Tonguthai & J.L. Maclean), pp. 259–66. Department of Fisheries, Bangkok, Thailand, and International Center for Living Aquatic Resources Management, Manila, Philippines.

Pullin, R.S.V., Bhukaswan, T., Tonguthai, K. & Maclean, J.L. (eds) (1988). *The Second International Symposium on Tilapia in Aquaculture, ICLARM Conference Proceedings*, **15**. Department of Fisheries, Bangkok, Thailand, and International Center for Living Aquatic Resources Management, Manila, Philippines.

Rab, M.A. (1989) *Intensive nursing of Nile tilapia* (Oreochromis niloticus) *fry*. M.Sc. thesis, Asian Institute of Technology, Bangkok.

Rana, K.J. (1986a) *Parental influences on egg quality, fry production and fry performance in* Oreochromis niloticus *(Linnaeus) and* O. mossambicus *(Peters)*. PhD thesis, Institute of Aquaculture, University of Stirling.

Rana, K.J. (1986b) An evaluation of two types of containers for the artificial incubation of

Oreochromis eggs. *Aquaculture & Fisheries Management*, **17**, 139–45.
Rana, K.J. (1990a) Influence of incubation temperature on *Oreochromis niloticus* (L.) eggs and fry. I. Gross embryology, temperature tolerance and rates of embryonic development. *Aquaculture*, **87**, 165–81.
Rana, K.J. (1990b) Influence of incubation temperature on *Oreochromis niloticus* (L.) eggs and fry. II. Survival, growth and feeding of fry developing solely on their yolk reserves. *Aquaculture*, **87**, 183–95.
Rana, K.J. (1990c) Reproductive biology and the hatchery rearing of tilapia eggs and fry. In *Recent Advances in Aquaculture*, Vol. 3 (eds J.F. Muir & R.J. Roberts), pp. 343–406. Croom Helm Ltd, London & Canberra.
Rana, K.J. & Macintosh, D.J. (1988) A comparison of the quality of hatchery-reared *Oreochromis niloticus* and *O. mossambicus* fry. In *The Second International Symposium on Tilapia in Aquaculture, ICLARM Conference Proceedings*, **15** (eds R.S.V. Pullin, T. Bhukaswan, K. Tonguthai & J.L. Maclean), pp. 497–502. Department of Fisheries, Bangkok, Thailand, and International Center for Living Aquatic Resources Management, Manila, Philippines.
Rana, K.J., Muiruri, R.M., McAndrew, B.J. & Gilmour, A. (1990) The influence of diluents, equilibration time and prefreezing storage time on the viability of cryopreserved *Oreochromis niloticus* (L.) spermatozoa. *Aquaculture & Fisheries Management*, **21**, 25–30.
Rifai, S.A. (1988) Growth and production of *Oreochromis niloticus* of three strains in cage culture with and without caudal fin cutting. In *The Second International Symposium on Tilapia in Aquaculture, ICLARM Conference Proceedings*, **15** (eds R.S.V. Pullin, T. Bhukaswan, K. Tonguthai & J.L. Maclean), p. 589. Department of Fisheries, Bangkok, Thailand, and International Center for Living Aquatic Resources Management, Manila, Philippines.
Ross, B. & Jauncey, K. (1981) A radiographic estimation of the effect of temperature on gastric emptying time in *Sarotherodon niloticus* × *S. aureus* hybrids. *Journal of Fish Biology*, **19**, 333–44.
Rothbard, S. & Hulata, G. (1980) Closed-system incubator for cichlid eggs. *Progressive Fish Culture*, **42**, 203–4.
Rothbard, S. & Pruginin, Y. (1975) Induced spawning and artificial incubation of *Tilapia*. *Aquaculture*, **5**, 315–21.
Sampson, D.R.T. (1986) *An evaluation of the 'arena method' of tilapia fry production in Kenya (with observations of fingerling grow-out and the potential for producing all-male tilapia)*. M.Sc. thesis, Institute of Aquaculture, University of Stirling.
Santiago, C.B., Banes-Aladaba, M. & Laron, M.A. (1982) Dietary crude protein requirement of *Tilapia niloticus* fry. *Kalikasan, Philippine Journal of Biology*, **11**, 255–65.
Santiago, C.B., Aladama, M.B. & Offilia, S. (1987) Influence of feeding rate and diet form on growth and survival of Nile tilapia (*Oreochromis niloticus*) fry. *Aquaculture*, **64**, 277–82.
Sarig, S. (1971) Diseases of warmwater fishes. In *Diseases of Fishes* (eds S.F. Snieszko & H. Axelrod). TFH Publications Nepture City, New Jersey.
Schoenen, P. (1982) *A Bibliography of Important Tilapias (Pisces: Cichlidae) for Aquaculture*. *ICLARM Bibliographies*, **3**. International Center for Living Aquatic Resources Management, Manila, Philippines.
Scott, A.G., Penman, D.J., Beardmore, J.A. & Skibinski, D.O.F. (1989) The 'YY' supermale in *Oreochromis niloticus* (L.) and its potential in aquaculture. *Aquaculture*, **78**, 237–51.
Shaw, E.S. & Aronson, L.R. (1954) Oral incubation in *Tilapia macrocephala*. *Bulletin of the American Museum of Natural History*, **103**, 379–415.
Siddiqui, A.Q., Howlader, M.S. & Adam, A.A. (1988) Effects of dietary protein levels on growth, feed conversion and protein utilization in fry and young Nile tilapia. *Aquaculture*, **70**, 63–74.
Sipe, M. (1992) Tilapia marketing in the USA. *INFOFISH International*, No. 3/92, 23–5.
Siraj, S.S., Smitherman, R.O., Castillo-Gallusser, S. & Dunham, R.A. (1983) Reproductive traits for three year classes of *Tilapia nilotica* and maternal effects on their progeny. In *International Symposium on Tilapia in Aquaculture* (eds L. Fishelson & Z. Yaron), pp. 210–8. Tel Aviv University Press, Tel Aviv.
Soliman, A.K., Jauncey, K. & Roberts, R.J. (1986) The effect of varying forms of dietary ascorbic acid on the nutrition of juvenile tilapias (*Oreochromis niloticus*). *Aquaculture*, **52**, 1–10.
Srisakultiew, P. & Wee, K.L. (1988) Synchronous spawning of Nile tilapia through hypophysation and temperature manipulation. In *The Second International Symposium on Tilapia in Aqua-*

culture, *ICLARM Conference Proceedings*, **15** (eds R.S.V. Pullin, T. Bhukaswan, K. Tonguthai & J.L. Maclean), pp. 275–84. Department of Fisheries, Bangkok, Thailand, and International Center for Living Aquatic Resources Management, Manila, Philippines.

Subasinghe, R.P. & Sommerville, C. (1985) Disinfection of *Oreochromis mossambicus* (Peters) eggs against commonly occurring potentially pathogenic bacteria and fungi under artificial hatchery conditions. *Aquaculture & Fisheries Management*, **16**, 121–7.

Subasinghe, R.P. & Sommerville, C. (1986) Possible cause of mortality of *Oreochromis mossambicus* eggs under artificial incubation. In *The First Asian Fisheries Forum*, (eds J.L. Maclean, L.B. Dizon & L.V. Hosillos), pp. 337–40. Asian Fisheries Society, Manila, Philippines.

Subasinghe, R.P. & Sommerville, C. (1988) Scanning electron microscope study of the causes of mortality in *Oreochromis mossambicus* (Peters) eggs under artificial incubation. *Journal of Fish Diseases*, **11**, 409–16.

Subasinghe, R.P. & Sommerville, C. (1992) Effects of temperature on hatchability, development and growth of eggs and yolksac fry of *Oreochromis mossambicus* (Peters) under artificial incubation. *Aquaculture & Fisheries Management*, **23**, 31–9.

Subasinghe, R.P., Sommerville, C. & Rana, K.J. (1990) In *The Second Asian Fisheries Forum*, (eds R. Hirano & I. Hanyu), pp. 69–72. Asian Fisheries Society, Manila, Philippines.

Suresh, A.V. & Kwei Lin, C. (1992) Tilapia culture in saline waters: a review. *Aquaculture*, **106**, 201–26.

Tave, D. (1988) Genetics and breeding of tilapia: a review. In *The Second International Symposium on Tilapia in Aquaculture, ICLARM Conference Proceedings*, **15** (eds R.S.V. Pullin, T. Bhukaswan, K. Tonguthai & J.L. Maclean), pp. 285–94. Department of Fisheries, Bangkok, Thailand, and International Center for Living Aquatic Resources Management, Manila, Philippines.

Trewavas, E. (1966) A preliminary review of fishes of the genus *Tilapia* in the eastward-flowing rivers of Africa, with proposals for two new specific names. *Revue de Zoologie et Botanique Africaine*, **74**, 394–424.

Trewavas, E. (1982) Tilapias: taxonomy and speciation. In *The Biology and Culture of Tilapias, ICLARM Conference Proceedings*, **7** (eds R.S.V. Pullin & R.H. Lowe-McConnell), pp. 3–14. International Center for Living Aquatic Resources Management, Manila, Philippines.

Trewavas, E. (1983) *Tilapiine Fishes of the Genera* Sarotherodon, Oreochromis *and* Danakilia. British Museum (Natural History), London.

Uchida, R.N. & King, J.E. (1962) Tank culture of tilapia. *Fish and Wildlife Service Fishery Bulletin*, **62**, 21–52.

Watanabe, W.O., Kuo, C.M. & Huang, M.C. (1984) Experimental rearing of Nile tilapia fry (*Oreochromis niloticus*) for saltwater culture. *ICLARM Technical Report*, **14**. Council for Agricultural Planning and Development, Taipei, Taiwan and International Center for Living Aquatic Resources Management, Manila, Philippines.

Watanabe, W.O., Kuo, C.M. & Huang, M.C. (1985) Salinity tolerance of the tilapias *Oreochromis aureus*, *O. niloticus* and an *O. mossambicus* × *O. niloticus* hybrid. *ICLARM Technical Report*, **16**. Council for Agricultural Planning and Development, Taipei, Taiwan and International Center for Living Aquatic Resources Management, Manila, Philippines.

Wee, K.L. & Tuan, N.A. (1988) Effects of dietary protein level on growth and reproduction of Nile tilapia (*Oreochromis niloticus*). In *The Second International Symposium on Tilapia in Aquaculture, ICLARM Conference Proceedings*, **15** (eds R.S.V. Pullin, T. Bhukaswan, K. Tonguthai & J.L. Maclean), pp. 401–10. Department of Fisheries, Bangkok, Thailand, and International Center for Living Aquatic Resources Management, Manila, Philippines.

Woiwode, J.G. (1977) Sex reversal of *Tilapia zilli* by ingestion of methyltestosterone. *Technical Paper Series, Bureau of Fisheries and Aquatic Resources*, Philippines, **1**, 1–5.

Yater, L.R. & Smith, I.R. (1985) Economics of private hatcheries in Laguna and Rizal provinces, Philippines. In *Philippine Tilapia Economics, ICLARM Conference Proceedings*, **12** (eds I.R. Smith, E.B. Torres & E.O. Tan), pp. 15–32. International Center for Living Aquatic Resources Management, Manila, Philippines.

Chapter 13
Carps (Cyprinidae)

13.1 Introduction and taxonomy
13.2 Carp culture in Israel
13.3 Chinese carps in Israel
13.4 Broodstock management
    13.4.1 Required numbers of common carp broodstock
    13.4.2 Pre-spawning maintenance of broodstock
    13.4.3 Selection and transport of broodfish
    13.4.4 Chinese carp broodstock
    13.4.5 Indoor holding tanks
13.5 Propagation and spawning
    13.5.1 Gonadal recrudescence and hormonal cycles in common carp
    13.5.2 Induced spawning
    13.5.3 Endocrine changes during induced spawning
    13.5.4 Duality of gonadotropins
    13.5.5 Pond spawning of common carp
    13.5.6 Indoor spawning: collection of gametes and artificial fertilization
    13.5.7 Incubation of eggs
13.6 Nursing of fry
    13.6.1 The nursery pond
    13.6.2 Predators of larvae and fry
    13.6.3 Algae: *Prymnesium*
    13.6.4 Primary nursing of larvae and fry
    13.6.5 Secondary nursing
    13.6.6 Nursing of mixed-sized carp
13.7 Propagation and culture of ornamental cyprinids
    13.7.1 Koi
    13.7.2 Goldfish
13.8 Future research priorities for cyprinids
    13.8.1 Genetic temperaments and ploidy manipulations
    13.8.2 Maturation and spawning
    13.8.3 Cryopreservation of sperm and cold storage of eggs and embryos
    References

Carps are farmed in the northern and the central parts of Israel, characterized by mild winters (10–12°C) and hot (28–32°C) dry summers. Silver, bighead and grass carp are grown together with common carp (a crossbreed of the genetically improved Dor-70 and Yugoslavian strains), tilapia and mullets in polyculture. Various combinations of these fish and black carp are introduced also into reservoirs to control water quality. Common carp reach puberty within 1 year while silver, bighead and grass carp mature in 2–3 years.

Once the genetic lines of the broodstock are established, the problem is how to maintain the broodfish in order to attain gonadal maturity in time and eventually gametes of high quality. Broodfish are maintained in 0.3–0.5 ha ponds at densities $<2000\,kg\,ha^{-1}$. Common carp sexes are separated to prevent uncontrolled spawning. Fish are fed a protein-rich diet (30% protein) at the rate of 0.5–1.0% of their biomass 2 months prior to spawning.

Another problem is the synchronization of spawning in order to obtain populations of uniform fry ready for nursing. This problem has been solved by spawning induction using a calibrated carp pituitary extract or ethanol-preserved carp pituitaries. Recently, the super active GnRH analogue has been used in combination with a dopamine antagonist. Spawning is carried out either in groups on mats immersed in spawning ponds or in hatcheries.

While the first stage of primary nursing (to 0.5–1.0 g) conducted in hatcheries mainly faces problems of water quality and sanitation, which are easily controlled in recirculating systems, nursing in earthen ponds is more complex. Overestimated numbers of larvae stocked into nursing ponds, because of difficulty in accurate counting, may lead to starvation, retarded growth and even mortality. Such a situation is circumvented by frequent examination of fry condition and control of water quality.

13.1 Introduction and taxonomy

The increasing importance of fish culture has made it essential for the commercial fish culturist to secure the reliable production of sufficient numbers of high quality fry for stocking by steadily improving methods and techniques of propagation. These are met by adopting methods of genetic improvement of existing fish stocks and by application of artificial breeding under controlled environmental condition to increase the survival and quality of fry.

This chapter deals with broodstock maintenance, spawning and fry nursing of fish belonging to three sub-families of Cyprinidae. These are:

- the common carp. *Cyprinus carpio* with its two variants, the mirror carp and the Japanese ornamental carp (nishikigoi, commonly known as 'Koi') and the goldfish varieties, all belonging to the sub-family Cyprininae;
- the silver carp, *Hypophthalmichthys molitrix* and the bighead carp, *Hypophthalmichthys* or *Aristhichthys nobilis*, both belonging to the sub-family Hypophthalmichthynae; and
- the grass carp, *Ctenopharyngodon idella* and the black carp, *Mylopharyngodon piceus* of the sub-family Leuciscinae (Makeeva 1969).

The common carp was naturally spread in Asia and in continental Europe during the last post-glacial period. Fish culture has been practised in ancient China for millennia, and the common carp was the first fish species to be domesticated owing to its easy maintenance and propagation in captivity (Wohlfarth 1984). Chinese isolates of common carp developed genetically diverse races, characterized by their unique genetic traits, responding to local culture techniques. Various genetic races of the common carp developed in other eastern and southern Asian countries under specific environmental conditions and aquacultural practices (Moav et al. 1975, Wohlfarth et al. 1975).

Carps were imported into ancient Rome and stored in pools, *piscinae*, to supply fresh tablefish (Balon 1974). The domestication of common carp in central and eastern Europe started in the seventh century AD in monastery ponds, 'to supply fresh fish for days on which the consumption of meat was prohibited by religious regulations' (Balon 1974). In the thirteenth century fish farming was established in western Poland, and in the sixteenth century a practical fish culture manual was officially published to help fish farmers (Strumienski 1573).

Most of the common carp world production (>95%) is concentrated in 11 countries producing annually above 10 000 t each (FAO 1990). In 1988, the combined production of common carp in these countries exceeded 1 000 000 t, 75% of which was being produced in China and in the USSR. Indonesia and Japan produced 88 000 and 18 000 t, respectively and the rest of the total production came from eastern Europe, where carp production forms 45–98% of inland fish culture (Table 13.1). Carp is the major component of Israeli fish culture, making up 55% to 65% of total yields (Sarig 1991).

The Chinese carps, which originated from central and eastern Asia, are widely distributed in the basins of large rivers, such as the Yangtze, Hsiang, Han and Si-kiang. The natural reproduction of these species is stimulated there by river floodings, caused by summer monsoons. The natural spawning grounds

Table 13.1 Production of common carp in countries producing over 10 000 t per year (FAO 1990)

	Fish yields from inland waters (1000 t per year)		
Country	Common carp	Total fish	Percentage of common carp from total fish production
China	584.6	3930.2	14.9
USSR	252.9	359.5	70.3
Indonesia	87.8	328.0	26.8
Japan	18.1	113.5	15.9
Poland	21.8	26.0	83.8
Germany	19.4	41.5	46.7
Romania	17.0	38.0	44.7
Czechoslovakia	17.0	20.9	81.3
Yugoslavia	13.1	13.4	97.8
Hungary	12.5	18.1	69.1
Iran	10.0	11.0	90.9

of Chinese carps still provide a great part of the fry for stocking in ponds (Committee 1981). Since the 1960s the advantages of Chinese carp for improving water quality and rapid growth were recognized by both aquaculturists and agencies dealing with management of fresh water. Consequently, Chinese carp have been introduced into many countries throughout the world (Welcomme 1988). Nevertheless, China is still the leading country in the production of these species (Table 13.2).

The following points should be considered in any carp propagation and breeding programme:

(1) Since the carp is an extremely fecund fish, and a single female can produce hundreds of thousands of eggs, the rate of inbreeding may increase if small numbers of broodfish are used to produce fry for stocking. Cross breeding of genetically selected lines may alleviate this problem by heterosis and by the production of progenies characterised by desired traits (Sin 1982).

(2) The advantages of pond spawnings using special substrates for eggs as an alternative to artificial propagation in controlled environment should be considered. The use of environmental stimuli or spawning inducing agents (HCG, pituitary extract, GnRH) is another point to be considered.

(3) Methods of incubation may consist of hapas, pond-immersed mats for sticky eggs, flow-through incubators of various shapes for degummed carp eggs or semi-buoyant eggs of Chinese carps. The most suitable technique should be selected with regards to the local situation. Nevertheless, in order to increase the survival of embryos and larvae, temperature and water quality should be monitored and controlled using any of these methods.

(4) Properly constructed small size ponds, that can be easily controlled, drained and refilled with fresh water are usually to be preferred. Density of fry should be frequently controlled during the nursing period to estimate growth and survival. Overcrowding in nursery ponds results in low survival rates. Size frequency distribution of fry maintained at high densities tends

Table 13.2 Production of Chinese carps in mainland China. Yields for 1988 reported by FAO (1900)

Species	1988 Production (t)	1988 World production (t)
Silver carp	1 481 000*	1 598 446
Grass carp	584 600	599 226
Bighead	701 500	705 267
Black carp	117 000	117 000

* The production of silver carp in the USSR (the second country producing this species) in 1988 only amounted to 98 342 t

to be asymmetric and skewed to the right. Such a phenomenon, often known as 'shoot-carp' is typical of carp populations (Wohlfarth 1977a).

13.2 Carp culture in Israel

Carp culture in Israel was initiated in marshes near Acre in 1934 (Hornell 1935); since then, common carp is the predominant fish cultured in Israel.

The 'Dor-70' line is an up-selected group of common carp developed at the Fish and Aquaculture Research Station, Dor. The progeny generated by crossing Dor-70 with carp of the Yugoslavian strain exhibits superior performances under Israeli conditions (Wohlfarth 1977b, Wohlfarth et al. 1980). At present, this crossbred is the only common carp cultured commercially in Israel. Crosses between the European and Chinese common carp have also been evaluated in genetic improvement programmes; crosses involving the Chinese strains of the common carp were found to have advantageous growth performance and high fecundity, especially under extreme conditions of low oxygen levels, high temperatures and high fish densities (Wohlfarth et al. 1980).

The sub-tropical climate of Israel (about 32°N, 35°E) is characterized by long, warm and dry summers and rather short winters with moderate temperatures. The seasonal multi-annual average of the temperature in a pond of 0.2 ha and 1.5 m depth is shown in Fig. 13.1. Carp tolerate a wide range of temperatures (4–35°C) and oxygen levels as low as $0.5\,\mathrm{mg}\,O_2\,l^{-1}$ (Sarig, 1966).

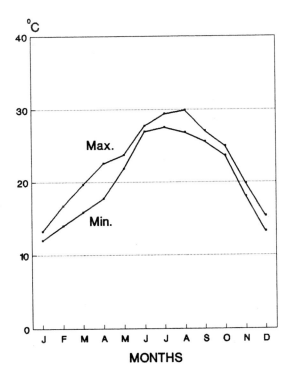

Fig. 13.1 Minimal and maximal temperatures in a 0.2 ha, 1.5 m depth, earthen pond used for rearing common and Chinese carp broodfish.

The high water temperatures in Israel accelerate growth and result in yields three or more times that in most European countries. Shortages of water and land resources limit the area for aquaculture in Israel, encouraging intensification and introduction of polyculture systems. Common carp is propagated in Israel either in ponds or in indoor facilities (Rothbard 1981). At present, the majority of farms depend on fry supplied by several fish hatcheries. Carp pituitaries calibrated pituitary extract manufactured in Israel, or GnRH analogues combined with dopamine antagonists, are used to induce spawning.

13.3 Chinese carps in Israel

Chinese Carps when reared in polyculture, form a biological complex that can control water quality in lakes and reservoirs by utilizing different ecological niches. By filtering unicellular algae or zooplankton, silver carp and bighead carp eliminate excess plankton blooms (Hepher & Pruginin 1981). The grass carp is able to restore water flow in channels covered with excess vegetation (Hickling 1971), and the black carp can eliminate snail populations, reducing the problems of water flow through pipes, filters and other installations (Leventer 1979), and can alleviate problems caused by the golden snail infesting rice fields (Acosta & Pullin 1991).

Chinese carps were introduced to Israel probably in the late 1960s (Welcomme 1988) in conjunction with the construction of the Israeli National Water Carrier (INWC). This enterprise was aimed at carrying water from the Sea of Galilee in the northern part of Israel, via a system of aqueducts, pipes, tunnels and operational reservoirs, to the Negev desert in the south of the country (Leventer 1987). The Sea of Galilee water is rich in nutrients; these combined with strong solar illumination result in a succession of food chains and a deterioration in water quality. There are two methods of controlling water quality, chemical and biological. The biological option was found by INWC authorities to be a more effective treatment. A biological complex, consisting of four species of Chinese carps, common carp and tilapias, able to control various stages of the biological food chain, was selected to achieve this goal. Certain characteristics of the four Chinese carp species, the silver carp (*Hypophthalmichthys molitrix*), the bighead (*Aristichthys nobilis*), the grass carp (*Ctenopharyngodon idella*) and the black carp (*Mylopharyngodon piceus*), are presented in Table 13.3.

Soon afterwards, the Chinese carp were introduced into fish ponds to improve water quality and to increase fish production. At present all species are propagated artificially in hatcheries and grown in polyculture (except black carp) with the common carp, tilapias and grey mullets. A hybrid between silver carp and bighead is preferred by some farms, owing to its good catchability and intermediate feeding habits (Spataru *et al.* 1983).

Table 13.3 Feeding habits and age at maturation of Chinese carps, used as a biological control system in Israel

Species	Feeding habits	Age of maturation*
Hypophthalmichthys molitrix	Planktophagous (unicellular algae	2 years for males
Aristichthys nobilis	Zooplanktonphagous (small invertebrates)	3 years for females
Ctenopharyngodon idella	Macrophytophagous (water vegetation)	
Mylopharyngodon piceus	Benthophagous (bottom feeder, e.g. snails)	5–6 years for males 6–8 years for females

* In Israel

13.4 Broodstock management

13.4.1 Required numbers of common carp broodstock

The number of spawners to be taken for propagation depends on the quantity of fry required for stocking. Stocking rates of common carp in Israel vary with the management system and the stocking rate of other species grown in polyculture. In general, average yield of common carp grown together with other fish is about $2.8\,t\,ha^{-1}$, while the total average yield is more than $5\,t\,ha^{-1}$ (Sarig 1991). Stocking 100 ha of ponds requires approximately 3 million carp fry. Producing this number of fish in an indoor hatchery requires about 30 female spawners and half the number of males. Owing to losses, about 25% of the broodstock has to be replaced each year. The number of spawners must be higher if spawning is carried out in ponds. Table 13.4 presents differences in propagation methods under indoor and field conditions and the number of fry expected with each approach.

13.4.2 Pre-spawning maintenance of broodstock

Temperature plays a major role in gonadal recrudescence and timing of ovulation in common carp (Horvath 1986), although under experimental conditions photoperiod may modulate ovarian maturation (Davies *et al.* 1986). In Israel, ovarian recrudescence starts at the end of summer (Yaron & Levavi-Zermonsky 1986) and is complete after the female has accumulated 1000–1200 temperature degree days (DD). Calculation of the DD required to achieve mature females begins in the spring, when water temperatures are above 15°C and increasing gradually. At the end of DD accumulation, females are usually ready for spawning. This calculation is in line with the results reported for the common carp females maintained for 2–3 years in temperate climates, where broodfish in the third year of their lives require 2000 DD to reach maturity (Horvath

Table 13.4 Comparison between hatchery and field propagation of common carp as practised in Israel. Conditions required for the production of 1 000 000 fry of 1 g, in one breeding cycle

Data compared	Indoor propagation	Field propagation
Equipment	Special equipment, installation, tanks	Properly designed earthen ponds
Broodstock	10 females (2 kg) 5 males	200 females (2 kg) 200–300 males
Ovulation timing	Complete control	Environment-dependent
Hormone treatment	Indispensable	Not required
Manpower	1–2 persons	3–5 persons
Water system	Closed, recirculated	Open pond
Treatment of eggs	Degumming	None
Incubation	Conical incubators	Special substrate
Larvae per kg female	c. 80 000	c. 10 000
Primary nursing	Indoor and in pond	None
Survival (1 g fry)	~70%	~50%
Knowhow	Technical skills	Field-pond experience

1985, 1986). However, in order to select precisely the females suitable for spawning induction, the stage of oocyte maturation is determined by the position of the germinal vesicle in ovarian biopsy of several females sampled from the broodstock (see [13.4.3] below).

After spawning, fish are usually stocked in earthen ponds until winter. Two months prior to spawning (January–February), fish are removed from the pond and the sexes stocked separately into 0.2–0.5 ha ponds, at densities of about $2 t ha^{-1}$. Each pond is equipped with an aerator and a source of good quality water. Although carp reach puberty in one year, 2–3 year old spawners, of 1.5–4.0 kg, are the most suitable for transportation, handling and propagation. Biological data concerning common carp propagation are presented in Table 13.5.

Owing to low stocking rates of broodfish, the broodstock ponds are rich in natural food. However, to enhance gonadal development and to avoid accumulation of fat, protein rather than carbohydrate-rich supplementary feed should be used (Hepher & Pruginin, 1981). Spawners are fed 30% protein pellets all-year round; in summer, a daily ratio of 1% of the biomass is given which is reduced to 0.5% in winter. The natural food, abundant in the spring, may enhance gonadal maturation.

13.4.3 Selection and transport of broodfish

Broodfish are easily stressed by poor pond conditions and careless handling. This needs to be considered in every step prior to spawning induction. Special attention should be given to avoid overcrowding spawners in nets, while seining the pond.

Table 13.5 Biological data on common carp reproduction and nursing of fry in Israel

Age of spawners	At least 2 years
Initial weight of spawners	>1 kg
No. of eggs per kg female	100 000–200 000
Spawning season	February–July/August
Method of spawning	Pond or indoor spawning
Natural spawning site	Shallow water vegetation
Eggs diameter	Pre-hydrated: 1.0–1.5 mm
	Activated: 1.5–2.5 mm
Pre-hydrated eggs per kg	600 000–1 000 000
No. of eggs per kg spawner	80 000–120 000
Hatching time (days)	$2-4\frac{1}{2}$ (temperature dependent)
Yolk sac larva	3 days post-hatching
Start of independent fry	Primary nursing: 7–9 mm
Initial prey size	100–300 µm
2–3 weeks old fish	c. 0.5–1.0 g at a density of 1 million per ha

Fish are examined for their stage of ripeness while in net enclosures. This is carried out, without removing the fish from water, only by palpating the female's belly. Ripe females may be recognized by their reddish and swollen genital papilla. The males are gently turned over in the water and the ripeness is indicated when some drops of milt can be squeezed out.

In order to stock the ponds at as early a date as possible, spawning induction is attempted at the end of February or early March. However, at this early season only a proportion of the females are ready for induction, which may result in frequent spawning failures. Stimulation of milt production occurs in male common carp if kept together with females induced to ovulate by pituitary homogenate. It has been suggested that this effect is pheromonal (Billard et al. 1989). In order to avoid wastage of broodfish, a simple biopsy technique has been adopted for the selection of the females ready for spawning induction. A Tygon tubing of 4 mm internal diameter is inserted via the genital duct about 6 cm into the ovary, and ovarian samples containing approximately 40–50 follicles are aspirated. The ovarian samples are cleared in Serra's fluid (ethanol: formalin (40%) acetic acid; 6:3:1). The opaque oocytes become translucent within 3 min and remain so for 6–7 min longer. During this interval, the position of the germinal vesicle is determined in at least 40 oocytes. In experiments performed early in the season it has been established that successful spawning induction can be achieved only in fish in which at least 65% of the oocytes have migrating germinal vesicles (Yaron & Levavi-Zermonsky 1986, reviewed by Shelton 1989). Females with oocytes which have not yet reached this stage are returned to the ponds to be used at a later date.

For short distance (pond–hatchery) transportation, selected fish are placed in soft cloth sacks in a tank with well oxygenated water. Fish transported in sacks remain calm during handling. While still in the wet sacks, females are weighed individually and held in separate tanks. Three to five males may be grouped in one holding tank, separate from the females.

13.4.4 Chinese carp broodstock

Chinese carps (excluding the black carp), reach sexual maturity in Israel at the age of 2 (males) or 3 years (females). In temperate climates, these fish mature at the age of 5–6 years. Black carp mature at the age of 6–7 years in Israel, while in China, where the fish are cultured as a table fish, black carp mature at the age of 8–10 years (Hickling 1971).

The Chinese carps grow fast and mature at large sizes. Black carp mature at a weight of 7–10 kg (80–90 cm). Other species mature at 3–5 kg (50–70 cm). Silver carp and bighead carp ripe males are easily recognized by a sandpaper-like roughness of the pectoral fins. In ripe male black carp nuptial tubercules cover 3–4 of the anterior rays of the dorsal fin and over ten rays of the pectoral fins; these are absent in the female. In ripe grass carp such tubercules are present in males but only variously found in females (Committee 1981).

Chinese carps can be maintained in the same pond as the common carp broodstock because of their different breeding habits. Both sexes may be kept together since they do not spawn spontaneously. Nevertheless, to avoid overcrowding of spawners, the total standing crop must not exceed 1500–2000 kg ha^{-1}.

Handling, transportation and selection of spawners of Chinese carps is performed by methods similar to those described for the common carp.

13.4.5 Indoor holding tanks

Most fish hatcheries in Israel are installed with a thermoregulated, recirculating water system to prevent waste of water, and to provide a low and uniform hydrostatic pressure in the pipelines. Gas-supersaturated water, due to high pressures, can be fatal for eggs and larvae.

Fish introduced into the hatchery are placed in holding tanks supplied with running water, disconnected from the hatchery water system to avoid contamination. Tanks of various capacities, from 0.5 m^3 to 1.2 m^3, are used as holding tanks. These are aerated, shaded and covered by a net to keep the broodfish calm and to prevent their jumping out.

Fish remain in the tanks for several hours before treatment against ectoparasites. A 4–5 h disinfection consists of a mixture of 0.2 ppm malachite green (Alderman 1985) and 30 ppm formalin (technical grade; 37%). After disinfection, the tanks are connected to the hatchery water system. Hormone treatment starts the next day to reduce faeces in the holding tank. Ripe females may be kept in the hatchery at water temperatures of 20–23°C for several weeks prior to initiation of spawning. Males with running milt may be maintained in similar conditions all the year round.

Chinese carp spawners are introduced into the hatchery, just prior to induced reproduction, because of their size and extreme sensitivity to handling. Immediately after collection of gametes, the broodfish are returned to their ponds.

13.5 Propagation and spawning

13.5.1 Gonadal recrudescence and hormonal cycles in common carp

Sexual maturation of the common carp depends mainly on the duration of its exposure to high temperatures (>18°C; Horvath 1986), but may be affected by day length (Davies *et al.* 1986). Under the climatic conditions in Israel, puberty of carp is reached within one growing season (or less), considerably faster than European carp populations (Horvath 1986).

The details of ovarian recrudescence in the local population of the common carp, and the associated fluctuations in hormone level, were studied in samples of 15 female carp taken monthly from the commercial harvest of Kibbutz Gan-Shmuel fish farm during the season of 1984–5 (Yaron & Levavi-Zermonsky 1986). Blood was sampled for hormone determinations and the ovaries were excised, weighed and fixed. The maximal diameter of the follicles was determined under the microscope in fresh ovarian tissue, and details of the oocytes' structure studied histologically. The pituitaries were removed for the determination of the gonadotropin (cGTH) content. The radioimmunoassay for cGTH was performed using a standard purified by Dr B. Breton (Rennes) and an anti cGTH serum developed in rabbits by Dr Lichtenberg (Hamburg) (Ribeiro *et al.* 1983, see [13.5.4] for comment on the duality of common carp GTHs).

In the vicinity of Gan-Shmuel, high water temperatures (28–30°C) are maintained in the fish-ponds from July through to September and only start to decline in October. The lowest temperatures (13–14°C) are recorded in December through to February (Fig. 13.2A).

Two populations of follicles can usually be discerned in the ovarian samples using a dissecting microscope; small, dormant follicles not exceeding 0.15 mm in diameter and developing follicles which occur in the ovary from August to May. In June and July only small follicles are found in the ovary; peripheral vacuoles are rarely seen in histological sections of the ovaries sampled in these months. The number of oocytes with peripheral vacuoles increases in August, but only in September are yolk granules deposited in the ooplasm. Up to this time these changes are not reflected in the gonadosomatic index (GSI; Fig. 13.2B,C). In subsequent months, vitellogenesis continues, and follicular diameters gradually increase in parallel with elevations in GSI. In February, the oocytes reach their maximal diameter of 1.10 ± 0.01 mm and the ovaries enter their pre-ovulatory phase. The values of oocyte diameter and GSI presented for May in Fig. 13.2B,C were taken from 15 females, five fish which had not spawned with GSI of 9.35 ± 2.2, and 10 postspawning fish with GSI of $0.87 \pm$

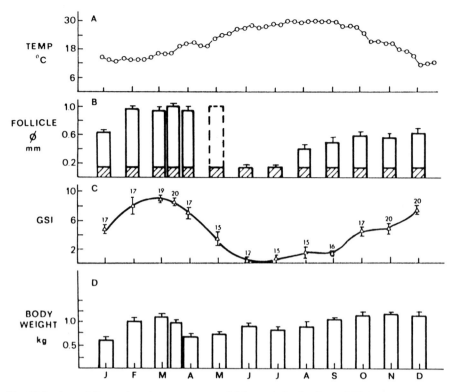

Fig. 13.2 Annual fluctuations in temperature of fish ponds in the Coastal Plain of Israel and parameters associated with reproduction in female carp (Dor-70 × Yugoslavian). Fish were obtained monthly from the commercial harvest of kibbutz Gan Shmuel; two samples were taken in April. (A) Weekly means of water temperature. (B) Diameter of ovarian follicles; at least 40 follicles were examined in each female, number of fish as in (C).

Open bars denote vitellogenic oocytes; hatched bars denote the presence of non-vitellogenic (dormant) oocytes throughout the year. Among the fish sampled in May some were postovulatory with dormant follicles only while others were preovulatory (dashed line bar). (C) Gonadosomatic index (GSI). (D) Mean body weight of fish in each sample; number of fish as in (C). (From Yaron & Levavi-Zermonsky 1986.)

0.07. The latter exhibited only small follicles in the ovarian biopsies. It is concluded that under the conditions prevailing in the ponds in the Coastal Plain of Israel, natural spawning takes place in May.

A prominent peak in circulating cGTH occurs in August at the onset of vitellogenesis; a second, less pronounced, but extended peak may be seen in April–May (Fig. 13.3D). There are two peaks in the pituitary content of cGTH, one in October–December and the other in February–March (Fig. 13.3E). Association of elevated GTH in the circulation and vacuolation of oocytes has also been reported in carp in Poland (Bieniarz et al. 1991).

Oestradiol levels are low in August but exhibit a transient peak in September, and a gradual increase in November and December during the active phase of vitellogenesis (Fig. 13.3A). Fish examined in January probably include a

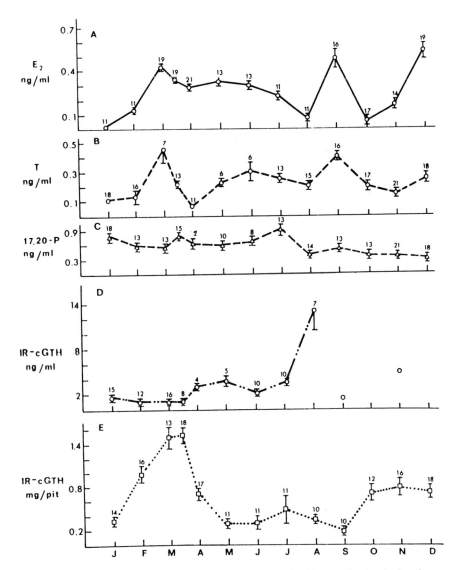

Fig. 13.3 Annual fluctuations in certain hormones associated with reproduction in female common carp. (A) Circulating oestradiol-17β (E_2). (B) Circulating testosterone (T). (C) Circulating 17α,20β-dihydroxy-4-pregnen-3-one (17,20-P). (D) Circulating cGTH. (E) cGTH content in the pituitary. (From Yaron & Levavi-Zermonsky 1986.)

younger year class (Fig. 13.2D), and showed a very low estradiol level which gradually increased in the subsequent months and remained high till July.

Testosterone levels fluctuate in a pattern similar to that of oestradiol (Fig. 13.2B). However, the level of the maturation-inducing progestogen (17α,20β dihydroxy-4-pregnene-3-one; 17,20-P) is extremely low (not exceeding 1 ng ml^{-1}) throughout the year (Fig. 13.3C). The only rise in the level of this steroid occurs during the induction of spawning (see [13.5.3]).

The increase in GTH and oestradiol in late summer and autumn, together with the growth in follicular diameter and GSI indicate that, under the conditions prevailing in Israeli aquaculture, the vitellogenic process in the common carp commences 2 months after spawning. Female carp studied in Poland showed high GTH levels in the circulation during and after the spawning season. These high levels were correlated with the vitellogenic phase which commenced immediately after spawning and was completed by December (Bieniarz et al. 1991). Female carp studied in France showed a peak in circulating GTH in July, at the normal time of spawning. This peak was followed by elevated oestradiol which lasted until November. Vitellogenesis in these fish was completed by October (Billard et al. 1987). Both European populations of carp exhibit a short vitellogenic phase starting immediately after spawning in June–July which is completed by October–December. Spawning in Israeli populations of carp occurs 2 months earlier in May, but vitellogenesis starts in August as in their European counterparts, Hence, Israeli carp enjoy 2 months of gonadal rest before the beginning of a new cycle. Furthermore, vitellogenesis in the Israeli carp extends throughout the winter months, and maximal oocyte diameters are not reached until February.

The stages of the testicular recrudescence of the common carp and the underlying endocrine changes under the conditions prevailing in Israeli aquaculture are poorly known. Nevertheless, milt can be expressed from most males at any time of the year. It is assumed, therefore, that spermatogenesis occurs at a certain rate all year round. The presence of ovulating females possibly augments spermatogenesis and milt production by the male when stocked together in earthen ponds (Billard et al. 1989).

13.5.2 Induced spawning

Hypophysation using crude pituitary homogenates was practised in Israel until the mid-1980s. Carp pituitaries were harvested from carp in fish-processing plants as a side product of the line producing carp fillets or ground carp meat. Pituitaries were collected from carp of 1 kg or more during early summer. The harvested glands were stored in ethanol and distributed to the farms by the Fish Breeders Association. The glands were blotted and homogenized in saline, and injected at the recommended dose of 0.2 gland equivalent per kg for priming at noon, and 1 gland equivalent per kg as a resolving dose, given 10–12 h later. Males were injected once with the equivalent of 0.5 gland per kg, 7–20 h prior to milt collection.

The resulting spawnings were generally satisfactory. However, when spawning induction failed, it was difficult to determine whether the recipient fish were not at the right stage or whether the glands did not contain sufficient gonadotropin. Indeed, the pituitary content of GTH in the common carp fluctuates throughout the year. The pituitary may contain as little as 300 µg per gland, or as much as 1.4 mg per gland, depending on season (Fig. 13.3E). Pituitary GTH content also depends on fish size (Yaron *et al.* 1984). The variability in the pituitary GTH content may explain the relatively poor (60% or less) and variable success of hypophysation recorded in fish farms by Billard *et al.* (1987)

A concerted effort of fish endocrinologists and the Dag Shan Fish Processing Plant resulted in the commercial production of a GTH-calibrated carp pituitary extract. The glands harvested in the plant are extracted and treated prior to lyophilization to ensure their biological activity after storage for more than 3 years. The gonadotropin content of each batch is determined by radioimmunoassay, and biological activity calibrated by an *in vitro* bioassay (Yaron *et al.* 1985) followed by a trial under routine hatchery conditions (Yaron *et al.* 1984).

The amounts of gonadotropin required for the priming and resolving doses of the pituitary extract were calculated in spawning experiments carried out under routine hatchery conditions and procedures. It was concluded that the priming dose should contain $50-100 \mu g\, kg$ cGTH and the resolving dose $250-500 \mu g\, kg^{-1}$ as determined by radioimmunoassay. Higher doses have proved to be less effective (Yaron *et al.* 1984). Accordingly, the commercial product is calibrated to contain the recommended doses for both priming and resolving injections ($450-600 \mu g\, kg^{-1}$ in total). The product is distributed together with instructions for dilution and injection volumes. Currently, most fish farms in Israel use the calibrated extract for spawning induction in common and fancy (koi) carp, grass carp, silver carp, bighead carp, black carp and goldfish.

It should be admitted that the popularity of the calibrated carp pituitary extract among Israeli aquaculturists has postponed the introduction of the hypothalamic approach for spawning induction in carp, namely the use of GnRH superactive analogue combined with dopamine antagonists. This approach has proved successful in spawning induction in carp and other species in many countries (Weil *et al.* 1986, Lin *et al.* 1986, 1988, Sokolowska *et al.* 1988; the reader is referred to the Fish Breeding Workshop on Induced Spawning of Asian Fishes, Singapore, 1987, in *Aquaculture* **74**, 1–4, 1988, for extensive reviews of methods). Research is now in progress to adapt and optimize this contemporary method to the conditions prevailing in Israeli carp hatcheries in terms of GnRH analogue type and dose, the adequate dopamine antagonist and its dose, and the effect of temperature on the timing of response.

13.5.3 Endocrine changes during induced spawning

In order to analyse the effects of the calibrated extract on the level of ovarian steroids and the kinetics of oocyte maturation, female carp were induced to spawn by a priming injection containing $70 \mu g\, kg^{-1}$ cGTH kg^{-1} given at noon, and a resolving dose containing $350 \mu kg^{-1}$ of the gonadotropin given 11 h later. Ovaries were biopsied and blood samples taken at intervals throughout the experiment (Figs 13.4,5, Levavi-Zermonski & Yaron 1986). Circulating level of cGTH increased from 2.3 ± 0.41 to $81 \pm 5.0\, ng\, ml^{-1}$ following priming, and to

$231 \pm 22.9\,\text{ng ml}^{-1}$ after the resolving dose was given. This high level is quite similar to that determined in naturally spawning carp ($256\,\text{ng ml}^{-1}$, Zhao et al. 1984) and is within the range reported in a similar experiment carried out in Poland using an uncalibrated dose of pituitary homogenate (Bieniarz et al. 1980). Oestradiol levels increased in response to the priming dose but exhibited a transient decrease following the administration of the resolving dose (Fig. 13.4). No change in the level of 17,20-P was noted following the priming injection, but a dramatic tenfold increase occurred 4 h after the injection of the resolving dose. Four hours later the level of 17,20-P declined considerably. This prominent peak occurred concomitantly with a transient decrease in oestradiol levels, probably indicating a transitory shift in the pathway of steroidogenesis.

Before the onset of the experiment, the majority of the oocytes contained germinal vesicles (GV) at an eccentric position. Migration of the germinal vesicle towards the periphery of the oocyte was stimulated by the priming injection; however, germinal vesicle breakdown (GVBD) and ovulation were noted (Fig. 13.5), only after the injection of the resolving dose and the dramatic increase in 17,20-P (Fig. 13.4). Oocytes of fish injected only with the priming

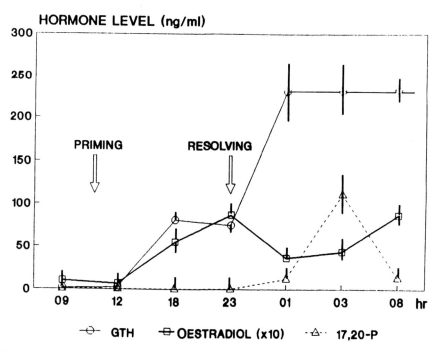

Fig. 13.4 Fluctuations in cGTH, E_2 and 17,20-P in the circulation of female common carp induced to spawn by a calibrated pituitary extract. Priming dose of the extract containing $17\,\mu\text{g cGTH kg}^{-1}\,\text{bw}$ was administered at noon (left arrow) and the resolving dose containing $350\,\mu\text{g cGTH kg}^{-1}\,\text{bw}$ was given 11 h later (right arrow). All females ovulated and were stripped the next morning. (Data from Levavi–Zermonsky & Yaron 1986.)

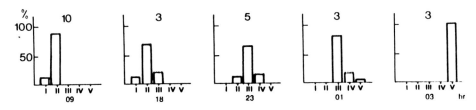

Fig. 13.5 Percentage of oocyte maturational stages in ovarian biopsies taken at intervals from carp before and during spawning induction. Fish and treatment as in Fig. 13.4. Figures above the histograms denote number of fish biopsied at each point. Maturational stages of the oocyte are defined as: I – central germinal vesicle (GV); II – migrating GV; III – peripheral GV; IV – GV breakdown; V – ovulated eggs. (Data from Levavi-Zermonsky & Yaron 1986.)

dose, although showing a progress in GV migration, reached neither GVBD nor ovulation till the end of the experiment (Levavi-Zermonsky & Yaron, 1986). The transient nature of the increase in 17,20-P in the carp has also been reported by other investigators, although the absolute values of the peaks differ between reports (Kime & Dolben 1985, Weil *et al.* 1986, Santos *et al.* 1986). The brevity of the peak might also explain the low and constant levels of the hormone which have been recorded during the annual cycle (Fig. 13.30, Yaron & Levavi-Zermonski 1986).

13.5.4 Duality of gonadotropins

It should be emphasized that the research described above was carried out prior to the findings of two gonadotropic hormones in salmonids (Suzuki *et al.* 1988a,b, Kawauchi *et al.* 1989). Only recently two distinct gonadotropins (GTH I and GTH II) have been isolated from mature female carp (Van Der Kraak *et al.* 1992). The two gonadotropins differ markedly in their molecular weight and amino acid composition; GTH I is of 51 kDa while GTH II is only 35 kDa. The biological activity of the two GTHs, as tested by the stimulation of oestradiol secretion from ovarian follicles of the goldfish, and testosterone from carp testis *in vitro*, was found to overlap considerably. Moreover, both GTHs were found to be equipotent in stimulating oocyte maturation *in vitro*. It is most probable that the GTH as measured in our work resembles GTH II which is more abundant than GTH I in the pituitary of mature fish.

13.5.5 Pond spawning of common carp

Pond spawning is carried out only with the common carps. Ripe broodfish of both sexes, sometimes injected with pituitary extract to stimulate synchronous spawning, are introduced into a pond freshly filled with water. Dykes in the spawning pond are kept clean of vegetation to reduce hiding places for potential pests which would prey on the eggs and larvae (see below). Spawning nests made of pine tree branches or mats made of synthetic brushes, are placed in the pond, 1 m² of nest being sufficient for each kg of female broodfish. Twenty to thirty females (30–50 kg) can spawn in a 1 ha pond. The recommended broodfish sex ratio is three males to two females. Fry of 0.5–1 g are usually removed at the age of 2–4 weeks and transferred to secondary nursing ponds.

Two sizes of ponds are used, 0.2–1 ha and 2–5 ha.

Small ponds, 0.2–1 ha

These are filled with fresh well water just prior to the introduction of broodfish. The nests, tied to metal frames, are located 70–80 cm below the water surface and 20–30 cm above the bottom. Spawning occurs within 24 h of the introduction of the spawners. Nests with fertilized eggs attached are transferred to larger ponds for incubation and primary nursing. Sometimes the broodfish are left for another day to spawn and only then removed, but the nests are replaced so that fry of two different ages will not mix. This procedure improves the survival of the fry and prevents suppression of the younger progeny by the older fry.

Large ponds 2–5 ha

These are deep ponds (c. 2 m) with a harvesting sump. The nests are placed in the sump, which is filled with water. Spawning mainly occurs within 24 h. When survival of eggs is high (above 80%), spawners are removed before hatching occurs. After larvae start free swimming, the nests are removed. The water is elevated to the desired level a few days later, when yolk sac absorption is complete and the free-swimming larvae appear searching for prey. The whole pond then serves for both primary and secondary nursing.

13.5.6 Indoor spawning: collection of gametes and artificial fertilization

Injection of the resolving dose of the pituitary extract results in final oocyte maturation and ovulation in the female and increases in milt production in the male. These processes are temperature-dependent, and therefore, for accurate timing of gamete collection, the latent period from the injection to ovulation should be considered to avoid overripening of eggs (Shelton 1990). The relationship between the duration of the latent period and water temperature for common carp is shown in Fig. 13.6. The broodfish in this experiment were kept at 19–22°C till the resolving dose was injected, then the temperature was maintained or raised to 24–28°C. Eggs were collected at the end of the latent period and recorded for each individual female. Similar relationships between ovulation and water temperature were also reported by Horvath (1978) in common carp and Chinese carps. Horvath expressed the latent period in hour-degrees (duration in hours between the resolving dose and ovulation × temperature in °C) and found that the optimal latency is 205–215, 210–220, 235–245 and 240–260 hour-degrees, for grass-carp, silver-carp, bighead and the common carp, respectively.

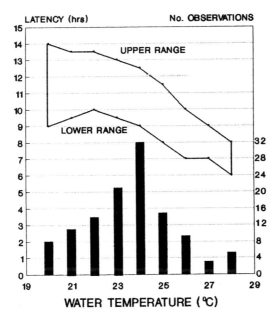

Fig. 13.6 The latency period between administration of the resolving dose and ovulation in the common carp, as a function of water temperature. Cumulative data from several hatchery spawnings of common carp induced by the calibrated pituitary extract according to the manufacturer's instructions.

The 'dry method' of fertilization used is as follows. Milt from several males is collected, mixed and stored at 4°C until the eggs are collected. A sample of the milt is activated by water and examined, in order to estimate spermatozoan motility. Broodfish are anaesthetized with benzocaine (ethyl-4-aminobenzoate) dissolved in water at the concentration of 1000 ppm, and eggs are stripped from each female into separate bowls, by applying a gentle pressure on the inflated belly (Fig. 13.7). The quantity of eggs is estimated by weighing the stripped eggs prior to fertilization. Egg size in common carp like most fish is correlated with female body weight and age. Eggs from yearling females are about 50% smaller than the eggs of older females (Hulata et al. 1974). The gonadosomatic index (GSI) of a ripe common carp female may each 25–30%. The estimated number of eggs produced by a 2–3 kg female is about 600 000.

Fertilization and degumming of common carp eggs is performed using the modified Woynarovich method. Eggs are activated with a solution of 4 g NaCl and 3 g urea/l^{-1} and 3–5 min later they are transferred into a solution of 4 g NaCl and 20 g urea/l^{-1} (Woynarovich & Woynarovich 1980, Rothbard 1981). A shaking table, operated by a low-gear motor with an excentric axis, keeps the eggs in constant shaking during degumming, enabling simultaneous treatment of many egg batches.

13.5.7 Incubation of eggs

Eggs are incubated as described by Rothbard (1981). Fertilized eggs are

Fig. 13.7 Stripping of 'Dor 70' carp.

incubated in conical funnel-shaped PVC-incubators in running water supplied from a thermo-regulated, biofiltered, recirculating water system (Fig. 13.8). Each incubator of 60–80 l can hold 200 000–300 000 eggs. The recirculated water enables the control of the incubation time and prevents gas supersaturation in the water, which may cause floating of the eggs in the funnel or mortalities of larvae owing to gas-bubble-disease.

At incubation temperatures of 25–28°C, Chinese carp eggs hatch after 23–28 h, and those of the common carp after 50–60 h (Fig. 13.9). Survival rates of common carp are recorded 24 and 48 h after fertilization. Those of Chinese carps are examined after 5–8 h, at the beginning of morula stage. The hatchlings swim through the overflow of the funnels to the primary nursing tanks. Feeding is initiated immediately after yolk sac absorption (c. 48 h after hatching). Incubation of black-carp eggs takes a few hours longer; the hatchlings are bigger than other Chinese carps and have larger yolk sacs; consequently, yolk sac absorption takes about 3 days.

The phase of yolk-sac absorption, initiation of external feeding until the fry reach the size of about 0.5–1 g, is termed 'primary nursing' irrespective of whether it is carried out in indoor tanks or outside ponds.

Carps (Cyprinidae) 341

Fig. 13.8 Battery of incubation funnels (Zoug jars).

13.6 Nursing of fry

13.6.1 The nursery pond

Owing to favorable temperatures of 21°C or more, nursing cycles are short, and small nursery ponds are the most effective for continuous utilization. Ponds of 0.1–1 ha and average depth of 1–1.5 m are preferred for secondary nursing (Rothbard 1982). These ponds can be drained readily, disinfected and refilled for the next nursing cycle. Ponds are stocked with fry, starting with common carp in spring (March–April) and continuing with Chinese carps in summer (May–July). In some hatcheries, late spawnings of common carp are carried out in July–August, further extending the utilization of the nursery ponds.

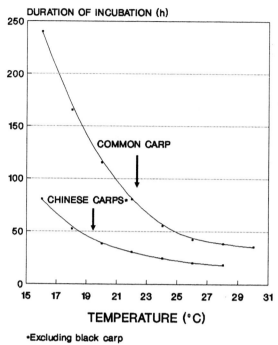

Fig. 13.9 The dependence of incubation time (between fertilization and hatching) on temperature in the common and Chinese carps (excluding the black carp).

Advantages of small nursery ponds are as follows:

(1) Shallow ponds warm up fast and allow penetration of sunlight, stimulating natural food chains which provide natural food for the small fry.
(2) Small ponds are well-oxygenated owing to their high surface: water ratio.
(3) Small ponds are easier to manage during routine activities such as seining, sampling, fertilization and disinfection of the pond.
(4) More economical utilization of water during short nursing cycles. This is important in conditions of chronic water shortage as in Israel.
(5) Risks of fry losses are minimized.
(6) More economic and more effective use of drugs and chemicals.

Nursery ponds are constructed with a bottom slope of 0.3–0.5% to enable concentrating fry during drainage and harvest. The wind direction is taken into account, especially on the coastal plains, to avoid precipitation of silt in the fishing sump, which may damage the gills of small fry during harvest. The combination of high temperatures (28–32°C) and the accumulation of organic wastes in nursery ponds may be fatal to small fry. Therefore, each nursing pond is provided with a constant supply of fresh water; this is particularly important when large numbers of fry are concentrated at harvest. Each nursery pond is also equipped with a paddle-wheel aerator which is operated at night.

The inlet pipe of the nursery pond is protected by a fine mesh (1 mm), saran net, preventing other fish from entering the nursery. The most harmful invader fish is the cichlid, *Tilapia zilli*, which matures at a small size, reproduces in large numbers and may compete with the nursing carp for the pond resources. The outlet from the nursery pond is located at the deepest part of the fishing sump, ensuring complete drainage of the pond. The overflow is provided by a screen with adequate-sized openings to prevent the escape of fry.

13.6.2 Predators of larvae and fry

The most dangerous predators of larvae in the nursery pond in Israel are water bugs (e.g. *Nepa rubra*, *Naucoris cimicoides*, *Ranatra linearis* and *Notonecta glauca*). These appear in large numbers during warm summers, preying directly on larvae or competing with them for other organisms. Heavy losses of eggs, larvae and even fry caused by these predators are common. To eradicate these pests, edible oil or diesel fuel is sprayed over the pond-surface at $20-30\,l\,ha^{-1}$, on windless early mornings. The oily fluid forms a film covering the pond surface and causes the air-breathing insects to suffocate. Usually, this treatment is repeated two or three times, on alternate mornings, till the fry are sufficiently strong to elude predators. Although copepods appear in the nursery ponds during the spring, they are not considered as pests, because the ponds are periodically drained and refilled with fresh water, usually 2–3 days before stocking the fry. A dense population of *Daphnia* spp. which competes with fish for oxygen, is eradicated by spraying Bromex at a concentration of 0.2–0.35 ppm.

Tadpoles and water snakes (*Natrix tessellata*) also prey on fry, usually those smaller than 20 g, (Hepher & Pruginin 1981). Losses in carp spawning ponds up to 80% due to frog-predation have been reported in Uganda (Pruginin 1967). Pond dykes kept clean of vegetation are effective in preventing the frogs from spawning in nursery ponds. Snake traps located in various sites of the pond reduce the number of snakes in the pond and protect the fry sufficiently from water snake predation.

13.6.3 Algae: *Prymnesium*

Prymnesium parvum is a worldwide free-swimming phytoflagellate, flourishing in brackish waters. *Prymnesium* ichthyotoxin (prymnesin) is lethal to gill-breathing organisms, including invertebrates. Heavy losses of fish, especially of carp, caused by prymnesin have been reported in the USA, England, Japan, USSR and other countries (Sarig 1971).

Prymnesium appeared in Israel for the first time in 1947, resulting in total mortality of fish populations in affected ponds within 3–5 h. Fry are particularly sensitive to the prymnesin and suffer mass mortalities. In affected ponds the fish gather in corners covered by vegetation or close to water inlets, become apathetic and are easily captured by birds. Mortality can be prevented by providing fry with fresh supplies of water.

Effective control of *Prymnesium* requires daily water examination to detect the presence of the algae; the mosquito fish, *Gambusia affinis*, is used as a test fish owing to its high sensitivity to the algal toxin. Ammonium sulphate (10 ppm) or copper sulphate (2−3 ppm) is applied to lyse the algal cells without harming the fish. Methods of detection, identification and eradication of *Prymnesium* are reviewed by Sarig (1971).

13.6.4 Primary nursing of larvae and fry

The primary nursing, which lasts for 3−4 weeks, may be carried out in indoor tanks or outdoor nursery ponds. At the end of this phase, fish reach 0.5−1 g and can be seined by fine-mesh nets, counted and stocked into secondary nursing ponds, where they are grown to the size required for stocking into ongrowing ponds. Fry stocking rates in nurseries of various size, as practised in the Gan Shmuel fish hatchery, are presented in Table 13.6.

After the yolk-sac has been absorbed, the larvae are fed with dry or live protein-rich food. The preferred live food is *Artemia salina* nauplii, incubated from decapsulated cysts (Sorgeloos *et al.* 1977), and provided mainly during the first days of feeding. Live 1 day old *Artemia* nauplii are added to tanks of larvae at concentrations of 1000−1500 per l. Later on, the feed consists of protein rich (56%) commercially available granulated food (Larvastart C-20, EWOS, Sweden). After 10−14 days of indoor nursing, the fry are transferred for an additional 2−3 weeks to an earthen pond, which had been filled with

Table 13.6 Nursing of fry of various cyprinids, propagated in Gan Shmuel Fish Breeding Center and nursed in different sizes of ponds

Species	Pond area (ha)	Stocking rate/ha ($\times 10^6$)	Nursing (days)	Survival (%)**	Mean final weight (g)
Silver carp	0.06	1.66	20	80	0.4
	0.06	2.0	18	66	0.3
	0.4	2.2	20	50	0.3−0.4
	0.06	2.5	25	91	~1.0
	0.06	3.3	30	60	1.0
	0.06	3.3	26	76	0.5
	0.06	5.0	20	84	0.3
	0.9	5.5	21	60	0.6
Common carp	5.0	0.7	24	74	1.4
	0.2	1.5	35	83	1.2
	0.5	3.0	15−35*	78	0.2−0.3
Grass carp	0.4	2.5	35	77	2.0−3.0
	0.06	4.0	47	10	0.2
	0.15	5.0	15−36*	50	0.1−0.2
Bighead	0.04	2.8	18 days	70	0.1−0.2

* Fry were harvested continuously during nursing
** Survival of fry according to estimated stocking rates

Table 13.7 Nursing of 0.3 g silver carp fry in secondary nursing ponds in three fish farms in Israel. Ponds were also stocked with 1000 common carps per ha to control filamentous algae (Sarig 1970)

Pond area (ha)	Number of fry per ha at harvest	Nursing (days)	Mean weight at harvest (g)	Individual weight gain (g per day)	Daily weight gain (g ha^{-1})
0.25	41 200	34	7	0.21	8 650
0.30	27 000	35	10	0.29	8 060
0.40	29 800	57	27	0.47	13 600

fresh water 2–3 days in advance: this completes the primary nursing. Fry numbers are estimated during transfer to the nursery ponds. The transfer procedure is carried out by means of syphons to reduce injury. Similar methods are applied in primary nursing of common carp and Chinese carp. Sometimes 100–200 g common carps are stocked in Chinese carp nursing ponds to control filamentous algae (Table 13.7, Sarig 1970).

When propagation and primary nursing are carried out in the same pond, the hatchlings subsist exclusively on natural food. Development of natural food is sometimes enhanced by a single addition of superphosphate and ammonium sulphate, each at the rate of 60 kg ha^{-1} (Rothbard 1982). Density of fry, produced in pond spawnings and grown to 1 g, should not exceed 1 000 000 fish per ha. At this density, fish reach 1 g after 2–3 weeks. Faster growth may indicate a smaller number of fish, possibly due to early mortality. Stocking rates of larvae higher than optimal may cause starvation of the whole population.

The density of fry is estimated routinely by sampling in order to control fish size and to determine whether their growth is inhibited. In the case of overcrowding, as indicated by retarded growth, supplementary feed is added by hand-scattering of fish meal or crushed fish pellets into the pond. When the fish reach 0.5 g, their density is reduced by selective seining of the pond and transferring the larger fry to another pond for secondary nursing.

13.6.5 Secondary nursing

Initial numbers of common carp fry, transferred (at the size of 0.5–1 g) into the secondary nursing ponds, are controlled at the time of stocking. Usually, stocking rates do not exceed 150 000–300 000 fish per ha. Growth is estimated by weekly sample weighing to establish the daily ration of feed according to biomass. During secondary nursing, fish are nursed to 50–100 g and fed according to the feeding schedule presented in Table 13.8.

Secondary nursing of Chinese carps is carried out at higher initial densities, and the growth of fingerlings is reduced by their higher biomass when they reach 7–12 cm. Fingerlings are stocked in polyculture with other fish, or in lakes and reservoirs for controlling water quality.

Table 13.8 Feeding schedule for individual common carp fry in secondary nursing ponds (from Sarig and Marek, 1974)

Fish BW (g) weight range	Final biomass kg ha^{-1}* (No. of fish)	Food per fish (g per day)	Feeding rate (%)	Feeding kg ha^{-1} day^{-1}
1–3	1000–1500 (1 000 000)	0.4	20.0	400
3–5	1500–2500 (500 000)	0.6	20.0	300–400
5–10	2500–5000 (500 000)	1.0	15.0	500–700
10–20	~6000 (400 000)	1.5	10.0	700–800
20–40	~10 000 (200 000)	2.5	9.0	400–600
40–70	~10 000 (180 000)	4.0	8.0	700–800
70–100	~10 000 (150 000)	6.0	7.0	800

* The biomass is controlled by weekly weighings. The growth is controlled by feeding and/or by reduction of the population density in the nursery

13.6.6 Nursing of mixed-sized carp

This type of nursing by which different sizes of common carp are co-cultured was popular in Israel until recently and is still practised on some farms.

The method is based on the concept that every pond serves both as a nursery and as a grow-out pond. In mixed-size nursing, secondary nursing of the small fish starts while the larger fish are growing to market size (~1 kg) in the same pond. When the market-sized fish are harvested, the small fish are removed and restocked in the same pond a few days later. For example, in Gan Shmuel Fish Farm, 3500–4000 carps per ha of 50–60 g are stocked together with 3000 carps per ha of 250–300 g. Fish are stocked in March and harvested 4 months later. The small fish reach the size of 250 g and are restocked for continuation of growth, while the large fish, which have reached the size of ~900 g, are marketed.

The advantages of mixed-size nursing are (1) that the pond is managed more efficiently, since each group of fish utilizes different feeds from a different ecological niche, and (2) that the risk involved in nursing large numbers of fry in a single pond is reduced to a minimum. The method is, however, labour intensive, because it requires sorting of fish at various stages of growth.

13.7 Propagation and culture of ornamental cyprinids

The culture of ornamental cyprinids in Israel was initiated at the beginning of the

1980s. The concept was to utilize the climatic conditions as well as the existing fish production facilities and knowledge and to exploit the close proximity of the various European markets. The Japanese ornamental carp (*nishikigoi*), popularly known as koi, and several varieties of goldfish were selected for this purpose.

13.7.1 Koi

The culture of nishikigoi started in the Niigata Prefecture, Japan, in the nineteenth century and spread to Singapore, Taiwan and other parts of the Far East, America, Europe and Israel. The variation of colours and colour patterns in the koi is enormous, and may discourage investigations of their inheritance. The published information concerning colour inheritance is limited and restricted to a Russian investigation (Katasonov 1978) and a preliminary investigation recently performed in Israel (Wohlfarth & Rothbard 1991).

Breeding of koi is similar to breeding of other carps. Broodfish are selected according to their coloration and colour pattern. The multi-coloured koi are presumably not true-breeding fish, and most of the progeny consists of fish with colour patterns which are of little value. Hand-sorting of fish is required to cull the less valuable individuals. Much effort is invested in selecting the fish of desirable colours and, consequently, the value of the fish is proportional to the intensity of sorting and culling.

Koi spawners are maintained, sexes separated, in small ponds and are transported to the hatchery prior to spawning. Group spawnings are performed either in small ponds (0.2–0.5 ha) or in the hatchery, on mats or by induction with pituitary extract. Incubation of eggs is carried out in funnels similar to those used for common carp. In pair spawnings, attempts are made to pair the fish according to their coloration and pattern. Usually the fish are induced to spawn twice a year, the males being utilized more frequently.

Fry are nursed and grown in earthen ponds to marketable sizes (3–70 cm). During the growing cycle, koi are sorted several times and only the highly valued fish are retained for marketing.

13.7.2 Goldfish

The goldfish or golden crucian carp (*Carassius auratus*) is a traditional Chinese fish, cultured for centuries in China. Its orange-golden ancestor was discovered during the Northern Song Dynasty (960–1127 AD) in Hangzhou and Jaxing, Zheijang Province (Li Zhen 1988). In 1502, goldfish were introduced into Japan as pets and later spread over the world, to Britain in 1794, to Italy at the end of the eighteenth century (Melotti 1986), and to the United States in 1878. Large scale commercial culture of goldfish in Israel started in the beginning of the 1980s, together with the koi.

Three varieties of goldfish are cultured on a large scale in Israel, the common goldfish, the red comet and the blue shubunkin. The common goldfish is the original golden crucian carp in which the colours have been intensified through selective breeding. The red comet is a long-tail goldfish with orange to red colouration. Red-white coloured individuals (sarasa-comets) are bred through selection or hand-selected from the red comet population. The blue shubunkin has a comet-like shape and variegated coloration. It appears in combinations of blue-red-black colorations. When blue shubunkins are inbred they yield a population phenotypically segregated into wild-type, variegated and white (transparent) fish, with a clear Mendelian proportion of 1:2:1. Breeding transparent and wild-type fish results in a 100% variegated population. The heredity of colours and scale patterns of this variety was reported by Matsui (1934) and Vaughan (1986).

Male and female goldfish spawners are maintained in small ponds (200–1000 m^2) lined with plastic, or in concrete basins (30–100 m^3), to prevent the development of submerged vegetation and spontaneous spawning. Usually, 1 year old fish (>20 cm), are used as broodfish. Spawning occurs when spawning substrates (mats or pine tree branches) are placed in the pond. The substrates carrying the sticky eggs are removed the next day and transferred to a primary nursing pond, where the fry are nursed to 0.5–1.0 g. Since the broodfish continue to spawn for another 1–2 days, new substrates are placed and the cycle is repeated.

13.8 Future research priorities for cyprinids

13.8.1 Genetic improvements and ploidy manipulations

The common carp, although domesticated for many generations, is still amenable to further genetic improvement. Genetic research, such as that carried out in Dor-70, may yield superior lines of fish adapted to the local conditions in the respective countries.

Despite the popularity of using Chinese carps to control water quality problems in various parts of the world, such operations do carry important environmental risks. An attractive solution to circumvent such hazards is stocking water bodies with sterile triploid fish (Cassani & Caton 1986) or with monosex populations produced through gynogenesis and sex-inversion (Shelton 1986). Gynogenesis can also serve as a powerful tool for the production of inbred lines of ornamental cyprinids such as the koi, characterized by desired colour patterns that can serve as ancestors of koi clones (Rothbard 1991).

Additional approaches for the improvement of koi broodstock requires investigation of colour inheritance. Some steps in this direction have already been initiated (Wohlfarth & Rothbard 1991).

13.8.2 Maturation and spawning

Maturation of cyprinids is highly dependent on ambient temperatures. Even at high temperatures, Chinese carps, especially the black carp, mature at a late age, and keeping broodstock for many years is costly and risky. In addition, late maturation constrains genetic research in these fish. Shortening the time to first maturation needs intensive research on the endocrine mechanisms controlling puberty in fish.

13.8.3 Cryopreservation of sperm and cold storage of eggs and embryos

The research on methods for sperm cryopreservation is in an advanced stage and may shortly become a potentially useful tool in aquaculture management. Adaptation of techniques developed for storage of mammalian sperm, eggs and embryos to freshwater fish species (koi, black-carp, grass-carp etc.) is still required. Such investigations will allow breeding of fish where maturation of the sexes does not coincide, a problem often encountered in crossing related species. It may also allow year-round availability of gametes from fish with a restricted spawning season.

Cryopreservation of gametes or embryo may facilitate transport of genetic material and reduce the hazard of transferring parasites and pathogens together with the transported fish.

Last, but not the least, is the establishment of a 'sperm bank' of highly valued individual fish to preserve genetic material of known quality for the initiation of selective breeding programmes, and that of endangered species ('gene bank') for future renovation.

References

Acosta, B.O. & Pullin, R.S.V. (eds) (1991) Environmental Impact of the Golden Snail (*Pomacea* sp.) on Rice Farming System. *ICLARM Conference Proceedings*, **28**, 34.
Alderman, D.J. (1985) Malachite green: a review. *Journal of Fish Diseases*, **8**, 289–98.
Balon, E.K. (1974) *Domestication of the Carp, Cyprinus carpio L.* Royal Ontario Museum of Life Sciences, Miscellaneous Publication.
Bieniarz, K., Epler, P., Thuy, L.N. & Breton, B. (1980) Changes in blood gonadotropin level in mature female carp following hypophyseal homogenate injection. *Aquaculture*, **20**, 65–9.
Bieniarz, K., Weil, C., Epler, P., Mikolajczyk, T., Bougoussa, M. & Billard, R. (1991) Maturational gonadotropin hormone (GtH) and gonadotropin releasing hormone (GnRH) changes during growth and sexual maturation of female carp. In *Proceedings of the Fourth International Symposium on the Reproductive Physiology of Fish, Norwich*, (eds A.P. Scott, J.P. Sumpter, D.E. Kime, & M.S. Rolfe), p. 61.
Billard, R., Bieniarz, K., Popek, W., Epler, P., Breton, B. & Alagarswami, K. (1987) Stimulation of gonadotropin secretion and spermiation in carp by pimozide-LRH-A treatment: effect of dose and time of day. *Aquaculture*, **62**, 161–70.
Billard, R., Bieniarz, K., Popek, W., Epler, P. & Saad, A. (1989) Observations on a possible pheromonal stimulation of milt production in carp, *Cyprinus carpio* L. *Aquaculture*, **77**, 387–92.

Cassani, J.R. & Caton, W.E. (1986) Efficient production of triploid grass carp (*Ctenopharyngodon idella*) utilizing hydrostatic pressure. *Aquaculture*, **55**, 43–50.

Committee for the Collection of Experiences in the Culture of Fresh Water Fish Species in China (1981) *Science of the culture of fresh water fish species in China*. International Development Research Centre, Canada (microfiche IDRC-TS16E).

Davies, P.R., Hanyu, I., Furukawa, K. & Nomura, M. (1986) Effect of temperature and photoperiod on sexual maturation and spawning of the common carp III. Induction of spawning by manipulating photoperiod and temperature. *Aquaculture*, **52**, 137–144.

FAO (1990) *Aquaculture Production (1985–1988)*. Prepared by Fishery Information, Data Statistics Service, Fisheries Department. FAO Fisheries Circular No. 815. REV. 2. FAO, Rome.

Hepher, B. & Pruginin, Y. (1981) *Commercial Fish Farming (with Special Reference to Fish Culture in Israel)*. John Wiley & Sons, New York.

Hickling, C.F. (1971) *Fish Culture*. Faber & Faber, London.

Hornell, J. (1935) *Report on the Fisheries of Palestine*. Crown Agents for the Colonies, London.

Horvath, L. (1978) Relation between ovulation and water temperature by farmed cyprinids. *Aquacultura Hungarica*, **1**, 58–65.

Horvath, L. (1985) Egg development in the common carp (*Cyprinus carpio* L.). In *Recent Advances in Aquaculture*, Vol. 2 (eds J.F. Muir & R.J. Roberts), pp. 31–77. Westview Press, Colorado.

Horvath, L. (1986) Carp oogenesis and the environment. In *Aquaculture of Cyprinids*, (eds R. Billard & J. Marcel), pp. 109–17. INRA, Paris.

Hulata, G., Moav, R. & Wohlfarth, G. (1974) The relationship of gonad and egg size to weight and age in the European and Chinese races of the common carp *Cyprinus carpio* L. *Journal of Fish Biology*, **6**, 745–58.

Katasonov, V. Ya. (1978) Color in hybrids of common and ornamental (Japanese) carp III. Inheritance of blue and orange types. *Soviet Genetics*, **14**, 1522–8.

Kawauchi, H., Suzuki, K., Itoh, H., Swanson, P., Naito, N., Nagahama, Y., Nozaki, M., Nakai, Y. & Itoh, S. (1989) The duality of teleost gonadotropin. *Fish Physiology and Biochemistry*, **7**, 29–38.

Kime, D.E. & Dolben, I.P. (1985) Hormonal changes during induced ovulation in the carp, *Cyprinus carpio*. *General and Comparative Endocrinology*, **58**, 137–49.

Levavi-Zermonsky, B. & Yaron, Z. (1986) Changes in gonadotropin and ovarian steroids associated with oocyte maturation during spawning induction in the carp. *General and Comparative Endocrinology*, **62**, 89–98.

Leventer, H. (1979) *Biological Control of Reservoirs by Fish*. Mekoroth Water Co., Nazareth, Israel.

Leventer, H. (1987) *The Contribution of Silver Carp*, Hypophthalmichthys molitrix, *to the Biological Control of Reservoirs*. Mekoroth Water Co., Nazareth, Israel.

Lin, H.R., Kraak, G.V.D., Liang, J.-Y., Peng, C., Li, G.-Y., Lu, L.-Z., Zhou, X.-J., Chang, M.-L. & Peter, R.E. (1986) The effects of LHRH analogue and drugs which block the effects of dopamine on gonadotropin secretion and ovulation in fish cultured in China. In *Aquaculture of Cyprinids*, (eds R. Billard & J. Marcel), pp. 139–50. INRA, Paris.

Lin, H.R., van der Kraak, G., Zhou, X.-J., Liang, J.-Y., Peter, R.E., Rivier, J.E. & Vale, W.W. (1988) Effects of [D-Arg^6, Trp^7, Leu^8, Pro^9 NEt]-lutenizing hormone-releasing hormone (sGnRH-A), and [D-Ala^6, Pro^9 NEt]-luteinizing hormone-releasing hormone, LHRH-A), in combination with pimozide or domperidone, on gonadotropin release and ovulation in the Chinese loach and common carp. *General and Comparative Endocrinology*, **69**, 31–40.

Li Zhen (1988) *Chinese Goldfish*. Foreign Languages Press, Beijing, China.

Makeeva, A.P. (1969) Characteristics of embryonal and fry development of some pond Cyprinidae. In *Proceedings of a Conference on Genetic selection and Hybridization of Fish, Leningrad*, (ed. B.I. Czerfas), pp. 148–74. (Translated by the Israel Program for Scientific Translations, Jerusalem, 1972.

Matsui, Y. (1934) On the inheritance of the scale transparency of goldfish. *Journal of the Imperial Fisheries Institute (Tokyo)*, **30**, 47–66.

Melotti, P. (1986) Goldfish (*Carassius auratus* L.) farming in Italy. In *Aquaculture of Cyprinids*, (eds R. Billard & J. Marcel), pp. 369–76. INRA, Paris.

Moav, R., Hulata, G. & Wohlfarth, G. (1975) Genetic differences between the Chinese and

European races of the common carp.: I. Analysis of genotype-environment interaction for growth rate. *Heredity*, **34**, 323–40.
Pruginin, Y. (1967) The culture of carp and *Tilapia* hybrids in Uganda. *FAO Fisheries Report*, **44**(4), 223–9.
Ribeiro, L.P., Ahne, W. & Lichtenberg, V. (1983) Primary culture of normal pituitary cells of carp (*Cyprinus carpio*) for the study of gonadotropin release. *In Vitro*, **19**, 41–5.
Rothbard, S. (1981) Induced reproduction in cultivated cyprinids – The common carp and the group of Chinese carps: I. The technique of induction, spawning and hatching. *Bamidgeh*, **33**, 103–21.
Rothbard, S. (1982) Induced reproduction in cultivated cyprinids – The common carp and the group of Chinese carps: II. The rearing of larvae and the primary nursing of fry. *Bamidgeh*, **34**, 20–32.
Rothbard, S. (1991) Induction of endomitotic gynogenesis in the *nishiki-goi*, Japanese ornamental carp. *Israeli Journal of Aquaculture – Bamidgeh*, **43**, 145–55.
Santos, A.J.G., Furukawa, K., Kobayashi, M., Bando, K., Aida, K. & Hanyu, I. (1986) Plasma gonadotropin and steroid hormone profiles during ovulation in the carp *Cyprinus carpio Bulletin of the Japanese Society of Scientific Fisheries*, **52**, 1159–66.
Sarig, S. (1966) Synopsis of biological data on common carp *Cyprinus carpio* Linnaeus, 1758 (Near East and Europe). FAO *Fisheries Synopses*, **31**(2), (various pages in synopsis).
Sarig, S. (1970) Initial results of silver carp (*Hypophthalmichthys molitrix*) breeding in Israel fishponds in 1969. *Bamidgeh*, **22**, 95–100.
Sarig, S. (1971) *The Prevention and Treatment of Diseases of Warmwater Fishes under Subtropical Conditions, with Special Emphasis on Intensive Fish Farming*. TFH Publications, Neptune City, NJ.
Sarig, S. (1991) The fish culture industry in Israel in 1990. *Israeli Journal of Aquaculture – Bamidgeh*, **43**, 103–11.
Sarig, S. & Marek, M. (1974) Results of intensive and semi-intensive fish breeding techniques in Israel in 1971–1973. *Bamidgeh*, **26**, 28–48.
Shelton, W.L. (1986) Broodstock development for monosex production of grass carp. *Aquaculture*, **57**, 311–19.
Shelton, W.L. (1989) Management of finfish reproduction for aquaculture. *Reviews in Aquatic Sciences*, **1**, 497–535.
Shelton, W.L. (1990) Sex control in carps. *Symposium on Carp Genetics, Szarvas, Hungary*, (in press).
Sin, A.W. (1982) Stock improvement of common carp in Hong Kong through hybridization with introduced Israeli race 'Dor-70'. *Aquaculture*, **29**, 299–304.
Sokolowska, M., Mikolajczyk, T., Epler, P., Peter, R.E., Piotrowski, W. & Bieniarz, K. (1988) The effects of reserpine and LHRH or salmon GnRH analogues on gonadotropin release, ovulation and spermiation in common carp, *Cyprinus carpio* L. *Reproduction, Nutrition and Development*, **28**, 889–98.
Sorgeloos, P., Bossuyi, E., Baeza-Mesa, M. & Persoone, G. (1977) Decapsulation of *Artemia* cysts: a simple technique for the improvement of the use of brine shrimp in aquaculture. *Aquaculture*, **12**, 311–15.
Sparatu, P., Wohlfarth, G.W. & Hulata, G. (1983) Studies on the natural food of different fish species in intensively polyculture ponds. *Aquaculture*, **35**, 283–98.
Strumienski, O. (1573) *O sprawie sypaniu, wymierzaniu i rybieniu stawow, takze o przekopach, o wazeniu i prowadzeniu wody. Ksiazki wsystkim gospodarzom potrzebne.* W Krakowie, Lazarz Andrysowic drukowal (Reproduced by Institute Slaski, Opole 1987).
Suzuki, Y., Kawauchi, H. & Nagahama, Y. (1988a) Isolation and characterization of two distinct gonadotropins from chum salmon pituitary glands. *General and Comparative Endocrinology*, **71**, 292–301.
Suzuki, Y., Nagahama, Y. & Kawauchi, H. (1988b) Steroidogenic activity of two distinct gonadotropins. *General Comparative Endocrinology*, **71**, 452–8.
Vaughan, M.H. (1986) Fancy, oriental goldfish. *Carolina Tips*, **49**, 33–5.
Van Der Kraak, G., Suzuki, K., Peter, R.E., Hoh, H. & Kawauchi, H. (1992) *General and Comparative Endocrinology*, **85**, 217–29.
Weil, C., Fostier, A. & Billard, R. (1986) Induced spawning (ovulation and spermiation) in carp

and related species. In *Aquaculture of Cyprinids* (eds. R. Billard J. Marcel), pp. 119–37. INRA, Paris.

Welcomme, R.L. (compiler) 1988. International introductions of inland aquatic species. *FAO Fisheries Technical Paper*, **294**.

Wohlfarth, G. (1977a) Shoot carp, *Bamidgeh*, **29**: 35–56.

Wohlfarth, G.W. (1977b) Israel. In *World Fish Farming: Cultivation and Economics*, (ed E.E. Brown), pp. 359–70. The AVI Publishing Company Inc., Westport, Conn.

Wohlfarth, G.W. (1984) Common carp. In *Evolution of Domesticated Animals*, (ed I.L. Mason), pp. 375–80. Longman, London and New York.

Wohlfarth, G.W. & Rothbard, S. (1991) Preliminary investigations on color inheritance in Japanese ornamental carp (*nishiki-goi*). *Israeli Journal of Aquaculture – Bamidgeh*, **43**, 62–8.

Wohlfarth, G., Moav, R. & Hulata, G. (1975) Genetic differences between the Chinese and European races of the common carp: II. Multicharacter variation – a response to the diverse method of fish cultivation in Europe and China. *Heredity*, **34**, 341–50.

Wohlfarth, G.W., Lahman, M., Hulata, G. & Moav R. (1980) The story of 'Dor-70', a selected strain of the Israeli common carp. *Bamidgeh*, **32**, 3–5.

Woynarovich, E. & Woyharovich, A (1980) Modified technology for elimination of stickiness of common carp (*Cyprinus carpio*) eggs. *Aquacultura Hungarica*, **2**, 19–21.

Yaron, Z. & Levavi-Zermonsky, B. (1986) Fluctuations in gonadotropin and ovarian steroids during the annual cycle and spawning of the common carp. *Fish Physiology and Biochemistry*, **2**, 75–86.

Yaron, Z., Bogomolnaya A. & Levavi B. (1984) A calibrated carp pituitary extract as a spawning-inducing agent. In *Research in Aquaculture*, (eds H. Rosenthal & S. Sarig). *European Mariculture Society Special Publication*, **8**, 151–68.

Yaron, Z., Bogomolnaya, A. & Donaldson, E.M. (1985) The stimulation of estradiol secretion by the ovary of *Sarotherodon aureus* as a bioassay for fish gonadotropin. In *Current Trends in Comparative Endocrinology*, Vol. 1. (eds B. Lofts & W.N. Holmes), pp. 225–8. Hong Kong University Press, Hong Kong.

Zhao, W.-X., Jiang, R.-L., Huang, S.-J. & Zou, H.-Q. (1984) The annual cycle of gonadotropin (GTH) in the pituitary and blood serum of the carp (*Cyprinus carpio* L.). *General and Comparative Endocrinology*, **53**, 457–8 (Abstract).

Chapter 14
Origins and Functions of Egg Lipids: Nutritional Implications

14.1 Energetics of gonad formation
    14.1.1 Seasonal aspects
    14.1.2 Monounsaturated fatty acids as energy sources
14.2 Lipids and vitellogenesis
14.3 Lipid metabolism in early development
    14.3.1 Lipid composition of eggs
    14.3.2 Lipid utilization in fertilized eggs
14.4 Essential fatty acids in early development
    14.4.1 Conversion of C18 to C20 and C22 polyunsaturated fatty acids
    14.4.2 Membrane functions of polyunsaturated fatty acids
    14.4.3 Polyunsaturated fatty acids as precursors of eicosanoids
14.5 Conclusions
Acknowledgements
References

Fatty acids are mobilized from the neutral lipid reserves of fish adipose tissue during gonadogenesis and transferred via the serum to the liver where they are assembled into the egg-specific lipoprotein, vitellogenin. Up to 60% of the free fatty acids mobilized, preferentially saturated and monounsaturated fatty acids, can be catabolized to provide metabolic energy for egg lipoprotein biosynthesis. The remainder, preferentially (n-3) polyunsaturated fatty acids (PUFA) and especially 22:6(n-3), are incorporated into the phospholipid-rich vitellogenin which is transferred via the serum to the developing eggs. The major egg phospholipid is invariably phosphatidylcholine, and eggs with short and long incubation times have low and high levels of triacylglycerols respectively.

Phospholipids and triacylglycerols in eggs have levels of (n-3)PUFA of circa 50% and 30% respectively, composed principally of 22:6(n-3) and 20:5(n-3) in a ratio of circa 2:1. The fatty acids of both phospholipids and triacylglycerols, including their (n-3)PUFA, are catabolized to provide metabolic energy for the developing egg and early larva, but the chief role of (n-3)PUFA is in the formation of cellular membranes. Because of the unusual richness of 22:6(n-3) in neural cell membranes this fatty acid has a critical role in the formation of the brain and the eyes, which constitute a large fraction of the embryonic and larval body mass. The small quantities of 20:4(n-6) in fish eggs are located almost exclusively in phosphatidylinositol and a specific role for this fatty acid in eicosanoid formation is indicated. From considerations of juvenile fish and analyses of fish eggs, an optimal level of (n-3)/(n-6)PUFA in broodstock diets and in fish eggs of

5:1–10:1 is indicated. The importance of using high quality marine fish oils in broodstock nutrition for successful normal embryonic and early larval development is stressed.

14.1 Energetics of gonad formation

14.1.1 Seasonal aspects

The fundamental importance of lipids in animal growth, namely their roles as sources of metabolic energy and sources of essential materials for the formation of cell and tissue membranes, is clearly seen in fish. This is because lipid rather than protein or carbohydrate is the favoured source of metabolic energy in most species of fish, and also because fish lipids are generally much richer in long chain polyunsaturated fatty acids (PUFA) of the (n-3) series, i.e. eicosapentaenoic acid or 20:5(n-3) and docosahexaenoic acid or 22:6(n-3), than most other animal lipids. As a consequence the majority of fish species can contain, at least at certain stages of their life cycles, high levels of lipids rich in (n-3)PUFA, i.e. oils.

Seasonal variations in the levels of lipids in fish are related fundamentally to the reproductive cycle, it being the norm in the natural environment that fish accumulate large lipid reserves during the period spring–late summer when food is plentiful prior to their developing gonads during late winter–early spring. The majority of marine fish spawn in spring, and so the fertilized eggs hatch when there is a plentiful supply of food available from the plankton blooms of spring–early summer. Consequently, marine fish generally contain their highest and lowest levels of lipid in early winter and early spring respectively (see e.g. Lovern & Wood 1937 for herring). Normally, food availability during winter is generally low, so that some of the lipid accumulated during spring–summer will be used for metabolic energy for maintenance purposes when fish overwinter.

However, such maintenance costs, whether they be associated with short-term swimming activity, osmoregulation or tissue turnover, are likely to be low compared with the metabolic costs of growth. An exception to this is when long distance migrations associated with spawning take place, when the metabolic costs of sustained swimming can be high. Nonetheless, growth requires a particularly large investment of metabolic energy in the formation of gonads and their resulting gametes, which is fundamentally a specialized form of tissue growth often occurring in the wild in the absence of external food, and is a major energy drain in fish as in all animals. It is the norm, therefore, that mature marine fish enter the winter rich in lipid and emerge in the following spring post-spawning with their lipid stores severely depleted.

The foregoing generalization is well illustrated by studies in both capelin (*Mallotus villosus*) and herring (*Clupea harengus*) in the wild. Capelin can routinely accumulate between 10% and 20% of their wet body weight as oil, stored largely in perivisceral and subcutaneous adipose tissue in the form of triacylglycerols. In excess of 70% of the oil in the muscle fillet, which includes the subcutaneous adipose tissue, of the female capelin is mobilized during gonadal development with some 40% of the mobilized lipid being accounted for by lipid deposited in the roe (Table 14.1) (Henderson *et al*. 1984). Therefore 60% of the lipid mobilized from female muscle fillet is catabolized to provide metabolic energy reflecting the high energy cost of producing the female gonad. The male capelin mobilizes the same proportion of its fillet oil as the female but deposits very small amounts of the mobilized oil in its milt, reflecting the much higher investment of metabolic energy in physical activity in the male compared with the female (Henderson *et al*. 1984).

Table 14.1 Mobilization of lipid from muscle into gonad of male and female capelin, *Mallotus villosus*, during gonadogenesis*

	Males	Females
Lipid initially present in muscle (mg)	442	382
Lipid finally present in muscle (mg)	108	88
Lipid mobilized from muscle (mg)	334	294
Lipid deposited in gonad (mg)	0	112
Lipid catabolized from mobilized lipid (mg)	334	182
% initial muscle lipid mobilized	76	77
% of mobilized lipid deposited in gonad	0	38
% of mobilized lipid catabolized	100	62

* Data are for a standardized fish of 10 g wet weight (from Henderson *et al*. 1984)

14.1.2 Monounsaturated fatty acids as energy sources

The importance of lipid as metabolic energy sources in gonad formation in female fish is clearly seen when individual fatty acids are considered. Zooplankton-consuming fish in northern latitudes, for example herring and capelin, accumulate their body triacylglycerol depots very largely from the large wax ester depots of their zooplankton diet, so that the fish triacylglycerols are rich in 20:1(n-9) and 22:1(n-11) fatty acids derived from the corresponding 20:1(n-9) and 22:1(n-11) fatty alcohols in their dietary wax esters (Sargent *et al*. 1979, Sargent & Henderson 1986). The abundance of 20:1(n-9) and 22:1(n-11) fatty acids is a characteristic of virtually all commercial fish oils produced from northern Atlantic fisheries which are currently used extensively in aquaculture. Therefore, farmed fish fed such fish oils (and indeed the accompanying fish meals derived from the same fisheries) are themselves rich in these fatty acids.

However, because of their relatively long chains, 20:1(n-9) and 22:1(n-11) fatty acids cannot be easily accommodated in tissue phospholipids and are instead confined almost exclusively to triacylglycerols. It is noteworthy, therefore, that negligible amounts of the 20:1 and 22:1 fatty acids abundant in the large amounts of body oil triacylglycerols mobilized during roe formation in both capelin (Henderson et al. 1984) and herring (Henderson & Almatar 1989) are deposited in the eggs of these species, whose lipids, like those of many marine fish eggs, are composed principally of phospholipid. Thus, the 20:1 and 22:1 fatty acids are largely catabolized to provide the metabolic energy necessary for the formation of gonad and eggs.

The converse holds for long chain (n-3)PUFA in that, because the phospholipids that constitute the major lipids of fish eggs are richer in (n-3)PUFA than the triacylglycerols that are virtually the only lipid in reserve adipose tissue, there is a preferential transfer of (n-3)PUFA from the body adipose reserves to the eggs (Henderson et al. 1984, Henderson & Almatar 1989). An interesting outcome of the foregoing is that the lipid mobilized from the muscle fillets of both male and female fish is replaced largely by water. This stems directly from the stoichiometry of fatty acid catabolism namely:

$$C_{16}H_{32}O_2 + 23O_2 \rightarrow 16CO_2 + 16H_2O + 137ATP,$$

That is, 256 g of fatty acid (palmitic acid, 16:0) generates 288 g of water, or one volume of fat with a specific gravity of circa 0.90 generates circa one volume of water. This inverse relationship between oil and water in fish flesh ensures that the total volume change of the fish is minimized during spawning and it accounts for the poor condition of spawned-out fish. In the present context it emphasizes the energy demand of reproduction in fish and the extent to which it is met from lipid catabolism.

Therefore, the first requisite for successful broodstock nutrition is to provide a sufficient lipid in the diet to meet the metabolic energy requirements of reproduction. It is unlikely that the fatty acid composition of the lipid required for energy production during gonadogenesis is unduly demanding. However, in cold water fish including the salmonids, it is met largely from a blend of medium chain saturated and long chain monounsaturated fatty acids, principally 16:0, 18:1(n-9), 20:1(n-9) and 22:1(n-11). This is easily satisfied by commercial fish oils currently available in north-west Europe. It is less easy to be categorical about the amount of dietary lipid required to meet the energy requirements of reproduction, since this will depend on the species and, in particular, on the amount of lipid present in its eggs. A theoretical estimation might be about two to three times the amount of lipid that is present in the eggs produced in the optimal situation. This can probably be met in practice by feeding broodstock diets containing 10–20% of their dry weight as fish oil.

14.2 Lipids and vitellogenesis

It has long been known that injection of the hormone oestradiol into sexually immature fish, both male and female, induces the appearance in serum of vitellogenin, the egg-specific lipoprotein (Plack et al. 1971). The mechanistic basis of this phenomenon is:

(1) the activation by oestradiol of a hormone-sensitive lipoprotein lipase in adipose tissue;
(2) hydrolysis within the adipocytes of their triacylglycerols by lipoprotein lipase to generate free fatty acids;
(3) export of free fatty acids from the adipocytes to the serum where they are transported, adsorbed on to serum albumin, to the liver;
(4) uptake of free fatty acids by the liver, where they are incorporated principally into phospholipids and to a lesser extent into triacylglycerols; and
(5) association of the newly synthesized lipids with egg specific apoproteins, themselves biosynthesized in the rough endoplasmic reticulum of the liver from amino acids mobilized to a large extent from muscle protein, to form the vitellogenin that is exported to the eggs via the serum (Fig. 14.1).

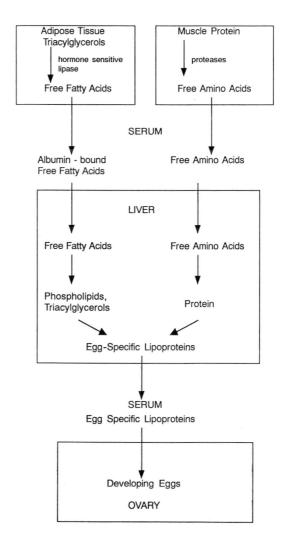

Fig. 14.1 Tissue relationships in the biosynthesis of egg-specific lipoproteins.

The quantitative importance of this major transformation of amino acids and fatty acids, derived principally from adipose tissue reserves and muscle, into egg lipids and proteins is seen by the marked but transient enlargement of the liver mass, e.g. 3–5 fold in the capelin (Henderson et al. 1984), that occurs in the early stages of oogenesis in fish (Fig. 14.2). This reflects a marked proliferation in the endoplasmic reticulum of the liver where the biosynthesis of vitellogenin occurs. Large amounts of metabolic energy are required for the formation of the expanded endoplasmic reticulum itself and more so for the biosynthetic reactions it performs.

Vitellogenin has been demonstrated in the serum of numerous teleost fish. It is a specialized very high density serum lipoprotein (VHDL) consisting of approximately 80% protein and 20% lipid. Phospholipids and triacylglycerols account for about two-thirds and one-third of the lipids respectively, and because of its abundance of phospholipids, the lipid of vitellogenin is rich in (n-3)PUFA, especially 22:6(n-3). During egg formation vitellogenin is taken up by developing oocytes by a process of pinocytosis and is cleaved in the egg to generate the egg yolk proteins, lipovitellin and phosvitin. Therefore the bulk of the (n-3)PUFA-rich phospholipids in eggs is located in lipovitellin. In addition to the foregoing, it is probable that eggs rich in triacylglycerols as well as phospholipids take up lipid-rich serum lipoproteins, e.g. very low density

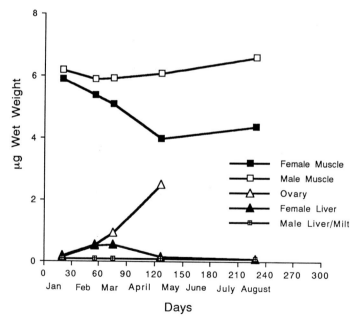

Fig. 14.2 Changes in the wet weights of muscle, liver and gonad of female and male capelin, *Mallotus villosus*, during gonadogenesis and spawning. Gonads begin to develop in January and fish spawn in early May. Changes in the weights of male liver and gonad are negligible when plotted with changes in weights of female liver and gonad. (From the data of Henderson et al. 1984.)

lipoproteins (VLDL) whose high lipid content is mainly due to triacylglycerols.

The foregoing processes have now been confirmed in a large number of fish species. Some recent studies include those for sea trout (Copeland & Thomas 1988), catfish (Smith et al. 1988), winter flounder (Nagler & Idler 1990), eel (Petersen & Korsgaard 1989), common carp (Carnevali & Belvedere 1991), stickleback, sea bass and roach (Covens et al. 1987) and rainbow trout (Babin 1987). A general account of lipid metabolism in teleost fish is contained in the review by Henderson & Tocher (1987) and a comprehensive review of plasma lipoproteins including vitellogenin in fish is that by Babin & Vernier (1989).

14.3 Lipid metabolism in early development

14.3.1 Lipid composition of eggs

Lipids are major sources of metabolic energy throughout embryonic development in fish (Terner 1979, Boulekbache 1981) and, in line with this, the amount of lipid in eggs generally correlates with the time interval between spawning and egg hatching or larval first feed (Blaxter 1969, Kaitaranta & Ackman 1981). Thus, freshwater spawners such as the salmonids generally shed relatively large eggs with large lipid reserves containing substantial amounts of triacylglycerols as well as phospholipids, and with long incubation periods up to 20 weeks. They generate large larvae that are easy to feed with defined diets. In contrast, many marine fish generally shed small eggs with relatively modest reserves of lipid composed principally of phospholipids, and with short incubation periods of around 20 days. They generate small larvae that are very difficult to feed with defined diets.

Typical of the foregoing are the eggs of Atlantic salmon, *Salmo salar*, which contain approximately 30% of their dry weight as lipid, 48% and 44% of which consists of triacylglycerols and phospholipids respectively (Cowey et al. 1985). The eggs of a typical marine gadoid, the cod *Gadus morhua*, contain 13% of their dry weight as lipid, 13% and 72% of which consists of triacylglycerols and phospholipids respectively (Tocher & Sargent 1984).

The major saturated and monounsaturated fatty acids in the triacylglycerols of fertilized salmon eggs are 16:0 (11.4%) and 18:1(n-9) (31%) respectively; the same is true for fertilized cod eggs where 16:0 and 18:1(n-9) account for 7.5% and 18% respectively of the total fatty acids in the triacylglycerols. These differences reflect the higher content of (n-3)PUFA in the triacylglycerols of cod eggs, where 22:6(n-3) and 20:5(n-3) account for 16% and 11% of the total fatty acids respectively. In salmon eggs the corresponding figures for 22:6(n-3) and 20:5(n-3) in triacylglycerols are 11.4% and 6.0% respectively. The fatty acids of the phospholipids of the eggs of salmon and cod contain 42% and 46% respectively as (n-3)PUFA, and in both cases the ratio of 22:6(n-3) to 20:5(n-3) is approximately 2:1 (Cowey et al. 1985, Tocher & Sargent 1984). A summary of lipid and fatty acid compositions of marine fish eggs is given in Table 14.2.

Table 14.2 Lipid and fatty acid compositions of fish eggs*

	Cod	Haddock	Herring	Capelin
Egg diameter (mm)	1.35	1.30	1.35	1.10
Moisture content (%)	74	86	74	70
Lipid content (% dry weight)	13.2	10.7	14.6	26.3
Polar lipid (% total lipid)	71.7	71.3	69.0	50.7
Neutral lipid (% total lipid)	28.3	28.7	31.0	49.3
Phosphatidyl choline (% total lipid)	45.6	45.8	57.6	37.7
Triacylglycerol (% total lipid)	12.5	8.3	14.8	30.4
Cholesterol (% total lipid)	6.1	9.5	8.3	3.1
Fatty acids in polar lipid (% total)				
saturates	28.1	25.5	32.7	27.4
monounsaturates	20.3	20.4	14.5	18.1
20:4(n-6)	1.9	3.7	1.0	1.1
20:5(n-3)	15.3	12.6	13.7	19.0
22:6(n-3)	28.6	27.6	31.4	24.6
Fatty acids in triacylglycerols (% total)				
saturates	21.3	20.4	30.6	24.9
monounsaturates	41.5	32.9	33.5	35.9
20:4(n-6)	1.2	2.8	0.6	0.5
20:5(n-3)	10.9	14.0	9.7	13.1
22:6(n-3)	16.0	14.5	17.1	14.1

* Data assembled from Tocher & Sargent (1984)

14.3.2 Lipid utilization in fertilized eggs

The above data illustrate the quantitative importance of (n-3)PUFA, especially 22:6(n-3), and of both triacylglycerols and phospholipids in embryonic and subsequent early larval development in fish. *A priori*, it might be expected that triacylglycerols would be largely consumed as a source of metabolic energy in early development and that phospholipids with their high concentrations of essential (n-3)PUFA would be conserved for the formation of new tissue in the growing embryo and early larva.

This simple prediction, however, is not borne out in practice. Triacylglycerols are indeed consumed as a metabolic energy source during embryonic development, as is clearly seen in the detailed study of Cowey *et al.* (1985), who showed that triacylglycerols in salmon eggs decreased from 3.4 mg per ontogenetic unit at fertilization to 2.7 mg in eyed eggs, 2.4 mg in yolk sac fry and 1.4 mg in swim-up fry. However, phosphatidylcholine (lecithin), which is by far the major phospholipid in salmon eggs, decreased from 2.9 mg in fertilized eggs to 2.5 mg in eyed eggs, 2.0 mg in yolk sac fry and 1.1 mg in swim-up fry. Thus, the net consumption of phosphatidylcholine (1.8 mg) is effectively the same as the net consumption of triacylglycerols (2.0 mg).

The situation is even more striking in developing herring eggs, where the marked preferential utilization of phosphatidylcholine, the major lipid component of the egg by far, is reflected by its decreasing from 62% of the total lipid in newly shed, fertilized eggs to 40% at hatch and 34% at 15 days post hatch (Tocher *et al.* 1985a). Over the same period the percentage of triacylglycerols in total egg lipid actually increases from 14% at fertilization to 23% at hatch and then falls to 13% at 15 days post-hatch. An analogous situation, i.e. preferential utilization of egg phosphatidylcholine over egg triacylglycerols, occurs during egg development in the cod (Fig. 14.3) (Fraser *et al.* 1988) and also in the halibut, *Hippoglossus hippoglossus*, where phosphatidylcholine is again the major egg lipid (Falk-Petersen *et al.* 1986, 1989).

Two aspects of the foregoing deserve consideration. First, the preferential utilization of phosphatidylcholine in fish egg development, seen especially in marine species, implies a relatively large catabolism of (n-3)PUFA. Detailed analyses by Tocher *et al.* (1985b) established that the modest amounts of triacylglycerols in herring eggs become progressively enriched with (n-3)PUFA as embryogenesis proceeds, consistent with a preferential retention of (n-3)PUFA mobilized from phosphatidylcholine in triacylglycerols. However, this sparing mechanism is unlikely to conserve fully the (n-3)PUFA mobilized during the consumption of phosphatidylcholine during egg development and it must be deduced that substantial amounts of (n-3)PUFA are catabolized to provide metabolic energy during embryogenesis. It can equally be deduced, however, that saturated and monounsaturated fatty acids are quantitatively more important than (n-3)PUFA in providing metabolic energy for egg development.

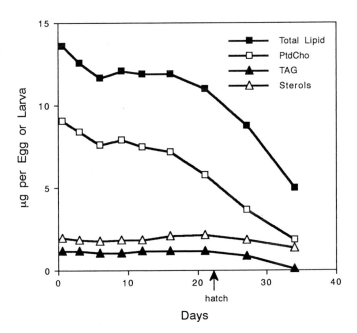

Fig. 14.3 Lipid utilization by fertilized eggs and early larvae of cod, *Gadus morhua*. (From the data of Fraser *et al.* 1988.) PtdCho, phosphatidylcholine; TAG, triacylglycerols; sterols, cholesterol + cholesteryl esters.

Second, given that the developing egg is a closed system, the mobilization of phosphatidylcholine will generate elevated internal levels of both inorganic phosphate and choline. Presumably the inorganic phosphate adds to the phosphate reserves of the eggs located, *inter alia*, in phosvitin and, to that extent, phosphatidylcholine may be regarded as contributing to the phosphate stores in eggs. The fate of the choline mobilized from phosphatidylcholine is less clear, especially given the relatively large quantities involved. It has been speculated (Falk-Petersen *et al.* 1989) that the choline may be metabolically oxidized to glycine betaine, which may serve to osmoregulate the water space produced from the oxidation of fatty acids to generate metabolic energy during embryogenesis. There is, however, no experimental evidence for this speculation and the fate of the choline mobilized from phosphatidylcholine remains unresolved.

This issue has considerable significance for broodstock nutrition because it underlines the importance of phosphatidylcholine and particularly choline biosyntheses in the maternal fish during oogenesis. This is likely to link in turn to methionine metabolism, because of the importance of this amino acid as a source of methyl groups for choline biosynthesis, and underlines the central role vitellogenin plays in linking lipid and protein nutrition in broodstock.

14.4 Essential fatty acids in early development

The preferential transfer of (n-3)PUFA from the adipose reserves of broodstock fish into egg lipids, particularly phosphatidylcholine, has already been referred to. This process underlines the importance of these fatty acids as essential constituents of the phospholipids of cell membranes, and it is vital that the egg contains a sufficient supply of these essential nutrients to allow growth and development of the embryo and subsequent larva from fertilization until first feeding. Any deficiency in the essential fatty acid content of the egg will obviously have profound consequences for embryonic and larval development.

14.4.1 Conversions of C18 to C20 and C22 polyunsaturated fatty acids

Much research has been conducted on the essential fatty acid requirements of freshwater and marine fish. From the available evidence (reviewed by Sargent *et al.* 1990, 1993) it can be stated that:

(1) Fish have an absolute requirement for (n-3)PUFA and almost certainly also for (n-6)PUFA.
(2) Freshwater fish are generally capable of converting the C18 PUFA 18:3(n-3) and 18:2(n-6) to their higher homologues 22:6(n-3) and 20:4(n-6) respectively.
(3) Those marine fish so far studied are incapable of these conversions and require the performed end products 20:4(n-6) and 22:6(n-3) (Fig. 14.4).

However, the differences between freshwater and marine fish in their ability

(n-6) SERIES

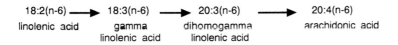

(n-3) SERIES

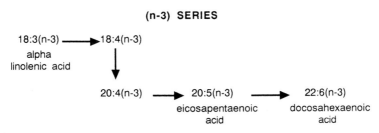

Fig. 14.4 Pathways of conversion of C18 (n-6) and (n-3) polyunsaturated fatty acids to their higher C20 and C22 derivatives.

to convert C18 PUFA to C20 and C22 homologues is more apparent than real in that such differences can be more readily accounted for by the natural feeds of the fish. Thus, much of the evidence for marine fish is based on detailed studies of the turbot, *Scophthalmus maximus*, an extreme carnivorous piscivore, whereas the majority of the freshwater species studied are more omnivorous in their feeding habits and include insectivores and herbivores. In general, the natural diets of marine fish are smaller marine fish which invariably contain a luxus of 22:6(n-3), whereas many of the insect and plant diets of freshwater fish contain predominantly 18:3(n-3) (Sargent *et al.* 1989).

In line with the above notion, recent studies in this laboratory strongly suggest that the pike, *Esox lucius*, an extreme freshwater carnivore, cannot convert 18:3(n-3) to 22:6(n-3) at significant rates so that the species almost certainly has an absolute requirement for 22:6(n-3) (R.J. Henderson, unpublished data). Emphasis on differences in feeding habit rather than environmental salinity as the major determinant of essential fatty acid requirements of fish, and indeed of animals generally, also emphasizes the extent to which the natural diets and therefore the PUFA intake of fish can change both qualitatively and quantitatively throughout development from first feeding larvae to mature adults. For example, rapidly growing salmon fry in fresh water will require large amounts of (n-3)PUFA for rapid tissue growth and will rapidly convert 18:3(n-3) ingested from ingested aquatic insects into 22:6(n-3), as well as avidly utilizing 22:6(n-3) ingested in other constituents of their diet. A late juvenile salmon in sea water with a lesser growth rate will require smaller quantities of (n-3) and has a ready dietary source of 22:6(n-3) in its piscine diet, obviating the requirement to convert any dietary 18:3(n-3) to 22:6(n-3).

Therefore, essential fatty acid requirements determined experimentally on juvenile and maturing fish cannot be reliably extrapolated to early developmental forms, above all to deduce quantitative requirements. Seen from this standpoint, it is notable that the major (n-3)PUFA of all fish eggs are 22:6(n-3) and 20:5(n-3) in a ratio of approximately 2:1, so that the potential problem of converting C18 PUFA to their C20 and C22 homologues has been prudently avoided in the egg. However, what constitutes optimal levels of 22:6(n-3) and 20:5(n-3), far less 20:4(n-6), in fish eggs, is not known and, in the continuing absence of direct experimental evidence, deductions can be made only by considering the amounts and the functions of essential fatty acids in the developing embryo and larvae up to first feeding.

14.4.2 Membrane functions of polyunsaturated fatty acids

As the (n-6)PUFA, principally 20:4(n-6), have a generalized role in maintaining the structural and functional integrity of cell membranes in higher vertebrates, so the (n-3)PUFA, principally 22:6(n-3), have a generalized role in maintaining the structural and functional integrity in cell membranes in fish. This is illustrated by the high levels of molecular species of phosphoglyceride rich in 22:6(n-3) in fish tissues (Table 14.3). In addition, 22:6(n-3), has a specific and important structural role in neural cell membranes, i.e. the brain and the eyes, of all vertebrates including fish. Some 40% of the dry matter of fish brain is lipid, some 10% of which is phosphatidylethanolamine (cephalin) (Tocher & Harvie 1988, Mourente et al. 1991). Moreover, 22:6(n-3) accounts for some 40% of the total fatty acids in phosphatidylethanolamine from-fish brain with the single molecular species di-22:6(n-3) phosphatidylethanolamine accounting for some 15% of all of the molecular species of this phospholipid in trout brain (Bell & Tocher 1989, Bell & Dick 1991a).

In trout retina the di-22:6(n-3) molecular species accounts for over 40% of the molecular species of phosphatidylethanolamine (Bell & Tocher 1989) and

Table 14.3 Molecular species composition (mol %) of phosphatidylcholine (PC) and phosphatidylethanolamine (PE) in tissues of the cod, *Gadus morhua**

	Muscle		Liver		Brain		Retina		Roe	
	PC	PE	PC	PE	PC	PE	PC	PE	PC	PE
22:6/22:6	6	20	2	4	1	14	29	72	1	2
20:5/22:6	8	10	2	4	0	1	1	2	1	1
16:0/20:5	18	4	16	8	2	2	3	1	15	9
16:0/22:6	22	13	25	14	18	19	23	6	31	15
18:1/20:5	4	5	6	9	0	3	1	1	6	13
18:1/22:6	4	13	6	25	2	15	2	6	15	26
16:0/18:1	6	0	6	0	17	3	13	0	5	1
18:1/24:1	0	0	0	0	13	0	0	0	0	0

* Data from Bell & Dick (1991a) for muscle, liver, brain and retina and from Bell (1989) for roe

the corresponding value in the cod retina is over 70% (Table 14.3) (Bell & Dick, 1991a). Neural tissue, especially the eyes, can form a relatively large proportion of the total body mass of embryonic and larval fish, so that a relatively large fraction of the total 22:6(n-3) required for embryonic and larval growth is directed towards the formation of cell membranes vital for normal development and functioning of the visual and neural systems of fish larvae. Therefore, any deficiency in 22:6(n-3) during embryonic and larval development will have serious consequences for the successful performance of sophisticated predatory larvae, particularly during and immediately following first feeding.

At the present time there is no direct evidence of the possibility that a deficient or even sub-optimal broodstock nutrition could perturb levels of 22:6(n-3) in fish eggs to the point of influencing predatory behaviour in the emergent larvae. However, in a recent study (Navarro & Sargent 1992) it has been noted that separate and distinct behavioural patterns can develop in a single batch of herring larvae reared from eggs under conditions where the larvae ingested very limited amounts of the live food offered. One group of these effectively starving larvae developed shiny eyes and an abnormal swimming behaviour, another developed shiny eyes and lay listlessly on the bottom of the tank, and a third group behaved normally. The behaviour of the abnormal groups is characteristic of essential fatty acid deficiency in fish (Castell *et al.* 1972, Watanabe *et al.* 1980, Navarro *et al.* 1988).

One interpretation of these findings is that the normal group resisted starvation more effectively than the two abnormal groups, a situation that could arise from differences in the starting levels of essential nutrients in different eggs within the batch. Since the single batch of eggs hatched in this experiment was derived from more than one female (and more than one male) the possibility exists that different individual females may produce eggs with different nutrient levels.

The concerns that arise from the above findings and interpretations should be counterbalanced by the probability that the development of neural and visual systems is maintained at all costs during embryonic and larval development, and that abnormalities in these systems are likely to occur only under conditions of severe nutritional deprivation. More troublesome is the concern that sub-normalities can develop in these systems through subnormal embryonic nutrition and that such subnormalities persist in later life. Visual subnormalities from sub-optimal provision of 22:6(n-3) during early development have recently been established in higher primates and man (Neuringer *et al.* 1986, Uauy *et al.* 1990) and associated mental subnormalities persist in later life (Lucas *et al.* 1992).

14.4.3 Polyunsaturated fatty acids as precursors of eicosanoids

The second important function of essential fatty acids is their role as precursors of eicosanoids. It is well established in mammals that arachidonic acid, 20:4(n-6), is the major precursor of the eicosanoids, and it is generally held that these short-lived, highly biologically active molecules are produced in trace amounts only. It is now also accepted that (n-3)PUFA, principally 20:5(n-3), have a physiological role in mammals in modulating the formation of eicosanoids from 20:4(n-6) by competing with the enzyme systems converting 20:4(n-6) to eicosanoids (see Fig. 14.5) (Anon. 1992). It is well established that biologically active eicosanoids in fish are formed principally from 20:4(n-6) and, on this ground alone, it can be concluded that (n-6)PUFA are essential nutrients for fish (Sargent et al. 1993).

Such reasoning is strongly supported by the preferential location of the generally very small amounts of 20:4(n-6) in fish phospholipids in phosphatidylinositol (Table 14.4) (Bell & Dick 1991b), a lipid class with a specialized and major cell membrane role in signal transduction. Thus phosphatidylcholine, which accounts for some 46% of the total lipid in cod eggs, has as its major molecular species 16:0/22:6 (31%), 18:1/22:6 (15%) and 16:0/20:5 (15%), and broadly the same holds for phosphatidylethanolamine which accounts for some 20% of the total lipid in the egg (Bell 1989, Tocher & Sargent 1984) (Table 14.3).

In contrast, phosphatidylinositol, which accounts for only 3% of the total lipid in cod eggs, has as its major molecular species 18:0/20:4 (37%) and 18:1/20:4 (17%) (Table 14.4) (Bell 1989, Tocher & Sargent 1984). Such data point

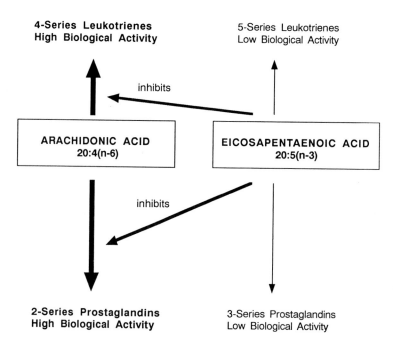

Fig. 14.5 Interactions between arachidonic acid and eicosapentaenoic acid in the production of eicosanoids (prostaglandins and leukotrienes)

Table 14.4 Molecular species composition (mol %) of phosphatidylinositol from tissues of the cod, *Gadus morhua**

	Muscle	Liver	Brain	Retina	Roe
16:0/22:6	5	2	5	26	6
18:0/20:5	13	5	41	17	8
18:0/22:6	40	2	8	18	5
18:0/20:4	11	49	9	24	37
18:1/20:5	3	3	14	2	5
18:1/22:6	10	2	4	3	9
18:1/20:4	4	18	3	3	17

* Data from Bell & Dick (1991b) for muscle, liver, brain and retina and from Bell (1989) for roe

strongly to the essentiality of 20:4(n-6) for embryonic and larval development, with a specialized role in eicosanoid production.

Because of the modulating effect of (n-3)PUFA in eicosanoid production from 20:4(n-6) in mammals, there has been much debate in recent years as to what constitutes an optimal dietary ratio of (n-6)/(n-3)PUFA, especially in human diets. A recent independent report advises a desirable ratio of (n-6)/(n-3)PUFA of 5:1 for the UK national diet (Anon. 1992). This issue is relevant to aquaculture because it is not unknown for commercial fish feeds to contain substantial amounts of 18:2(n-6) derived from vegetable oils, as well as fish oils rich in 22:6(n-3) and 20:5(n-3).

Research in the NERC Unit of Aquatic Biochemistry, Stirling, has established that heavy supplementation of salmon diets with vegetable oils rich in (n-6)PUFA, under conditions when there is a sufficiency of (n-3)PUFA in the diet, can lead to severe cardiac disorders in salmon (Table 14.5) (Bell *et al.* 1990). These disorders became apparent only when the fish were stressed by transporting and they correlated with increased eicosanoid production from 20:4(n-6) in the fish (Bell *et al.* 1990, 1992). Eicosanoids are, of course,

Table 14.5 Incidence of cardiac muscle necrosis in salmon post-smolts fed diets containing fish oil or sunflower oil*

Weeks on diet	Fish oil diet			Sunflower oil diet		
	Total fish	Normal fish	Necrosed fish	Total fish	Normal fish	Necrosed fish
16	5	5	0	5	1	4
18	6	6	0	10	1	9
20	10	9	1	9	5	4
Total	21	20	1	24	7	17

* Data of Bell *et al.* (1990)

produced physiologically in response to a range of external stressors, and it is logical to anticipate that diets with an over high ratio of (n-6)/(n-3)PUFA exaggerate normal stress responses, in fish no less than in humans. Therefore the undesirability of over-supplementing fish diets with vegetable oils rich in 18:2(n-6) is self evident, particularly in broodstock since the eicosanoid prostaglandin E3 is a physiological inducer of egg shedding in ripe fish.

There is now accumulating evidence in man that many of the cardiac pathologies appearing later in life and associated with acknowledged risk factors, including an over-high dietary ratio of (n-6)/(n-3)PUFA, originate *in utero* where sublethal pathologies are triggered by the presence of a risk factor during pregnancy (Barker et al. 1989, 1990). This is to say that the embryo is vulnerable to an over-high maternal ratio of dietary (n-6)/(n-3)PUFA during pregnancy. Although there are no experimental studies on fish in this area, all of our current knowledge of fish lipid nutritional biochemistry leads us to believe that an excess of (n-6)PUFA in broodstock fish will be readily transferred to the egg and result in an elevated production of eicosanoids during subsequent embryogenesis and larval development. This is likely to increase the susceptibility of the resulting larvae to external stressors.

The undoubted involvement of essential fatty acids in eicosanoid production both in broodstock fish and developing eggs and larvae and their fundamental involvement in stress reactions, demands consideration of what constitutes an optimal or even desirable dietary ratio of (n-3)/(n-6)PUFA in broodstock and their resulting fish. Unfortunately there is little or no direct experimental evidence on this in mature fish, far less in developing embryos and larvae. At present we can only venture the advice that 'natural is best' in which case we advocate a dietary ratio of (n-3)/(n-6)PUFA in the region of 5:1–10:1 as an ideal for marine fish, including broodstock. We would be concerned with a ratio much less than 5:1 and we would studiously avoid a ratio of (n-3)/(n-6)PUFA of less than 1:1.

As important, in the absence of definitive evidence to the contrary, we strongly advocate use of only the end-products of PUFA biosynthesis, i.e. 22:6(n-3), 20:5(n-3) and 20:4(n-6), in diets and avoidance as far as possible of substantial amounts of their precursors, 18:2(n-6) and 18:3(n-3), irrespective of whether the fish species in question is capable of elongating C18 PUFA to their C20 and C22 homologues or not. This is because the pathways of elongating and further desaturating C18 PUFA to C20 and C22 PUFA are highly complex, to the point that it remains virtually impossible to predict with confidence for any animal species the blend of C20 and C22 end-products that will be generated from a given blend of C18 precursors (Sargent et al. 1992).

An important aspect of the recommended use of C20 and C22 PUFA is that it is prudent to use 22:6(n-3) and 20:5(n-3) in the ratio of about 2:1 found in most fish eggs. This is because of ongoing uncertainty of the extent to which 20:5(n-3) can be converted to 22:6(n-3) by fish and the concern that excess 20:5(n-3) may over-modulate eicosanoid production from 20:4(n-6).

14.5 Conclusions

The importance and apparent complexities of lipid nutritional biochemistry revealed in some of the foregoing should not obscure the fact that our present knowledge is consistent with the view that lipid nutritional problems need not commonly arise in practical broodstock management. It is important to recognize the dual but inseparable roles of lipids in animal metabolism, namely their role as a source of metabolic energy and their role as a source of essential components for cellular membranes.

An alternative expression of this duality is that lipid nutrition is fundamentally a matter of balance between fatty acids. This review has highlighted the importance of the dietary ratio of (n-3)/(n-6)PUFA in lipid nutrition, which, although important, is but one aspect of the overall ratio of saturated fatty acids: monounsaturated fatty acids: (n-6)PUFA: (n-3)PUFA. The theoretical complexities in lipid nutrition largely arise from the fact that changes in the absolute amounts of any one of these families of fatty acids inevitably alter the relative dietary contributions of the others. Thus, a diet very rich in saturated fatty acids runs the risk of being partly deficient in essential fatty acids, whereas a diet containing a sufficiency but no more of essential fatty acids is likely to be deficient in energy.

Moreover, competitive metabolic interactions can occur between the four families of fatty acids, particularly in the case of chain elongation and further desaturation reactions when competitions exist between monounsaturated fatty acids, (n-6)PUFA and (n-3)PUFA, and possibly also between saturated fatty acids, for the same limited number of enzymes. This last consideration makes it prudent to use, wherever possible, end-products of fatty acid biosynthesis in practical nutrition.

Fortunately, the theoretical problems of fish lipid nutrition become largely academic in practice, especially in broodstock nutrition, where, in the light of our present understanding, currently available commercial fish oils closely approach what may be regarded as a very satisfactory balance of the four fatty acid families. This conclusion ought to be self-evident on the grounds that commercial fish oils are largely derived from late summer−early winter fish that will shortly become active in gonadogenesis, i.e. the oils are effectively derived from wild broodstock fish in the first place.

On the basis of the above it is concluded that, given sensible and judicious selection of commercial fish oils, broodstock lipid nutrition *per se* should not be a serious cause of excessively poor egg and larval quality, certainly up to the stage of first feeding larvae. One cannot conclude, however, that problems with egg and larval quality can be entirely divorced from broodstock lipid nutrition. It is stressed in particular that interactions of broodstock diet and other environmental factors, including disease organisms and environmental pollutants, with genetic variations in individual broodstock remains largely unexplored. Of these, the genetic variations are likely to have the biggest and perhaps the most subtle impact. Solutions to the problems of defining and controlling egg and larval quality may continue to be elusive until basic knowledge of developmental fish biology is substantially improved.

Acknowledgements

I readily acknowledge my colleagues in the Lipid Group of NERC Unit of Aquatic Biochemistry, Michael Bell, Gordon Bell, Jim Henderson and Douglas Tocher, for their generous help in preparing this text, for their major contributions to many of the original findings referred to, and their ongoing stimulus in developing many of the views expressed.

References

Anon. (1992) *Unsaturated Fatty Acids. Nutritional and Physiological Significance. The Report of the British Nutrition Foundation's Task Force.* Chapman & Hall, London.

Babin, J. (1987) Apolipoproteins and the association of egg yolk proteins with plasma high density lipoproteins after ovulation and follicular atresia in the rainbow trout (*Salmo gairdneri*). *Journal of Biological Chemistry*, **262**, 4290−6.

Babin, P.J. & Vernier, J.-M. (1989) Plasma lipoproteins in fish. *Journal of Lipid Research*, **30**, 467−89.

Barker, D.J.P., Winter, P.D., Osmond, C., Margetts, B. & Simmonds, S.J. (1989) Weight in infancy and death from ischaemic heart disease. *Lancet*, **ii**, 577−80.

Barker, D.J.P., Bull, A.R., Osmond, C. & Simmonds, S.J. (1990) Fetal and placental size and risk of hypertension in adult life. *British Medical Journal*, **301**, 259−62.

Bell, M.V. (1989) Molecular species analysis of phosphoglycerides from the ripe roes of cod (*Gadus morhua*). *Lipids*, **24**, 585−8.

Bell, M.V. & Dick, J. (1991a) Molecular species composition of the major diacyl glycerophospholipids from muscle, liver, retina and brain of cod (*Gadus morhua*). *Lipids*, **26**, 565−73.

Bell, M.V. & Dick, J. (1991b) Molecular species composition of phosphatidylinositol from the brain, retina, liver and muscle of cod (*Gadus morhua*). *Lipids*, **25**, 691−4.

Bell, M.V. & Tocher, D.R. (1989) Molecular species composition of the major phospholipids in brain and retina from rainbow trout (*Salmo gairdneri*). *Biochemical Journal*, **264**, 909−15.

Bell, J.G., McVicar, A.H., Park, M.T. & Sargent, J.R. (1990) Effects of high dietary linoleic acid on fatty acid compositions of individual phospholipids from tissues of Atlantic salmon (*Salmo salar*): association with a novel cardiac lesion, *Journal of Nutrition*, **121**, 1163−72.

Bell, J.G., Sargent, J.R. & Raynard, R.S. (1992) Effects of increasing dietary linoleic acid on phospholipid fatty acid composition and eicosanoid production in leucocytes and gill cells of Atlantic salmon (*Salmo salar*). *Prostaglandins, Leukotrienes and Essential Fatty Acids*, **45**, 197−206.

Blaxter, J.H.S. (1969) Development: eggs and larvae. In *Fish Physiology*, Vol. 3 (eds W.S. Hoar & D.J. Randall), pp. 177−252. Academic Press, New York.

Boulekbache, H. (1981) Energy metabolism in fish development. *American Zoologist*, **12**, 377−89.

Carnevali, O. & Belvedere, P. (1991) Comparative studies of fish, amphibian and reptilian vitellogenins. *Journal of Experimental Zoology*, **259**, 18−25.

Castell, J.D., Sinnhuber, R.O., Wales, J.H. & Lee, D.J. (1972) Essential fatty acids in the diet of rainbow trout (*Salmo gairdneri*): growth, feed conversion and some gross deficiency symptoms. *Journal of Nutrition*, **102**, 77−86.

Copeland, P.A. & Thomas, P. (1988) The measurement of plasma vitellogenin levels in a marine teleost, the spotted sea trout (*Cynoscion nebulosus*) by homologous radioimmunoassay. *Comparative Biochemistry and Physiology*, **91B**, 17−24.

Covens, M., Covens, L., Ollevier, F. & DeLoof, A. (1987) A comparative study of some properties of vitellogenin (Vg) and yolk proteins in a number of freshwater and marine teleost

fishes. *Comparative Biochemistry and Physiology*, **88B**, 75–80.
Cowey, C.B., Bell, J.G., Knox, D., Fraser, A. & Youngson, A. (1985) Lipids and antioxidant systems in developing eggs of salmon (*Salmo salar*). *Lipids*, **20**, 567–72.
Falk-Petersen, S., Falk-Petersen, I.-B., Sargent, J.R. & Haug, T. (1986) Lipid class and fatty acid composition of eggs from the Atlantic halibut (*Hippoglossus hippoglossus*). *Aquaculture*, **52**, 207–11.
Falk-Petersen, S., Sargent, J.R., Fox, C., Falk-Petersen, I.-B., Haug, T. & Kjorsvik, E. (1989) Lipids in Atlantic halibut (*Hippoglossus hippoglossus*) eggs from planktonic samples in northern Norway. *Marine Biology*, **101**, 553–6.
Fraser, A.J., Gamble, J.C. & Sargent, J.R. (1988) Changes in lipid content, lipid class composition and fatty acid composition of developing eggs and unfed larvae of cod (*Gadus morhua*). *Marine Biology*, **99**, 307–13.
Henderson, R.J. & Almatar, S.M. (1989) Seasonal changes in the lipid composition of herring (*Clupea harengus*) in relation to gonad maturation. *Journal of the Marine Biological Association of the UK*, **69**, 323–34.
Henderson, R.J. & Tocher, D.R. (1987) The lipid composition and biochemistry of freshwater fish. *Progress in Lipid Research*, **26**, 281–347.
Henderson, R.J., Sargent, J.R. & Hopkins, C.C.E. (1984) Changes in the content and fatty acid composition of lipid in an isolated population of the capelin *Mallotus villosus* during sexual maturation and spawning. *Marine Biology*, **78**, 255–63.
Kaitaranta, J.K. & Ackman, R.G. (1981) Total lipids and lipid classes of fish roe. *Comparative Biochemistry and Physiology*, **69B**, 725–9.
Lovern, J.A. & Wood, H. (1937) Variations in the chemical composition of herring. *Journal of the Marine Biological Association of the UK*, **22**, 281–93.
Lucas, A., Morley, R., Cole, T.J., Lister, G. & Leeson-Payne, C. (1992) Breast milk and subsequent intelligence quotient in children born preterm. *Lancet*, **339**, 261–4.
Mourente, G., Tocher, D.R. & Sargent, J.R. (1991) Specific accumulation of docosahexaenoic acid (22:6n-3) in brain lipids during development of juvenile turbot *Scophthalmus maximus* L. *Lipids*, **26**, 871–7.
Nagler, J.J. & Idler, D.R. (1990) Ovarian uptake of vitellogenin and another very high density lipoprotein in winter flounder (*Pseudopleuronectes americanus*) and their relationship with yolk proteins. *Biochemistry and Cell Biology*, **68**, 330–5.
Navarro, J.C. & Sargent, J.R. (1992) Behavioural differences in starving herring *Clupea harengus* L. larvae correlate with body levels of essential fatty acids. *Journal of Fish Biology*, **41**, 509–13.
Navarro, J.C., Hontoria, F., Varo, I. & Amat, F. (1988) Effect of alternate feeding a poor long chain polyunsaturated fatty acid *Artemia* strain and a rich one on sea bass (*Dicentrarchus labrax*) and prawn (*Penaeus kerathurus*) larvae. *Aquaculture*, **74**, 307–17.
Neuringer, M., Connor, W.E., Lin, D.S., Barstad, L. & Luck, S. (1986) Biochemical and functional effects of prenatal and postnatal omega-3 fatty acid deficiency on retina and brain in rhesus monkeys. *Proceedings of the National Academy of Science of the USA*, **83**, 4021–5.
Petersen, I. & Korsgaard, B. (1989) Experimental induction of vitellogenin synthesis in eel (*Anguilla anguilla*) adapted to sea-water or freshwater. *Comparative Biochemistry and Physiology*, **93B**, 57–60.
Plack, P.A., Pritchard, D.J. & Fraser, N.W. (1971) Egg proteins in cod serum. Natural occurrence and induction by injections of oestradiol 3-benzoate. *Biochemical Journal*, **121**, 847–56.
Sargent, J.R. & Henderson, R.J. (1986) Lipids. In *The Biological Chemistry of Marine Copepods*, (eds E.D.S. Corner & S.C.M. O'Hara), pp. 59–108. Clarendon Press, Oxford.
Sargent, J.R., McIntosh, R.M., Bauermeister, A.E.M. & Blaxter, J.H.S. (1979) Assimilation of the wax esters of marine zooplankton by herring (*Clupea harengus*) and rainbow trout (*Salmo gairdneri*). *Marine Biology*, **51**, 203–7.
Sargent, J.R., Henderson, R.J. & Tocher, D.R. (1989) The lipids. In *Fish Nutrition*, (ed. J. Halver), 2nd edn, pp. 153–218. Academic Press, New York.
Sargent, J.R., Bell, M.V., Henderson, R.J. & Tocher, D.R. (1990) Polyunsaturated fatty acids in marine and terrestrial food webs. In *Comparative Physiology*, Vol. 5, *Animal Nutrition and Transport Processes: Nutrition in Wild and Domestic Animals*, (eds J. Mellinger, J.P. Truchot & B. Lahlou), pp. 11–23 Karger, Basel.

Sargent, J.R., Bell, J.G., Bell, M.V., Henderson, R.J. & Tocher, D.J. (1993) The metabolism of phospholipids and polyunsaturated fatty acids in fish. In *Aquaculture: Fundamental and Applied Research*, (eds B. Lahlou & P. Vitiello). *Coastal and Estuarine Studies*, **43**, 103–24. American Geophysical Union, Washington DC.

Smith, M.A.K., McKay, M.C. & Lee, R.F. (1988) Catfish plasma lipoproteins: *in vivo* studies of apoprotein synthesis and catabolism. *Journal of Experimental Zoology*, **246**, 223–35.

Terner, C. (1979) Metabolism and energy conversion during early development. In *Fish Physiology*, Vol. 8 (eds W.S. Hoar, D.J. Randall & J.R. Brett), pp. 261–78. Academic Press, New York.

Tocher, D.R. & Harvie, D.G. (1988) Fatty acid compositions of the major phosphoglycerides from fish neural tissues; (n-3) and (n-6) polyunsaturated fatty acids in rainbow trout (*Salmo gairdneri*) and cod (*Gadus morhua*) brains and retinas. *Fish Physiology and Biochemistry*, **5**, 229–39.

Tocher, D.R. & Sargent, J.R. (1984) Analyses of lipids and fatty acids in ripe roes of some northwest European marine fish. *Lipids*, **19**, 492–9.

Tocher, D.R., Fraser, A.J., Sargent, J.R. & Gamble, J.C. (1985a) Lipid class composition during embryonic and early larval development in Atlantic herring (*Clupea harengus* L.). *Lipids*, **20**, 84–9.

Tocher, D.R., Fraser, A.J., Sargent, J.R. & Gamble, J.C. (1985b) Fatty acid composition of phospholipids and neutral lipids during embryonic and early larval development in Atlantic herring (*Clupea harengus* L.). *Lipids*, **20**, 69–74.

Uauy, R.D., Birch, D.G., Birch, E.E., Tyson, J.E. & Hoffman, D.R. (1990) Effect of dietary omega-3 fatty acids on retinal function of very-low-birth-weight neonates. *Pediatric Research*, **28**, 485–92.

Watanabe, T., Oowa, F., Kitakima, C. & Fujita, S. (1980) Relationship between dietary value of brine shrimp *Artemia salina* and their content of omega-3 highly unsaturated fatty acids. *Bulletin of the Japanese Society of Scientific Fisheries*, **46**, 35–41.

Chapter 15
Larval Foods

15.1 Introduction
15.2 Algal culture
15.3 Rotifer culture
15.4 *Artemia* culture
    15.4.1 Optimization of the nutritional status
    15.4.2 Use of stress tests for larval quality assessment
    15.4.3 Use of on-grown *Artemia* of specific size
15.5 Artificial diets
15.6 Disease control through live food
15.7 Future prospects
    References

Larval ontogeny in fish is characterized by important anatomical and physiological phases with consequent changes in nutrient requirements. Depending on the size of the larva and the specialization of its digestive system at the initiation of exogenous feeding, live food remains an essential requirement for many marine and some freshwater fish species. This paper reviews the latest developments in the selection, production and/or use of the most commonly used live feeds in larviculture, i.e. micro-algae, rotifers, and brine shrimp.

Species-specific dietary requirements are met by application of live food enrichment techniques with selected nutrients. It is very likely that broodstock nutrition and conditioning might influence the nutritional requirements of the larvae, especially at start feeding. Co-feeding and/or early weaning with formulated feeds should allow hatcheries to gradually reduce the live food requirements. Fish larval quality, expressed as its resistance against stress conditions, can also be influenced by the dietary regimes applied in the hatchery.

15.1 Introduction

One of the main reasons for the improvements in success of fish and crustacean culture in the 1970s and 1980s was the development of reliable techniques for the mass-production of quality fry and fingerlings. This breakthrough was achieved by domestication of species, mainly involving appropriate techniques for broodstock management and controlled reproduction in captivity, zootechnical management, production of adequate (live) food, and disease control (Sorgeloos & Léger 1992).

Closing the life cycle was relatively easy for a few selected species such as salmonids, mainly because their hatched larvae carry a big yolk sac with enough food reserves for the first weeks of their development. After this period the fingerlings are already sufficiently developed and of a size to accept formulated feeds readily.

However, most marine fish with aquaculture potential have very limited yolk reserves and at start feeding still have small mouths and primitive digestive systems. Nutrition at commercial production level, especially at the early larval stages, is far more critical here, and live food production techniques and feeding strategies needed to be developed before industrial scale production can grow. In shrimp larvae, food size is not the only problem: larvae moult into different stages, eventually changing from herbivorous filter-feeders to carnivorous hunters.

Over the past two decades adequate feeding strategies have been developed for marine fish and crustacean larvae, resulting in the worldwide use of various species of microalgae (approximate size 2–20 μm) the rotifer *Brachionus plicatilis* (50–200 μm) and the brine shrimp *Artemia* (420–8000 μm).

At the 'Larvi '91 – Fish and Crustacean Larviculture Symposium' in Gent, Belgium (see Lavens *et al*. 1991), it was stated that, although standard practices are established for several species around the world, further optimization is required in the commercial production of fry or post-larvae in order to guarantee predictable production outputs, improve its cost-effectiveness and to realize a commercially-feasible production of all potential aquaculture candidates.

In these respects special attention still needs to be paid to improving feeding strategies, egg quality and hygienic conditions. The goal of this paper is not to review what has been achieved in the 1980s, for example live food HUFA-enrichment (see review of Léger *et al*. 1986, Bengtson *et al*. 1991, Sorgeloos & Léger 1992, Sorgeloos *et al*. 1993) but to report on new developments and prospects with regard to the selection, production and use of the most commonly used live food, as well as artificial diets.

15.2 Algal culture

Algal culture for fish and shellfish hatcheries is generally recognized as being very labour intensive and expensive. Producing monocultures of selected species of microalgae requires stock cultures of several species kept under sterile conditions, and batch cultures of increasing size (flasks, then carboys, and finally plastic bags of 100 l content or even tanks up to a size of a few cubic metres) using micron-filtered and UV-treated water. Furthermore, problems still exist with respect to contamination as well as consistent nutritional quality, e.g. batch variations in the (n-3) highly unsaturated fatty acid (HUFA) content (Olsen 1989), which affects the consistency of hatchery-outputs.

This provided the rationale for investigations into alternatives or supplements to live algae. Today, various approaches and formulations are being applied on experimental and industrial scales with varying success. Examples of such substitute products are freeze-dried heterotrophically grown microalgae (Laing & Verdugo 1991), manipulated yeasts (Coutteau *et al*. 1990a), and various microparticulate and encapsulated diets (Jones *et al.* 1979, Langdon & Waldock 1981, Kanazawa *et al.* 1982). Algal replacement diets for rotifer culture are discussed in the next section.

In shrimp larviculture, algal substitutes were successfully introduced in the mid-1980s (Léger *et al.* 1985) and are now in routine use. Bivalve hatcheries are still the largest consumers of algae. Although full substitution has not yet been accomplished, there is experimental proof of successful supplementation up to 80% and 50% of the algae for the Manila clam (*Tapes semidecussata*) and the hard clam (*Mercenaria mercenaria*), respectively (Coutteau *et al.* 1990b, 1991) (Fig. 15.1). Further optimization is expected, especially with respect to the fulfilment of the nutritional requirements of the bivalve spat.

Fish larvae do not filter-feed but are carnivorous hunters, as a result of which microalgae do not constitute a major food source at the start of feeding. Nevertheless, in many commercial operations, algae are introduced in the larval rearing tanks at concentrations of 50 000−150 000 cells per ml. Several attempts have been made to explain possible beneficial effects of this green water technique. In some cases, e.g. cod (Meeren 1991) and halibut (Reitan *et al.* 1991), it has been shown that the larvae take up substantial numbers of microalgae during the first days after hatching ('green stomachs') which may support the idea that they are used as a direct food source at the start of feeding. Fortuitous ingestion, furthermore, may be a source of micro-nutrients, which are not available through the administered rotifers or brine shrimp larvae. Also, they may supply exogenous enzymes which could assist in the digestion of zooplankton food ingested by the fish larvae, which at the start of feeding have only a primitive digestive system. Indirectly they may stimulate enzymatic synthesis (Hjelmeland *et al.* 1988) and onset of feeding (Naas *et al.*

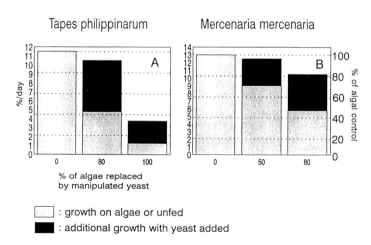

Fig. 15.1 Comparison of the daily growth rate of *Tapes philippinarum* (A) and *Mercenaria mercenaria* (B) cultured on algae and replacement diets. (Adapted from Coutteau *et al.* 1990b, 1991.)

1992). In addition, recent findings at SINTEF, Trondheim, Norway (Y. Olsen, pers. comm.) reveal that algal polysaccharides may act as non-specified immunological stimulants in the larval fish and in this way may contribute to more stable patterns of production.

Another effect may be created by conditioning the dietary quality of the rotifers fed to the fish culture tanks. Rotifers that are not immediately taken up by the fish larvae may maintain their optimal nutritional status by feeding on the microalgae in the tank. Microalgae may further act as water quality conditioners by stripping nitrogenous substances as well as by suppressing the development of harmful bacteria.

In Europe this green water technique is used in several sea bass and bream hatcheries and in most production systems for turbot fry. At the Laboratory of Aquaculture & Artemia Reference Center, a standard clear-water culture procedure in 100 l tanks has been successfully developed for turbot, yielding survival rates of 46% at day 13 (P. Dhert et al., pers. comm.). However, when the culture tanks were illuminated with halogen lamps (wavelength 450–750 nm) instead of fluorescent tubes (wave length 450–630 nm) lower survival rates were observed (28% ± 5% against 41% ± 8% at day 22). The increased stress of fluorescent lights noted in turbot larvae (i.e. black coloured larvae with higher sensitivity for salinity stress) could be correlated with sub-optimal feeding regimes. The fact that this light effect has never been reported in other turbot rearings (all green water technique) may be related to the light shading effect when microalgae are added to the culture tanks.

15.3 Rotifer culture

Bottlenecks in the optimal use of rotifers, which are especially administered as a starter diet in marine fish larviculture (Fukusho 1989), are mainly related to reliable and cost-effective techniques for continuous mass production.

A recent break-through in production technology has been the development of an artificial diet (Culture Selco, Artemia Systems NV, Belgium) which completely replaces algae and at the same time eliminates the need of an extra enrichment period for enhancement of the rotifers' dietary value. This dry product needs to be suspended in water prior to feeding. Provided it is continuously aerated and cold-stored, the food suspension of Culture Selco can be used in automatic feeding for as long as 48 h. A standard feeding protocol has been developed (Table 15.1) and tested on several rotifer strains in 100 l tanks (Fig. 15.2).

Under these feeding conditions a doubling time for different rotifer strains is generally obtained every 3 days. Under optimal conditions, i.e. strain-adapted temperatures, twice-a-week water renewals, ciliate removal and culture under shaded conditions, doubling of the population may even be expected every 24 h (Lavens et al. 1994).

Although various rearing tanks for mass production of rotifers can be used,

Table 15.1 Feeding protocol for rotifer culture using Culture Selco (CS)

Rotifer density per ml	Administered CS in g per 10^6 rotifers per day
100–150	0.80
150–200	0.70
200–250	0.60
250–300	0.55
300–350	0.50
350–400	0.45
400–450	0.40
450–500	0.35

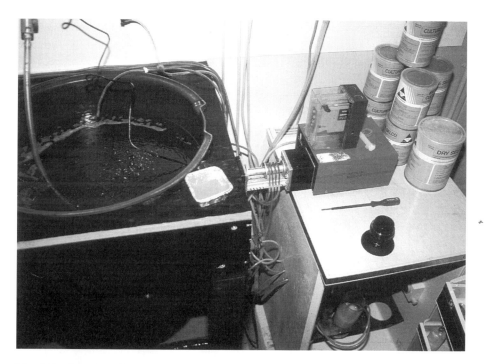

Fig. 15.2 Overview of the procedure for rotifer production using an artificial diet and regular automated feedings from cold-stored feed suspensions.

optimal food conversion ratios are achieved only in tanks with a smooth surface and a cylindro-conical shape, which ensures minimal sedimentation. Since this type of tank shape is not always available, slight alterations to the proposed feeding scheme may be required. Excellent results in using this diet as an alternative to the traditional mixture of algae and yeast were also obtained on a commercial scale in different bream and bass hatcheries, which reported two to three times higher growth rates; there were also much improved production outputs, with initial inoculation densities of 200 rotifers per ml in four 800 l tanks producing in total 23×10^9 rotifers over a two month rearing period and allowing an average daily harvest of 50% (Komis et al. 1991).

Not only are the 'off the shelf' availability and predictable rotifer outputs important features of this artificial diet, possibly more importantly they allow constant and predictable levels of (n-3) HUFA enrichment which cannot be reached by any other microalgal enrichment procedure (Fig. 15.3). Rotifers grown on the Culture Selco replacement diet have constant levels of eicosapentaenoic acid (EPA) and docosahexaenoic acid (DHA) of $6\,\text{mg}\,\text{g}^{-1}\,\text{DW}$ and $4\,\text{mg}\,\text{g}^{-1}\,\text{DW}$, respectively. Recent investigations on gilthead seabream larviculture (Mourente et al. 1993) demonstrated that during first feeding the best growth rate was achieved with a diet of Culture Selco-reared rotifers, which combined a high (n-3) HUFA content with a high DHA:EPA ratio.

Study of the economics of rotifer production in operational hatcheries reveals that the exclusive use of artificial diets in rotifer culture reduces the unit production costs of rotifers by more than 60% (Fig. 15.4).

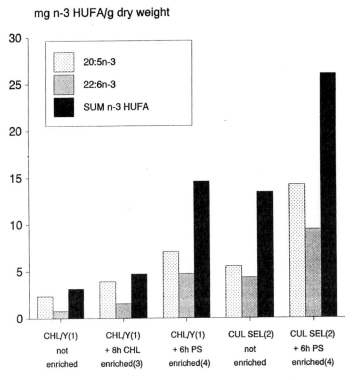

Fig. 15.3 Enrichment levels in rotifers following different culture techniques. (1) rotifers cultured for 5 days on *Chlorella* sp. (CHL) + yeast (Y); (2) rotifers cultured for 5 days on Culture Selco (CUL SEL) only; (3) harvested rotifers enriched with *Chlorella* sp. for 8 h; (4) harvested rotifers enriched with Protein Selco (PS) for 6 h.

Larval Foods 379

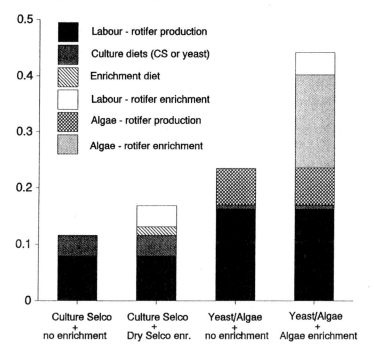

Fig. 15.4 Rotifer production cost for different culture techniques.

15.4 Artemia culture

Of all the live feeds used in larviculture, brine shrimp *Artemia* constitutes the most widely used genus. For this reason considerable progress has been made in the past decade in characterizing and improving its value as a larval diet; this includes selective use of specific strains, hatching quality control, optimal hatching and disinfection procedures, cold storage of freshly-hatched nauplii and enrichment procedures for essential fatty acid (EFA) incorporation (see review of Sorgeloos and Léger 1992). The most recent developments involve a further improvement of the nutritional composition of *Artemia* through enrichment with special components (DHA, phospholipids, vitamin C) in order to fulfil the nutritional requirements of the predator, and the use of *Artemia* preparations other than the nauplii (e.g. juveniles and decapsulated cysts).

15.4.1 Optimization of the nutritional status

Docosahexaenoic Acid (DHA)

The essential nature of the fatty acid docosahexaenoic acid 22:6n-3 (DHA) in (marine) aquaculture is now without question (Watanabe 1991, Sargent *et al.*

1993). As a result, considerable efforts have been made to incorporate high levels of DHA and high ratios of DHA/EPA (20:5n-3, eicosapentaenoic fatty acid) in live foods. So far the best results have been obtained with the self-emulsifying product High DHA-Super Selco (Artemia Systems, Belgium). Compared with the results obtained with traditional products like Super Selco, the boosting of freshly-hatched *Artemia* with this product under standard enrichment practices results in a 45% increase of DHA and a DHA/EPA ratio of 1.60 instead of 0.75 (Figs 15.5, 15.6).

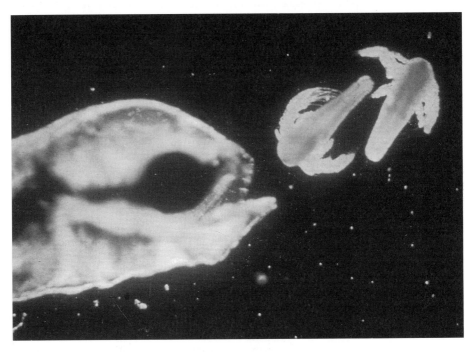

Fig. 15.5 Feeding one-day-old *Artemia* nauplii enriched with HUFA has proved to be essential for the commercial production of marine fish larvae (here Asian sea bass).

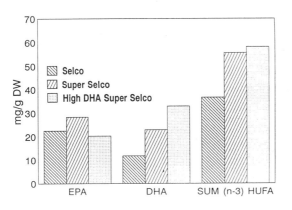

Fig. 15.6 (n-3) HUFA levels in *Artemia* after a 24h enrichment period using different self-emulsifying concentrates.

The necessity of incorporating DHA in the diet of Japanese flounder and red sea bream has recently been proven (Devresse *et al.* 1992). Fish from the control treatment received unenriched rotifers and *Artemia*; the fish in the DHA enrichment treatment were fed rotifers and *Artemia* boosted with High DHA-Super Selco. The use of DHA resulted in much higher survival rates than in the controls; it also improved resistance to stress conditions (Table 15.2). Identical experiments conducted on red sea bream were even more conclusive. In this experiment the growth on day 38 of DHA-fed larvae was 50% better than those in the control treatment.

DHA might also be involved in the pigmentation of turbot and other flatfishes. In this respect, the preliminary results of Devresse *et al.* (1992) suggest that the concentration of the DHA component might be of less importance than the EPA/DHA ratio (Fig. 15.7). A typical recirculation system used for turbot larviculture experiments is shown in Fig. 15.8.

On turbot, recent investigations on isolated cells have demonstrated that the conversion from EPA to DHA is very slow (Sargent *et al.* 1993). In this respect direct supplementation with DHA might be beneficial for turbot larvae. However, methods involving the dosage and boosting with DHA during the early

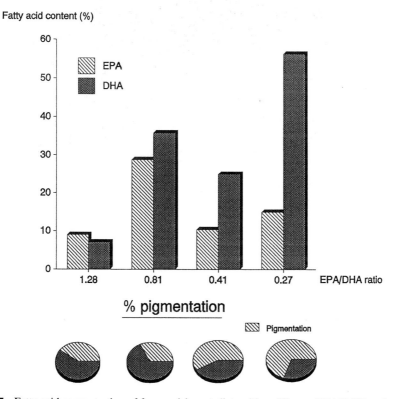

Fig. 15.7 Fatty acid concentration of four enrichment diets with a different EPA/DHA ratio and their effect on turbot pigmentation. (Adapted from Devresse *et al.* 1992.)

Table 15.2 Results of growth, survival, and stress resistance of the Japanese flounder *Paralichthys olivaceus* on day 50 fed either unenriched rotifers and *Artemia* (= control) or high DHA Super Selco enriched live food (= DHA) (from Devresse *et al.* 1992)

	Control	DHA
Growth (length in mm)	19.1	28.7
Survival (%)	1.8	21.5
Vitality test (survival %)	40.0	93.0

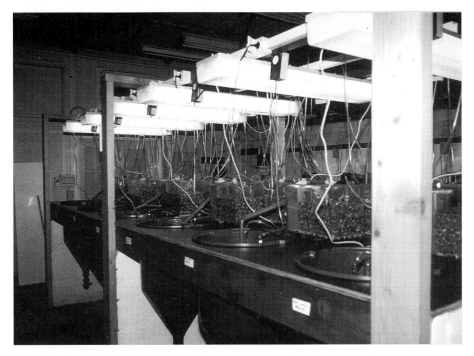

Fig. 15.8 Experimental aquaria for turbot larviculture, using cylindro-conical tanks equipped with individual recirculation systems.

larval stages should be used only with extreme care, since the requirements of the larvae may depend not only on their ontogenetic stage but also on the fatty acid reserves in their yolk-sacs, which may be highly variable (Lavens & Sorgeloos 1991a), depending on prior conditioning of the broodstock. For instance, a modest administration period with DHA-boosted rotifers of 2 consecutive days was enough to improve the physiological condition of the larvae. By contrast, extending the feeding period with the same diet to 4 days was so detrimental to the larvae that it produced similar larval quality to that obtained with HUFA – deficient rotifers reared on bakers' yeast (Fig. 15.9) (Dhert *et al.* 1992a).

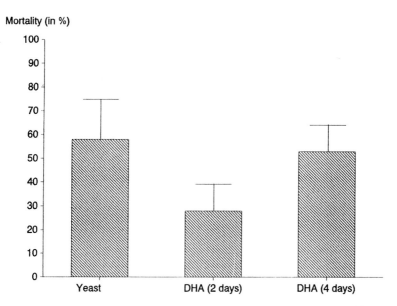

Fig. 15.9 Percentage mortality of turbot larvae exposed to a salinity stress test (48‰) at the end of the rotifer stage. (Adapted from Dhert et al. 1992a.)

It is clear that, depending on the extra supplementation with essential fatty acids during the rotifer stage, larvae with various nutritional and maybe physiological characteristics can be obtained. In this respect, the nutritional manipulation of *Artemia* has to be adapted according to the feeding strategy followed during the rotifer stage. If the requirements for (n-3) fatty acids have been satisfied during the rotifer stage, the enrichment of *Artemia* with high levels of essential fatty acids can become optional (Table 15.3).

Therefore, the formulation of 'blue print' feeding strategies for a single species may be extremely difficult. Too many nutritional, ontogenetic and even zootechnical factors can interact and create periodic imbalances in the fatty

Table 15.3 Relative length and stress sensitivity of turbot larvae fed on EPA- and DHA-containing *Artemia*. The results of the length and stress sensitivity are relative figures, expressing how much the parameter varied from the control treatment (newly-hatched GSL *Artemia*)

Age of the fish	Relative stress sensitivity		Age of the fish	Relative length	
	EPA	DHA		EPA	DHA
Day 8	–	0.48			
Day 10	–	0.92	Day 10	1.02	1.15
Day 14	0.80	0.80			
Day 21	0.67	0.72	Day 25	1.07	0.97
Day 28	0.94	0.70			
Day 35	1.36	1.23	Day 35	1.02	0.99

acid requirements of the larvae; this results in alternative periods with deficiencies or excesses. Only daily monitoring of larval quality (as performed for instance by several stress test procedures) can guarantee a state of nutritional well-being in fish. Experimentation with different diets containing various fatty acid ratios during the complete ontogenetic development may finally help to determine an optimal feeding strategy.

As stated above, the dietary requirements of fish larvae may also be significantly influenced by their biochemical composition at the start of feeding, a factor which it is well-known can be modified by alteration in the nutritional status of the broodfish; for example Watanabe & Miki (1993) demonstrated that red seabream broodfish fed with raw krill shortly before or during spawning produced eggs and larvae of significantly better quality than fish fed the control diet. The active components for the improvement of the larval quality were detected as an asthaxanthin isomer and phosphatidylcholine.

The transfer of fatty acid nutrients from the broodstock diet of the mother to the offspring has also been reported for *Epinephelus tauvina* (Dhert et al. 1991). Feeding the female broodfish with a lipid-rich diet (Marila, Artemia Systems, Belgium) resulted in a 21% increase of total lipid concentration in the eggs and significantly larger oil globules. Although the broodstock enrichment diet had an initial EPA/DHA ratio of 1:3, the gain in the EPA component was more pronounced (46% increase compared with 9% for DHA) (Table 15.4). Further investigations are, however, required to detect the possible effects on egg and larval quality and subsequent nutritional requirements at the start of feeding.

Table 15.4 Egg-quality characteristics in grouper broodstock fed with a maturation booster (from Dhert et al. 1991)

	Control broodfish	Enriched broodfish
Egg diameter (in µm)	792 ± 5	801 ± 23
Oil globule diameter (in µm)	179 ± 7	189 ± 6*
Total lipids (%)	20.2 ± 2.2	24.4 ± 2.4*
EPA (mg.g^{-1})	4.6 ± 0.4	6.7 ± 0.5*
DHA (mg.g^{-1})	23.5 ± 2.2	25.7 ± 2.0*
Σ(n-3) (mg.g^{-1})	30.7 ± 2.4	35.9 ± 1.9*

* Significantly different at the $P < 0.05$ level with the control

Ascorbic acid

In recent years vitamin C has also been considered as an important dietary component in larviculture (Dabrowski 1992). As a consequence, tests have been conducted at the Laboratory of Aquaculture, University of Gent, to incorporate this nutrient into live feeds but in a stable and bio-available form. Using the standard enrichment procedure (Léger et al. 1987) and experimental self-emulsifying concentrates containing 10%, 20% and 30% (on a DW basis) of ascorbyl palmitate (AP) in addition to triglycerides, high levels of free ascorbic acid (AA) can be incorporated into brine shrimp nauplii (Fig. 15.10). A 10% AP inclusion in the emulsion only slightly enhances (900 µg g^{-1} DW) the natural levels occurring in freshly-hatched nauplii (600 µg AA g^{-1} DW). However, 20 or 30% additions increased AA levels in *Artemia* respectively 3-fold and 6-fold after 18 h enrichment at 27°C, while (n-3) HUFA levels remained at similar levels to those produced by normal enrichment procedures.

When the total amount of enrichment product (0.6 g l^{-1}) is split over three administrations (24 h, 36 h and 42 h after the start of hatching) instead of two (24 h and 36 h), even higher levels of AA are incorporated into the *Artemia*, respectively 3.4 and 2.4 mg AA g^{-1} DW (Fig. 15.8). Moreover, these AA concentrations do not decrease when the 24 h enriched nauplii are stored for 24 h at densities of 300 per ml in normal sea water of 4°C and 28°C, respectively (Fig. 15.11, G. Merchie *et al.* unpublished data). In fact, an increase is observed even after 12 h storage when the nauplii were three times enriched; this may mean that there is a continued assimilation of AA from the AP remaining in the gut, which so far cannot be detected by the applied analytical method.

An effect of vitamin C enrichment into *Artemia* nauplii on larviculture success has been documented in only one experiment, with *Macrobrachium rosenbergii*. No significant differences in survival and growth were found until day 28 for the two groups fed on Great Salt Lake *Artemia* enriched with an experimental emulsion E50 containing respectively 0 (= control) or 20% AP. However, the physiological condition of the AP-fed larvae, measured by means of a stress test (Romdhane *et al.* 1994), was much higher than in the control group (Table 15.5). Since much higher levels of AA are incorporated in the

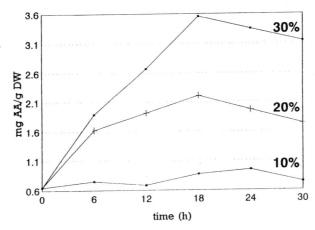

Fig. 15.10 Ascorbic acid enrichment in *Artemia* nauplii: effect of different emulsion concentrations.

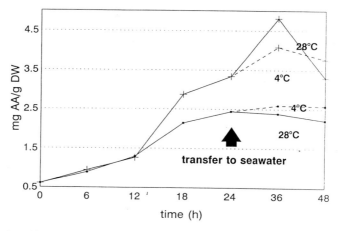

Fig. 15.11 Ascorbic acid enrichment and storage in *Artemia* nauplii: effect of 2 × (0, 12 h) against 3 × (0, 12, 18 h) distributions of a total amount of 0.6 g l^{-1} emulsion containing 20% AP; storage after enrichment at respectively 4 and 28°C.

Table 15.5 Effect of vitamin C enrichment in *Artemia* nauplii on *Macrobrachium* larviculture success (until day 28)

	Control	Control + 20% vitamin C
Survival (%)	65	43
Ind. length (mm)	9.31	9.43
Ind. weight (µg)	831	888
Metamorphosis (%)	11	15
Stress resistance (%S)	9	33
Vitamin C in larvae (µg AA g^{-1} DW)	365	552

body tissue of *Macrobrachium* larvae when fed a vitamin C-rich food source (Table 15.5), it may be assumed that a positive effect on stress resistance is created by feeding vitamin C-enriched live foods.

This hypothesis is supported by Dabrowski (1992), who states that stress creates increased ascorbate requirements for larval fish. Nevertheless, more experiments are required to determine the requirements and role of ascorbic acid in the larval stages of fish and crustaceans. The enrichment procedure for *Artemia*, and possibly rotifers, where various levels of AA can be incorporated into the live food, may be an important method of improving larval quality.

15.4.2 Use of stress tests for larval quality assessment

Immediate measurement of larval vitality is difficult and can often not be evaluated by parameters describing growth or survival. It was the Japanese who first developed techniques to evaluate vitality. For this purpose they exposed the fish to stress conditions and measured recovery from this induced stress. The simplest method to induce stress consists of lifting the fish from the water for a few seconds and measuring the mortality induced by the exposure to air. Standardization in this testing procedure is not easy and results in a high variability among replicates.

In order to improve the reliability of the stress test procedure a salinity stress test has been designed (Fig. 15.12) in which the fish are exposed to a high salinity (55–65‰), and consequent survival of the animals is measured at regular time intervals (Dhert et al. 1992b,c). This stress test has the advantage that the stress sensitivity can be expressed as an index, i.e. the mathematical sum of cumulative mortality, noted at regular time intervals. Moreover, the index allows statistical interpretation of the data. In this way valuable information can be obtained which allows, for example, the selection of proper feeding strategies and the quality/price evaluation of post-larval shrimp and fish species.

The stress test procedure has been successfully applied in fish larval nutrition studies because it enables early detection of nutritional deficiency and offers the possibility of anticipating culture failures at an early stage (Dhert et al. 1990). Other applications of the stress test have been reported in the field of larval biomedication where the test allows the evaluation of the physiological condition of the fish fry immediately after challenging them with pathogenic bacteria (Dhert et al. 1992b,c).

15.4.3 Use of on-grown *Artemia* of specific size

Although the advantages of using gradually increasing sizes of *Artemia* are fully recognized, for they contain a higher individual protein and energy content and thus improve the predator's energy budget, they are easily boosted with various nutrients, they enable considerable savings to be made in numbers of *Artemia* cysts and they reduce cannibalistic behaviour (Lavens & Sorgeloos 1991b), the complexity and high cost of production make integration of super-intensive *Artemia* rearing systems in fish hatcheries up to now prohibitive. As a consequence, a simplified and flexible but reliable technique for the short-term culture of brine shrimp juveniles up to 3 mm in length (27 µg DW) for use as a nursery diet in fish and shrimp farming, has recently been developed (Dhont et al. 1993).

Cultures are performed in rectangular tanks of 100–500 l containing sea water of 35‰ and 25°C but with no water renewal. Aeration is provided by PVC tubes fixed to the tank bottoms. Food consists of a suspended micronized product (YM20: Artemia Systems NV, Belgium) which is semi-continuously distributed to the cultures following a direct feeding regime (Table 15.6); nevertheless each culture requires daily adjustment of the feed ration in order to keep transparency levels at 15–20 cm. Feeding rations are independent of *Artemia* density, and these may be selected to provide the desired growth rate (Fig. 15.13). Moreover, uniform sizes are obtained at any moment during culture and survival rates remain above 70%.

Fig. 15.12 A simple stress test consisting of the capture and concentration of 10 fish larvae (above) before bringing them into a high-salinity test solution (below).

Table 15.6 Directive feeding regime with YM20 and 10 animals per ml (g per 1000 l culture volume) and enrichment ratios added daily to the food suspension (mg DRY SELCO l^{-1} culture volume)

Day	0	1	2	3	4	5	6	Total
YM20	200	225	250	300	325	350	350	2000
DRY SELCO	–	37.5	62.5	87.5	112.5	137.5	162.5	600

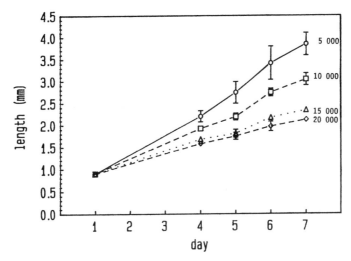

Fig. 15.13 Growth curves of *Artemia* cultured over 7 days under standard conditions but at different densities: (○) 5000 animals per litre; (□) 10000 per litre; (△) 15000 per litre; (◊) 20000 per litre.

Cultured *Artemia* juveniles can subsequently be enriched with (n-3) HUFA in the same way as for nauplii, resulting in higher HUFA levels, which are achieved within shorter enrichment periods (Table 15.7). Furthermore, Dhont et al. (1991) have developed an alternative boosting method. Instead of adding the enrichment emulsion at the end of the culture period, a dry booster (Dry Selco, Artemia Systems NV, Belgium) is distributed together with the food over the entire culture period. Daily increasing doses of Dry Selco are administered so that the total amount equals $0.6 \text{ g} l^{-1}$ by the end of the culture period (Table 15.6). This method has several advantages over the first method: five times higher levels of (n-3) HUFA are incorporated in 7 days old *Artemia* (Table 15.7), the risks for oxygen depletion are avoided, and partially-enriched *Artemia* of a specific size class can be harvested and fed to the predator at any time during the culture period (Dhont et al. 1991).

It is expected that this technique, which has only a minimal investment cost and which is relatively easy to apply as a routine procedure by hatchery personnel, will be integrated in the near future in hatchery operations of

Table 15.7 Comparison of enrichment results of *Artemia* juveniles and nauplii

Artemia	Procedure	Duration	20:5n-3 (mg g^{-1} DW)	22:6n-3 (mg g^{-1} DW)	Σ n-3 HUFA (mg g^{-1} DW)
Nauplii[a]	after hatching	12 h	7.9	4.4	14.4
Juveniles	after culture	4 h	5.8	4.4	14.2
	during culture	7 days	44.2	16.5	64.3

[a] Data compiled from Léger *et al.* (1987)

marine species with specific nutritional requirements before weaning, for example halibut and mahi-mahi.

15.5 Artificial diets

It is generally accepted that in fish larviculture a reduction in the requirements for live foods would contribute significantly to further optimization of this sector of aquaculture production as it would allow a more standardized production protocol and consequently a more reliable and cost-effective production output. Whereas for several freshwater fish species this has been achieved without major difficulties, many problems are encountered with marine fish, principally because the larvae are mostly very tiny, sensitive and rather poorly developed at hatching. Generally, development continues progressively up to metamorphosis and includes significant changes in morphology and enzymology of the digestive system. The nutritional requirements of marine fish, e.g. for HUFA, are also more specific. Replacement diets therefore require a lot of special characteristics with respect to particle size, physical performance in water, attractability, digestibility and nutritional composition. They must also remain cost effective.

It may, therefore, not be surprising to find that complete replacement diets for marine larviculture are still more an illusion than a reality. Nevertheless, in recent years much work has been devoted to the development of artificial diets with partially optimized characteristics and which therefore can be fed at earlier larval stages (early weaning diet) or which can partially replace live food (co-feeding diet). Up to the present, however, only preliminary data are available.

Person-Le Ruyet *et al.* (1993) reported maximal survival rates and no differences in growth when good quality *Dicentrarchus labrax* larvae were weaned at day 22 instead of day 35 onto a commercial weaning diet (FFK, Sevbar, Japan). Larvae of lower quality, however, performed less well as far as survival and growth were concerned when early weaning was applied.

Co-feeding experiments on an experimental scale in 30 l aquaria with *D. labrax* during the hatchery period (days 21–36) showed that partial replacement (50%) of *Artemia* by the commercial diet R1 (Artemia Systems NV,

Belgium) resulted in a higher growth rate compared with the control group (6.17 against 4.41 mg DW), and a slightly higher stress resistance (39 against 55% mortality in the stress test). This experiment has been repeated under pilot scale conditions (500 l) to include the weaning period (days 36–50). At day 50 co-fed fry performed equally well as far as growth and survival were concerned as fry fed the standard feed regime (10.2 against 9.2 mg DW, respectively 30 against 35%).

With the aim of improving our knowledge of the quantitative nutritional requirements of larval stages of various aquaculture species, several investigators have tried to develop purified particulate diets which can be used as standard reference diets in nutritional studies. These diets not only have to meet the same requirements as mentioned above but, moreover, should utilize selected ingredients of defined chemical composition; this makes formulation of a nutritionally-adequate diet even more difficult and far more expensive. Attempts in this respect for crustaceans have been made by Kanazawa *et al.* (1982) and Castell *et al.* (1989).

Our laboratory recently initiated work to develop a reference diet for marine fish fry (>30 d). In order to meet the pre-requisites of particle size, physical performance in water and digestibility, the same process technology is to be applied as for the commercial microparticulate diets (MPD). This means, however, that relatively large batches (>10 kg) of experimental product need to be prepared, making trials very costly. It was decided, therefore, to divide the preparation in two steps:

(1) Extrusion of a basic (nucleus) diet formulated with (semi-) purified proteins, carbohydrates, vitamin + mineral mix, and an attractant (this is the non-variable fraction of all trial diets).
(2) Spray-dried coating of selected purified lipid fractions of this basic diet (This would include the fat soluble components).

This approach, however, has the disadvantage that the MPD can be only partially manipulated (proteins and carbohydrates cannot be altered) which make it unsuitable for nutritional studies of non-lipid components. Testing on a laboratory scale of several formulated diets on the basis of different (semi-) purified protein sources (casein/gelatin, cod protein) are under way for *Dicentrarchus labrax* (Fig. 15.14). Once its suitability is proven, this reference diet could be made available for nutrition studies of different marine fish species (Fig. 15.15).

Another reference diet containing decapsulated *Artemia* cysts is also being developed at our laboratory for use with different species of freshwater fish.

15.6 Disease control through live food

At present mortalities due to microbial infections in larval cultures of fish and shrimp are mainly treated or prevented by dissolving relatively high doses of

Fig. 15.14 A random block arrangement of tanks is used for the experimental facility where nutritional requirements of sea bass are tested using an artificial reference diet.

REFERENCE DIET
FOR MARINE FISH LARVAE
Purified ingredients + attractant
- * proteins
- * carbohydrates
- * vitamin-mix
- * mineral-mix

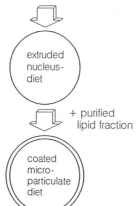

Fig. 15.15 Production scheme for the reference diet for marine fish larvae.

broad-spectrum antibiotics in the culture water. However, success of this method is not guaranteed; other important constraints are the (uncontrolled) use of high quantities of expensive drugs and concern over their subsequent discharge into the environment and possible development of resistant bacteria (Brown 1989).

Improved disease control may be obtained by oral delivery of the drug to the predator larva. In this respect it has recently been demonstrated on a laboratory scale that live food enrichment techniques through bioencapsulation in *Brachionus* and *Artemia* may be an excellent tool for the prophylactic and therapeutic treatment of larvae with drugs as well as vaccines. Verpraet *et al.* (1992) have demonstrated that high doses of more than $290\,\mu g\,g^{-1}$ DW of trimethoprim and sulphamethoxazole can be bioencapsulated in *Artemia*. Moreover, as shown in Fig. 15.16, the drugs are directly incorporated in sea bass fry upon administration of these enriched *Artemia* nauplii: high therapeutic levels in the fish tissue of $22.6\,\mu g\,g^{-1}$ DW were reached after only 3 h of feeding (Chair *et al.* 1991). Challenge tests to confirm the therapeutic value of this technique are ongoing.

Vaccination through oral administration has also been shown to be feasible. A method for optimal uptake of *Vibrio anguillarum* vaccine by *Artemia* nauplii has been developed by Campbell *et al.* (1993). 80-days old sea bass fry orally vaccinated via *Artemia* show a significant increase in growth (DW) compared with the bath-vaccinated or non-vaccinated groups 5 weeks after treatment (Fig. 15.17) (M. Dehasque, pers. comm.).

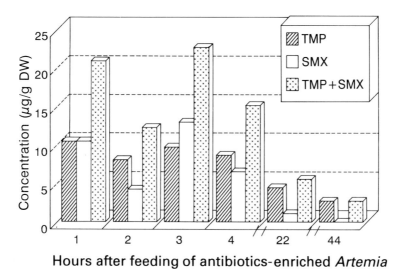

Fig. 15.16 Accumulation of TMP and SMX in *Dicentrarchus labrax* fry (60 days old) after feeding once (at 0 h) antibiotic-enriched *Artemia*. TMP—trimethoprim; SMX—sulphamethoxazole.

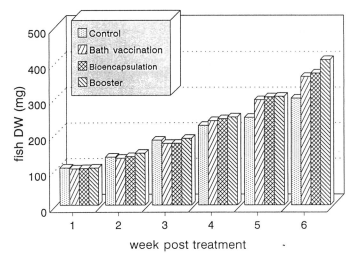

Fig. 15.17 Effect of oral vaccination on *Dicentrarchus labrax* larval growth.

15.7 Future prospects

Production of sufficient quantities of quality fry remains a cornerstone in the further development of aquaculture. Optimization of the larval feeding strategy will play an important role in achieving this. In this respect research efforts in the near future should concentrate on:

- much more user-friendly and cost-effective applications;
- a better definition of the nutritional requirements of specific stages, including the larval quality status at start feeding; and
- the impact of feeding strategy on the engineering and design of culture systems (e.g. automation and creation of water flow patterns).

Wherever possible, maintenance of live food stocks (microalgae, rotifers, copepods) should be reduced or eliminated from the hatchery operation. Even the use of the contemporary cheap and easy to handle *Artemia* will be reduced in favour of formulated microparticulate diets. For some species, such as penaeids, the moment when hatcheries will use only artificial diets is fast approaching. However, full live food substitution is less likely to be achieved in marine fish larviculture in the near future. Although appropriate process technologies will be developed to meet the pre-requisites of the larval feed, their production costs may be too high to be competitive against live food. Consequently, their commercial applicability is uncertain. The cost-effectiveness of live and formulated feeds will be the key to their proportional use in co-feeding regimes.

Elucidation of species and ontogenic stage-specific nutritional requirements will furthermore facilitate the choice of feeding strategy. These requirements may be less stringent in cases where essential nutrients are endogenously supplied through the parental broodfish. Improvements of broodstock conditioning, more specifically broodstock nutrition, and its effect on larval quality should, therefore, be ranked as a high-priority research topic.

Finally, curative and preventive treatment of diseases of larvae through the food chain may be far more effective than via the relatively large volume of culture water surrounding them.

References

Bengtson, D.A., Léger, Ph. & Sorgeloos P. (1991) Use of *Artemia* as a food source for aquaculture. In *Artemia Biology*, (eds R.A. Browne, P. Sorgeloos, & C.N.A. Trotman), pp. 255–85. CRC Press, Boca Raton, Florida.

Brown, J. (1989) Antibiotics: their use and abuse in aquaculture. *World Aquaculture*, **20**(2), 34–43.

Campbell, R., Adams, A., Tatner, M.F., Chair, M. & Sorgeloos, P. (1993) Uptake of *Vibrio anguillarium* vaccine by *Artemia salina* as a potential delivery system to fish fry. *Fish & Shellfish Immunology*, **3**, 451–59.

Castell, J.D., Kean, J.C., D'Abramo, L.R. & Conklin, D.E. (1989) A standard reference diet for crustacean nutrition research. I. Evaluation of two formulations. *Journal of the World Aquaculture Society*, **20**(3), 93–9.

Chair, M., Romdhane, M., Dehasque, M., Nelis, H., De Leenheer, A.P. & Sorgeloos, P. (1991) Live-food mediated drug delivery as a tool for disease treatment in larviculture. II. A case study with European seabass. In *Larvi '91 – Fish & Crustacean Larviculture Symposium*, (eds P. Lavens, P. Sorgeloos, E. Jaspers & F. Ollevier), pp. 412–14. European Aquaculture Society, Special Publication, **15**. Gent, Belgium.

Coutteau, P., Lavens, P. & Sorgeloos, P. (1990a) The use of yeast as single-cell protein in aquacultural diets. *Medical Faculty Landbouww Rijksuniversity, Gent*, **54**(4b), 1583–92.

Coutteau, P., Lavens, P., Léger, P. & Sorgeloos, P. (1990b) Manipulated yeast diets as a partial substitute for rearing bivalve molluscs: laboratory trials with *Tapes semidecussata*. In *Book of Abstracts, World Aquaculture '90, 10–14 June 1990, Halifax, NS, Canada*, p. 111. Imprico, Quebec, Canada.

Coutteau, P., Hadley, N., Manzi, J. & Sorgeloos, P. (1991) Manipulated yeast diets as a partial substitute for the nursery culture of the hard clam *Mercenaria mercenaria*. In *Aquaculture and the Environment*, pp. 77–8. Dublin, Ireland, 10–12 June 1991. Short Communications and Abstracts. European Aquaculture Society Special Publication, **14**. Bredene, Belgium.

Dabrowski, K. (1992) Ascorbate concentration in fish ontogeny. *Journal of Fish Biology*, **40**, 273–9.

Devresse, B., Léger, P., Sorgeloos, P., Murata, O., Nasu, T., Ikeda, S., Rainuzzo, J.R., Reitan, K.I., Kjorsvik, E. & Olsen, Y. (1992) Improvement of flatfish pigmentation through the use of DHA enriched rotifers and *Artemia*. In *Book of Abstracts, Fifth International Symposium on Fish Nutrition and Feeding, Santiago, Chile, September 1992*. Fundacion Chile, Santiago, Chile.

Dhert, P., Lavens, P., Duray, M. & Sorgeloos, P. (1990) Improved larval survival at metamorphosis of Asian seabass (Lates calcarifer) using ω3-HUFA-enriched live food. *Aquaculture*, **90**, 63–74.

Dhert, P., Lim, L.C., Lavens, P., Chao, T.M., Chou, R. & Sorgeloos, P. (1991) Effect of dietary essential fatty acids on egg quality and larviculture success of the greasy grouper (*Epinephelus tauvina*, F.): preliminary results. In *Larvi '91 – Fish & Crustacean Larviculture Symposium*, (eds P. Lavens, P. Sorgeloos, E. Jaspers & F. Ollevier), pp. 58–62. European Aquaculture Society, Special Publication, **15**. Gent, Belgium.

Dhert, P., Lavens, P. & Sorgeloos, P. (1992a) Fatty acid requirements for European seabass (*Dicentrarchus labax*) and turbot (*Scophthalmus maximus*) in function of their ontogenetic development. In *Book of Abstracts, Third Asian Fisheries Forum, Singapore, October 1992*. Asian Fisheries Society, Philippines.

Dhert, P., Lavens, P. & Sorgeloos, P. (1992b) A simple test for quality evaluation of cultured fry of marine fish. *Proceedings of the Forum of Applied Biotechnology, Mededelingen, van de Faculteit Landbouwwetenschappen,* **57**(4A), 2135–43.

Dhert, P., Lavens, P. & Sorgeloos, P. (1992c) Stress evaluation: a tool for quality control of hatchery-produced shrimp and fish fry. *Aquaculture Europe,* **17**(2), 6–10.

Dhont, J., Lavens, P. & Sorgeloos, P. (1991) Development of a lipid-enrichment technique for *Artemia* juveniles produced in an intensive system for use in marine larviculture. In *Larvi '91 – Fish & Crustacean Larviculture Symposium,* (eds P. Lavens, P. Sorgeloos, E. Jaspers & F. Ollevier), pp. 51–5. European Aquaculture Society, Special Publication, **15**, Gent, Belgium.

Dhont, J., Lavens, P. & Sorgeloos (1993) Preparation and use of *Artemia* as food for shrimp and prawn larvae. In *CRC Handbook of Mariculture,* Vol. 1, *Crustacean Aquaculture,* (ed J.P. McVey), pp. 61–93. CRC Press, Boca Raton, Florida.

Fukusho, K. (1989) Biology and mass production of the rotifer *Brachionus plicatilis. International Journal of Aquatic Fisheries Technology,* **1**, 232–40.

Hjelmeland, K., Pedersen, B.H. & Nilssen, E.M. (1988) Trypsin content in intestines of herring larvae, *Clupea harengus,* ingesting inert polystyrene spheres or live crustacea prey. *Marine Biology,* **98**, 331–5.

Jones, D.A., Kanazawa, A. & Ono, K. (1979) Studies on the nutritional requirements of the larval stages of *Penaeus japonicus* using microencapsulated diets. *Marine Biology,* **54**, 261–7.

Kanazawa, A., Teshima, S. & Sasada, H. (1982) Culture of prawn larvae with microparticulate diets. *Bulletin of the Japanese Society of Scientific Fisheries,* **48**(2), 195–9.

Komis, A., Candreva, P., Franicevic, V., Moreau, V., Van Ballaer, E., Léger, P. & Sorgeloos, P. (1991) Successful application of a new combined culture and enrichment diet for the mass cultivation of the rotifer *Brachionus plicatilis* at commercial hatchery scale in Monaco, Yugoslavia, France, and Thailand. In *Larvi '91 – Fish & Crustacean Larviculture Symposium,* (eds P. Lavens, P. Sorgeloos, E. Jaspers & F. Ollevier), pp. 102–3. European Aquaculture Society, Special Publication, **15**. Gent, Belgium.

Laing, I. & Verdugo, C.G. (1991) Nutritional value of spray-dried *Tetraselmis suecica* for juvenile bivalves. *Aquaculture,* **92**, 207–18.

Langdon, C.J. & Waldock, M.J. (1981) The effect of algal and artificial diets on the growth and fatty acid composition of *Crassostrea giga* spat. *Journal of the Marine Biological Association of the UK,* **61**, 431–48.

Lavens, P. & Sorgeloos, P. (1991a) Variation in egg and larval quality in various fish and crustacean species. In *Larvi '91 – Fish & Crustacean Larviculture Symposium* (eds P. Lavens, P. Sorgeloos, E. Jaspers & F. Ollevier), pp. 221–2. European Aquaculture Society, Special Publication, **15**. Gent, Belgium.

Lavens, P. & Sorgeloos, P. (1991b) Production of *Artemia* in culture tanks. In *Artemia Biology,* (eds R.A. Browne, P. Sorgeloos & G.N.A. Trotman), pp. 317–50. CRC Press, Boca Raton, Florida.

Lavens, P., Sorgeloos, P. Jaspers, E. & Ollevier, F. (1991) *Book of Short Communications & Abstracts, of Larvi '91 – Fish & Crustacean Larviculture Symposium.* European Aquaculture Society, Special Publication, **15**. Gent, Belgium.

Lavens, P., Dhert, P., Merchie, G., Stael, M. & Sorgoloos, P. (1994) A standard procedure for the mass production of an artificial diet for rotifers with a high nutritional quality for marine fish larvae. *Proceedings of the Third Asian Fisheries Forum, Singapore, 1992.* Asian Fisheries Society, Philippines (in press).

Léger, P., Bieber, G.F. & Sorgeloos, P. (1985) International study on *Artemia* XXXIII. Promising results in larval rearing of *Penaeus stylirostris* using a prepared diet as algal substitute and for *Artemia* enrichment. *Journal of the World Mariculture Society,* **16**, 354–67.

Léger, Ph., Bengtson, D.A., Simpson, K.L. & Sorgeloos, P. (1986) The use and nutritional value of *Artemia* as a food source. *Oceanography and Marine Biology, an Annual Review,* **24**, 521–623.

Léger, P., Naessens-Foucquaert, E. & Sorgoloos, P. (1987) International study on *Artemia.* XXXV. Techniques to manipulate the fatty acid profile in *Artemia* nauplii and the effect on its nutritional effectiveness for the marine crustacean *Mysidopsis bahia* (M.). In *Artemia research and its applications,* Vol. 3 (eds P. Sorgeloos, D.A. Bengtson, W. Decleir & E. Jaspers), pp. 411–24. Universa Press, Wetteren, Belgium.

Meeren, T.v.d. (1991) Algae as first food for cod larvae, *Gadus morhus* L.: filter feeding or ingestion by accident? *Journal of Fish Biology*, **39**, 225–37.

Mourente, G., Rodriguez, A., Tocher, D.R. & Sargent, J.R. (1993) Effects of dietary docosahexaenoic acid (DHA; 22:6n-3) on lipid and fatty acid compositions and growth in gilthead seabream (*Sparus aurata* L.) larvae during first feeding. *Aquaculture*, **112**, 79–98.

Naas, K.E., Naess, T. & Harboe, T. (1992) Enhanced first feeding of halibut larvae (*Hippoglossus hippoglossus* L.) in green water. *Aquaculture*, **105**, 143–56.

Olsen, Y. (1989) Cultivated microalgae as a source of omega-3 fatty acids. In *Fish, Fats and Your Health, Proceedings of the International Conference on Fish Lipids and Their Influence on Human Health*, pp. 51–62. Svanoy Foundation, Norway.

Person-Le Ruyet, J., Fischer, C. & Thébaud, L. (1993) Sea bass (*Dicentrarchus labrax*) weaning and ongrowing onto Sevbar. In *Proceedings of the Fourth International Symposium on Fish Nutrition and Feeding, Biarritz, France, June 1991*, (eds S.J. Kaushik & P. Luquet), pp. 623–8. INRA Editions, Versailles, France.

Reitan, K.I., Bolla, S. & Olsen, Y. (1991) Ingestion and assimilation of microalgae in yolk sac larvae of halibut, *Hippoglossus hippoglossus* (L.). In *Larvi '91 – Fish & Crustacean Larviculture Symposium*, (eds P. Lavens, P. Sorgeloos, E. Jaspers & F. Ollevier), pp. 332–4. *European Aquaculture Society, Special Publication*, **15**. Gent, Belgium.

Rhomdane, M.S., Devresse, B., Léger, P. & Sorgeloos, P. (1994) Effects of feeding nutritionally enriched *Artemia* during a progressively increasing period on the larviculture of the freshwater prawn *Macrobrachium rosenbergii*. *Journal of the World Aquaculture Society* (in press).

Sargent, J.R., Bell, J.G., Bell, M.V., Henderson, R.J. & Tocher, D.J. (1993) The metabolism of phospholipids and polyunsaturated fatty acids in fish. In *Aquaculture: Fundamental and Applied Research*, (eds B. Lahlou & P. Vitiello), pp. 103–24. *Coastal and Estuarine Studies*, **43**. American Geophysical Union, Washington, DC.

Sorgeloos, P. & Léger, P. (1992) Improved larviculture outputs of marine fish, shrimp and prawn. *Journal of the World Aquaculture Society*, **23**, 251–64.

Sorgeloos, P., Lavens, P., Léger, P. & Tackaert, W. (1993) The use of *Artemia* in marine fish larviculture. In *Proceedings of Finfish Hatchery in Asia '91, Tungkang, Taiwan, December 1991*, (eds C.S. Lee, M.S. Su & I.C. Liao), pp. 73–86.

Verpraet, R., Chair, M., Léger, P. Nelis, H., Sorgeloos, P. & De Leenheer, A.P. (1992) Live-food mediated drug delivery as a tool for disease treatment in larviculture. The enrichment of therapeutics in rotifers and *Artemia* nauplii. *Aquaculture Engineering*, **11**, 133–9.

Watanabe, T. (1991) Importance of Docosahexaenoic acid in marine larval fish. In *Larvi '91 – Fish & Crustacean Larviculture Symposium*, (eds P. Lavens, P. Sorgeloos, E. Jaspers & F. Ollevier), p. 19. *European Aquaculture Society Special Publication*, **15**. Gent, Belgium.

Watanabe, T. & Miki, W. (1993) Astaxanthin: an effective dietary component for red seabream broodstock. In *Proceedings of the Fourth International Symposium on Fish Nutrition and Feeding, Biarritz, France, June 1991*, (eds S.J. Kaushik & P. Luquet), pp. 23–36. INRA Editions, Versailles, France.

Chapter 16
Red Sea Bream (*Pagrus major*)

16.1 Introduction and taxonomy
16.2 Management of broodstock
16.3 Egg incubation
16.4 Quality of eggs
16.5 Broodstock nutrition
16.6 Larval rearing
16.7 Quality of larvae
16.8 Larval nutrition and the role of essential fatty acids (EFA)
16.9 Future prospects
References

Three decades after the initiation of red sea bream *Pagrus major* mariculture, the technology has undergone various stages of refinement, and of late there has been an overproduction of seedlings. Among the unresolved problems, the seasonal limitation of spawning seems to be the forerunner. The necessity of broodstock diets is yet to be established. Even though the larval technology has been commercially successful, the suitability and availability of live and inert larval diets merits further research. As lipids are vital to larval survival, docosahexanoic acid-rich live food has to be identified and mass cultured. The larval production techniques continue to depend on a lot of manpower, thereby demanding more automation. A problem-solving approach would further consolidate bream culture as an industry standard.

16.1 Introduction and taxonomy

Sparid fish are found in temperate and tropical oceans, inhabiting the coastal and continental shelf areas where water currents are warm. Some species of this family of bony fish, classified under order Perciformes, attain great size and are important both as sources of food and for sport. Out of the 41 species grouped into 22 genera and belonging to four sub-families, many are hermaphroditic but later differentiate to males or females as they get older (Smith & Heemstra 1986). The high and well-compressed bodies of the Sparidae are generally coloured either red or silver-grey. The former groups live in relatively deep sea and the latter in shallow waters. The Sparidae in Japan comprise three sub-families, Denticinae, Pagrinae and Sparinae (Masuda *et al.* 1984).

Among the Sparids, the red sea bream *Pagrus major* (Temminck and Schlegel), of the sub-family Pagrinae, is perhaps one of the most important food fish, especially for the Japanese. This fish, which grows to about 100 cm SL (Fig. 16.1), has been established as a prime aquaculture species over the decades.

Fig. 16.1 Broodstock red sea bream.

Fish belonging to sub-family Sparinae, commonly called porgies, are also being cultivated. Among them, the black sea bream *Acanthopagrus schlegeli* (Bleeker) is the most popular.

Tracing the development of red sea bream culture in Japan, artificial breeding and larval production were first attempted in Hiroshima in 1887 (Davy 1991). Subsequently, a hatchery established there in 1902 encountered a variety of problems. However, the breakthrough came very late in 1962, when 22 juveniles were produced from 150 000 eggs at a fisheries centre in Kanagawa. Later, at Ehime in 1965, mass propagation of red sea bream was established using *Brachionus plicatilis* as live food. Research carried out since then has eventually established the technology for larval mass-culture (Fukusho 1989). The main factors contributing to successful seed production were considered to be the introduction of rotifers as live food, the establishment of natural spawning systems in tanks, and the better environmental management of the larval rearing systems.

Larval care greatly depended on the availability and nutritional quality of food organisms such as rotifers, *Artemia* nauplii, wild copepods, and bivalve and echinoderm larvae. Investigations on broodstock have helped in extending the spawning period for more than 6 months by controlling water temperature and feeding frozen krill.

Commercial husbandry of red sea bream was initiated in 1962, almost a decade later than yellowtail, adapting the prevalent culture systems. The rapid increase in production during the 1970s was due to the introduction of floating net cage technology. Aquacultural production of red sea bream in Japan has recently risen to about 61 000 t (1991), corresponding to a rapid increase in seedling production, which was supported by technical developments related to egg-taking, larval-rearing and mass-culture of quality live foods. Over the last few years, in response to the strong demand for seedlings, there has been a several-fold increase from the 110 million juveniles produced in 1989 (Fig. 16.2). Of this, 77 million juveniles in the size range 10−306 mm were used for on-growing in commercial farms. The remaining numbers falling in the group 12−154 mm were restocked into the coastal waters to augment the natural resource. The consumption of pelleted feed rose from 30 000 t (1990) to 41 000 t (1991).

Red sea bream intensive farming procedures were recently reviewed by Foscarini (1988). He indicated that the success and popularity of this fish are due to good knowledge of its biology and behaviour in the wild and of the nutritional requirements at different stages of its culture; these together with fast growth, tolerance to a relatively wide temperature range and ease of spawning make the red sea bream an ideal species for culture.

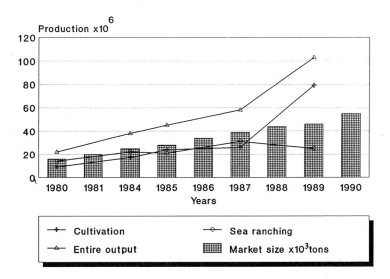

Fig. 16.2 Production trends of seedling and market size red sea bream in Japan.

16.2 Management of broodstock

The red sea bream spawns easily in captivity with 3−6 year-old fish being considered the ideal reproductive ages. Broodstock are usually kept in floating net cages (Fig. 16.3) and fed frozen raw fish such as sardine, mackerel, sand lance, saury, shrimps and commercial dry pellets. Moist pellets prepared by mixing raw fish with formulated mash in a ratio of 8:2 or 9:1 are also used. When the spawning season approaches, special feeds are frequently fed to broodstock. These are fresh fish, frozen krill and moist pellets prepared from

Fig. 16.3 Floating net cages for broodstock red sea bream.

these materials supplemented with vitamin mixtures. Feeding frozen krill has been empirically known to result in stable spawning and the production of good quality eggs. Among the dry feeds, squid meal was a better protein source than white fish meal. Spawning and egg quality were later found to be greatly affected by the nutritional quality of diets given to broodstock shortly before or during spawning.

Broodstock are generally transferred to the spawning tanks on land from floating net cages about 4–6 weeks before spawning. The spawning season can be recognized by the initiation of spawning in net cages. The onset is earlier in the Pacific region than in the Seto Inland Sea and the Sea of Japan. It starts from late March–early April in the former region and late April–late May in the sea areas. Water temperature during that period also varies with the locality, ranging from 11–17°C. The spawning period ends in June or July. The National Research Institute of Aquaculture recently succeeded in early spawn-taking by employing heated water circulation tanks (Fukusho 1989). Increasing the water temperature of the tanks, by about 4°C higher than ambient, advanced the spawning by 15–20 days compared with the times of natural spawning (Fig. 16.4). This technique has been adopted in some districts where warm waste water from atomic and steam power stations is available.

Spawning tanks are commonly rectangular or round, with a depth of 1–2 m and capacities ranging from 50 to 100 m^3. Equal numbers of male and female fish, 3–7 years old, are maintained at densities of 0.7–1.5 individuals per m^3 (Table 16.1). Spawning occurs at nightfall and lasts for several nights. Higher

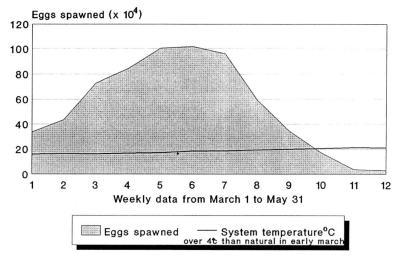

Fig. 16.4 Spawning of red sea bream on manipulation of temperature in broodstock tank. (Modified from Fukusho 1989.)

Table 16.1 Spawn-taking conditions at some fish farming centres in Japan[*1]

	Fish farming centre		
	Hyogo	Kyoto	Ohita
Spawning season	21 Apr–29 Jun	14 Apr–08 Jul	09 Apr–15 Jun
Water temperature (°C)	14–21	–	–
Spawning tank (m^3)	120	50	100 × 2
Number of parents	180	220	371
Age (years)	4	4	4
Body weight (kg)	1–3	1–2	1–2
Male:Female ratio	1:1	1:1[*2]	1:1[*2]
Eggs produced (×10^6)	90	258	6262
per female	1	2.35	3:38
per day	1.1	3.04	9.21

[*1] After Fukusho (1989) [*2] assumed

stocking densities, a decrease in salinity and unequal sex ratios adversely affect the success of spawning. Small fresh fish such as anchovy, sand lance and mackerel are preferred feedstuffs. Of late, raw fish has been replaced by dry feeds without any ill effects on spawning and egg quality.

16.3 Egg incubation

Although the exact techniques vary, generally the pelagic eggs of red sea bream are collected in fine-meshed (200–400 μm) nets by draping these across the overflow water from the spawning tanks or by siphoning the eggs into the nets. The eggs are then transferred to incubation nets and kept in a flow-through system until they hatch. Alternatively, they are introduced directly into a smaller rearing tank at a density of 30 000–90 000 eggs per m^3 (1800 eggs per g fertilized eggs). Eggs spawned during the earlier stages of the season are poorer in quality in terms of hatching rate. At the peak of spawning, about 90% of fertilized eggs hatch. A female parent of 3–7 years of age can produce 100–400 × 10^4 eggs during one spawning season. Employing 50–100 m^3 tanks holding about 40–80 fish in equal male:female ratio, several million eggs could be obtained every day for a month or more.

16.4 Quality of eggs

In red sea bream the percentage of buoyant eggs floating on the water surface is considered very important for the evaluation of egg quality. The hatchability of floating eggs is high and they have a normal development. The unfertilized and dead eggs are deposited at the bottom of the tank. Normal eggs of cultured *Pagrus major* have diameters ranging from 0.66 to 1.03 mm and contain a 0.25 mm oil globule (Fukuhara 1985). When there are more than two oil globules development is abnormal. Poor quality eggs also have wrinkles on the egg membrane, minute black spots on the egg cortex or a swollen cortical ooplasma (Sakai *et al.* 1985). The above characteristics are used as criteria for quality assessments of red sea bream eggs.

16.5 Broodstock nutrition

As nutritional factors have been shown to play a decisive role in breeding programmes, this has been a significant area for investigation of methods for improving the production cycle of red sea bream. Several experiments (Watanabe *et al.* 1984a,b,c) have revealed the importance of pre-spawning nutritional regimes on egg quality. One of the components, raw krill, seems to have a distinct quality-enhancing effect compared with fish meal, which is a common dietary ingredient. Assessing egg and larval quality based on the percentage of buoyant eggs, of total hatch and normal larvae, it was found that viable offspring production was more than doubled when krill or krill lipids were included in the diets (Table 16.2).

In order to identify the effective components in krill which were able to improve egg quality, krill meal was separated into defatted meal and oil and these were substituted for the corresponding fish products in a normal diet. Subsequently, it was shown that the lipid portion could induce the production of larger numbers of healthy larvae. The krill oil contains polar (PL) and non-polar (NL) fractions, mainly constituted by phosphatidylcholine (85%) and triglyceride (81%) respectively. On offering fish meal-based control diets and test diets comprising krill polar and non-polar lipids, to 4-year old broodstock red sea bream held in land-based spawning tanks (Fig. 16.5), it was found that

Table 16.2 Progeny-quality[1] from red sea bream broodstock maintained on normal fish meal ration compared to raw krill and krill lipid-based rations

Stage	Diets			
	Fish meal[2]	Raw krill[3]	Krill PL[4]	Krill NL[5]
Buoyant eggs	64	98	95	96
Hatch-out	67	92	83	93
Normal larva:				
% of viable eggs	75	85	88	87
% of spawn	32	77	71	79

[1] Average of 30–60 days' spawning as percentage
[2] Data of 5 years; diet contained 45% crude protein and 10% lipid
[3] Data of 2 years
[4] Krill polar lipid
[5] Krill non-polar lipid

the lipid fractions effectively improved the number of viable larvae (Table 16.2). The non-polar lipid seemed to be slightly better than the polar (Watanabe *et al.* 1991a,b). This led to the postulation that the attributive components might be phosphatidylcholine in the polar and astaxanthin in the non-polar lipid fraction.

Astaxanthin (3,3'-dihydroxy-β, β-carotene-4,4'-dione) occurs widely as a red animal pigment especially in crustaceans. The purified product is a mixture of three stereoisomers. The 3S,3'R, a meso-isomer which is lowest in proportion

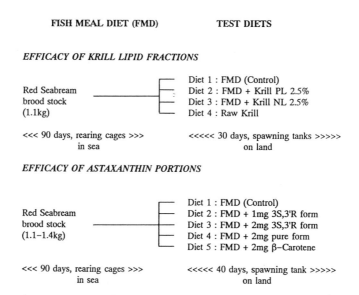

Fig. 16.5 Feeding programmes for identifying the effective components that enhance the quality of fish eggs.

of natural astaxanthin was used as a tracer to investigate the metabolism of carotenoid and its transportation to eggs. This fraction separated by HPLC and the unfractionated pure astaxanthin were added as supplements to a fish meal based diet (Fig. 16.5). In addition a β-carotene containing diet was compared with the fish meal control diet.

The results (Fig. 16.6) indicate that the percentage of buoyant eggs was significantly increased up to 97% when 2 mg of the 3S,3'R isomer or the full profile stereoisomeric astaxanthin were used. The final productivity of viable larvae from the total eggs produced was only 36% in the control group whereas the inclusion of 2 mg of the unseparated isomers or the 3S,3'R isomer resulted in a two-fold improvement in numbers of larvae. Therefore, the principal active component in the non-polar lipid fraction was astaxanthin. The fact that egg quality was enhanced by both forms of astaxanthin demonstrated that the metabolic involvement was unrelated to the chemical configuration.

On the other hand, β-carotene did not reveal any specific role as supported by analytical data of egg carotenoids (Table 16.3). No carotenoids were detected in the eggs produced by broodstock fed the control diet or the diet containing β-carotene, suggesting its low availability to fish and minimal transport to the eggs through the diet. The total carotenoid content was 50–80 μg per 100 g eggs from broodstock fed the isomeric mixture or the 3S,3'R meso-isomer. More than 90% of the total egg carotenoid was astaxanthin, apart from a small amount of idoxanthin presumably derived from astaxanthin. The configuration of astaxanthin in the eggs was found to be the same as that supplemented to the diet.

In another experiment on red sea bream broodstock, Miki et al. (1984) supplemented diets with 0.1% β-carotene and 0.3% canthaxanthin or krill oil or frozen raw krill, shortly before spawning. Levels of carotenoids incorporated in

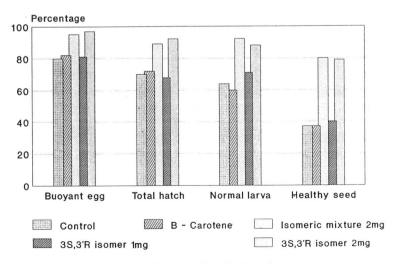

Fig. 16.6 Influence of dietary carotenoids on quality of red sea bream eggs.

Table 16.3 Content and composition of carotenoids in eggs of red sea bream

Group	Astaxanthin content (μg per 100 g)	Carotenoid composition*[1] (%)	
		Astaxanthin	Idoxanthin
2 mg isomeric mixture fed	56–74	90	4–7
1 mg 3S,3'R isomer fed	28–31	90	4–6
2 mg 3S,3'R isomer fed	62–80	90	3–6

*[1] Also contained traces of 4-Ketozeaxanthin

the good eggs were approximately 60 μg per 100 g irrespective of the amount and kind of carotenoids supplemented to the diets. Canthaxanthin was transferred to the eggs but not β-carotene. Egg quality was very high in the krill group as well as in the other two groups.

Other investigations on red sea bream broodstock revealed that vitamin E and phospholipids also improved the quality of eggs (Watanabe et al. 1991a,b). Thus, the egg quality enhancing effect could be attributed to the physiological mechanisms involving vitamin E and phospholipid or astaxanthin. Carotenoids generally have a strong ability as quenchers or scavengers of active free radicals like singlet oxygen and hydroxyl ions; among them astaxanthin is the strongest. This could be the reason why astaxanthin seems to be the bio-active component responsible for producing superior quality fish eggs. Free radicals, initiated by various factors including active oxygen, attack lipids and protein in biomembranes, leading to a deterioration of egg quality, the situation being averted when they are stabilized by quenchers or scavengers like vitamin E and C, phospholipids and carotenoids.

16.6 Larval rearing

Three larval stages have been described (Kohno et al. 1983) based on the mode of swimming and feeding, as follows:

(1) Early larval stage with a notochord up to 5.2 mm in length, during which stage there is little swimming activity and the larvae feed by swallowing.
(2) Transitional stage up to about 7 mm notochord length, during which the larvae develop basic adult structures, bite food and resort to caudal movements.
(3) An advanced larva above 7 mm in length which has defined swimming and feeding abilities.

The management of the fry can be separated into two parts: the first is the caring of larvae from 3 up to 12 mm in size, and the second concerns juveniles from 12 to 25 mm. Newly-hatched larvae are either kept in the same egg incubation tanks or transferred to an indoor or outdoor larval rearing tank. Currently large water tanks with capacities of 50 to 100 m^3 are being employed. These well-aerated tanks can either be rectangular or circular (Fig. 16.7), usually fabricated in concrete or fibreglass. Sea water is supplied to individual tanks from a reservoir which receives filtered water through a pumping system. Water is changed as required during the first 2–3 weeks, the rate being gradually increased with the growth of larvae. Marine *Chlorella* ('green water') at a concentration of 300–400 cells per ml of water is added to the tanks for stabilizing the water quality and as a food source for rotifers.

The larval population density depends on the mode of primary rearing, but ranges between 10 000 and 50 000 per m^3. A recommended optimum stocking density for experimental rearing is around 5000 larvae for a 500 l tank. Outdoor tanks are covered with shade nets to reduce the light intensity at the water surface to a maximum of 5000 lux.

Almost all larvae of marine finfish are initially fed on live foods such as rotifers and marine copepods. The most popular feeding schedule is shown in Fig. 16.8. The live foods chosen depend on the size of the larvae. Hatched larvae with body lengths greater than 2.3 mm are given rotifers as the initial diet and this is continued for about 30 days after hatching. When larvae attain a length of 7 mm or more, marine copepods such as *Tigriopus*, *Acartia*, *Oithona* and *Paracalanus*, or in their absence *Moina* and *Daphnia* of freshwater origin, are fed to larvae, together with rotifers which on their own are too small for

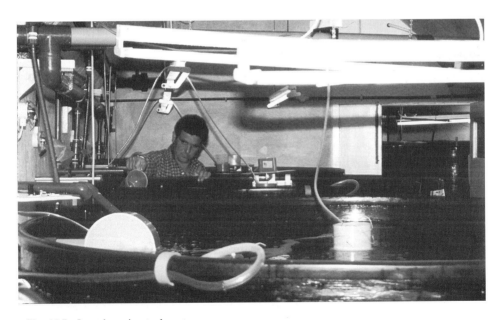

Fig. 16.7 Larval rearing tanks.

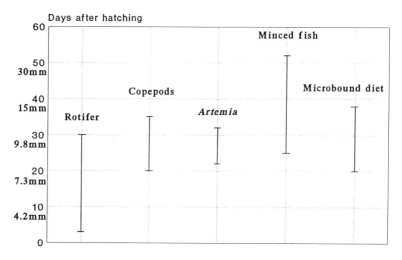

Fig. 16.8 Live food schedule adopted for marine fish larvae.

7 mm larvae. The recommended minimum density of rotifers which would prevent larval starvation is five individuals per ml.

The brine shrimp, *Artemia*, is frequently used as food for many marine fish larvae, especially when there is a shortage of marine copepods and because it is commercially available. The amount of live feed required to supply a rearing tank is estimated to be 1.3−1.4 times more than that consumed by the larvae, thus allowing for the escape of these organisms out of the rearing system.

Fish larger than 10−11 mm which are in the secondary rearing phase are fed on minced fish, shellfish or artificial micro-diets. The feeding schedule and organism used vary according to the type of hatchery and its geographic location. However, during the initial change-over period, most production centres mix minced fresh fish with short-neck clams and krill in equal ratios. Later on they are put on fresh fish alone, occasionally supplemented with vitamins.

The quality of water in the larval tanks has to be properly maintained. The bottom of the tanks must be cleaned by removing unconsumed feed, dead fish and debris. This cleaning has to be done specifically during and after the initiation of feeding with microparticulate diets. Care has also to be taken not to allow any floating debris which may prevent the larvae from gulping air during swim-bladder inflation. Very strong aeration must also be avoided for similar reasons. There has been a great deal of automation in larval production. Apart from tank cleaning and feeding equipment, a bio-farming system linking *Chlorella*, rotifer and the larvae is currently being operated in Japan.

16.7 Quality of larvae

Lordosis has been a serious problem in the rearing of red sea bream larvae.

Lately, studies have demonstrated that lordosis occurs in fish with deflated swim bladders owing to their failure to gulp air at the water surface when they reach a body length of 4–4.5 mm i.e. approximately 10–15 days after hatching (Kitajima *et al.* 1977, 1981). Such fish with malformed swim bladders have difficulty in maintaining their position in the upper or middle layers of the water column and have to swim at oblique angles with rapid fin strokes. This situation induces lordosis, which compensates for the oblique direction of the body axis by a distortion in the spinal column. Additionally, it has been found that swift water currents caused by excessive aeration make it impossible for larvae to swim up toward the water surface and to gulp air (Iseda *et al.* 1982).

In diet-induced lordosis, it was observed that feeding larval red sea bream, rotifers with a low content of (n-3) highly unsaturated fatty acids (n-3 HUFA) resulted in low swimming activity, a lack of endurance and improper reflex responses (Fig. 16.9, Watanabe *et al.* 1980). Consequently, they could not swim up and inflate the swim bladder. Lordosis has also been reported in other species of hatchery-reared fish with deflated swim bladders.

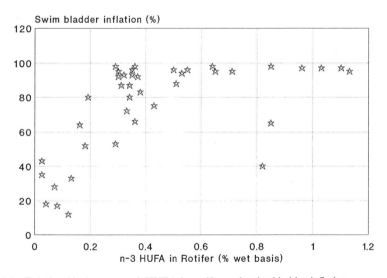

Fig. 16.9 Relationship between n-3 HUFA in rotifer and swim bladder inflation.

16.8 Larval nutrition and the role of essential fatty acids (EFA)

As the hatchery phase depends largely on live foods, they have been the subject of much research. Knowledge of the essential fatty acids (EFA) indicate that marine fish larvae require (n-3) HUFA such as eicosapentaenoic acid (EPA) and docosahexaenoic acid (DHA) as EFA for their normal growth. The requirement of (n-3) HUFA is about 0.5% for both larval and juvenile red sea bream (Watanabe *et al.*, 1989). Therefore, the concentration of n-3 HUFA in live foods is the principal determinant of its dietary value.

However, it was not clear which of the HUFA, EPA or DHA was more important for the larvae. In freshwater fish in which linolenic acid (18:3n3) is converted to DHA via EPA, the two may not vary much in EFA efficiency, unlike the marine species in which conversion from EPA to DHA is very limited (Kanazawa 1985). This is also supported by the fact that DHA, which is usually high in the eggs of marine fish species, is quickly reduced during larval development, although it is not known whether DHA is utilized as an energy source during development or converted to other physiologically important substances such as prostaglandins (Watanabe 1978).

Recent experiments have evaluated the comparative importance of EPA and DHA in marine fish. Larval and juvenile red sea bream were offered rotifers enriched by the direct method with methyl oleate (OA), EPA and DHA (purity > 99%) and a mixture of (n-3) HUFA (Watanabe *et al.* 1989). The EFA-deprived fish had poor growth, survival and vitality in addition to a high incidence of hydrops (Table 16.4). Improved results were obtained on incorporation of EPA, DHA or (n-3) HUFA mixtures into rotifers.

However, only the larvae in the DHA fed group showed significant reductions in hydrops and better performance in the vitality tests. Even though the (n-3) HUFA fed rotifers provided DHA, its dietary availability is probably limited because certain proportions of the EFA might be crucial to the larval bream. The fatty acid analysis on the total lipids of the larvae has shown that the level of assimilation of DHA was much higher than that of EPA; although EPA was partly converted to 22:5n-3, no retroconversion of DHA was observed, indicating the importance of DHA.

More recently Takeuchi *et al.* (1991) also attempted a relative estimation of DHA and EPA requirements by providing enriched *Artemia* to larval red sea bream. Ethyl esters of EPA and DHA (purity 99.9%) and ethyl oleate (purity 95%) were used for enrichment. After an 11-day feeding period, survival rates

Table 16.4 Comparative performance indices of larval red sea bream fed rotifers enriched with methyl oleate (OA), eicosapentaenoic acid (EPA), docosahexaenoic acid (DHA) and ester-85[*1] (E-85) for 8 days

Rotifers fed on	Total body length (mm)	Survival rate (%)	Vitality test[*2]		Hydrops (%)
			5 s	2 min	
OA(0.3)[*3]	4.4	32	0	0	47
EPA(4.7)	4.7	59	17	6	21
DHA(4.0)	5.0	57	63	52	7
E-85(7.0)	4.7	23	29	7	20

[*1] Methyl esters of n-3 HUFA containing 37% EPA and 42% DHA
[*2] Survival rate at vitality tests in which 50 larvae were scooped out, held in air for 5 s or 2 min, transferred to tank and mortality recorded
[*3] Values in parentheses are the n-3 HUFA content of rotifers in percentage dry matter basis

indicated the importance of EFA. There were hardly any mortalities in either the EPA- or DHA-enriched *Artemia*-fed fish. Growth was similar in all groups but vitality tests placed DHA as the superlative essential fatty acid for sea bream larvae (Table 16.5). In some way DHA seems to be able to alleviate various forms of stress in fish larvae.

In red sea bream, (n-3) HUFA such as EPA and DHA are actively incorporated into phosphatidylethanol amine(PE) and phosphatidylcholine(PC) and various other phospholipids (Table 16.6). EPA was also converted to 22:5n-3 in both PC and PE as observed in the former experiment with the larvae fed EPA-rotifer. It is interesting to note that EPA levels were markedly increased in the larvae fed EFA-deficient *Artemia*, probably mobilized from PL or neutral lipids, thus indicating functional differences between PC an PE. Poor inviable larvae resulting from being fed EPA alone in the diet may occur because excessive amounts of EPA promote an imbalance of EPA and DHA in the phospholipids of biomembranes. In order to facilitate membrane fluidity or various functions of phospholipids, it appears that a specific ratio of EPA:DHA has to be maintained. Further experimentation should clarify what proportions are required.

Table 16.5 Comparative performance indices of larval red sea bream fed *Artemia* nauplii enriched with OA, EPA and DHA for 11 days

Artemia fed on	Total body length (mm)	Survival rate (%)	Vitality test (%)[*1]	
			30 s	2 min
OA(0.5/ND)[*2]	20.9	60	0	–
PA(12.3/ND)	21.0	99	3	–
DHA(1.5/5.7)	20.5	100	100	57

[*1] See footnote of Table 16.4
[*2] Values in parentheses are the EPA/DHA content of *Artemia* in percentage dry matter basis; ND = not detected

Table 16.6 Incorporation of EPA and DHA into phosphatidylcholine and phosphatidylethanolamine in larval red sea bream fed *Artemia* enriched with EPA or DHA (area %)

Fatty acid	Phosphatidylcholine					Phosphatidylethanolamine				
	Initial					Initial				
	I	II*	OA	EPA	DHA	I	II*	OA	EPA	DHA
18:1n9	14.8	17.6	30.2	16.7	14.9	7.2	8.2	19.2	8.2	7.5
20:5n3	5.6	3.0	3.5	17.7	4.5	5.9	8.2	11.3	22.8	3.0
22:5n3	1.2	1.0	0.5	3.5	0.5	3.5	3.3	2.0	10.7	1.0
22:6n3	6.0	3.8	0.9	1.1	19.0	20.6	15.0	3.4	4.2	37.1
n3 HUFA	14.7	9.4	6.3	23.4	25.2	33.2	29.4	20.4	40.2	43.2

* Larvae fed the newly-hatched *Artemia* nauplii without enrichment for 3 days

Table 16.7 The n-3 HUFA requirements of marine larval fish and relative evaluation of EFA

Fish species and food organism	Requirement (% dry matter)		EFA value
Red sea bream			
Rotifer	n-3 HUFA	3.5	DHA > EPA
Artemia	n-3 HUFA	3.0	DHA > EPA
Yellow tail			
Artemia	n-3 HUFA	>3.9	DHA > EPA
Striped knifejaw			
Rotifer	n-3 HUFA	>3.0	
Artemia	n-3 HUFA	>3.0	
Fish (26 mg)	DHA	1	DHA > EPA
Flounder			
Artemia	n-3 HUFA	>3.5	
Fish (5 mm)	DHA/EPA	>1.0	DHA > EPA
Turbot			
Rotifer	n-3 HUFA	1.2–3.2	

The EFA requirement has been established for several marine fish at the live food dependant stage. In all species hitherto examined, DHA has a superior EFA value (Table 16.7). In the mass propogation techniques for marine larval fish, the enrichment of live foods with DHA has to become a standard procedure.

16.9 Future prospects

Much work has been carried out on the successful domestication of red sea bream. Though the choices in culturing this species are varied, difficulties are encountered particularly in the initial hatchery phases. Further developments could be:

- systematization of larval mass propagation through automation;
- establishment of stable methods for enrichment and bulk production of live foods;
- formulation of microparticulate diets to reduce the dependence of the larvae on live organisms; and
- identification of dietary factors affecting quality of eggs and spermatozoa before and after spawning.

This would enable the red sea bream to be an aquaculturist's model fish.

References

Davy, F.B. (1991) *Mariculture Research and Development in Japan. An Evolutionary Review.* IDRC. Ottawa.

Foscarini, R. (1988) A review: Intensive farming procedure for red seabream (*Pagrus major*) in Japan. *Aquaculture*, **72**, 191–246.

Fukuhara, O. (1985) Functional morphology and behaviour of early life stages of red seabream. *Nippon Suisan Gakkaishi*, **51**, 731–43.

Fukusho, K. (1989) Fry production for marine ranching of red seabream. *International Journal of Aquaculture and Fisheries Technology*, **1**, 109–17.

Iseda, H., Ishihara, M., Sumida, S., Owaki, M. & Tabata, T. (1982) Prevention of lordosis in the juvenile red seabream, *Pagrus major* reared in ponds – II. Relationship between the initial rearing conditions and the lordotic deformity. *Bulletin of Kumamoto Prefectural Fisheries Experimental Station*, **2**, 25–45.

Kanazawa, A. (1985) Essential fatty acid and lipid requirements of fish. In *Nutrition and Feeding in Fish*, (eds C.B. Cowey, A.M. Mackie & J.G. Bell), pp. 281–98. Academic Press, London.

Kitajima, C., Iwamoto, H. & Fujita, S. (1977) Relationship between curvature of vertebral column and undeveloped swim bladder in hatchery-reared red seabream, *Pagrus major*. *Bulletin of Nagasaki Prefectural Institute of Fisheries*, **8**, 137–40.

Kitajima, C., Tsukashima, S., Fujita, S., Watanabe, T. & Yone, Y. (1981) Relationship between uninflated swim bladder and lordotic deformity in hatchery-reared red seabream, *Pagrus major*. *Nippon Suisan Gakkaishi*, **47**, 1289–94.

Kohno, H., Taki, Y., Ogasawara, Y., Sirojo, Y., Taketomi, Y. & Inoue, M. (1983) Development of swimming and feeding functions in larval *Pagrus major*. *Japanese Journal of Ichthyology*, **30**, 47–59.

Masuda, H., Amaoka, K., Araga, C., Uyeno, T. & Yoshino, T. (eds) (1984) *The Fishes of the Japanese Archipelago*. Tokai University Press, Tokyo.

Miki, W., Yamaguchi, K., Konosu, S. & Watanabe, T. (1984) Metabolism of dietary carotenoids in eggs of red seabream. *Comparative Biochemistry and Physiology*, **77B**, 665–8.

Sakai, K., Nomura, M. & Takashima, F. (1985) Characteristics of naturally spawned eggs of red seabream. *Nippon Suisan Gakkaishi*, **51**, 1395–9.

Smith, M.M. & Heemstra, P.C. (eds) (1986) *Smith's Sea Fishes*. Springer Verlag, Berlin.

Takeuchi, T., Toyota, M. & Watanabe, T. (1991) Dietary value to red seabream of *Artemia* nauplii enriched with EPA and DHA. *Abstracts of the Annual Meeting of Japanese Society of Scientific Fisheries, Tokyo*, p. 243.

Watanabe, T. (1978) Nutritional quality of live foods. In *Dietary Lipids in Aquaculture*, (ed. Japanese Society of Scientific Fisheries), pp. 93–111. Koseisha-Koseikaku, Tokyo.

Watanabe, T., Ohashi, S., Kitajima, C., Seikai, T., Tsukashima, Y., Fujita, S. & Yone, Y. (1980) Relationship between the n-3 HUFA content in the rotifer and aeration on the swim bladder inflation in larval red seabream. *Abstracts of the Annual Meeting of Japanese Society of Scientific Fisheries, Hakata*, p. 242.

Watanabe, T., Arakawa, T., Kitajima, T. & Fujita, S. (1984a) Effect of nutritional quality of broodstock diets on reproduction of red seabream. *Nippon Suisan Gakkaishi*, **50**, 495–501.

Watanabe, T., Itoh, A., Murakami, A., Tsukashima, Y., Kitajima, C. & Fujita, S. (1984b) Effect of nutritional quality of diets given to brood stock on the verge of spawning on reproduction of red seabream. *Nippon Suisan Gakkaishi*, **59**, 1023–8.

Watanabe, T., Itoh, A., Satoh, S., Kitajima, C. & Fujita, S. (1984c) Effect of dietary protein levels and feeding period before spawning on chemical components of eggs produced by red seabream. *Nippon Suisan Gakkaishi*, **51**, 1501–9.

Watanabe, T., Izquierdo, M.S., Takeuchi, T., Satoh, S. & Kitajima, C. (1989) Comparison between eicosapentaenoic and docosahexaenoic acid in terms of essential fatty acid efficiency in larval red seabream. *Nippon Suisan Gakkaishi*, **55**, 1635–40.

Watanabe, T., Lee, M., Mizutani, J., Yamada, T., Satoh, S., Takeuchi, T., Yoshida, N., Kitada, T. & Arakawa, T. (1991a) Effective components in cuttlefish meal and raw krill for improvement of quality of red seabream *Pagrus major* eggs. *Nippon Suisan Gakkaishi*, **57**, 681–94.

Watanabe, T., Fujimura, T., Lee, M., Fukusho, K., Satoh, S. & Takeuchi, T. (1991b) Effect of polar and non-polar lipids from krill on quality of eggs of red seabream *Pagrus major*. *Nippon Suisan Gakkaishi*, **57**, 695–8.

Index

aeration
  excessive, larval lordosis, 409
  oxygen requirements, red drum, 132
Aeromonas, on eggs, 19
Aeromonas salmonicida, furunculosis, 202, 228
Africa, aquaculture *see* tilapia, Nile
algae, toxicity, *Prymnesium parvum*, 343
algal culture, 374–6
  green water techniques, 375–6, 407–8
  microalgae
    in fish nutrition, 375–6
    shading effect, 376
algal substitutes, shrimp larviculture, 375
Alteromonas
  antibiotic tolerance, 186
  on eggs, 19
ambisexuality, sea bream, 96
Amyloodinium ocellatum infections, 119
anaesthetic, quinaldine sulphate, 123
ANCOVA, 7–9
androgenesis, 79, 82
antibiotics
  addition to hatchery waters, 19
  bioencapsulation, 393–4
  sperm diluents, 64
  tolerance, bacterial pathogens, 186
aquaculture
  breeding goals, 86–90
    disease resistance, 88
    extended environmental tolerance, 87
    growth rate, 88–9
    sex control, 86–7
  world production (1990), 280
Aquazine, 225
Arachidonic acid, 20:4(n–6) pathway, 366
Aristichthys nobilis, 322
Artemia culture, 132, 162–3, 379–90, 408
  ascorbic acid incorporation, 384–6
  bioencapsulation, 393–4
  bream diet, 113
  decapsulation, 163
  optimization of nutritional value by DHA/EPA-enrichment, 161, 379–84, 410–12

  PUFA-enriched, 159
    DHA/EPA-enrichment, 380
  size of nauplii, 387–8
  substitute diet for, 132–3
artificial diets, 390–1
artificial insemination procedure, 71
  delay, rainbow trout, 71
  salmonids, 47, 71
ascorbic acid
  diet supplementation, 107–8, 298
  incorporation into *Artemia* culture, 384–6
  seminal fluid, 30
asthaxanthin pigment, role in nutrition, 15, 384, 404–6
Atlantic cod, croaker, halibut *see these names*
Atractoscion nobilis see bass, white sea

β-carotene, 405–6
bacterial kidney disease (BKD), *Renibacterium salmoninarum*, 202
bacterial pathogens
  antibiotic tolerance, 186
  control methods through live food, 391–4
  on egg surface, 19
  on eggs and larvae, 185–7
bass, Australian (*Macquaria novemaculeata*), egg cryopreservation, 65
bass, sea (*Dicentrarchus labrax*), 138–68
  diet
    early weaning, 390
    effects of quality, 154–6
    effects of ration, 154
  egg and larval quality, 156–9
  holding tanks, 160–3
  reproductive strategies, 140–4
  seasonal endocrine cycles, 142–4
  spawning
    endogenous rhythms, 152–4
    environmental control, 148–52
    induction, 144–8
  sperm, cooling and warming rates, 71

bass, white sea (*Atractoscion nobilis*), 118, 126
  optimum culture temperature, 131
  see also sciaenids
biological oxygen demand (BOD), emergency lowering measures, 227
biotechnological approaches, 76–93
  broodstock improvement technologies, 78–86
    chromosome manipulations, 79–82
    gene transfer, 82–3
    marker-assisted selection, 83–6
  fish as suitable subjects, 76–8
black drum (*Pogonias chromis*), 118–19
Brachionus plicatilis (rotifer) culture, 132, 162, 376–9
  bioencapsulation of antimicrobials, 393
bream, black sea (*Acanthropagrus schlegeli*), 399
bream, gilt-head sea (*Sparus aurata*), 94–117
  diet, egg and larval quality, 106–11
  larval rearing, 113
  management, recommendations, 111–13
  sex reversal and sex ratio, 96–7
  spawning
    egg collection, 112–13
    induction, 97–105
      dopamine antagonists, 101–3
      GnRH, 98–101
      GnRH delivery, 101–3, 102
      management, 111–12
      normal season, 111
      year-round egg production, 105–6
  sperm
    cooling and warming rates, 71
    dilution and motility, 31
  taxonomy, 95–6
bream, red sea (*Pagrus major*), 398–413
  diet and fecundity studies, 107
  eggs
    incubation, 402–3
    quality, 403
  larval rearing, 406–8
    nutrition, 409–12
    quality, 408
  management of broodstock, 400–2
  nutrition, 403–6
  taxonomy, 398–400
broodstock management *see* management practice
bullhead, black (*Ictalurus melas*), 221
bullhead, brown (*Ictalurus nebulosus*), 221
bullhead, yellow (*Ictalurus natalis*), 221

calcium, effects on sperm, 36–8
calcium ions, motility of sperm, 36–7
canthaxanthin, 405–6
  see also asthaxanthin
capelin (*Mallotus villosus*), liver mass, 358
capelin oil, (n−3) PUFA source, diet supplementation, 107
cardiac pathologies, PUFA ratios, 367–8
carotenoids
  anti-oxidant effectiveness, 16
  composition, 404–6
carp, *see also* cyprinids
carp, bighead (*Hypophthalmichthys nobilis*), 322, 326–7, 344
carp, black (*Mylopharyngodon piceus*), 322, 326–7
carp, Chinese
  defined, 325
  Israel, culture, 325–6, 330
    fry rearing, 341, 345–6
    maturation, 349
    propagation and spawning, 338, 349
  triploidy, 348
carp, common (*Cyprinus carpio*), 321–52
  eggs, storage temperature, 62
  fry rearing, 341–3
    primary and secondary nursing, 344–6
  Israel, culture, 325–6
    biological data, 329
    hatchery versus field propagation, data, 328
  predators of fry, 343
  production
    China (1988), 324
    selected countries (1990), 323
  propagation and spawning, 331–40
    egg incubation, 339–40
    endocrine changes, 335–7

gonadal recrudescence and hormonal
  cycles, 331−4
 induced spawning, 334−5
 pond spawning, 337−8
 sperm
  activation, 33
  fertilizing capacity, 32, 34
  ionic composition, 29
  motility, beat frequency, 35
  viability (unfrozen), 59
 summary and future research prospects,
  348−9
carp, crucian (*Carassius carassius*), 347−8
carp, grass (*Ctenopharyngodon idella*),
  322, 326−7, 344−5
 eggs, storage temperature, 62
carp, ornamental, 346−8
carp, silver (*Hypophthalmichthys
  molitrix*), 322, 326−7, 344−5
 eggs, storage temperature, 62
catfish, African (*Clarias anguillaris*)
 key, 243−5
 strains used, 245
 systematic status, 245
catfish, African (*Clarias gariepinus*), 242−
  76
 egg size, larval survival rates, 269
 maintenance of pubertal broodfish,
  252−5
 reproduction, annual rhythms, 248−52
 selection, maintenance and breeding of
  broodfish, 255−61
 survival of larvae, 267−73
 taxonomy and systematics, 243−8
 triploidy and gynogenesis, 261−6
catfish, Asian (*Clarias batrachus*), 242,
  246−8
 egg size, larval survival rates, 269
catfish, blue (*Ictalurus furcatus*), 221
catfish, channel (*Ictalurus punctatus*),
  220−41
 broodstock management, 222−9
 eggs, incubation and hatching, 232−5
 production of fry and small fish, 236−40
 spawning, 229−32
catfish, flathead (*Pylodictus olivarius*), 221
catfish, white (*Ictalurus catus*), 221
channel catfish virus disease, 220, 228

Chilodonella, on tilapia fry, 289
Chlorella, role in aquaculture, 375−6,
  378, 407, 408
Chorulon *see* hCG
chromosome manipulations, 79−82
 see also gynogenesis; triploidy
clams, algal substitutes, 375
Clarias, African species, key, 243−4
Clarias batrachus see catfish, Asian
Clarias gariepinus see catfish, African
Clarius lazera, 243−5
Clarius macrocephalus, eggs, time for
  optimal quality, 20, 21
Clarius mossambicus, 243−5
Clarius senegalensis, 243−5
cod, Atlantic (*Gadus morhua*), 169−96
 biology and taxonomy, 171−2
 broodstock management, 175−83
 eggs
  composition, 359−60
  time for optimal quality, 20, 21
  viability, 184−7
 historical background, 173−4
 quality of seed, assessment, 187−8
 sperm
  ionic composition, 29
  respiration, 40
conservation
 and gene banks, 5−6, 54−6
 see also cryopreservation
conservation and rescue programmes,
  Pacific salmon, 211−14
copepods, in diet, 159
copepods in larval nutrition, 407
croaker, Atlantic (*Micropogonias
  undulatus*), 118−37
 endocrine control, 120−1
 GTH I and II release, 103, 121
 MIS (maturation-inducing steroid),
  stimulation by GTH II, 122
cryopreservation of gametes, 55−6, 65−
  71
cryopreservation of gynogenetic lines, 6
 dilution ratio, 42
 post-thaw mortality, 43
cryopreservation of sperm, 5, 42−4, 55−
  6, 65−71, 212, 349
 cooling and warming rates, 70−1

cryoinjuries, 66
cryoprotectants, 42, 68–70
diluents, 67–8
equilibration time, 70
extenders, 124
gene bank, 55–6
insemination of thawed milt, 71
physico–chemical objectives, 65–6
post-thaw motility, effects of diluents and equilibration time, 69
process, 66–71
Ctenopharyngodon idella, 62, 322, 326–7, 344–5
Cynoscion nebulosus see trout, spotted seatrout
Cynoscion regalis see weakfish
Cynoscion xanthulus see orangemouth corvina
cyprinids, 321–52
 homogametic females, 264–6
 seminal fluid, 30
Cytophaga, on eggs, 19

Daphnia, in diet, 159
daylength *see* photoperiod
DHA (docosahexaenoic acid)
 conversion from EPA, 381–2
 EPA/DHA ratios, and pigmentation of turbot, 381
 essential nature, 379
 incorporation into diet, 107–10, 380–4, 409–12
Dicentrarchus labrax see bass, sea
diet
 artificial diets, 390–1
 failure to spawn, 9–10
 good management practice, 15–16
 krill/cuttlefish addition, 107
 microdiets, 150
 PUFA supplementation, 107–10, 380–84, 409–12
 quality, effects, 106–11
  bream, 107–11
  cod, 177
  sea bass, 154–6
  trout, 7–8
 vitamin supplementation, 107–8
see also algal culture; *Artemia*; rotifers;
specific nutrients
17α,20β,dihydroxy–4–pregnen–3–one (17,20–P), 122, 333
 action on sperm, 37–8
diluents for sperm, 45, 67–8
dimethylacetamide, sperm cryopreservation, 68
disinfection
 Betadine, 235
 iodophor, 18
 malachite green/formalin, 330
DMSO
 cryoprotectant, 42, 69
 diluent for sperm, 68
DNA electroporation, 82
DNA markers, 83–6
 RAPD technique, 83
DOCA, side-effects, 258, 261
domperidone *see* dopamine antagonists
dopamine antagonists, 3, 101–3
 potentiation of GnRH
  bass, 146–8
  sea bream, 101–3
drum *see* black drum; red drum

EFA
 in early development, 362–8
  C18–C20, 22 polyunsaturated fatty acids, 362–4
  deficiency, symptoms, 365
 see also PUFA
egg lipids *see* lipids *and specific lipids*
eggs
 cell symmetry at 8–16 cell stage, 188
 cryopreservation, 65
 disinfection, iodophor, 18
 lipid composition and utilization, 359–62
 quality
  management practice, 11–15
  overripening, 19–21
 quality assessment, 13–14
  determinants of quality, 15, 269
 size
  and fecundity, 6–11
  weight–fecundity relationships, 7
 time after ovulation for optimal quality, 20

eicosanoids, 379–84, 409–12
eicosanoids *see also* DHA
    (docosahexaenoic acid); EPA
    (eicosapentaenoic acid)
endogenous rhythms
  catfish, 154
    rainbow trout, 154
    sea bass, 152, 154
environmental manipulation, 3–5
environmental tolerance, extended, 87
EPA (eicosapentaenoic acid), 365–8
  production from 20:4(n–6), 366–7
  PUFA as precursors, 365–8
ethylene glycol, cryoprotectant, 42
euryhaline species, spawning induction, 130
extended spawning, red drum (*Sciaenops ocellatus*), 128–30

fathead minnow (*Pimephales promelas*), forage fish, 225
fatty acids
  monounsaturated, as energy sources, 355–6
  polyunsaturated *see* PUFA
  stoichiometry, 356
  see also EFA; PUFA
fecundity, and size of eggs, 6–11, 107, 154
feeds *see* diet
fertilization rate
  index of quality, 13–14
  sole, 180
  various spp., 58–9
Flavobacterium
  antibiotic tolerance, 186
  on eggs, 19
Flexibacter ovolyticus, egg mortality, halibut, 185
fluosol, sperm, preservation, 65–6
forage fish, fathead minnow, 225
fungal disease, treatment, 19, 235
furunculosis, 202

gadoids
  egg composition, 359–60
  sperm, short-term preservation, 57–8
Gambusia affinis, test for ichthyotoxin, 343–4

gene banks, and conservation, 5–6
gene transfer, broodstock improvement technology, 82–3, 89
genetic drift, sea bass, 139–40
gill parasites *see Amyloodinium ocellatum* infections
glycerol, sperm cryopreservation diluent, 68
GnRH (gonadotrophin-releasing hormone)
  degradation pattern, salmon, 99
  potentiation by dopamine antagonists, sea bream, 101–3
  sciaenids, 121–2
  specific binding protein, 99
  structure–activity relationships, 98–101
  see also hypophysation
GnRHa (mammalian analogues)
  bioavailability, 100
  biodegradation in bream, 98
  cholesterol-containing implants, 103, 104
  delivery systems, bream, 103–5
  implantation, 104
  intraperitoneal injection, salmon, 206
  list, 99
    properties, 99
  low efficiency, 101
  overstimulation, 127, 144–5
  stimulation of GTH II, 121–2
  structure–activity relationships, 98–101
  superpotent analogues
    biological properties, 99
    bream, 99
  treatment
    carp, 335
    red drum, 126
    salmon, 205–7, 213
    sea bass, 144–8
    sea trout (oral), 127–8
  use of dopaminergic antagonists, 101–3
  year-round egg production, 105–6
  see also hypophysation
(s)GnRHa (salmon–GnRH analogue)
  bioassay, 100
  biological properties, 99
  effects, croaker, 121
  extended spawning, 129–30

oral administration, 126–30
  sea trout, 128
  single injection, species, 126
  see also hypophysation
goldfish (*Carassius auratus*), 347–8
  bioassay, 100
  eggs, time for optimal quality, 20, 21
  GnRH-binding protein in blood, 99
gonadotrophins
  GTH I (vitellogenic), 337
    carp, 337
    sea trout, croaker, 121, 126
  GTH II (maturational), 337
    Atlantic croaker, 121–2
    carp, 331, 337
    red drum, 126
    sea bream, 100–1
    sea trout and croaker, priming of oocyte maturation, 122
    stimulation of MIS, 122
  radioimmunoassay (RIA), 98, 331
grayling, cryopreservation of sperm. 43
green water techniques, 375–6
grouper, sperm pH, 34
growth hormone, sciaenids, 121
GSI, carp, 332
gynogenesis, 79, 81–2
  cryopreservation of gynogenetic lines, 6
  induced
    in catfish, 266
    in salmonids and cyprinids, 264–6, 348

haddock, egg composition, 360
halibut, Atlantic (*Hippoglossus hippoglossus*), 169–96
  biology and taxonomy, 172
  broodstock management, 175–83
  eggs
    storage temperature, 62
    time for optimal quality, 20, 21
    viability, 184–7
  historical background, 174
  quality of seed, assessment, 187–8
  sperm
    activation, 33
    cryopreservation, 43
      post-thaw mortality, 43

  motility, beat frequency, 35
  viability (unfrozen), 59
hatcheries, conservation and rescue programmes, Pacific salmon, 211–14
hCG
  administration, natural versus stripping dosage, 97
  spawning induction (hypophysation), 258–61
hCG (mammalian analogues of)
  administration, bream, 98
  overstimulation, 127, 144–5
herring, eggs
  composition, 360
  time for optimal quality, 20, 21
Heterobranchus longifilis, × *C. gariepinus*, 245
Hippoglossus hippoglossus see halibut, Atlantic
homogametic females, in salmonids and cyprinids, 264–6
hormonal sex reversal, tilapias, 311–13
HUFA (highly unsaturated fatty acids) see PUFA
hydrops, larval bream, 410
17α-hydroxyprogesterone, 258, 259
Hypophthalmichthys molitrix see carp, Chinese; carp, silver
Hypophthalmichthys nobilis see carp, bighead; carp, Chinese
hypophysation, 2–3
  carp, 334–5
  catfish, 258–61
  hCG dose, 97, 259–60
  sciaenids, 123–30
  sea bass, 144–8
  sea bream, 97–105
  see also specific hormones

ichthyotoxin, control, 343–4
Ictalurus see bullhead; catfish
Idoxanthin, 405
indoor systems, red drum (*Sciaenops ocellatus*), 132
infectious haematopoietic necrosis virus (IHNV), 88
iodophor, egg disinfection, 18

see also asthaxanthin
isobutyl−1−methylxanthine, sperm medium, 71

jacks, 200

11−ketotestosterone, oral administration, 124
kidney disease (BKD) *Renibacterium salmoninarum*, 202
koi, 346−8
kokanee salmon, 200
krill, analysis and utility, 384, 403−4

LDH, in seminal fluid, 40
'lek', 279
LHRHa
  spawning induction, 3, 126, 144−8
  see also GnRHa
light *see* photoperiod control
linolenic acid, conversion to DHA, 410
lipids, 353−72
  coating in artificial diet, 391
  metabolism, early development, 359−62
  vitellogenesis, 356−9
lipoproteins
  egg-specific
    biosynthesis, 357
    see also vitellogenin
lipovitellin, 358
liver, transient enlargement, 358
lordosis, causes, 159, 408−9

Macquaria novemaculeata see bass, Australian
Macrobrachium rosenbergii (prawn), fed ascorbic acid-enriched *Artemia* culture, 385−6
malachite green/formalin, disinfection, 330
mammalian hormone analogues *see* GnRHa
management practice, 1−24
  biotechnological approaches, 76−93
  control of reproduction, 1−2
  hypophysation, 2−3
  induction of spawning, 2−3

induction of spawning *see* hypophysation
nutritional aspects, 15−16
  see also diet
quality measurement, 13−15
marker-assisted selection, broodstock improvement, 83−6, 89
maturation-inducing steroid (MIS)
  bass, 142−4
  sciaenids, 122
methanol, cryoprotectant, 42, 69
17−α−methyl testosterone, hormonal sex reversal in tilapias, 311−13
MHC genes in fish, 88
Micropogonias undulatus see croaker, Atlantic
MIS (maturation-inducing steroid)
  17α,20β,dihydroxy−4−pregnen−3−one (17,20−P), 37−8, 122
  in salmonids, 142
  in sea bass, 142, 144
  stimulation by GTH II, 122
  17α,20β,21,trihydroxy−4−pregnen−3−one (20βS), croaker, sea trout, 122
mosquito fish (*Gambusia affinis*), test for ichthyotoxin, 343−4
mouthbrooders, artificial incubation methods, tilapias, 301−11
Mylopharyngodon piceus see carp, black

Nannochloropsis, in bream diet, 113
Naucoris, control, 343
Nepa, control, 343
neural cell membranes, (n−3) PUFA, 353
Notonecta, control, 343

17β-oestradiol
  action, induction of vitellogenin, 142, 356−8
  levels, cod and halibut, 175
Oncorhynchus spp. *see* salmon; salmonids
oocytes
  growth, sea trout, 120−1
  GVBD, 146
  maturation, priming, sea trout and croaker, 122
oral vaccination, 391−4

orangemouth corvina (*Cynoscion xanthulus*), 119–37
  hCG injection, death, 127
  see also sciaenids
Oreochromis mossambicus, eggs, storage temperature, 62
Oreochromis niloticus see tilapia, Nile
ozonization, 19

perch, yellow (*Perca flavescens*), DNA RAPD, 84
perfluorocarbon emulsions, 65–6
phosphatidylcholine
  functional comparison with phosphatidylethanolamine, 411
  major component of eggs, 360–1, 384
  fate in egg, 362, 366
  preferential utilization, 360–1
phosphatidylethanolamine
  cod retina, 365
  functional comparison with phosphatidylcholine, 411
  trout retina, 364–5
phosphatidylinositol
  20:4(n−6) essentiality, 366–7
  molecular species composition, cod eggs, 366–7
phosphocreatinine, 40
phosvitin, 358
photoperiod control, 3–5
  gonadal maturation, 125
    red drum, 125
    salmon, 206
    sea trout, 125
    sea bass, 148–50
photoperiodic species, light requirements, 4
pike (*Esox lucius*)
  PUFA, absolute requirement for (n−3) PUFA 22:6(n−3), 363
  sperm pH, 34
Pimephales promelas, forage fish, 225
pimozide see dopamine antagonists
pituitary harvest, GTH content, 334–5
pituitary hormones see GnRH; hypophysation; LHRH
Plecoglossus, eggs, time for optimal quality, 20, 21

poaching, control, 229
polylactic–polyglycolic copolymer (PLGA), GnRHa implants, 104
polyploidy, 79–81
predators, control, 228–9
progestagen see $17\alpha,20\beta$-dihydroxy-4-pregnen-3-one
17α-hydroxy-progesterone, 258, 259
propane–diol, cryoprotectant, 42
prostaglandins, production, 366
Prymnesium parvum (phytoflagellate), control, 343
Pseudomonas, antibiotic tolerance, 186
Pseudomonas disease, 19, 228
PUFA
  absolute requirement, 16
  bream, 409
  cod, 362
  C18–C20, and C22 in early development, 362–8
  DHA/EPA ratios, 16, 362–8, 409–12
  DHA/EPA-enrichment of *Artemia* culture, 161, 379–84, 410–12
  diet supplementation, bream, 107–10
  in early development, membrane functions, 364–5
  membrane functions of PUFA, 364–5
  (n−3) PUFA 22:6(n−3), 362–8
  (n−3) PUFA, capelin oil, 107
  (n−6) PUFA 20:4(n−6), 362–8
  in vegetable oils, cardiac disorders, 367–8
  as precursors of eicosanoids, 365–8

quinaldine sulphate, anaesthetic, 123

radioimmunoassay (RIA), maturational gonadotrophin, 98, 331
Ranatra, control, 343
RAPD technique, 83–4
red drum (*Sciaenops ocellatus*), 118–37
  culture methods
    broodstock, 123–30
    eggs and larvae, 130–3
    extended spawning, 128–30
    indoor systems, 132
    precocious maturation, 124
  reproduction, 120–2

redfish *see* red drum
Renibacterium salmoninarum, bacterial kidney disease (BKD), 202
rescue programmes, Pacific salmon, 211–14
RFLPs, DNA markers, 83–6
RIA, maturational gonadotrophin, 98, 331
rotifers
  artificial diet for, 376–8
  bioencapsulation of antimicrobials, 393–4
  culture, 132, 162, 376–9
  dispensing, 113
  microalgae in nutrition, 375–6

20β-S *see* maturation-inducing steroid (MIS)
salinity, euryhaline species, 130
salmon hormone analogues *see* (s)GnRHa
salmon, amago (*Oncorhynchus rhodurus*), 198–9
salmon, Atlantic (*Salmo salar*)
  egg composition, 359
  environmental manipulation, 4–5
  sperm
    ionic composition, 29
    osmolality of milt, 61
    potassium concentration of milt, 61
    preservation, fluosol, 65–6
    respiration, 40
    urine contamination, 60
    viability (unfrozen), 58
  trypsin-like isoenzymes as markers, 89
salmon, chinook (*Oncorhynchus tschawytscha*), 198–9
salmon, chum (*Oncorhynchus keta*), 198–9
  eggs, storage temperature, 62
  sperm, viability (unfrozen), 58
salmon, coho (*Oncorhynchus kisutch*), 20, 198–9
salmon, masu (*Oncorhynchus masu*), 198–9
salmon, Pacific (*Oncorhynchus* spp.), 196–219
  culture of broodfish, 16–18, 200–10
    capture and holding, 200–1
    genetic factors, 207
  judging ripeness, 203
  ocean ranching, 209–10
  spawning, 204–7, 213–14
  stress and disease prevention, 201
  supplementation hatcheries, 207–9
  hatcheries conservation and rescue programmes, 211–14
  taxonomy and life history, 198–200
salmon, pink (*Oncorhynchus gorbuscha*), 198–9
  eggs, storage temperature, 62
  sperm, viability (unfrozen), 58
salmon, sockeye (*Oncorhynchus nerka*), 198–9
  eggs, storage temperature, 62
  sperm, viability (unfrozen), 58
salmon, yamame, 200
salmonids
  artificial insemination procedure, 47
  DNA markers, 84–5
  egg composition, 359
  eggs
    time for optimal quality, 20, 21
    weight–fecundity relationships, 7
  homogametic females, 264–6
  intratesticular sperm, 40
  PUFA, 363
  seminal fluid, 30
  sperm
    cryopreservation, diluents, 68
    insemination post thaw, 71
    motility, 35–6
    short-term preservation, 41, 57–8, 62–3
Saprolegnia, treatment, 19
Sarotherodon
  sperm, viability (unfrozen), 59
  taxonomy, 278–9
Sarotherodon mossambicus, 255
sciaenids, 118–37
  growth hormone, 121
scoliosis, causes, 159, 408–9
seminal fluid
  composition, 30
  LDH in, 40
  proteases contained, 41
  quality in cryopreservation, 67
  see also cryopreservation of sperm
sex reversal, sea bream, 96

sex reversal in tilapias, 311–13
shrimp larviculture, algal substitutes, 375
shubunkin, 348
Simazine, 225
smoltification, timing, 5
sole, Dover (*Solea solea*), fertilization rate, 180
sparids, 398–413
Sparus aurata see bream, gilt-head sea
spawning, induction *see* hypophysation
sperm
  acrosome, presence/absence, 27
  chemical composition, 29–30
  concentration, 33
  cryoinjuries, 66
  cryopreservation, 5, 42–4, 124, 212
  diluents, 45, 67–8
  fertilization rates, various spp., 58–9
  fertilizing capacity, 32
  historical background, 26
  inactivation by radiation, 264–6
  intratesticular sperm, 40
  mitochondria, 28, 38–40
  morphology, 27–9
  motility, 31–2, 35–40
  physiology, 33–40
  production, 29
  progestagen action, 37–8
  quality, assessment, 31–3
  recommendations, handling brood males, 44–6
  sampling, 45
  short-term preservation, 41–2, 54–5, 56–7, 58–65
    dilution media, 64–5
    oxygen-enriched environments, 63–4
    storage temperature, 58–9, 61–3
  stripping, 44–6
  survival in vivo, 41–2
  see also cryopreservation of sperm
spottail bass *see* red drum
spotted seatrout (*Cynoscion nebulosus*), 118–37
squid, in diet, 107–11
Stizostedion see walleye
stressors, 16–17
  handling, 17, 127
  larval stress, 184
  sunlight, 202

test, larval quality assessment in nutritional studies, 386–7
sulphamethoxazole, bioencapsulation, 393–4
swim bladder function, 409

temperature effects
  maturation control, 3–4
  overripening, 20
  short-term preservation of sperm, 58–9, 61–3
  spawning frequency, 128–9
  see also specific fish
tetraploidy, 81
theophylline, sperm medium, 71
Thymallus, cryopreservation of sperm, 43
thyroxine, oral administration, 124
tilapia, Nile (*Oreochromis niloticus*), 277–320
  artificial incubation methods, 284–5, 301–11
  cannibalism, 289, 290
  eggs
    storage temperature, 62
    time for optimal quality, 20, 21
  genetics, 283–4
  hormonal production of all-male fry populations, 311–13
  parasites, 289
  predators, 289
  reproductive biology, 283–9
  saltwater tolerance, 282
  sperm
    morphology, 27–8
    post-thaw motility, 69, 70
    quality in cryopreservation, 67
  subject area reviewed, 282–3
  summary and future prospects, 313
  taxonomy and natural history, 278–93
  traditional fry production methods, 289–93
    improved management, 293–301
Tilapia spp.
  sperm, viability (unfrozen), 59
  taxonomy, 278–9
toxins
  bioassays, 17
  prymnesin (phytoflagellate), 343

trash fish (*Boops boops*), 154
triacylglycerols, as egg energy source, 360–1
Trichodina, on tilapia fry, 289
17α,20β,21,trihydroxy-4-pregnen-3-one (20βS), 122
trimethoprim, bioencapsulation, 393–4
triploidy, 79–81
  disease resistance, 88
  growth rate, 88–9
  hybrids
    brook trout, 80
    carp, 348
    catfish, 261–4
    rainbow trout, 80–1, 89
trout, brook (*Salvelinus fontinalis*)
  hybridization with rainbow trout (triploid), 80
  sperm, viability (unfrozen), 58
trout, brown (*Salmo trutta*)
  broodfish, genotype importance, 7
  sperm, viability (unfrozen), 57, 58
trout, rainbow (*Oncorhynchus mykiss*)
  artificial insemination procedure, 71
  broodfish, genotype importance, 7–11
  cryopreservation of sperm, 43, 57, 58, 70–1
  DNA fingerprint patterns, 85
  DNA RAPD, 83–4
  eggs
    quality assessment, 14–15
    time for optimal quality, 20, 21
    weight–fecundity relationships, 7–11
  environmental manipulation, 4–5
  feeding, 10
  fertilization, percentage rate, 12–13
  hatchery procedures and timings, 18
  sperm
    cooling and warming rates, 70–1
    insemination post thaw, 71
    preservation, perfluorocarbon emulsions, 65–6
    quality in cryopreservation, 67
    respiration, 40
    viability (unfrozen), 58
  triploid hybrids, 80–1
trout, spotted seatrout (*Cynoscion nebulosus*), 118–37

GnRHa injection, stimulation of GTH-II, 121–2
MIS (maturation-inducing steroid), stimulation by GTH-II, 122
oocyte growth, 120–1
see also sciaenids
turbot (*Scophthalmus maximus*), 170
  eggs, for optimal quality, 20
  fertilization rate, 180
  lipid nutrition, 363
  seminal fluid, 30
  sperm
    dilution and motility, 31
    ionic composition, 29

UV sterilization, 19

verapamil, effects on sperm, 36–8
Vibrio
  anguillarium vaccine, 393–4
  egg mortality, halibut, 185
viral haemorrhagic septicaemia virus (VHSV), 88
vitamin C
  diet supplementation, 107–8, 298
  incorporation into *Artemia* culture, 384–6
  seminal fluid, 30
vitamin E, diet supplementation, 107–8
vitellinogen, levels, cod and halibut, 175
vitellogenesis
  GnRH induction of GtH, 98–101
  regulation in sciaenids, 121
vitellogenin (egg-specific lipoprotein), 356–9
  composition, 258
  ELISA, sea bass, 142
  hepatic production, 121
  occurrence, 258
  oestradiol levels, 121
  see also lipids

walleye
  DNA RAPD, 83–4
  sperm, viability (unfrozen), 59
water bugs, control, 343
weakfish (*Cynoscion regalis*), 120
world production, 280